D1221301

PRINCIPLES

OF

MINERAL DRESSING

PRINCIPLES
OF
MINERAL DRESSING

BY

A. M. GAUDIN

Consulting Metallurgist; Richards Professor of Mineral Dressing, Massachusetts Institute of Technology; Formerly, Research Professor of Mineral Dressing, Montana School of Mines

McGRAW-HILL BOOK COMPANY, INC.

NEW YORK AND LONDON

1939

PRINTED IN THE UNITED STATES OF AMERICA

XIV

23030

To

FRANCIS ANDREW THOMSON

respected chief, wise counselor,
true friend

PREFACE

The past thirty years have witnessed the introduction of unparalleled changes in the art of dressing ores and minerals. These changes have included the natural evolution in machinery and also the introduction of radically new methods. Yet, in spite of the tremendous strides made by flotation and other adhesion processes, the older methods have retained a significant place in the art. The field is thus much wider than it was when mineral dressing was simply the mechanical preparation of ores.

College instruction in this subject has therefore had to meet a new problem. Treatment of mineral dressing in detail, within the time allotted for that subject in undergraduate curricula, has become an impossible task for the student and the instructor alike: I have found it best to modify the classical instruction in such a way as to lay emphasis on principles and to minimize machine descriptions, details of practice, and discussion of auxiliary operations.

This book is an outgrowth of lecture notes and is designed primarily for the use of seniors in chemical, geological, metallurgical, and mining engineering, but I hope that it may also be of service as a reference book.

Throughout this text it has been my aim to take the student behind the scenes, to examine what is going on in each process. To that extent the book is written from the *unit process* point of view. I am one of those who hold that there is no fundamental difference between the dressing of any of the ores or minerals, that is, that there is no such thing as coal dressing as a distinct art from the milling of gold or copper ore. I believe that this point of view, if properly nurtured, is bound to lead young college graduates to a wider adoption of good ideas and devices from one field of industry to another.

It has also been my aim to present mineral dressing as a living art that requires constant adjustment to its environment. If many issues have been left in a controversial form it has been in order to stimulate the imagination of the student by suggesting many a use for his services.

For the convenience of those who wish to take mathematical treatments for granted, those parts of the text are generally printed in finer type. References are placed at the ends of chapters; although numerous, they do not represent a complete record. They are, however, those references which I have found to be most directly related to the text at hand.

Many are those to whom I should express thanks in connection with the preparation of this text. In the first place, I wish to acknowledge the help that I have received from books and articles. I am particularly indebted to the works of Berthelot, Louis, Richards and Locke, Rittinger, Taggart, Truscott, and Wark. The debt that I owe to writers of technical articles is more difficult to acknowledge specifically as the articles are so numerous; but to an extent this is met by the use of references.

In the second place, I am indebted to several of my former colleagues, at Montana School of Mines, who have graciously applied themselves to the task of helping me with problems of one kind or another; more particularly to Dr. George L. Shue, Professor of Physics, Dr. Curtis L. Wilson, Professor of Metallurgy, and the Misses Marjorie Bedinger and Guinevere Crouch, librarians. I wish to say here that this book was prepared almost wholly in Butte, while I was associated with the Montana School of Mines.

In the third place, I am indebted to many of my professional friends who have read and criticized parts of the manuscript, more particularly Messrs. Will H. Coghill, Reginald S. Dean, George G. Griswold, Jr., Edward Menzies, Reinhardt Schuhmann, Jr., Arthur F. Taggart, Henry D. Williams, and S. R. Zimmerley. In the same way I am indebted to Mr. John C. Rabbitt, of Butte and Cambridge, who has edited the manuscript, and to Mr. A. W. Schlechten, of Anaconda, Mont., and Cambridge, who has assisted me in reading proof.

Last, but not least, I am indebted to those of my students who, by their questions in class, have made it possible to correct and enlarge my lecture notes, year after year, until they appeared sufficiently complete to form the basis for this book.

A. M. Gaudin.

Cambridge, Mass.,
November, 1939.

CONTENTS

PAGE

CHAPTER XIX

CHAPTER XX

CHAPTER XXI

CHAPTER XXII

PRINCIPLES OF MINERAL DRESSING

CHAPTER I

INTRODUCTION

Most mineral substances are wanted in a relatively high state of purity or in certain sizes only. For instance, traprock, which is used as ballast, is not wanted in very coarse chunks or as fine powder, but only in lumps of intermediate size. Again, coal is worthless unless it is sufficiently free of waste material, and copper ore containing less than about 5 per cent copper is not directly usable. It is therefore apparent that many ores and minerals require some preparation before use either to enhance their chemical purity or to better utilize their physical properties.

Mineral dressing is commonly regarded as the processing of raw minerals to yield marketable products and waste by means that do not destroy the physical and chemical identity of the minerals. Hydrometallurgy, pyrometallurgy, and oil refining may also deal with raw minerals, but those processes change the character of some or all of the constituents of the raw material.

Ore dressing is mineral dressing applied to ores.

Economic Justification. Compared with hydrometallurgical, pyrometallurgical, or refining processes, mineral dressing is inexpensive. Economic justification for its inclusion in the mineral industry is readily made.

The benefits resulting from dressing metalliferous ores prior to smelting derive from the following sources:

1. Savings in freight—since no freight has to be paid on the waste discarded by the dressing operations.
2. Reduced losses of metal at the smelter—because of reduction in amount of metal-bearing slag produced at the smelter.
3. Reduction in total smelting cost—because of reduction in tonnage to be smelted.

Against these advantages, of course, must be charged the losses incurred in dressing the ore and the cost of the dressing operation.

The following problems are typical:

Example 1. A lead ore containing on the average 15 per cent lead, 0.2 oz. of gold per ton, 5 oz. of silver per ton, 6 per cent iron, 2 per cent zinc, 12 per cent sulphur, and 40 per cent insoluble can be dressed to yield a concentrate containing 60 per cent lead, 0.8 oz. of gold per ton, 20 oz. of silver per ton, 5 per cent iron, 3 per cent zinc, 18 per cent sulphur, and 3 per cent insoluble.

The freight to the smelter costs $1.50 per ton for shipment of the ore and $2.25 for shipment of the concentrate. The smelter purchases the ore or concentrate according to the following schedule:

TYPICAL LEAD-SMELTER SCHEDULE
(Rocky Mountain District)

Payments:

Gold. If 0.02 oz. per ton and up to 5 oz. per ton, pay for all at the rate of $31.82 per ounce; if over 5.0 oz. per ton, pay at the rate of $32.67 per ounce. No payment if less than 0.02 oz. per ton.

Silver. Pay for 95 per cent of the silver (minimum deduction of 0.5 oz.) at government or open-market price, whichever is applicable. No payment if less than 1 oz. per ton.

Lead. Pay for 90 per cent of the lead, as determined by dry assay (wet assay less 1.5 unit*), at the New York price less 1.5 cts. per pound. No payment if less than 3 per cent, dry assay.

Copper. Pay for 90 per cent of the copper, as determined by wet assay, at the New York price less 5.5 cts. per pound.

Iron. Pay at the rate of 6 cts. per unit.

Charges:

Insoluble. Charged for at 10 cts. per unit.

Zinc. 6 per cent free; excess charged for at 30 cts. per unit.

Arsenic. 2 per cent free; excess charged for at 35 cts. per unit.

Sulphur. 2 per cent free; excess charged for at 25 cts. per unit. Maximum charge $2.50 per ton.

Treatment. $2.50 per ton on the basis of 30 per cent dry lead assay. Increase treatment charge by 10 cts. for each unit of lead under 30 per cent, and decrease it by 10 cts. for each unit of lead over 30 per cent.

* Each unit represents 1 per cent, or 20 lb. per ton.

What net saving per ton of ore must be credited to dressing if the cost of that operation is $0.75 per ton of ore treated and if the lead recovery is 90 per cent?

Example 2. A gold ore containing 0.45 oz. of gold per ton, 3.5 oz. of silver per ton, 0.6 per cent copper, 12 per cent iron, 40 per cent silica, and 6 per cent alumina can be treated by a combination of amalgamation and flotation. Fifty-five per cent of the gold is obtained in the amalgam; in addition,

40 per cent of the gold and 80 per cent of the silver are obtained as a flotation concentrate containing 3.6 oz. of gold per ton, 56.0 oz. of silver per ton, 10.2 per cent copper, 30 per cent iron, 10 per cent silica, and 5 per cent alumina. The ore (or concentrate) is shipped to a copper smelter which pays for the metals according to the following schedule:

TYPICAL COPPER-SMELTER SCHEDULE
(Western United States)

Payments:

Copper. Pay for 96 per cent of the copper content (minimum deduction of 10 lb. per ton) at New York price less 2.5 cts. per pound.

Silver. Pay for 95 per cent of the silver content (minimum deduction of 1 oz. per ton) at government or open-market price, whichever is applicable, less 2 cts. per ounce.

Gold. Pay for 95 per cent of the gold (minimum deduction of 0.01 oz. per ton) at $20 per ounce, plus 90 per cent of the premium in excess of $20.67 per ounce. At the present government price of $35 per ounce, this is equivalent to paying for all the gold at $31.82 per ounce.

Charges:

Treatment. $4 per ton, as a basis. If metal payments are in excess of $15 per dry ton, add 10 per cent of this excess to the treatment charge; add 12 cts. for each unit* of iron; deduct 2.5 cts. for each unit of silica in excess of alumina; the maximum treatment charge being $5.50 per ton.

* Each unit represents 1 per cent, or 20 lb. per ton.

If freight is by truck and costs $1.50 per ton of ore or concentrate and if the cost of the dressing operation is $1.50 per ton treated, what net saving per ton of ore must be credited to dressing?

Example 3. Waste has been accumulating from the operation of a coal washery. Space becomes valuable, so that a credit of $0.06 per ton can be had for removal of, or reduction in, the existing culm banks. If treated by flotation, this culm can be made to yield a coal having a market value of $0.50 per ton at the washery. One ton of this coal can be obtained from each 2 tons of culm. Cost of dressing the culm, excluding drying of the clean coal to market requirements, is $0.11 per ton of culm. Cost of drying to market requirements is $0.20 per ton of cleaned coal. What is the saving accomplished by cleaning the slack?

In a number of cases, as in the marketing of molybdenite, asbestos, feldspar, mica, and almost all nonmetalliferous mineral products, the products sold have to meet certain stringent requirements. If they do not meet them, they are either unsalable or so heavily penalized as to justify even the loss of a large amount of the material in order to acquire suitable grade.

Scope of Mineral Dressing. To a large extent, mining and quarrying operations are concentrating operations in that they bear only on those portions of mineral deposits whose content of the material wanted is the highest available. This was especially true of mining in antiquity, but it remained so to a large degree until very recent times because of the crudity of dressing methods. With improvements in mineral dressing, it has become possible to mine increasingly lower grade ores until the practice which may be termed "bulk mining" has come to supersede to a large extent that which by contrast is termed "selective mining." That is, it has been found more advantageous to mine ore carelessly, including in the mined product much of the waste that happens to be in the way, but at a much lower cost than would be required for careful mining, with the view of later purifying the ore by careful, large-scale, inexpensive dressing.

In a general way, the scope of mineral dressing is twofold, its object being to eliminate either (1) unwanted chemical species or (2) particles of unsuitable size or structure. The first of these objects is commonly considered to be the extent of ore dressing, and is indeed the more important of the two, but the other is of sufficient importance to warrant attention even in an introductory text.

The principal steps involved in the preparation of mineral particles from a chemical standpoint are (1) liberation of dissimilar particles from each other and (2) separation of chemically dissimilar particles. Likewise, the principal steps involved in the preparation of mineral products valued for their physical structure are (1) reduction in size and (2) separation of particles of dissimilar physical character. Thus, it will appear that in every instance mineral dressing involves a size-reducing or liberating operation or group of operations as a first step and a separating operation or group of operations as a second step. This generalization should not be emphasized too much since instances are common of operations in which liberation and separation stages are made to alternate in order to accomplish the desired ends most advantageously.

Historical Outline. In order to get a satisfactory perspective of the status of mineral dressing, it is desirable to review briefly the development of the art from its inception[33]. As in other fields of human endeavor, our information concerning the operations of the past is not so full as might be wished. We have to refer to Aristotle, Herodotus, and Pliny for information concerning antiquity. That information is incomplete, largely because educated Athenians and Romans alike would not lower themselves to such "common" duties as are implied in technological descriptions or discussions. The records from civilizations contemporary with, or older than, that of the Greeks and Romans are even less complete.

With the dawn of new thinking in the latter part of the medieval epoch, more attention was paid to matters pertaining to the mineral industry. A great record of medieval and early Renaissance practice is embodied in Agricola's "De re metallica," of which a translation into English by the Hoovers is available in most technical libraries.[2]

The oldest dressing method was unquestionably *hand sorting.* By this is meant the choosing of valuable lumps of ore from worthless lumps, basing such choice on appearance and "heft." The oldest metallurgical operations were so crude that only the highest grade material could be smelted. Naturally, hand sorting played a preponderant part in supplying the smelter with suitable feed. It was done underground, at the surface, or at both places. Hand sorting survives in modern dressing operations only where cheap labor is available and where certain peculiar characteristics of the mineral product are encountered. Underground hand sorting is still practiced where selective mining is employed. The wisdom of extensive underground hand sorting is not always defensible, and its economic worth should always be scrutinized.

Washing was, in all probability, the next process to be evolved. It is likely that even in very primitive civilizations man must have noticed that water exerts a cleansing action on fine mineral particles, effecting a separation between fine grains of sandy character and those of slimy character. Undoubtedly, washing was used, first without, and later with, conscious appreciation of chemical dissimilarity between coarse and fine particles. Remains of ancient works have recently been discovered which show that concentration by washing was practiced centuries before the Christian era. In recent times, washing has taken a "back seat" as a concentration method; it is nevertheless still used extensively for the beneficiation of some types of ores, especially oxide ores of iron.

Although the first steps to be evolved in the dressing of mineral products were concentrating procedures (hand sorting and washing), it was discovered quite early in the development of the art that valuable particles generally occur in relatively intimate aggregation with worthless material and that improvement in the operations could be obtained by first *crushing* the ore and then separating the dissimilar grains. The earliest crushing methods were of the sledge-hammer type, and the motive force was the brute force of the operator. This was superseded by crushing contraptions in which the work was done by horses or other draft animals. It is only in recent times that power-driven machinery was substituted for horses or man; it has produced such tremendous savings as to put an entirely different aspect on the whole art of mineral dressing.

From the washing operations of the ancients, it was only a step to the use (but not the understanding) of a combination of the washing force of streams with the force of gravity to segregate mineral particles of different specific gravities. Ruins of the works at the famous silver mines in Attica have yielded stone tables set at a slant; these were used by the Athenians for the concentration of the fines from the argentiferous lead ores of Laurium. Water was allowed to flow down the slope of the table, and it carried away the lighter and finer particles; at the same time, the heavier particles settled on the table. After a certain length of time, the valuable sediment was scraped from the table, and the operation was repeated as often as necessary on either the concentrated product or the waste. This was the beginning of *tabling*, a form of *gravity concentration.*

In medieval days, it was discovered that concentration of granular material could be effected by jiggling it on a submerged sieve. The machine in which this work was done was operated by hand; it was known as a jig and the operation as *jigging.* Much of this development occurred in the Harz mountains of north central Germany. Jigs have played a very important part in dressing, and although obsolescent for many purposes, they remain of great importance in coal cleaning and in the beneficiation of iron ores. They now are, of course, power driven.

From Agricola's time on, it became increasingly evident that the greater portion of the losses of valuable mineral was among the fine particles. To minimize this loss, devices called *buddles* and *vanners* were evolved, which were essentially modifications of the Athenian tables. Buddles and vanners were greatly improved in Cornwall and have remained in use there longer than in most other places. Toward the end of the nineteenth century, the *shaking table* devised by A. R. Wilfley of Denver succeeded in giving a new lease of life to gravity concentration by making the previously intermittent operation of tables a continuous operation[32,40].

This sketchy account of the historical growth of gravity concentration would not be adequate without a mention of those friends of the prospector the *sluice box* and the *rocker,* devices eminently suited to the handling of placer material.

Modern *grinding* machinery has been developed concurrently with the improvement of concentration technique for the treatment of particles of fine size. Early devices were essentially variants of equipment used in milling flour. They are now all historic monuments to man's ingenuity and have been replaced by more effective devices consisting of a grinding medium of tumbling stone pebbles, iron or steel balls, or steel rods contained within a rotating shell.

The development of crushing and grinding technique has necessitated the development of a technique for the *sizing* of the crushed or ground material—this, because crushing and grinding are most efficiently carried out on material from which have been removed the sizes finer than those to which it is desired to break or grind the material under treatment. The first sizing devices used were stationary *screens.* These are still used, although rotating cylindrical screens known as trommels have replaced them to a large extent; but even trommels are becoming obsolete as vibrating screens are being improved. For sizing fine particles, *classifiers* are preferred to screens. For their operation, classifiers depend on the difference in the settling velocity of particles of different sizes in water.

In recent years, forces other than that of gravity have been used to separate minerals. One of these is magnetism; for the concentration of some iron ores, *magnetic separation* is without a peer. Magnetic separation can also be used for the concentration of ores in which minerals occur that are only slightly magnetic. The application of this process is limited by the fact that the initial cost of the machinery is relatively high if the minerals are not very magnetic. It is limited, also, because the design of wet magnetic separators, *i.e.*, of machines operating on ores pulped with water, has lagged behind the development of magnetic separators operating on dry-crushed minerals.

Flotation, the newest method of mineral concentration, has been developed very recently, historically speaking. It utilizes forces characteristic of mineral surfaces in such a way that particles may be caused to adhere either to the water in which they are immersed or to air bubbles. The particles are collected in a froth or remain in the pulp, depending on whether they stick to the air or to the water.

In the treatment of sulphide ores, flotation is without a competitor, and its application to the treatment of nonsulphide ores is making great strides. A related process which seems due for startling advances in the field of nonmetallic minerals is that of *agglomeration*.

A number of other methods have been used to a varying extent and with local success. In these methods, use is made of characteristics other than chemical affinities, magnetism, or gravity. Among these should be mentioned methods of separation based on the electrical conductivity of mineral particles, their color, decrepitation upon heating, shape, and dielectric constant.

Operating Steps Involved. There are four principal types of operations. To these may be added a number of auxiliary operations which are not directly involved in effecting either liberation or separation. The four kinds of principal operations are comminution, sizing, concentration, and dewatering.

Comminution means reduction to a smaller size. It is accomplished either on the dry ore or in aqueous pulp. Depending upon the size of the material being comminuted, the operation is regarded as crushing or grinding. Crushing is almost always conducted on dry ore, and grinding may be wet or dry.

Sizing is the separation of a material into products characterized by difference in size. This can be accomplished by screening or by classifying, the latter being a sizing method depending upon the relationship existing between the size of mineral particles and their settling velocity in a fluid medium, generally water or air.

Concentration may be accomplished by washers, sluice boxes, shaking tables, flotation cells, magnetic separators, electrostatic separators, or other specific concentrating devices. It can also be achieved by screens, classifiers, or a combination of screens and classifiers.

Dewatering is generally carried only to the extent of producing a damp cake, in two steps: first in thickeners to remove most of the water; then in filters, which receive the thickened pulp and yield the damp mineral cake. If further dewatering is desired, driers requiring fuel for evaporation of moisture are essential.

Auxiliary operations are naturally quite diverse. They involve storing in bins, conveying (by conveyors, feeders, elevators, or pumps), sampling, weighing, reagent feeding, pulp distributing, etc.

To give an idea of usual arrangements of dressing machinery, a number of flow sheets are presented on the following pages. These flow sheets are qualitative in that no account is taken of the size or number of units required, and auxiliary operations are disregarded. The flow sheets are simply designed in order to give the reader who is being introduced to the subject of mineral dressing an idea of the most important machines needed for different treatment methods, of the general arrangement of the machines, and also a crude measure of the relative complexity of each type of treatment. Broadly speaking, the flow sheets are presented in order of increasing complexity.

In this description of flow sheets, the use of a certain amount of technical jargon is unavoidable. When in doubt, the reader is referred to other parts of the text, reference to which can be found in the index. A number of terms, however, recur so often that a brief definition is useful.

The *feed* to a concentrating machine, also known as heading, is the material received for separation. The products of the separation are the *concentrate* and *tailing*, if only two products are made, the concentrate being the valuable part and the tailing the discarded product. If more than two products are made, the other may be an additional concentrate or a *middling*. The additional product is termed a concentrate if it contains in concentrated form some other valuable constituent than the first concentrate; it is termed a middling if it contains the same valuable constituent as the concentrate but in a more diluted condition. Thus a shaking table may make a lead concentrate, a zinc concentrate, and a tailing; or else a zinc concentrate, a zinc middling, and a tailing.

When the concentrate that is obtained is not of sufficiently high grade and it is graded up by retreatment, the primary oper-

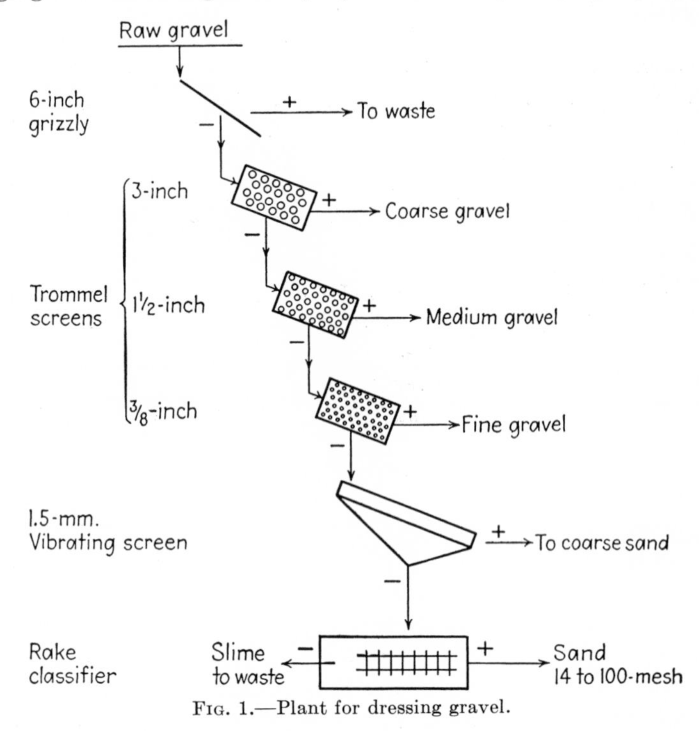

FIG. 1.—Plant for dressing gravel.

ation is known as a *roughing* operation and the retreatment operation as a *cleaning* operation. If several stages of retreatment are used, they may be termed recleaning, re-recleaning, etc. The concentrate from the primary operation is the rougher concentrate, that from the cleaning operation, the cleaner concentrate. The same terminology is adopted in connection with tailings or middlings.

Plant for Dressing Gravel.[4,5,36] Figure 1 represents a dressing plant for the preparation of concrete aggregate. It consists essentially of screens and classifiers, the screens (trommels) making coarse, medium, and fine gravel, and the classifier making sand. In addition, a vibrating screen removes coarse sand. The oversize from the coarsest screen and the fine over-

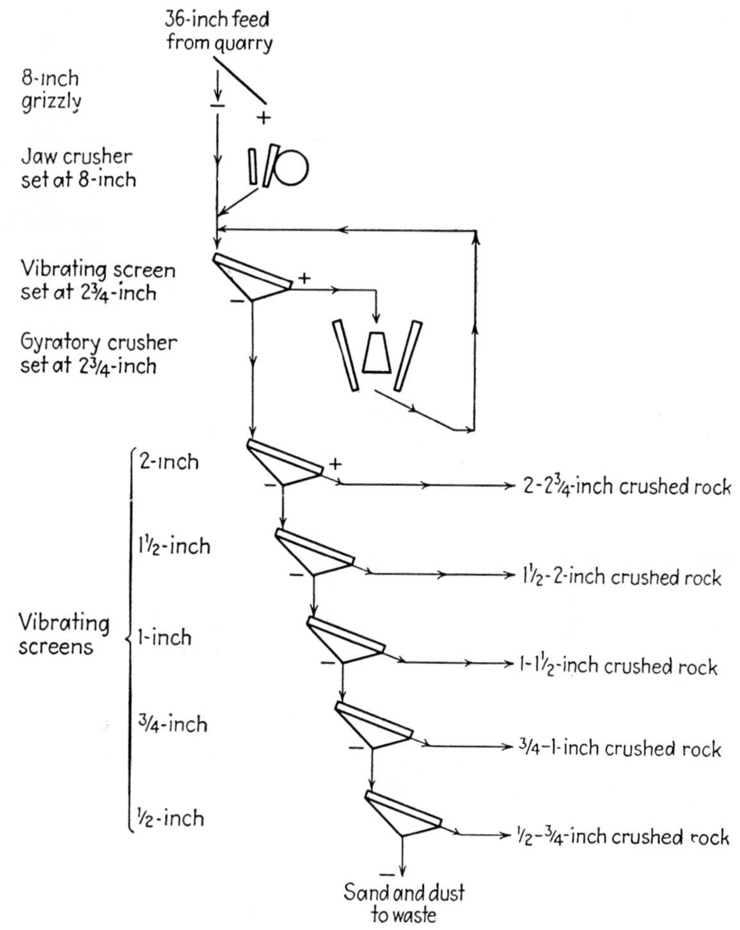

FIG. 2.—Plant for crushing and grading rock.

flow from the classifier (clay) are waste products. The sizing exerted by the various screens is useful as it permits the subsequent blending of aggregate of various sizes and of sand according to definite specifications.

Plant for Crushing and Grading Rock.(17,18,37) Figure 2 represents a rock-crushing plant. A jaw crusher crushes raw feed from which material finer than 8 in. has been removed by a

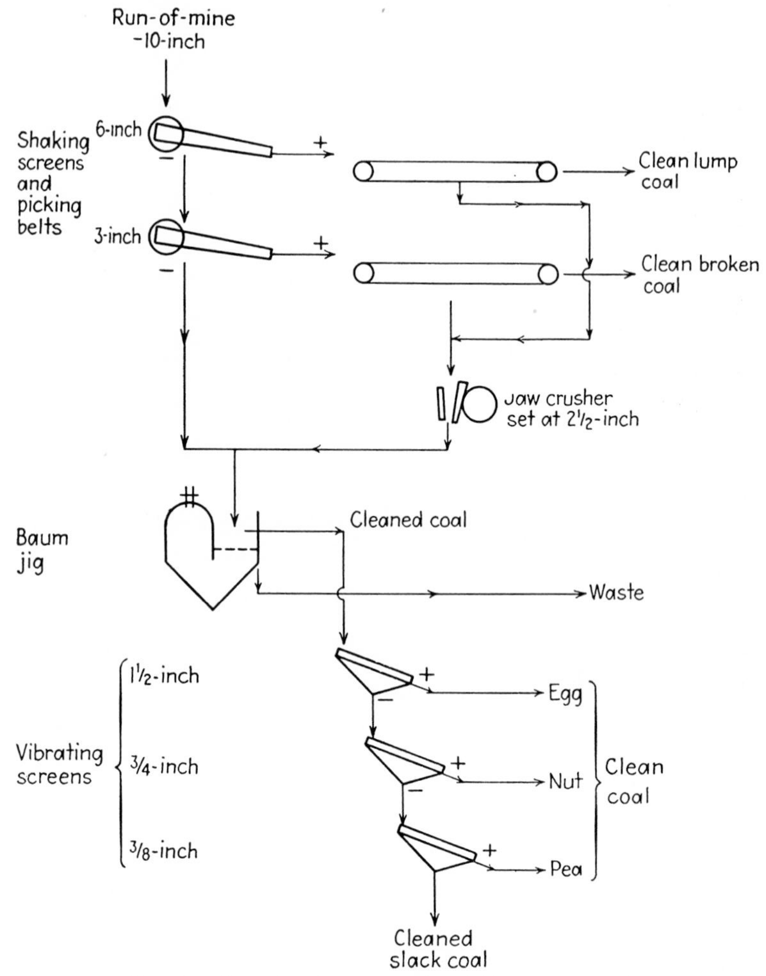

FIG. 3.—Plant for the preparation of bituminous coal.

grizzly screen. The product from the jaw crusher is reduced further to 2¾ in. by a gyratory crusher in closed circuit with a vibrating screen. Vibrating screens separate five sizes of gravel and waste fines.

It will be apparent that neither of the flow sheets outlined in Figs. 1 and 2 involves a concentrating operation.

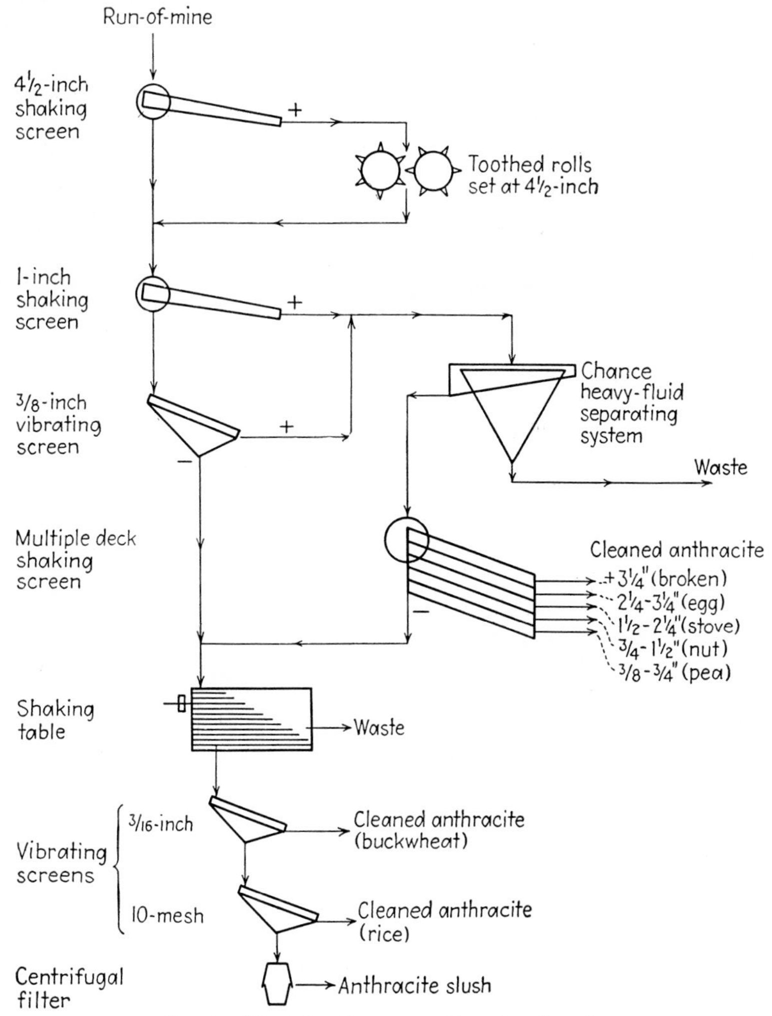

Fig. 4.—Plant for the preparation of anthracite.

Coal Plants. (6,14,15) Figure 3 represents a plant for the cleaning and preparation of bituminous coal. Concentration is by hand picking for the two coarse sizes, *viz.*, the 6- to 10-in. lump and 3- to 6-in. stove sizes, and by jigging for the minus 3-in. sizes.

Coarse sizing is done by shaking screens; fine sizing is done by means of vibrating screens.

Figure 4 represents an anthracite-preparation plant in which the coal is cleaned by a Chance heavy-fluid cone in the coarse

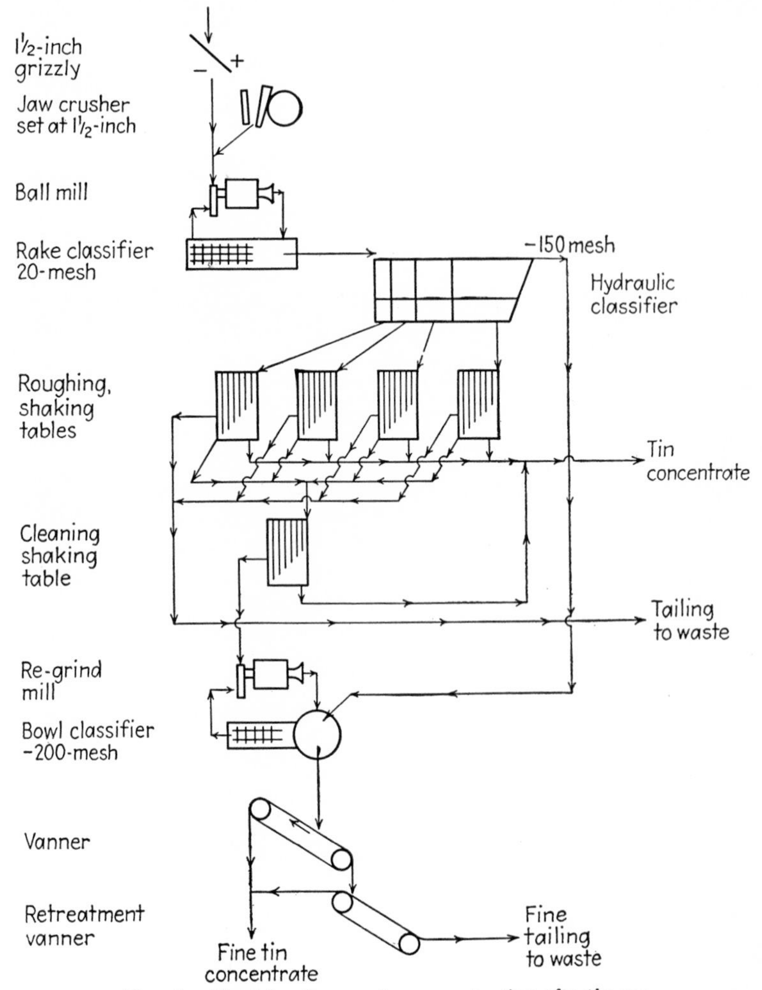

Fig. 5.—Plant for the gravity concentration of a tin ore.

sizes and a shaking table in the fine sizes. Reduction of coal to the largest commercial size ("broken") is obtained by toothed rolls. This type of crusher is selected because it keeps the formation of fines at a minimum; it receives only the plus 4½-in. lumps. Sizing to ⅜ in. is done in shaking screens to avoid disintegration, and the minus ⅜-in. coal is sized by vibrating screens. Dewatering is by natural drainage for the coarse sizes and by means of a centrifugal filter for the minus 10-mesh anthracite.

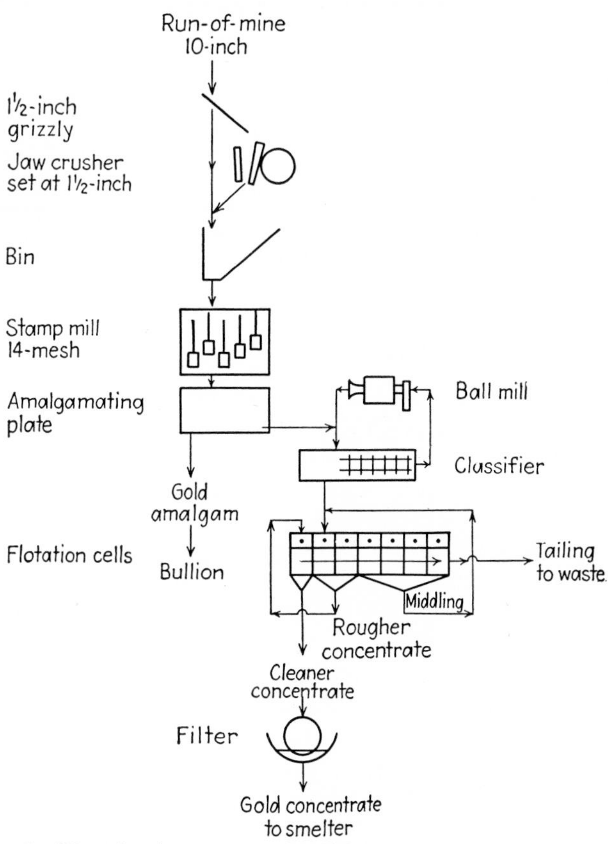

FIG. 6.—Plant for the recovery of gold by amalgamation and flotation.

Plant for the Gravity Concentration of a Tin Ore.(7,13) Figure 5 is a simplified flow sheet of a tin (cassiterite) concentration plant. The ore is reduced to 1½-in. by a jaw crusher then to 20 mesh by a ball mill, a grizzly screen and rake classifier being used to facilitate the task of the crusher and control the operation of the mill, respectively. The minus 20-mesh ore is classified into several grades by a hydraulic classifier, each grade being concentrated on a shaking table. The tables each make a concentrate, a middling, and a tailing. The middlings are retreated to yield additional concentrate and a product requiring further treatment. That product is crushed to 200 mesh in a regrind mill controlled by a bowl classifier which also received the overflow from the hydraulic classifier. Slime concentration is a two-stage vanner treatment.

Plant for the Recovery of Gold by Amalgamation and Flotation.[1,9,10,39] Figure 6 represents an amalgamation and flotation plant for the treatment of a gold ore. The run of mine, all finer than 10 in., passes over a grizzly screen having an aperture of 1½ in. The oversize is crushed in a jaw crusher set at 1½ in., and the crushed product, together with the undersize from the grizzly, is stored in a bin. The bin feeds a stamp mill crushing to 14 mesh (about 1/25 in.). The stamp mill discharge flows over an amalgamating plate into a classifier in closed circuit with a ball mill. The amalgamating plate removes the coarse ("free") native gold, yielding gold amalgam. The overflow from the classifier feeds to a battery of flotation cells, making a rougher concentrate, a middling, a cleaner concentrate, and a tailing, as shown in the flow sheet. The concentrate is dewatered in the suction filter.

Concentrating Plant for a Magnetite Ore.[26,35] Figure 7 represents a concentrating plant for the treatment of a magnetite ore. The ore is crushed to ¾ in. prior to concentrating, in three stages: jaw crusher, 42 to 10 in., gyratory crusher, 10 to 2 in., and spring rolls, 2 to ¾ in. The first two crushing steps are with by-pass of the fines by a screen, and the third step includes a screen in closed circuit. Dry magnetic concentration is carried out on three sizes, ⅜ to ¾ in., 3/16 to ⅜ in., and 1/16 to 3/16 in., the sizing being done on a multiple-deck vibrating screen. Each magnetic separator makes a finished tailing, part of which can be utilized as gravel. The separator working on the ¾- to ⅜-in. feed makes only a middling, but the others make concentrate and middling. The plus 3/16-in. middlings are crushed in spring rolls and returned to the multiple-deck vibrating screen. The minus 3/16-in. middling and the 1/16-in. undersize from the multiple-deck vibrating screen are ground to 65 mesh in a ball mill controlled by a rake classifier. A demagnetizer precedes the classifier to prevent magnetic flocculation in the classifier. Concentration is by roughing and cleaning in magnetic log washers. A demagnetizing step precedes the cleaning operation in order to

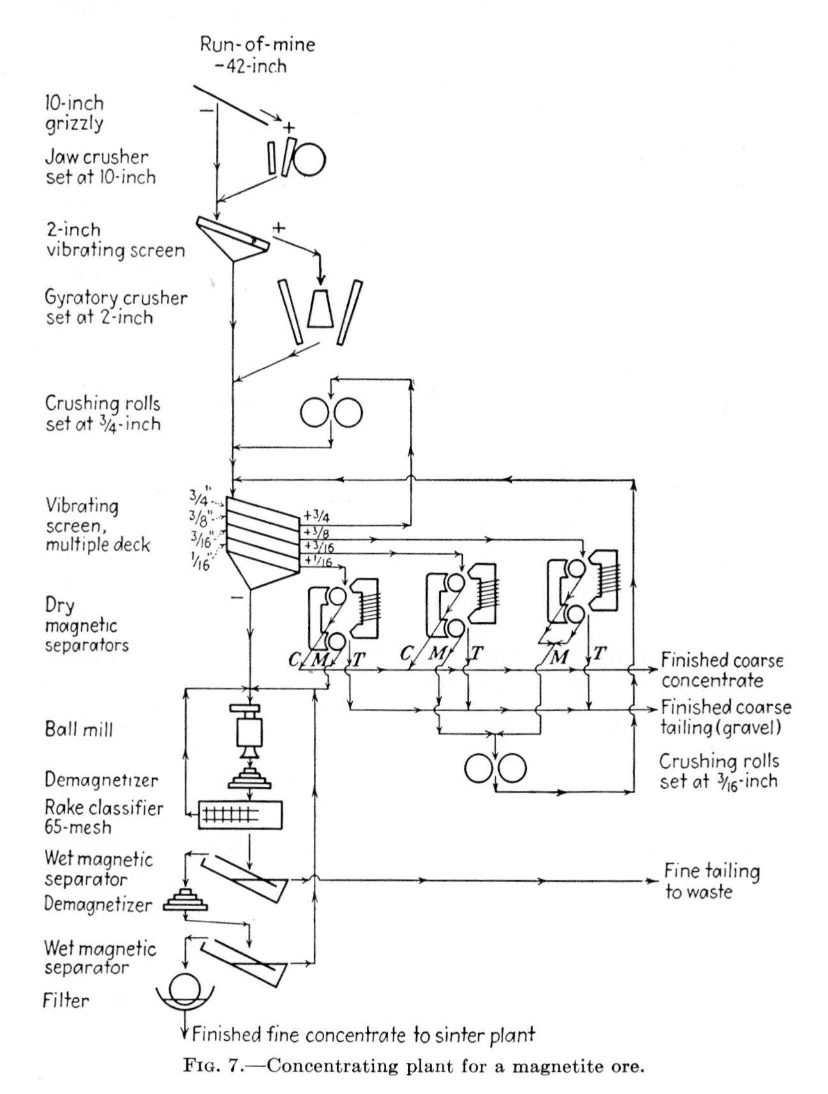

FIG. 7.—Concentrating plant for a magnetite ore.

eliminate the waste occluded mechanically within magnetic flocs. Filtering completes the dressing operation, and the concentrate is ready for sintering.

Flotation Plants. Figure 8 represents a concentrating plant for the treatment of phosphate rock.[21,22,28] The "rock" consists of phosphate nodules and sand loosely cemented with

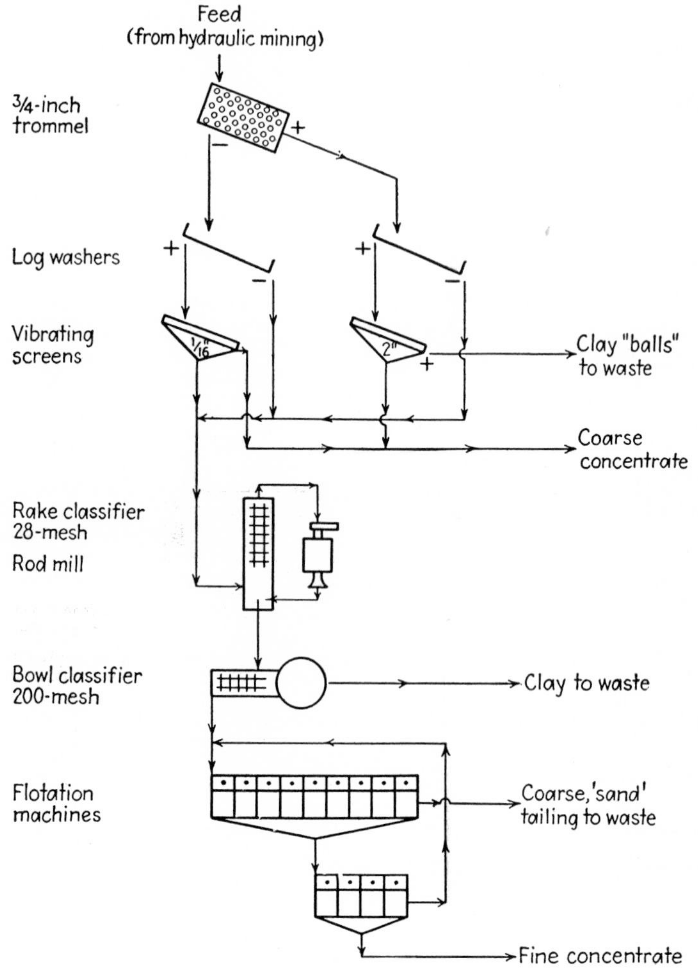

FIG. 8.—Plant for the concentration of phosphate rock.

clay. It is first disintegrated by tumbling in a trommel, and then in log washers. The material that passes through a ¾-in. trommel but settles in the log washer is passed over a vibrating screen making a separation at 1/16 in., the oversize being finished phosphate rock (coarse concentrate). The material retained by the trommel and settling in the subsequent log washer is further separated by a vibrating screen into "clay balls," which are dis-

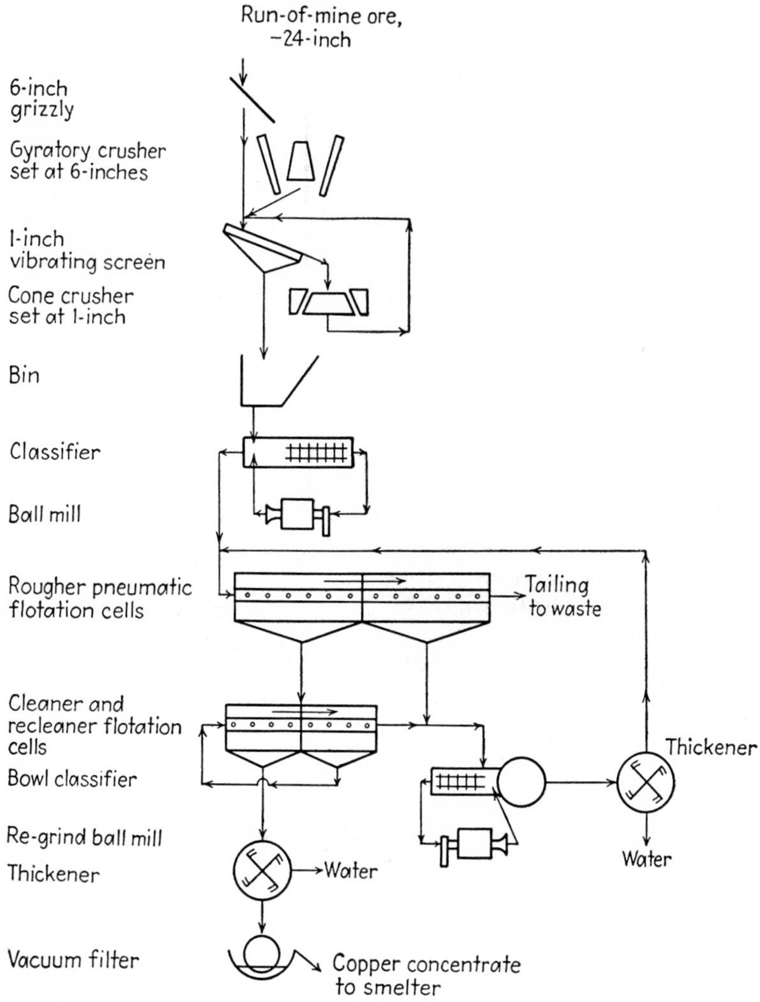

FIG. 9.—Flotation plant for the concentration of a copper sulphide ore.

carded, and cleaned rock. The undersize from the $\frac{1}{16}$-in. vibrating screen, together with overflow from the log washers, goes to a classifier and rod mill in closed circuit, the overflow of which feeds into a bowl classifier. The bowl classifier is set to overflow particles finer than 200-mesh (1⁄400 in.). The overflow (mostly clay) goes to waste, and the sands go to the flotation machine. This machine makes a rougher concentrate and a tailing. The rougher concentrate is cleaned once in another flotation machine which yields finished fine-size phosphate concentrate and a middling. This middling is returned for retreatment to the head of the primary or rougher flotation machine. The flotation concentrate is dewatered by sedimentation.

Figure 9 represents a plant for flotation of copper sulphide.(16,20,23,25,29) It features two stages of coarse crushing, the first in a gyratory crusher, and the second in a cone crusher, each with its screen. The crushed material, finer than 3⁄4 in. is stored in a bin. This part of the plant may be operated intermittently, *e.g.*, one shift per day. From the bin, the ore goes to a drag classifier in closed circuit with a ball mill. The overflow from the classifier goes to a pneumatic roughing flotation machine which makes a concentrate, a middling, and a tailing. The tailing is a finished waste product and is discarded. The rougher concentrate is cleaned twice in pneumatic machines. The recleaned concentrate is a finished product which is merely dewatered in the thickener and the vacuum filter on its way to the smelter. The tailing from the cleaner machine and the middling floated in the rougher machine are reground in a closed circuit consisting of a ball mill and a bowl classifier. The overflow from the bowl classifier is partly dewatered in a thickener before being returned to the head of the roughing flotation machine.

Figure 10 represents a flotation plant for the treatment of a complex lead-zinc ore(8,11,27,30,31,41,42). In this instance again, there are two stages of coarse crushing, arranged substantially as in the case of the plant diagrammed as Fig. 9. The crushed ore, finer than 5⁄8 in., is stored in a bin. From the bin, the ore passes to the grinding circuit consisting of a drag classifier and ball mill. The lead minerals are floated in the lead-roughing flotation machines. The deleaded pulp passes to a conditioning tank and thence to a zinc circuit which is practically a duplication of the lead circuit. The lead rougher concentrate from the lead rougher is cleaned and recleaned. The recleaner concentrate is dewatered in a thickener and filter on its way to the smelter. The recleaner tailing is returned to the head of the cleaner machine, and likewise the cleaner tailing is returned to the head of the rougher machine, countercurrent fashion. The cleaner tailing, however, is reground in a ball mill-bowl classifier circuit

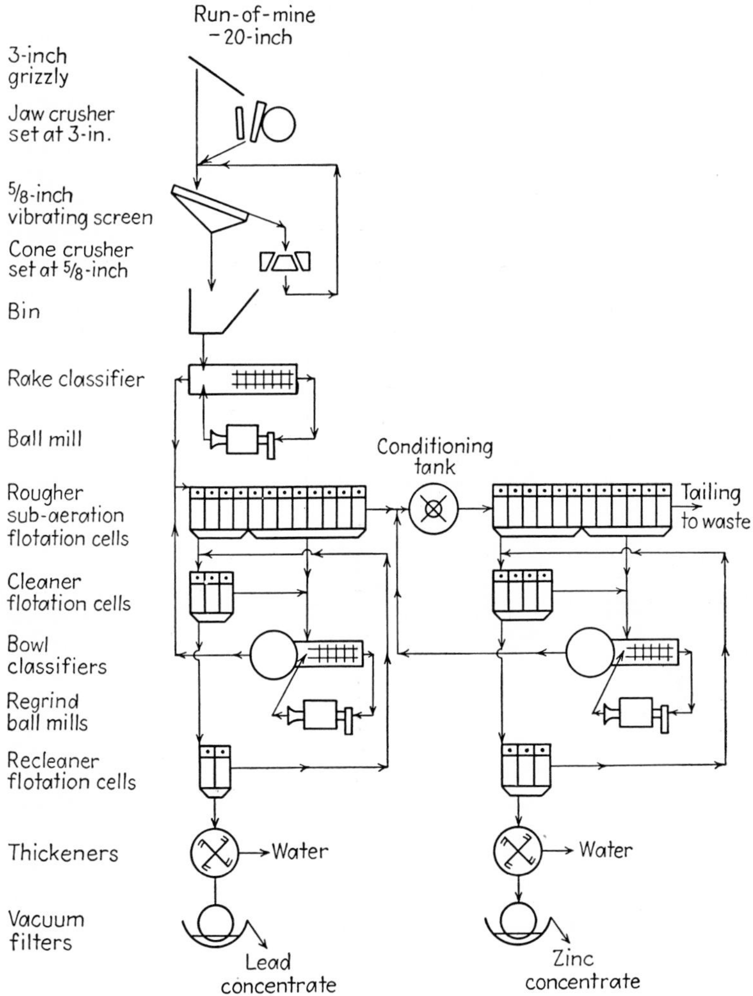

Fig. 10.—Plant for the concentration of a lead-zinc sulphide ore by selective flotation.

before being returned to the head of the lead circuit. In this treatment, it is joined by the middling floated in the rougher machine. This regrind of the rougher middling and of the cleaner tailing is necessitated by the presence in these products of a large proportion of locked particles.

Discussion of Type Flow Sheets. Table 1 summarizes the various types of principal operations in the flow sheets presented as Figs. 1 to 10. The number of times each type of operation occurs is also indicated. The student will find it instructive to examine each flow sheet with reference to Table 1.

TABLE 1.—PRINCIPAL OPERATIONS APPEARING IN FLOW SHEETS, FIGS. 1–10.

Flow sheet	Comminuting	Sizing	Concentration	Dewatering
1	0	6	0	0
2	2	7	0	0
3	1	5	3	0
4	1	10	2	1
5	3	4	7	0
6	3	2	3	1
7	5	7	5	1
8	1	5	4	0
9	4	4	3	3
10	5	5	6	4

Flow sheets descriptive of actual practice are usually detailed, and contain not only those devices which perform principal operations, but also those which perform auxiliary operations. They contain also a statement of the make of the machine, the size of the machine, and the number of units. It is a profitable exercise for the student to prepare a flow sheet reduced to essentials from a detailed flow sheet, and to count the number of times a principal operation of each type is performed.

Literature Cited

1. ADAM, H. R., and F. WARTENWEILER: Flotation of Banket Sands, *J. Chem. Met. Mining Soc. S. Africa*, **37,** 108–116, 205 (1936).
2. AGRICOLA: "De re metallica," trans. Herbert Clark Hoover and Lou Henry Hoover, The Mining Magazine, London (1912).
3. ALFORD, NEWELL G.: Bibliography of Coal Cleaning, Jan. 1, 1934 to Mar. 31, 1935, *1935 Yearbook Coal Mine Mechanization, Am. Mining Congr.*, 276–280.
4. ANABLE, ANTHONY: A Unique Sand and Gravel Plant, *Mining and Met.*, **15,** 403–406 (1934).
5. ANABLE, ANTHONY: Preparation of High-specification Sand at the Grand Coulee Dam, *Am. Inst. Mining Met. Engrs., Tech. Pub.* 715 (1936).

6. ANON.: United Electric Companies' Fidelity Mine and Washery, *Mining and Met.*, **17,** 416–420 (1936).
7. ANON.: A New Tin Producer in Siam, *Eng. Mining J.*, **137,** 42 (1936).
8. BEMIS, H. D.: Milling Methods and Costs at the Pecos Concentrator of the American Metal Co., Tererro, N. Mex., *U. S. Bur. Mines, Information Circ.* 6605 (1932).
9. BRADLEY, JAMES: Mining and Milling at the Spanish Mine, *Mining and Met.*, **12,** 435–439 (1931).
10. CAMPBELL, STEWART: Amalgamation-flotation of Gold Ore, *Eng. Mining J.*, **134,** 188–190 (1933).
11. DALTON, M. P.: Milling Methods and Costs at the Morning Concentrator of the Federal Mining and Smelting Company, Mullan, Idaho, *U. S. Bur. Mines, Information Circ.* 6587 (1932).
12. DAVIS, EDWARD: Magnetic Roasting of Iron Ore, *Univ. Minn., Mines Expt. Sta. Bull.* 13 (1937).
13. DERINGER, D. C., and JOHN PAYNE, JR.: Patiño—Leading producer of Tin: III. Ore Concentration at Catavi, *Eng. Mining J.*, **138,** 299–306 (1937).
14. HEBLEY, HENRY F.: Modernization of Coal-preparation Plants Continues without Striking Innovations, *Mining and Met.*, **17,** 36–37 (1936).
15. HOFFMANN, H.: Die Aufbereitung der Saarfettk ohlen, *Glückauf*, **72,** 945–960 (1936).
16. HUNT, H. D.: Milling Methods and Costs at the Concentrator of the Miami Copper Co., Miami, Ariz., *U. S. Bur. Mines, Information Circ.* 6573 (1932).
17. KEISER, H. D.: Crushed-stone and Ore-dressing Plant—A Profitable Combination, *Eng. Mining J.*, **132,** 4–7 (1931).
18. KEISER, H. D.: Barytes Producer Enters Washed-gravel Industry, *Eng. Mining J.*, **132,** 266–267 (1931).
19. LAWFORD, E. G.: An Example of Modern Practice in the Recovery of Tin from Tailing, *Eng. Mining J.*, **124,** 773–774 (1927).
20. MARTIN, H. S.: Milling Methods and Costs at the Arthur and Magna Concentrators of the Utah Copper Co., *U. S. Bur. Mines, Information Circ.*, 6479 (1931).
21. MARTIN, H. S.: Milling Methods and Costs at No. 2 Concentrator of the Phosphate Recovery Corporation, *Trans. Am. Inst. Mining Met. Engrs.*, **112,** 466–485 (1934).
22. MARTIN, H. S., and JAMES WILDING: "Industrial Minerals and Rocks," Chap. XXXIII, "Phosphate Rock," American Institute of Mining and Metallurgical Engineers, New York, (1937), pp. 543–570.
23. MORROW, BAYARD S.: Both Copper and Zinc Ores Treated by Selective Flotation in Concentrators at Anaconda, Montana, *Eng. Mining J.*, **128,** 295–300 (1929).
24. MORROW, J. B.: Coal Preparation, *Mining and Met.*, **15,** 56–58 (1934).
25. MUNRO, A. C., and H. A. PEARSE: Milling Methods and Costs at the Concentrator of the Britannia Mining and Smelting Co., Ltd., Britannia Beach, B. C., *U. S. Bur. Mines, Information Circ.* 6619 (1932).

26. Myners, T. F.: Magnetic Concentration Methods and Costs of Witherbee, Sherman and Co., Mineville, N. Y., *U. S. Bur. Mines, Information Circ.* 6624 (1932).
27. Page, W. C.: Trepca Mines Limited. IV. Milling the Ore, *Mining and Met.*, **17,** 584–585 (1936).
28. Pamplin, J. W.: Ore-dressing Practice with Florida Pebble Phosphates, Southern Phosphate Corporation, *Am. Inst. Mining Met. Engrs., Tech. Pub.* 881 (1938).
29. Pearse, H. A.: Three-product Flotation at the Britannia, B. C., Mill, *Mining and Met.*, **15,** 379–383 (1934).
30. Price, G. S.: Milling Methods and Costs at the Page Concentrator of the Federal Mining and Smelting Company, Kellogg, Idaho, *U. S. Bur. Mines, Information Circ.* 6590 (1932).
31. Read, T. A.: The New Lead and Zinc Flotation Sections of the Concentrator of Broken Hill South, Ltd., *Proc. Australasian Inst. Mining Met.*, **98,** 305–348 (1935).
32. Richards, Robert H.: The Anaconda Round Table, the Wilfley Table, and the Ten-spigot Classifier, *Mining and Met.*, **15,** 342–343 (1934).
33. Rickard, T. A.: "Man and Metals," Whittlesey House, McGraw-Hill Book Company, Inc., New York (1932).
34. v. Rittinger, P. Ritter: "Lehrbuch der Aufbereitungskunde," Ernst and Korn, Berlin (1867).
35. Roche, H. M., and R. E. Crockett: Evolution of Magnetic Milling at Scrub Oak, *Eng. Mining J.*, **134,** 241–244, 273–277 (1933).
36. Shaw, Edmund: Washing and Sizing Sand and Gravel, *Trans. Am. Inst. Mining Met. Engrs.*, **73,** 424–433 (1926).
37. Slattery, Louis A.: Large Crushing and Screening Plant Reflects Engineering Progress, *Eng. Mining J.*, **134,** 2–10 (1933).
38. Verhoeff, Jack R.: Coal Processing Plant of Peabody Coal Company, *Proc. Illinois Mining Inst.*, 1936, pp. 89–97.
39. Weekley, C. A., and S. W. Norton: Some Refractory Gold Ores of the Philippines, *Eng. Mining J.*, **138,** 412–413 (1937).
40. Wood, Henry E.: The Beginning of the Wilfley Concentrating Table, *Eng. Mining J.*, **125,** 975–977 (1928).
41. Young, A. B., and W. J. McKenna: Selective Flotation of Lead-zinc Ores at Tooele, Utah, *Eng. Mining J.*, **128,** 291–294 (1929).
42. Young, George J.: Selective Lead-zinc Flotation at Kimberley, B. C., *Eng. Mining J.*, **131,** 313–315 (1931).

CHAPTER II

CRUSHERS

Crushing may be defined as that operation or that group of operations in a mineral-dressing plant whose object is to reduce large lumps to fragments, the coarsest particles in the crushed product being $\frac{1}{20}$ in. or more in size. Although, in general, the object of crushing is *size reduction* only, this is occasionally coupled with a requirement for the production of a minimum amount of *fines*.

Crushers have been designed that will reduce rock so that all the fragments are finer than a definite maximum size. But no crusher has ever been devised that will produce only material exceeding a certain minimum size. There is produced, always, a substantial proportion of material that is crushed too fine for the purpose at hand. Although some types of machines produce a smaller proportion of fines than others, the difference between machines is small.

In describing a crushing operation, it is customary to state the maximum size of the particles in the feed and in the product. The ratio of these two sizes is known as the *reduction ratio* and is a convenient yardstick for comparing the performance of different machines. Other definitions of the reduction ratio are also sometimes given.

The crushing action in all crushing machines results from stresses that are applied to the particles to be crushed by some moving part in the machine, working against a stationary part or against some other moving part: a crusher is substantially a glorified nutcracker. The stresses set up strains within the particle to be broken which result in fracturing whenever they exceed the elastic limit of the material. It is therefore clear that if the crushing force is not intense enough, there may be no crushing effect. This explains in part why crushing machines, especially those designed for crushing effectiveness regardless of the overproduction of fines, are so very rugged and massive.

Crushers can be classified into four groups as follows: (1) for coarse or primary breaking, (2) for intermediate or secondary crushing, (3) for fine crushing, (4) for special uses. Jaw and gyratory crushers are used for coarse crushing; reduction gyratory crushers, cone crushers, disk crushers, and spring rolls are used for secondary crushing; and gravity stamp mills are used for fine crushing. There are a number of special machines for soft materials designed to limit the overproduction of fines or to crush selectively the more friable constituents of these soft materials. These include toothed rolls and hammer mills.

Modern practice tends to do without fine crushing as grinding devices have become improved and as new methods of concentration have supplanted older methods.

Coarse crushing is always conducted on dry material, and the tendency is also to conduct the intermediate crushing on dry material, although rolls are still used to crush wet.

Jaw Crushers

Jaw crushers[6] consist of two crushing faces or *jaws* one of which is stationary, being mounted rigidly in the crusher frame, and the other moving alternately toward the stationary face and away from it by a small throw.

Jaw crushers can be classified according to the point of minimum amplitude of motion on the moving face and according to the way in which the motion of the movable face is transmitted. In jaw crushers of the *Blake* type, the movable jaw is hinged at the top so that the greatest amplitude of motion is at the bottom of the crushing face. In the *Dodge* crusher, the reverse is true: the bottom of the jaw is the fulcrum, and the top moves by the greatest amplitude.

Blake Jaw Crusher. Blake crushers are made in the largest size and have the distinction of being the oldest type of coarse-crushing machinery still in current use. The Blake crusher dates back to 1858,[5] prior to which time coarse crushing was done by hand and hammer at a much higher cost.

In a typical Blake crusher, the crushing frame (Fig. 11) is made of cast steel ①. The jaws are made of cast steel lined with replaceable jaw plates ⑧ of alloy steel, generally manganese steel, and the sides of the crushing opening are made of manganese-steel cheek plates ⑨. Motion is transmitted from the main crusher shaft ④ by means of a pitman ③ working on an eccentric on that shaft, and of toggles ⑤. The pitman is nearly vertical and the toggles nearly horizontal. One of the toggles is set in steel bearings at one end on the frame and at the other on the pitman; the other toggle is set in steel bearings on the back of the movable jaw at one end and in the pitman at the other ⑥. As a result of this design, the rotation of the drive shaft causes an up-and-down translation of the pitman and consequently an increase and decrease of the distance between the back of the movable jaw and the frame. The movable jaw is kept pressed

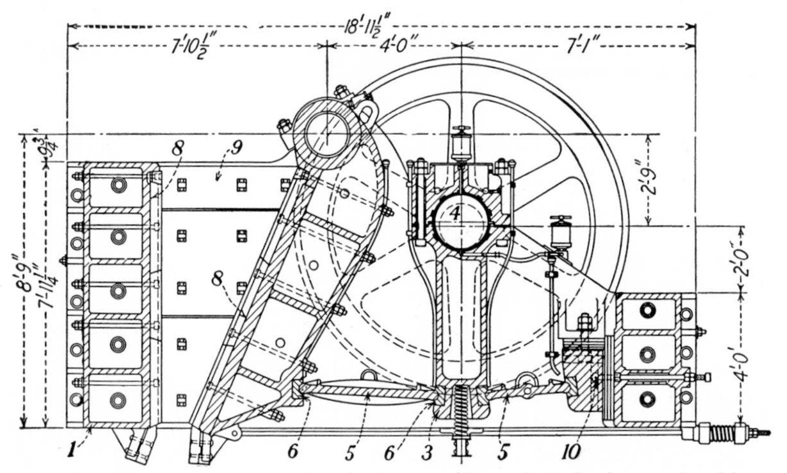

FIG. 11.—Blake-type jaw crusher (section). (*From C. G. Buchanan, Crushing Machinery Division, Birdsboro Steel Foundry and Machine Co.*)

against the toggles by a tension rod and spring. Provision is made for wear on the movable jaw by shims or by making it possible to move the setting block of the rear toggle ⑩. Forced-feed lubrication is the rule.

Because of the intermittent load on a jaw crusher, it is necessary to equalize this load by means of flywheels. One or two flywheels are generally mounted on the main shaft of the crusher, and the drive is by flat belt or V-belts.

Dodge Crusher. The Dodge crusher (Fig. 12) is made only in small sizes. It differs from the Blake crusher in that the fulcrum of the motion of the movable plate is at the bottom of the plate so that the width of the discharge opening remains practically constant. This results in the production of a more closely sized product. Also, the crusher has fewer parts: one toggle instead of two, and no pitman. The advantage of constant discharge opening and of more uniform size of product may be significant if the crusher is the only comminuting device used, but if the crusher is followed by other comminuting machinery, this minor advantage is of little practical value. The Dodge crusher, also, represents a less satisfactory design from a mechanical standpoint in that the displacement of the movable jaw is least where the crusher is required to disintegrate the finer lumps and greatest where the crusher has to deal with coarser lumps. The stresses are therefore uneven. This makes the machine weaker and is the reason why it is not built in large sizes. In large installations

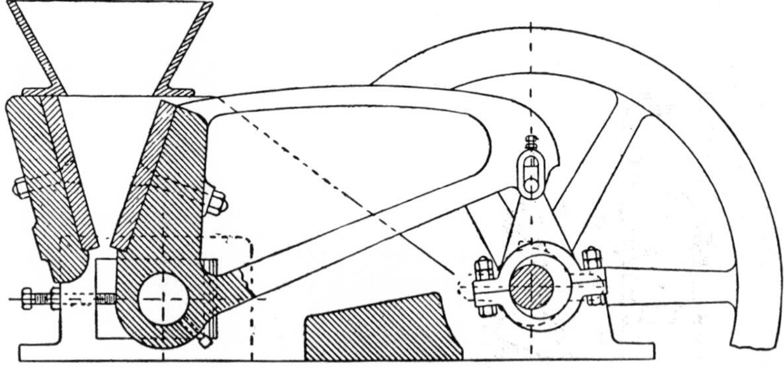

FIG. 12.—Dodge-type jaw crusher, generalized section. (*From Kennedy-Van Saun Mfg. & Eng. Corp*)

where elaboration of flow sheet by the introduction of screens is not a serious objection, the Blake crusher is preferred.

Telsmith Jaw Crusher. The Telsmith crusher (Fig. 13) resembles the Blake crusher in that motion of the swinging jaw is largest at the bottom, but it differs in that motion is transmitted to the movable plate, not by means of toggles and pitman, but directly from an eccentric on the main shaft by means of a horizontal, direct-stroke pitman.

The Telsmith jaw crusher is a new machine not made in sizes so large as standard Blake-type crushers but one that has met with considerable favor in practice.

A jaw crusher of new design is the Eimco-Fahrenwald crusher.[1] The principal feature of this machine is that the "stationary" jaw is mounted against heavy springs. This allows it to move away from the reciprocating jaw if an uncrushable object falls in the crusher.

Characteristics of Jaw Crushers. Jaw crushers are intended for use as primary crushers, *i.e.*, to receive the coarsest lump produced from the mine or quarry. Accordingly, jaw crushers have a relatively large *gape* (width of the receiving opening). The length of the receiving opening is generally somewhat

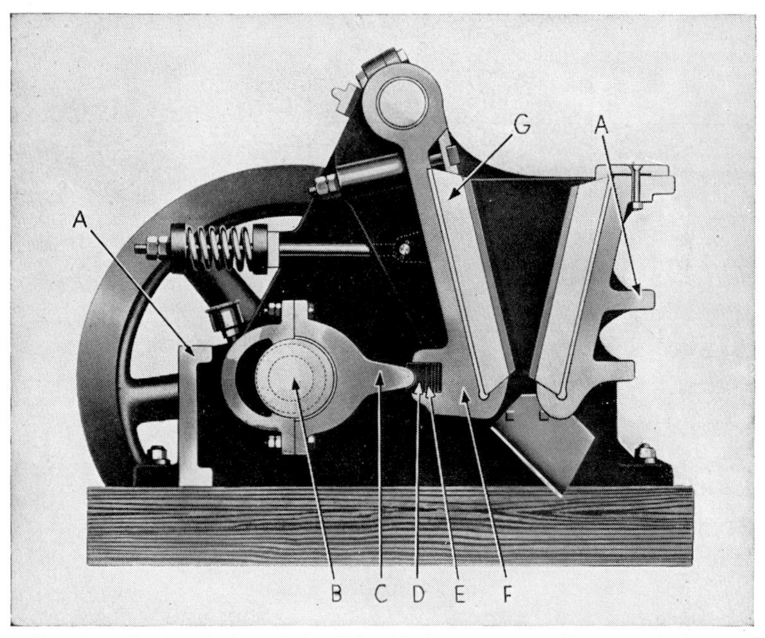

FIG. 13.—Sectional view of the Telsmith jaw crusher. Parts marked are as follows: (*A*) cast-steel main frame, (*B*) eccentric shaft, (*C*) horizontal direct-stroke pitman, (*D*) jaw block, (*E*) adjusting shims, (*F*) cast-steel swinging jaw, (*G*) reversible jaw die. (*From Smith Engineering Works.*)

greater than the width. The largest jaw crushers made to date have a receiving opening 84 by 120 in. They can accommodate a block 6 ft. in diameter and weighing many tons. All jaw crushers have an adjustable discharge opening (*set*) so that they can produce a product that is coarse or fine, within limits, depending upon the adjustment. In modern practice, it is not unusual for jaw crushers to yield a reduction ratio of 7 although in some instances the ratio is as small as 4. The tendency is to build and operate crushers so as to obtain a higher reduction ratio than heretofore.

The *capacity* of jaw crushers is very large. For instance, a crusher having a receiving opening 48 by 42 in. will reduce to 6 in. or finer nearly 4,000 tons per 24 hr. Table 2 (Allis-Chalmers

TABLE 2.—SIZES AND CAPACITIES OF TYPICAL JAW CRUSHERS

Size, inches	Capacity in tons per hour to size product stated				Size of pulley, inches	R.p.m.	Hp. required	Weight of heaviest piece, pounds	Total weight, pounds
	Min. size, inches	Tons	Max. size, inches	Tons					
15 × 10	1½	7	3½	21	30 × 9	235	15	3,650	10,000
24 × 15	2¼	22	5	60	42 × 13	210	35	13,600	27,000
36 × 24	3	45	6	110	76 × 12½	210	75	35,000	70,000
42 × 40	4½	90	8	190	84 × 18½	190	125	80,000	140,000
48 × 36	4½	110	8	225	84 × 18½	190	150	88,000	145,000
48 × 42	4½	110	8	225	96 × 18½	190	150	96,000	160,000
60 × 48	5	150	9	300	96 × 25	170	200	62,000	215,000
84 × 56	8	360	12	630	147 × 40¾	90	200	63,000	422,500
84 × 60	8	360	12	630	147 × 40¾	90	250	68,000	430,000
84 × 66	9	420	12	630	147 × 40¾	90	250	76,000	460,000

NOTE:—Product sizes given above correspond approximately to the discharge opening measured from tip of corrugation on one jaw plate to bottom of corrugation on the other plate when the jaws are open. Capacities as indicated are conservative and in some cases can be increased considerably, depending upon the character of material to be broken, operating conditions, etc.

Manufacturing Company) gives the sizes and capacities of "Superior" jaw crushers.

Taggart[16a] gives the following empirical formula for the capacity of crushers:

$$T = 0.6LS, \tag{II.1}$$

in which T is the capacity in tons per hour, L the length of the receiving opening, and S the width of the discharge opening (set), both in inches. This formula is said to be fairly accurate except for the smallest and largest crushers: for small crushers the answer is too high and for large crushers too small. That the Taggart formula represents only a first approximation is suggested by its disregard of such necessarily significant factors as specific gravity, hardness, toughness, moisture content, and structural weaknesses of the ore.

The *energy consumed* by jaw crushers varies considerably. It depends upon the following factors: size of feed, size of product, capacity of machine, properties of the rock, and percentage of idling time. Broadly speaking, the energy consumption decreases with increasing size of feed (at constant reduction ratio), with increasing size of product (at constant size of feed), and with increasing capacity of the machine. On the assumption that the idling loss is kept to a minimum, the energy consumed ranges from 0.3 to 1.5 kw.-hr. per ton crushed. Energy consumption of crushing devices is generally, though erroneously, called *power consumption.*

The *attendance* is usually one man per machine. The attendant has other duties besides that of watching the crusher operate, *viz.*, removing chips or tramp metal from feed belts, controlling bin discharge, barring down arching rock lumps, etc. In view of the fact that crushers of very large capacity are available at a first cost that is not unreasonable, it is generally considered good practice to install a crusher of such size that it will handle in one shift the whole input of a continuously operating mill.

Breakdowns are usually caused by chunks of metal jamming in the crusher and fracturing toggles, pitman, or eccentric; another source of breakdowns is the failure of the lubricating system of the crusher. Spare parts must be on hand to minimize lost time.

Wear is a substantial source of expense in jaw plates and cheek plates, both of which must be replaced from time to time. Wear cost consists not only of the cost of the new parts, but also of the labor of installation and of the time lost to operation.

GYRATORY CRUSHERS

Gyratory crushers[7,8] have been developed more recently than jaw crushers in order to supply a machine having an even greater capacity.

Reduced to its elements, a gyratory crusher consists of two substantially vertical, truncated, conical shells, the outer shell having its apex pointing down and the inner shell having its apex pointing up. The outer shell is stationary, and the inner shell is made to gyrate or to rotate, thus alternately receding from and approaching all the points on the periphery of the outer shell. There are several types: the best known are the supported-spindle gyratories (obsolete), the suspended-spindle gyratories, and the "parallel-pinch" crushers. Accurately speaking, the latter is not a gyratory crusher since the crushing head rotates eccentrically instead of gyrating.

Suspended-spindle Gyratory Crusher. Figure 14 is a section of a suspended-spindle-type gyratory crusher. The crusher consists of an outer frame (1) carrying a wearing surface known as *concaves* (2) and an inner *crushing head* (3) mounted on a spindle (4).

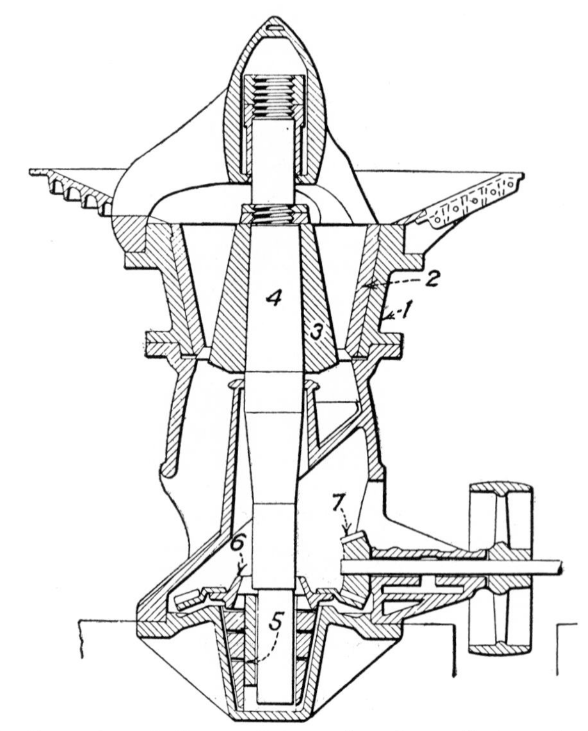

Fig. 14.—Suspended-spindle gyratory crusher (generalized section). (*After W. T. W. Miller.*)

This spindle is made to gyrate from a fixed fulcrum at the point of suspension by an eccentric sleeve (5) fastened to a beveled gear (6) which meshes into another beveled gear (7) driven from a horizontal drive shaft.

Parallel-pinch Crusher. Figure 15 represents the Telsmith parallel-pinch crusher. The essential difference between this and the suspended-spindle-type crusher is that the motion of the crushing head, instead of generating an acute cone with its apex at the point of suspension, generates a cylinder so that the extent of the pinch exerted on the rock to be crushed is the same at all points along the face of the crushing head.

A recent improvement[4] is the introduction of flaring or "bell-shaped" heads and concaves (Fig. 16) for gyratories used as secondary crushers (reduction gyratories). This improvement

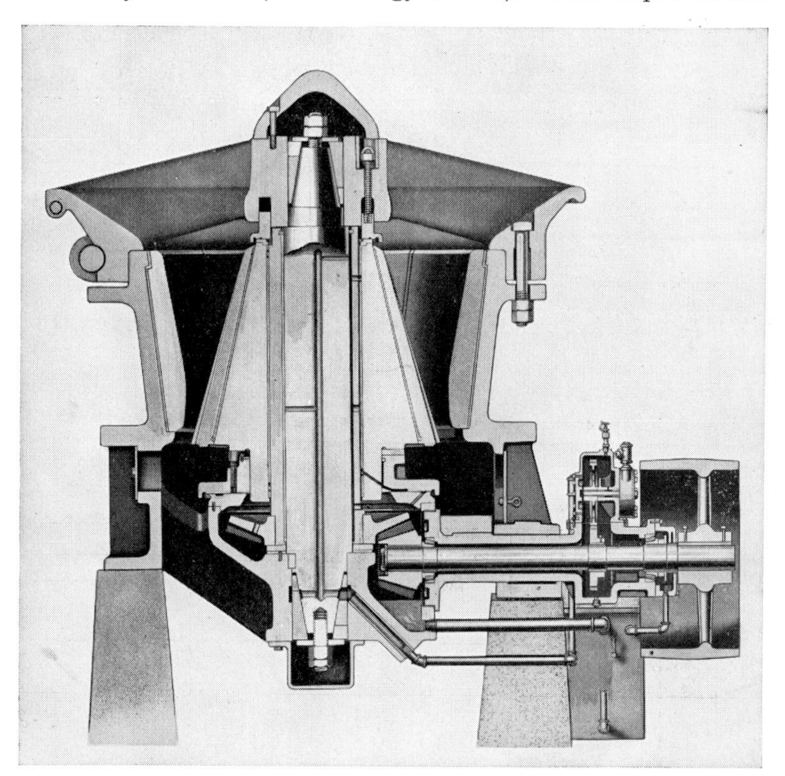

FIG. 15.—Sectional view of Telsmith parallel-pinch crusher. (*From Smith Engineering Works.*)

is said to prevent packing near the discharge, to reduce appreciably the energy consumption, and to cut down the wear.

To minimize gear trouble and reduce frictional energy consumption, modern gyratory crushers are built so that the gears are running in an oil bath. Oil pumps driven by the crusher itself circulate the oil through the gear mechanism, and the oil then passes to an outside cooler and filter. Some care is required to see that this cooling and filtering mechanism is in proper working order.

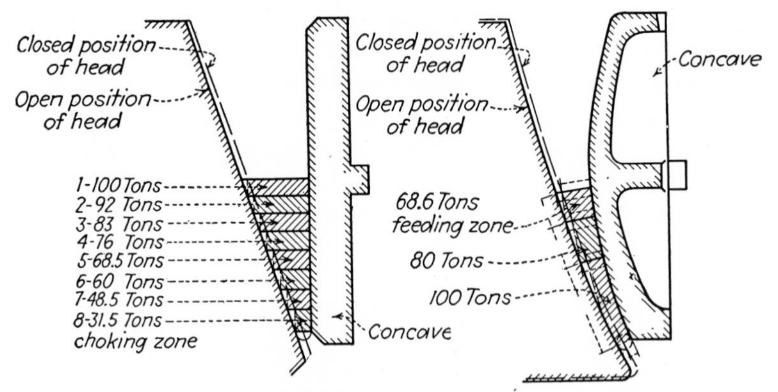

FIG. 16.—Conical and bell-shaped concaves and heads for gyratory crushers. (*After Bernhard.*)

TABLE 3.—SIZE AND CAPACITIES OF TYPICAL GYRATORY CRUSHERS

Size opening, inches	Approx. shipping weight, pounds	Size of each receiving opening, inches	Approximate capacity per hour, tons of 2,000 lb.†			
			Maximum set		Minimum set	
			Inches	Tons	Inches	Tons
2½	550	2½ × 14	½	¾	⅜	½
8	16,000	8 × 37	2	36	1	14
12	28,500	12 × 46	2¾	70	2	39
14	39,500	14 × 55	3½	95	2¼	60
16	60,000	16 × 63	4½	150	2½	100
20	104,000	20 × 80	5	275	3	150
26	160,000	26 × 100	6	400	3½	225
30	175,000	30 × 118	6½	450	4	265
36	255,000	36 × 136	7	600	4½	370
42	280,000	42 × 153	7½	700	5	410
42*	402,000	42 × 153	7½	1,000	5	600
48	520,000	48 × 166	9	1,890	5½	1,100
54	600,000	54 × 190	9½	2,100	6¼	1,350
60	950,000	60 × 210	10	2,400	7	1,600
72	1,400,000	72 × 242	12	3,400	9	2,500

* Extra heavy machine—made in standard drive only.

† Capacities are based on materials weighing 100 lb. per cu. ft. when crushed.

Characteristics of Gyratory Crushers. Gyratory crushers are available for the coarsest crushing. The largest crushers to date have a gape of 72 in. and weigh about 700 tons.

The capacity of gyratory crushers (Table 3) is much greater than that of jaw crushers handling the same size of feed; indeed, it is not uncommon to find that gyratory crushers having a gape suitable for the coarsest crushing step have at the same time an excessive capacity. As a result, one commonly has to use a jaw crusher for primary breaking. On the other hand, gyratory crushers for secondary crushing do not often have excess capacity. Many flow sheets show a single primary jaw crusher or gyratory crusher followed by two or more "reduction" gyratories.

The gyratory crusher has an advantage over the jaw crusher in that it has a more regular power draft. This is because the gyratory is crushing continuously, whereas the jaw crusher is working intermittently. In respect to reduction ratio, gyratory crushers do as well as jaw crushers; in respect to power consumption at equivalent capacity, the two types of machine are on a par, but at equivalent gape the gyratory has an advantage (as it has a higher capacity).

In comparing jaw crushers with gyratory crushers and choosing which to install, it may be concluded that if the quantity to be crushed is sufficiently small to be handled by one jaw crusher the latter is more economical, that if the quantity to be crushed is large enough to keep a gyratory crusher busy the latter is to be preferred. Taggart[(16b)] states it as a rule that "if the hourly tonnage to be crushed divided by the square of the gape expressed in inches yields a quotient less than 0.115, use a jaw crusher; otherwise, a gyratory."

CONE CRUSHERS

Relatively speaking, cone crushers are newcomers, but they have gained such wide popularity as an economical machine for intermediate crushing that they must be regarded as the standard.

Figure 17 is a sectional view of a *Symons cone crusher*. The principle of operation of the cone crusher is much like that of the gyratory crusher, but with two exceptions. In the first place, the outside crushing surface, instead of flaring in from top to bottom as it does in a gyratory crusher, flares out so as to provide an increased area of discharge for the crushed material (this is necessary if a substantial capacity is to be achieved). Secondly, the outer crushing surface, which becomes in the cone crusher an upper crushing surface, can be lifted away from the lower surface when an uncrushable obstruction enters the machine. This is accomplished by means of a nest of heavy coiled springs arranged in circular fashion around the crusher. It eliminates fracture from "tramp" iron. Symons cone crushers are made in two

Fig. 17.—Symons cone crusher, sectional view. (*From Nordberg Mfg. Co.*)

types: the standard crusher for the reduction of coarser feeds, and the short-head crusher for the reduction of finer feeds. The short-head crusher is shown in Fig. 17.

The cone crusher is a machine of very great capacity if compared with rolls or even with gyratory crushers set to deliver a product of comparable fineness. To do their best, cone crushers must be provided with a dry feed freed of fine particles. If this is not done, the crusher may clog. This requirement makes it necessary to use efficient screens in closed circuit with cone crushers. Lubrication of all moving parts in an enclosed oil bath is used, and a special oil-cooler is required.

TABLE 4.—SIZE AND CAPACITIES OF CONE CRUSHERS

A. "Standard" Crusher

Size of crusher, feet	Capacities in tons (2,000 lb.) per hour with discharge set, inches									
	¼	⅜	½	⅝	¾	1	1¼	1½	2	2½
2	14	20	25	30	35	45	50	60		
3	25	35	40	55	70	80	85	90	95	
4	..	60	80	100	120	150	170	177	185	
5½	..	100	130	160	200	275	300	340	375	450
7	..	...	225	280	330	450	560	600	800	900

B. "Short-head" Crusher

Size of crusher, feet	Approximate capacities in tons (2,000 lb.) per hour for closed-circuit operation, inches				
	⅛	3/16	¼	⅜	½
3	15	20	30	40	50
4	20	35	50	75	100
5½	..	65	100	135	175
7	..	120	150	240	300

Table 4 gives data as to size of crusher and capacity and Table 5 as to weight of crusher and size of motor.

TABLE 5.—WEIGHT AND POWER REQUIRED FOR CONE CRUSHERS

Size of crusher, feet	Weight, tons	Hp	Size of crusher, feet	Weight, tons	Hp
A. "Standard" Crusher			B. "Short-head" Crusher		
2	5.25	25 to 30	3	11.25	75
3	10.5	50 to 60	4	22.5	150
4	17.5	75 to 100	5½	44	200
5½	42.5	150 to 200	7	71.5	300
7	65	250 to 300			

The *Telsmith gyrasphere* (Fig. 18) is a crusher of recent introduction. It features a spherical steel head instead of a truncated cone, roller bearings instead of a main spherical bearing, and compression-yielding instead of tension-yielding coil springs.

ROLLS

Crushing rolls[12] consist of cylinders revolving toward each other so as to nip a falling ribbon of rock and discharge it, crushed, below. Rolls were invented more than 100 years ago, record of their use in Cornwall going back to 1806. In early

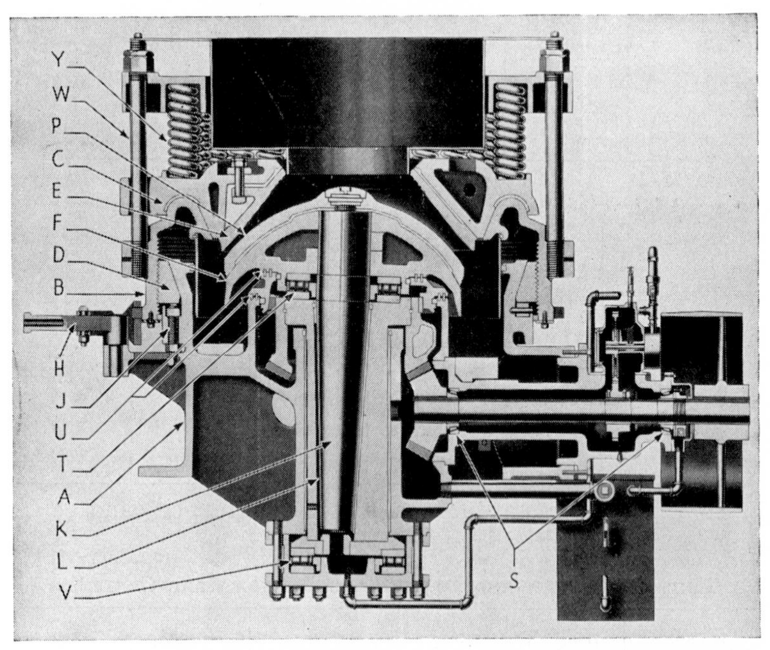

FIG. 18.—Telsmith gyrasphere crusher. Parts labeled are as follows: (*A*) frame, (*B*) adjusting collar, (*C*) concave retaining bowl, (*D*) sectional wedge ring screw, (*E*) manganese steel concave ring, (*F*) spherical head, (*H*) adjusting lever, (*J*) clamp screws for wedge ring, (*K*) main shaft, (*L*) eccentric, (*P*) manganese steel head mantle, (*S*) roller bearings for countershaft, (*T*) upper roller thrust bearing, (*U*) leather labyrinth seals, (*V*) lower roller thrust bearing, (*W*) main bolts for tightening springs, (*Y*) springs holding concave unit in place. (*From Smith Engineering Works.*)

machines (Fig. 19), the rolls were mounted on heavy shafts revolving in open bearings contained within cast-iron side frames. One of the rolls was driven positively and the other, or idler, by friction. To assure rotation of the idler and to prevent breakage, the idler was pressed sideways against the live roll by a heavy weight hung from a yoke.

In modern designs (Fig. 20), both rolls are positively driven, at much higher speeds, and breakage is prevented by mounting the bearings of one roll shaft against coil springs.

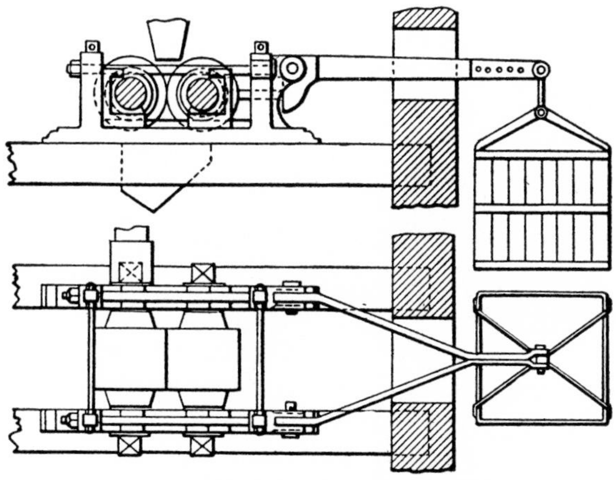

FIG. 19.—An early set of rolls after the Cornish style. These rolls were usually 18 to 30 in. in diameter and 14 to 20 in. wide. (*After Miller.*)

FIG. 20.—Modern Traylor crushing rolls, outside view. (*Traylor Eng. & Mfg. Co.*)

Rigid rolls, *i.e.*, rolls in which both roll shafts are revolving in fixed bearings, were used at one time but are now obsolete.

In spring rolls, the crushing surface consists of a suitably heat-treated steel shell shrunk on the main-roll casting. Roll shells are replaced when worn.

It has been observed that there is a tendency for grooving in the face of the roll. In modern machines, this is eliminated by providing one of the rolls with an automatic longitudinal motion which causes the opposing roll faces to change position. Figure 21 shows the lateral "fleeting" device installed to prevent angular corrugation and flanging of shells. This device causes points on opposite shells to move laterally with reference to each other.

Comparison with Other Crushers. Rolls are characterized by the production of a smaller proportion of fines than other crushers,

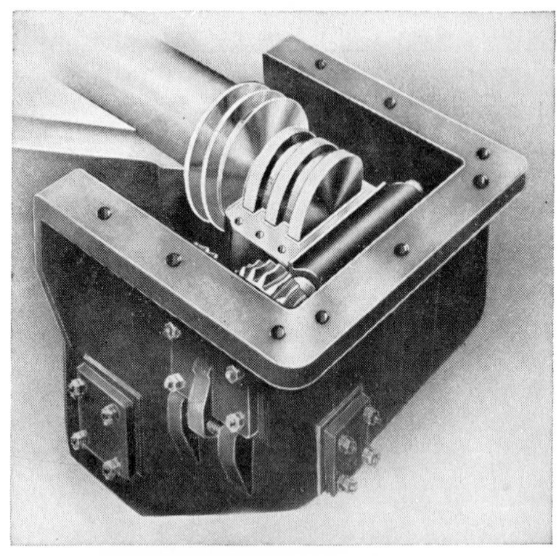

FIG. 21.—"Fleeting roll" mechanism for Traylor heavy-duty crushing rolls. (*Traylor Eng. & Mfg. Co.*)

and also by a smaller reduction ratio. The reason for the smaller proportion of fines seems to be related to the limited time during which material of a size already fine enough to fall through the rolls remains in the machine subject to further crushing (no "mastication").

The small reduction ratio is a result of the fact that it is not practicable to make rolls of a small curvature (and therefore of a great diameter) without unduly increasing the weight of the machine and the difficulties of construction. A given set of rolls could theoretically be made to crush with a very large reduction ratio if it were fed with only fine material, the reduction ratio thus decreasing with increasing size of feed, until it equals unity (no crushing) for the coarsest permissible feed. In practice, however, owing to the springing of the rolls when the machine is loaded, actual reduction ratio is much smaller.

Because of their inability to yield a high reduction ratio, rolls have not been installed so often in recent years as they used to be. With the increasing use of flotation as a concentrating method, the advantage offered by rolls of producing the minimum proportion of fines has ceased to be valued. The combination of one stage of size reduction in a cone crusher with one or two stages of size reduction in ball mills often suffices to reduce the ore to a fineness suitable for flotation. Whenever gravity concentration is used as the principal concentration method, rolls remain the best device for intermediate crushing.

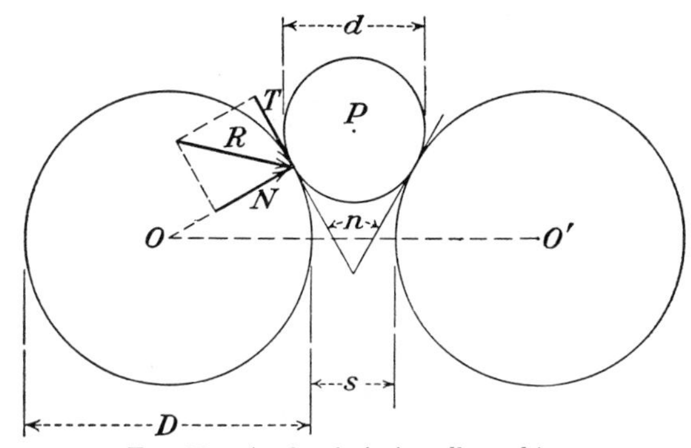

Fig. 22.—Angle of nip in roll crushing.

Angle of Nip. In choosing rolls for a certain duty, it is required to know the size of feed and the size of product that it is wished to make and the tonnage to be handled. Combination of the information relating to the size of feed and size of product determines the diameter and set of the rolls. Calculation is based on the coefficient of friction between the material crushed and the rolls and is as follows:

Let D and d be the diameters of the rolls and of the particle to be nipped, respectively (Fig. 22). Gravity being neglected, the forces acting on the particle are a normal force N and a tangential force T.

If the resultant of these forces R is directed downward, the particle will be nipped and crushed, otherwise it will ride in the trough formed by the rolls. Equating the vertical components of N and T describes the limiting condition for crushing.

$$N_v = N \sin\frac{n}{2}; \qquad T_v = T \cos\frac{n}{2}.$$

For $N_v = T_v$, the sufficient condition becomes

$$N \sin\frac{n}{2} = T \cos\frac{n}{2},$$

or

$$\tan\frac{n}{2} = \frac{T}{N}.$$

Under this limiting condition for crushing, the angle n is termed the *angle of nip*.

Since $T/N = \phi$ (coefficient of friction), it follows that the condition necessary and sufficient to ensure nip is

$$\tan\frac{n}{2} \leq \phi. \qquad \text{[II.2]}$$

If s is the distance apart of the roll faces, n, s, D, and d, are related as follows:

$$\frac{s}{2} + \frac{D}{2} = \left(\frac{D}{2} + \frac{d}{2}\right)\cos\frac{n}{2},$$

or

$$\cos\frac{n}{2} = \frac{D + s}{D + d}. \qquad \text{[II.3]}$$

Example. If the coefficient of friction between rock and steel is 0.4, what is the minimum diameter of roll to reduce 1½-in. pieces of rock to ½ in.?

The limiting value of $\frac{n}{2}$ is described by $\tan\frac{n}{2} = \phi = 0.4$; therefore

$$\frac{n}{2} = 21°48', \qquad \cos\frac{n}{2} = 0.9285.$$

Hence,

$$\frac{D + 0.5}{D + 1.5} = 0.9285, \qquad D = 12.5 \text{ in.}$$

Practically speaking, rolls of a diameter of 20 in. would probably be chosen.

Choke Feeding. In modern roll crushing, the aim is not so much to avoid production of fines as to utilize a machine that is well fitted for size reduction between the range of large cone crushers and that of rod and ball mills. This has resulted in widespread adoption of choke feeding. With choke feeding, there is no individual movement of particles and the classical theory of angle of nip is not applicable.

Advantages of choke feeding are as follows:

1. Possibility of using a very large circulating load (of the order of 400 per cent in one plant).

2. Reduction of wear by grinding "ore on ore" rather than "ore on metal."

3. Greater capacity.

Capacity. The capacity of rolls depends upon their speed, the width of their faces, their diameter, and their set. The theoretical capacity in tons per hour is

$$C = 0.0034\, NDWsG, \qquad [II.4]$$

in which N is the number of revolutions per minute, D the diameter of the rolls, W the width of face, s the set, all in inches, and G the specific gravity of the rock being crushed. Actual capacity is considerably less than this, averaging from 10 to 30 per cent of the theoretical. When dealing with rolls set "close," *i.e.*, without clearance between faces, the theoretical capacity is of course nil. But actual capacity may be large, as the choked feed results in the roll faces being held apart.

Rolls can be operated wet or dry. Dry crushing gives less wear and a lower output per machine. In one large plant when all the factors were weighed, it was found that wet and dry crushing were on a par.

GRAVITY STAMPS

Gravity stamp mills[14] are one of the oldest intermediate- and fine-crushing devices known. They operate on the principle of the mortar and pestle. Figure 23, reproduced from the Hoovers' translation of Agricola's book, presents a crude type of stamp mill consisting of four stamps crushing rock in an open mortar. This machine resembles the modern stamp mill, except for its smaller size, in that there are four stamps instead of five, and in that the crushing is done in the open instead of being done in a box enclosed by a screen. In modern stamp mills (Fig. 24), the crushing is accomplished by 5 (or 10) heavy blocks shod with a replaceable part and striking on a hard metal block. The stamp is raised by means of a cam working under a collar. The cam is designed so that after the stamp has been raised to the highest point it is allowed to drop on the crushing block. Ore is fed regularly in the mortar and falls on the block. The operation is

Fig. 23.—Medieval stamp mill. (*After Agricola's "De re metallica," as translated by H. C. Hoover and L. H. Hoover.*)

conducted in aqueous pulp, and the front, or the front and back, of the mortar box is closed by a screen allowing exit only to those particles fine enough to pass through the sieve. Material that is too coarse is compelled to stay in the mortar box until crushed. To get the best possible agitation of the pulp, one of several special sequences of stamp drops is desirable. Richards recommends the sequence 1, 3, 5, 2, 4—each number referring to one stamp in a battery of five, reckoned from one end of the battery to the other.

One of the essential requirements for the satisfactory operation of stamp mills is a uniform rate of feed. This is accomplished by feeders of the Challenge or similar type.

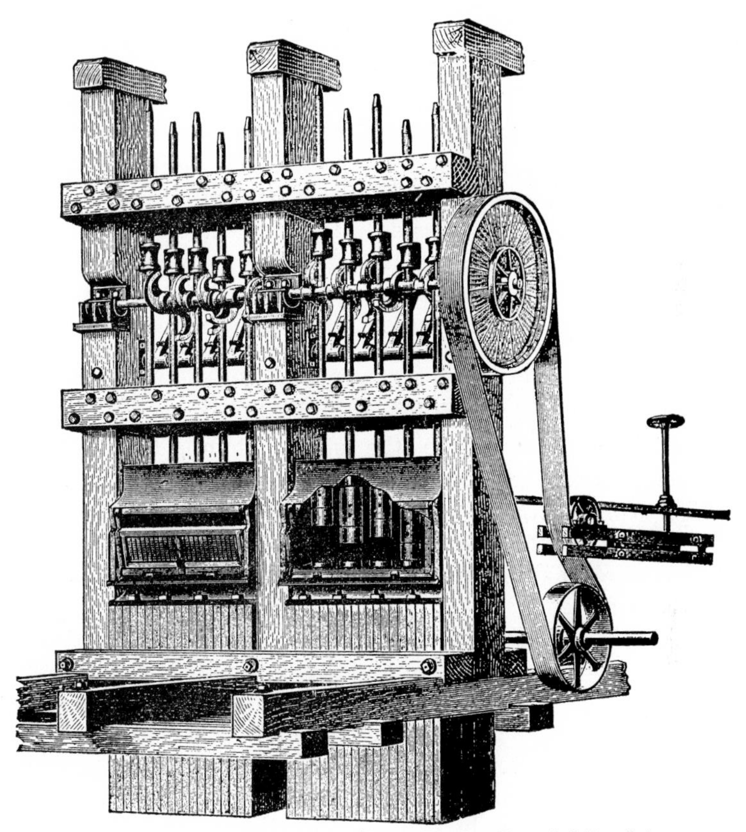

Fig. 24.—Recent stamp mill. (*From Traylor Eng. & Mfg. Co.*)

Stamp mills are especially well suited for crushing gold ores if it is desired at the same time to amalgamate the gold. This amalgamating operation can be conducted inside or outside of the mortar block. In the case of inside amalgamation, the violent ejection of the crushed rock from under the stamp favors encounter of gold with mercury; in the case of outside amalgamation, the wavelike surges of pulp flow induced by the pulses in the mill turn over settled particles and favor the encounter of gold with mercury. Stamp mills and amalgamation, until recently, have been viewed as nearly inseparable.

Gravity stamps are a flexible crushing device capable of taking a relatively coarse feed and of yielding a relatively fine product: it is possible to feed a stamp mill lumps 3 in. in diameter and to discharge particles finer than 1/50 in. This indicates a reduction ratio possibly as high as 150, as compared with 6 to 8 in jaw crushers and gyratory crushers, 7 to 10 in cone crushers, and 1.5 to 4 in rolls. On the other hand, the capacity of stamp mills is small in relation to the weight of the machine, to its cost, and to the space it occupies. Again, it causes much vibration and noise.

In summation, it can be said that gravity stamps are especially fitted to handle a small tonnage of coarse feed; to make a fine product, especially if the procedure is to be used in connection with amalgamation; and in cases where no grinding of the ore is required.

SPECIAL CRUSHERS

Special crushers include the steam stamp, Bradford coal breaker, disk crushers, toothed rolls, and hammer crushers.

The *steam stamp*[15a] resembles the gravity stamp in that a large weight strikes the ore. But in this case the blow is struck not by a freely dropping weight but by a weight driven down (and later lifted) by a steam piston. The complete stamp stem, shoe, piston, and piston rod weigh 5,500 to 7,900 lb.; the striking velocity is 20 to 24 ft. per sec.; and the blow is of the order of 25 to 50 tons and strikes on an area of about 2 sq. ft. The stamp drops on a crushing die mounted on a concrete foundation weighing 50 to 100 tons. The mortar is surrounded by punched-plate screens through which the crushed rock is discharged.

Steam stamps have been devised for the treatment of native-copper ore (Michigan). Copper finer than 3/8 to 5/8 in. is discharged through the peripheral screens; copper ranging from that size to 4 in. is caught in a device like a hydraulic classifier or plunger jig built in the foundation or to the side of the foundation. Pieces larger than 4 in. are removed manually at intervals.

Figure 25 represents a *Bradford breaker*.[15b] This machine is a combination crushing machine and screen which has been used for the elimination of coarse waste from coal. Its operation depends on the fact that coal is considerably softer and more friable than slate so that the breaker reduces the coal to a size fine enough to pass through the screen while the slate is retained

Fig. 25.—Bradford breaker. (*From Pennsylvania Crusher Co.*)

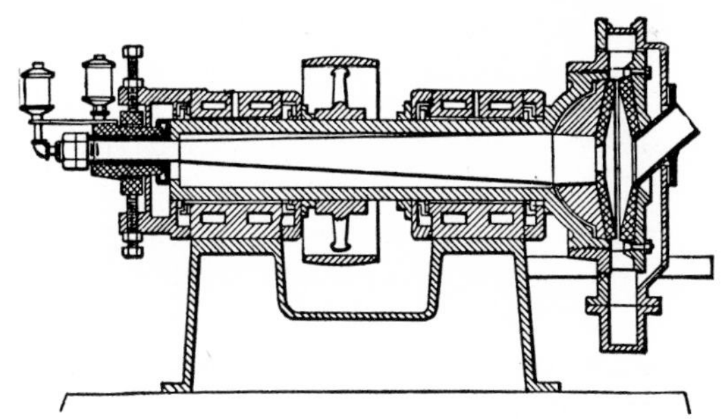

Fig. 26.—The disk crusher in its original form as introduced about 1908. (*After Miller.*)

in a size coarse enough to pass out of the machine as screen oversize. The Bradford breaker breaks by tumbling the coal on itself; in many respects, it resembles a ball mill without the balls.

Disk crushers(13) were much in favor a few years ago for intermediate crushing. In recent years, however, few installations have been made. This is traceable to the development of the superior cone crushers. Disk crushers consist of a pair of saucer-shaped disks revolving at substantially the same speed and in the

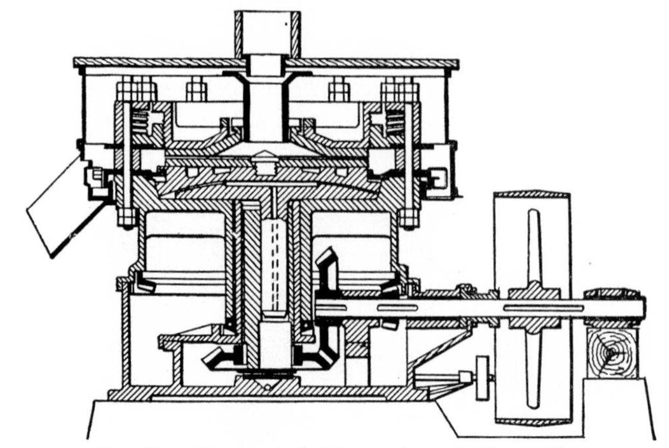

Fig. 27.—The vertical disk crusher. (*After Miller.*)

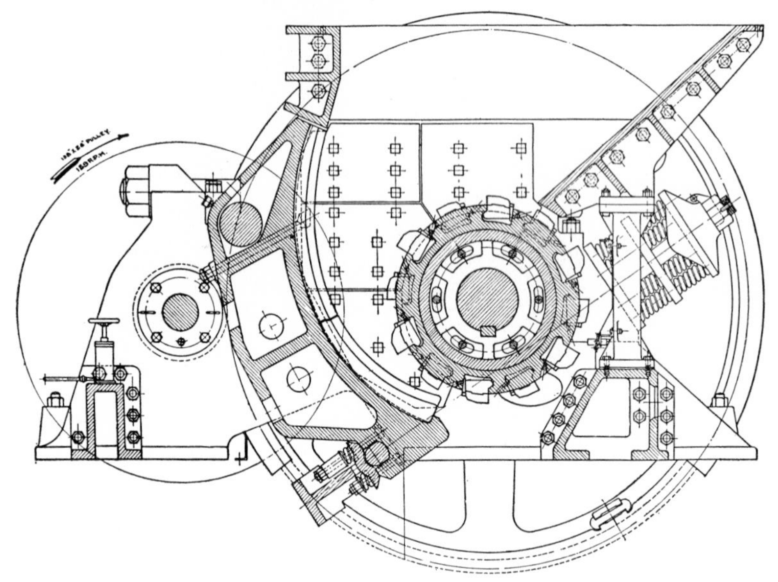

Fig. 28.—Single-roll, toothed crusher. (*From Allis-Chalmers Mfg. Co.*)

same direction. One of the disks has, also, a slight gyrating motion. Centrifugal force throws out the rock toward the periphery, where it is nipped and crushed. Horizontal-shaft disk crushers were first introduced (Fig. 26); vertical-shaft disk crushers (Fig. 27) came later and paved the way to the design of cone crushers as a happy compromise between them and reduction gyratories.

Toothed rolls are used extensively in coal preparation. In some machines, a pair of rolls is used[9]; in others, a single roll[10] (Fig. 28). Because of the friability of coal, it is possible to break

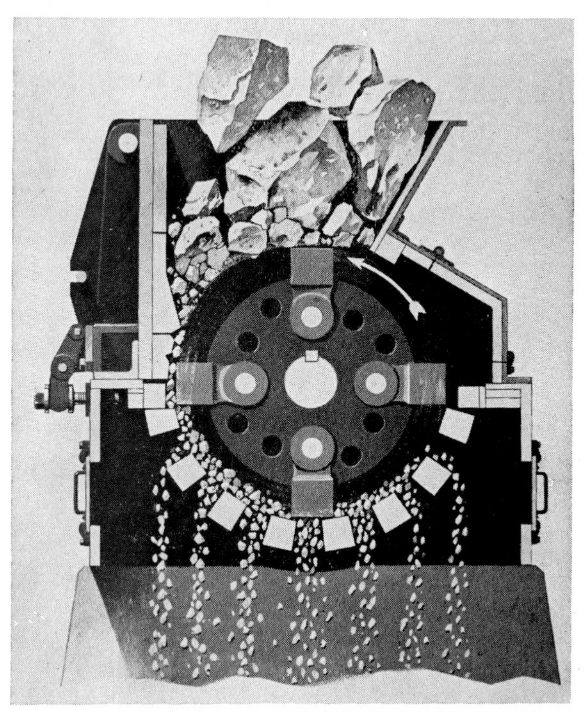

Fig. 29.—Jeffrey hammer crusher. (*From Jeffrey Manufacturing Co.*)

it with a minimum of fines in a machine such as toothed rolls provided with sharp teeth from which great pressure can be applied at some particular points. Toothed single-roll crushers are sometimes known as cubing rolls, because of the blocky appearance of the crushed product.

Swing-hammer crushers and *pulverizers*[11] are widely used in industry, but not for ore treatment. Figure 29 illustrates a typical design, the Jeffrey hammer crusher. This consists of four heavy blocks or hammers fastened by pins to a revolving disk, inside a crushing chamber. The centrifugal force of the hammers is employed to deliver the blow that reduces the rock in size. The rock remains in the crushing chamber until fine enough to drop through the openings in the bottom. Swing-hammer crushers are advantageous for the crushing of soft materials; if the silica content exceeds 10 per cent, some manufacturers think that the swing-hammer pulverizers are inadmissible.

Literature Cited

1. ANON.: Design of New Jaw Crusher Avoids Toggles, *Eng. Mining J.*, **135,** 580 (1934).
2. ANON.: England's Latest in Ore-crushing Machinery, *Mining and Met.*, **15,** 445–447 (1934).
3. ANON.: Novel Crusher Pitman and Toggle System, *Eng. Mining J.*, **135,** 236–237 (1934).
4. BERNHARD, R.: Bell-shaped Heads and Concaves for Gyratory Crushers, *Mining and Met.*, **13,** 107–108 (1932).
5. BLAKE, WILLIAM P.: The Blake Stone- and Ore-breaker: Its Invention, Forms, and Modifications and Its Importance in Engineering Industries, *Trans. Am. Inst. Mining Engrs.*, **33,** 988–1031 (1902).

5*a*. MILLER, W. T. W.: "Crushers for Stone and Ore," Mining Publications, Ltd., London (1935).

6. MILLER, W. T. W.: The Jaw Crusher as a Sledger, *Eng. Mining J.*, **123,** 876–882 (1927).
7. MILLER, W. T. W.: The Gyratory as a Sledging Crusher, *Eng. Mining J.*, **124,** 7–11 (1927).
8. MILLER, W. T. W.: The Gyratory as a Secondary Breaker, *Eng. Mining J.*, **124,** 131–136 (1927).
9. MILLER, W. T. W.: Giant Rolls—A Phase in the Development of Primary Breakers, *Eng. Mining J.*, **124,** 410–412 (1927).
10. MILLER, W. T. W.: Single-roll Crushers, *Eng. Mining J.*, **124,** 490–492 (1927).
11. MILLER, W. T. W.: Swing-hammer Crushers—Their Design and Application, *Eng. Mining J.*, **124,** 888–892, 930–933 (1927).
12. MILLER, W. T. W.: Crushing Rolls for Ore and Stone, *Eng. Mining J.*, **125,** 246–249, 331–336 (1928).
13. MILLER, W. T. W.: Disk Crushers—Their Development and Advantages, *Eng. Mining J.*, **125,** 449–451, 492–494 (1928).
14. MUNROE, H. S.: On the Weight, Fall, and Speed of Stamps, *Trans. Am. Inst. Mining Engrs.*, **9,** 84–99 (1880).
15. RICHARDS, ROBERT H., and CHARLES E. LOCKE: "Textbook of Ore Dressing," 2d ed., McGraw-Hill Book Company, Inc., New York, (1925); (*a*) pp. 33–37; (*b*) p. 462.
16. TAGGART, ARTHUR F.: "Handbook of Ore Dressing," John Wiley & Sons, Inc., New York (1927); (*a*) p. 255; (*b*) p. 280.

CHAPTER III

LABORATORY SIZING

Inasmuch as the effectiveness of practically all dressing operations is a function of the size of the particles treated, thorough knowledge of the size characteristics of the materials handled is of the greatest importance. As a result of this, a good command of laboratory sizing techniques is required.

There are several methods for sizing mineral particles, the most important of which is screening. Classification (elutriation) and microscope sizing are next in importance. With relatively large particles, it is often easiest to determine size by actual measurement with a yardstick of some kind, but this technique is restricted to the handling of particles several inches in diameter.

Effect of Particle Shape. Particles are generally irregular in shape so that it is difficult to define rigorously what is meant by particle size.[12,18] In spheres, for instance, the size of the particle is the diameter of the sphere. With cubes, it is debatable whether the size of the particle should be regarded as the edge, the long diagonal, or perhaps the diameter of a sphere whose volume is equal to that of the cube. With irregularly shaped particles, it is impossible to define strictly what is meant by the size of the particles; and yet no one will deny that he has a fair idea of the material under consideration if he is advised, for example, that it passes a screen with 3/8-in. openings and is retained on a screen with 1/4-in. openings.

Let a, b, c, stand for the length, breadth, and thickness of any particle. Different sizing techniques measure particles according to one or another or a combination of these three dimensions. Thus, when a particle is measured with a yardstick, its greatest dimension is generally considered; if the particle is put through a ring, as is customary in dealing with large lumps of coal, the second largest dimension of the particle is most significant. In screening particles with sieves made of woven wire having square openings, the second largest dimension of the particles is the most important, but the least dimension, and to some extent the greatest, also enter into the size determination.

In sizing by classification, all three dimensions have a bearing on the operation since the controlling factors are the mass of the particles (which is dependent equally on a, b, and c, and on the specific gravity of the particles) and the surface of the particle that is rubbing against the classification medium (also a function of a, b, and c). Since, however, the mass and surface of each particle cannot be described by the same function of a, b, and c, the settling of a particle in a resisting medium is dependent not only upon its mass, but also upon its shape.

In measuring particles under the microscope, it is customary to sprinkle them on a slide and to measure their diameter either in a random direction or else in two dimensions in the plane of vision at right angles to each other. In either case, the smallest dimension of the particles is neglected since they orientate themselves with the smallest dimension at right angles to the surface on which they are sprinkled.

From these considerations of the basic controlling factors in particle sizing, it follows that if sizing of a mineral product is accomplished by one technique down to a certain size, and then by a different technique from that point on, the data will necessarily show that different techniques do not actually measure the same thing: there will be a "jog" in the data. Since there is no universal sizing technique that will handle particles of all sizes, systematic irregularities in laboratory data must be accepted as being an unavoidable evil.

Sizing Scales. At first sight, it would appear that a most natural sizing scale—*i.e.*, a grouping of those sizes at which a splitting of the sample into grades is made—is one in which the various sizes are in arithmetic progression. For instance, it may be selected to separate a material so that consecutive grades differ from each other by 1.0 mm. Such a scale might include the following divisional sizes: 10, 9, 8, 7, 6, 5, 4, 3, 2, and 1 mm. It is obvious that a multitude of particles radically different from each other would be included in the grade from 0 to 1 mm., whereas all the particles in the 8- to 9- and 9- to 10-mm. grades would be very much alike.

A more satisfactory type of scale is one based on a geometric progression,[4,13] *i.e.*, one in which successive grades are in constant ratio. Such a scale preserves size relationships in due proportion. For example, such a scale might have a ratio of 3 and include the following divisional sizes: 9, 3, 1, $\frac{1}{3}$, $\frac{1}{9}$, $\frac{1}{27}$ mm., etc.

The geometric scale in which the ratio is $\sqrt[2]{2}$ has for some time been regarded as standard in the United States. A number of geometric scales having $\sqrt[2]{2}$ as ratio could of course be devised, by using as a reference point almost any size such as 1 mm., 0.1 mm., or 1 in. For practical manufacturing reasons, the reference screen opening is that of the so-called standard 200-mesh screen. This standard screen is a woven-wire screen having 200 wires per linear inch and likewise 200 openings per linear inch, in which the opening is a square having an edge of 74 microns, or 0.074 mm. Standard 200-mesh screens calibrated by the Bureau of Standards are available. In some foreign countries, a geometric screen scale has not yet been accepted. In dealing with particles finer than the finest screen available, an arithmetic size scale or even an arbitrary size scale is sometimes used. These procedures are definitely objectionable and should be discouraged.

A sizing scale based on the standard 200-mesh screen is presented in Table 6. For sizes coarser than 37 microns, screens are available as shown. For sizes finer than a few microns, the scale as presented is essentially theoretical since practical sizing cannot at present be extended beyond that point. It is interesting to observe that the range accessible to screening from 1 in. down covers 20 grades and the range not accessible to screening covers more than half again as many, or 32 grades. It is also of interest to note that screening and elutriation together cover but 26 grades, or only one-half of the total range. Another way of looking at this matter is to observe that one-third of the total size range is invisible, even with the help of the most powerful microscope. Table 6 indicates also the size of some common structural entities.

In describing a screen, it is customary to refer to the opening in either inches or millimeters or to the number of meshes per linear inch. Unless one is dealing with a standard screen, the latter description may be misleading as the opening of the screen is determined by both the number and the diameter of the wires of which it is made. In some countries, especially in continental Europe, it is customary to describe a screen by the number of openings per *square* centimeter, so that 9,000-mesh and 6,000-

TABLE 6.—A STANDARD SIZING SCALE BASED ON THE STANDARD 200-MESH SCREEN

Size		Mesh	Sizing method	Example
Millimeters	Microns			
26.67			↑ Screening	
18.85				River gravel
13.33				
9.423				
6.680	....	3		
4.699	....	4		
3.327	....	6		Pea gravel
2.362	....	8		
1.651	....	10	Screening	
1.168	....	14		
0.833	....	20		
0.589	....	28		
0.417	....	35		
0.295	....	48		Beach sand
0.208	....	65	↑ Classification (elutriation and sedimentation)	
0.147	....	100		
0.104	....	150		
0.074	....	**200**		
0.052	52	270		
0.037	37	400	Screening ↓	
	26		↑ Microscope	Fine silt
	18.5	(800)		
	13			
	9.25	(1,600)		Blood cells
	6.5			
	4.62	(3,200)	Classification ↓	
	3.25			
	2.32			
	1.62		Microscope	Many germs
	1.16			
	0.81			
	0.58			Wave length of visible light
	0.41		↑ Centrifuge	
	0.29			
	0.20		Microscope ↓	
	0.14			
	0.10			
	0.07		Centrifuge	
	0.05			
	0.035			Thinnest iridescent films visible by light interference
	0.025		Centrifuge ↓	
	0.017			
	0.012			
	0.008			
	0.006			
	0.004			Very large molecules
	0.003			
	0.002			
	0.0015			
	0.001			
	0.0007			
	0.0005			Average unit crystal

mesh screens are not unheard of. This, of course, is an entirely different way of describing a screen from that which is customary in America, and care should be exercised in reading foreign literature not to be led astray by so gross a discrepancy in meaning.

Graphical Representation of Sizing Analyses. There are a number of different methods of charting sizing tests. In one type of charts, the ordinate is the percentage of the total weight coarser than a given size. This type of graphical presentation, which is presented in Fig. 30 from the data in Table 7, is termed a *cumulative* type of plot. The abscissas may be either the actual screen opening expressed in millimeters or inches (as in Fig. 30) or some other function of size.

TABLE 7.—TYPICAL SCREEN ANALYSIS
(Crusher product obtained in crushing rolls)

Size, millimeters	Actual percentage	Cumulative percentage
+ 13.33	0.74	0.74
−13.33 + 9.423	2.60	3.34
− 9.423 + 6.680	11.41	14.75
− 6.680 + 4.699	32.12	46.87
− 4.699 + 3.327	18.34	65.21
− 3.327 + 2.362	11.05	76.26
− 2.362 + 1.651	7.13	83.39
− 1.651 + 1.168	4.83	88.22
− 1.168 + 0.833	3.33	91.55
− 0.833 + 0.589	2.30	93.85
− 0.589 + 0.417	1.71	95.56
− 0.417 + 0.295	1.33	96.89
− 0.295 + 0.208	0.94	97.83
− 0.208 + 0.147	0.70	98.53
− 0.147 + 0.104	0.52	99.05
− 0.104 + 0.074	0.33	99.38
− 0.074	0.62	100.00
	100.00	

In a second type of graphical presentation, the ordinates are the percentages of the total weight included within consecutive divisional sizes, *i.e.*, the amount retained by each screen and passing the screen immediately larger. The abscissas may again be either the actual screen opening expressed in millimeters

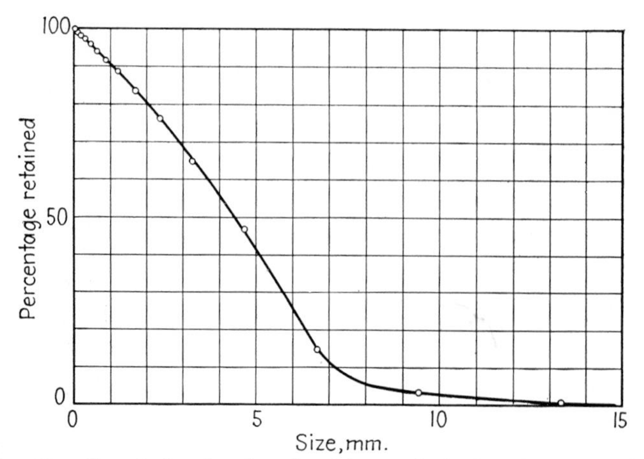

FIG. 30.—Cumulative charting of a sizing analysis (data from Table 7).

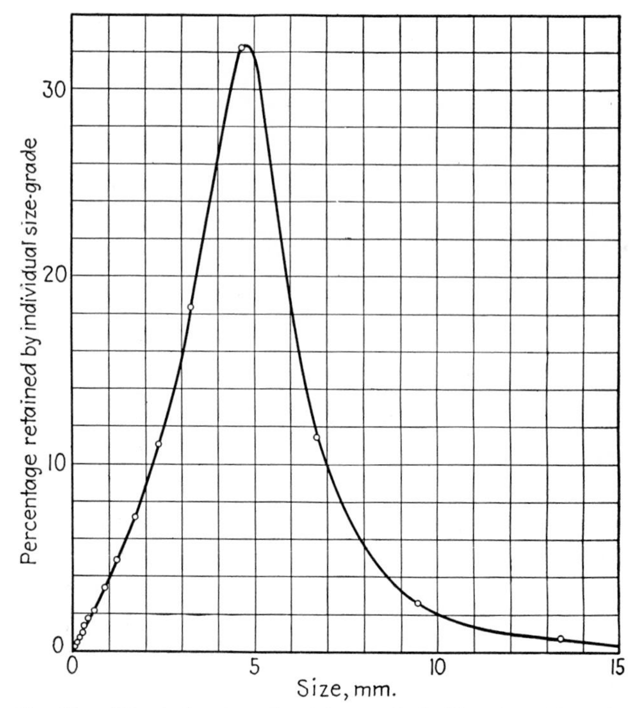

FIG. 31.—Direct charting of a sizing analysis (data from Table 7).

or inches, as in Fig. 31 or some other function of size. If this second type or *direct* plot is used, special care must be exercised; unless the standard screen scale has been used in making the screen analysis, this type of diagram is misleading in that uneven emphasis is placed on the various size ranges. This type of diagram, also, is embarrassing in dealing with the material finer than the finest size to which the sizing operation is conducted. Thus, if the finest screen used is the 200-mesh screen, the minus 200-mesh material includes particles ranging from 200- to 270-mesh, from 270- to 400-mesh, and so on down to particles of submicroscopic size. Indeed, on the basis of the standard scale,

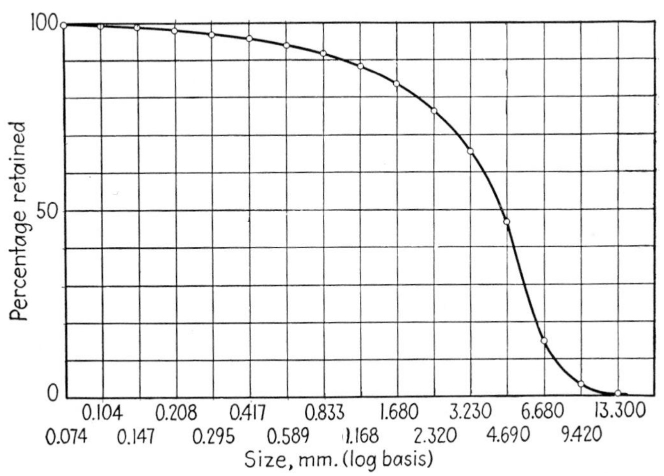

Fig. 32.—Cumulative charting of a sizing analysis, logarithmic scale (data from Table 7).

minus 200-mesh material is an association of particles belonging to some 36 different size grades (Table 6). The best course to follow in making a direct plot is to omit representation of the material passing the finest screen available. If it happens to be expedient to carry out the sizing test in the coarse range with a $\sqrt[2]{2}$ ratio and to continue this sizing test in the fine range with a ratio of 2, it is sufficient in plotting the data in the fine range to use as ordinate a scale one-half as large as that used to plot the data for which a closer size range has been used.

So far nothing has been said of the proper representation for the abscissas. It is, of course, obvious that a direct scale can be used, as in Fig. 30. This method crowds the points at the fine end of the scale. This is not a drawback if a cumulative ordinate is used (Fig. 30) since plotting of all the points is not necessary, but it becomes objectionable if a direct ordinate plot is used. In that case, the use of a logarithmic scale of abscissas is indicated. With that type of scale, since the divisional sizes are in constant ratio (and their logarithms in constant difference), the distances

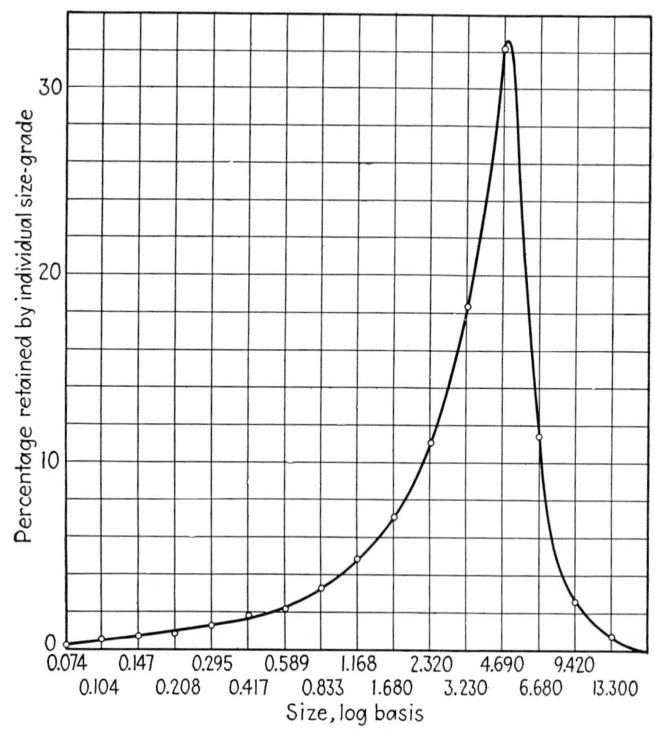

FIG. 33.—Direct charting of a sizing analysis, logarithmic scale (data from Table 7).

separating points corresponding to consecutive sizes are equal. This obviates the necessity of using logarithmic paper, or of calculating logarithms, if a logarithmic scale of abscissas is used. In direct-percentage *vs.* log-size plots (Fig. 33), the area under the curve is in all cases equal to unity, or 100 per cent. This is often useful.

For the sake of completeness, Fig. 32 has been added to Figs. 30, 31, and 33. The plotting is a cumulative percentage *vs.*

log-size plot. Although not so helpful as the plots typified by Figs. 30 and 33, it is used occasionally.

In studies of size distribution in crushed products, a particularly useful graph is obtained by plotting the logarithm of the percentage in each size grade against the logarithm of the size.[5] Another graph that has recently been adopted is one in which the ordinate is a log-log scale of the reciprocal of the cumulative percentage in each size grade, and the abscissa is the logarithm of the size.[7] Other graphs, still, have been proposed.[11,14,17]

Screening Technique. Testing screens are circular shells of brass 8 in. in diameter and about 2 in. high. The screen cloth is placed in the bottom of the shell so that material can be held on the screen. The screens are built to nest in each other in order of decreasing size; a complete set includes a cover and a bottom pan.

In screening by hand, the material is placed on the coarsest screen, used alone with a cover and a bottom. The assembly is held in the right hand and struck against the palm of the left hand which is used as a bumping block. This operation is repeated for a given length of time; after which the undersize is removed and the operation resumed to determine whether more material will pass through the sieve. Actually, there is no end to a screening operation as there always are particles whose shape is such that they will pass through the sieve only when they are presented in one particular direction. For practical purposes, however, after a certain length of shaking, there is very little material going through the screen so that a practical end point may be considered to have been reached.

Hand Screening. According to the American Standards Association,[1,2] in order to conduct a screen analysis by hand screening on a strictly accurate basis, it is necessary to proceed as follows. The lid on top of the screen is marked in six directions differing from each other by 60°, then, with the thumb opposite one of these marks, the screen is given 25 circular motions and bumps in 10 sec.; the direction of bumping is then changed by 60° by simply changing the grasp of the right hand on the screen assembly, and the operation is repeated until another 25 bumps have been administered, when the direction of bumping is changed by another 60°, etc. The cycle of 6 by 25, or 150, blows requires 1 min. At this stage, the undersize is removed from the pan and the operation is repeated for another minute. If the amount passing through the sieve during the second cycle is less than 0.05 per cent of the weight of the sample, the operation is complete; if not, the operation is repeated until that condition has been reached. It is patent from this description that the standard method may become extremely tedious. It is claimed that this method is more accurate than the method of screening

FIG. 34.—Ro-tap testing sieve shaker. In the illustration the machine is adjusted to hold one full-height sieve and pan and one half-height sieve and pan. (*From W. S. Tyler Company.*)

by machine, which will be described presently. This method is substantially that recommended in 1925 by Gross,[9a] except that Gross requires brushing of the sieves after each 1-min. sieving.

Automatic Screening. There are a number of machines for automatically screening material in testing sieves. One of the best known is the Ro-tap manufactured by the W. S. Tyler Company (Fig. 34). This machine causes a circular motion of the material on the screen; the frame in which the screens are placed is provided with a translating motion at one end and a circular motion at the other so that there is continual movement of the material on the screen. Besides, a blow is delivered to the nest of sieves, once for each revolution of the frame. The operation thus simulates the motion described in the preceding paragraph in connection with hand screening. A nest of sieves, generally up to 6 screens although occasionally up to 12 ("half-height"), is shaken in the machine for from 5 to 20 min., depending upon the extent to which it is desired to approach the end point of the screening operation. If proper care is taken not to overload the screens, and if the material is deslimed and well dried, machine screening gives excellent results.

The most satisfactory results in making a screen analysis are obtained when the amount of material on each screen, at the end of the screening operation, does not exceed that required to form a layer one particle deep. However, a bed of particles two deep can be handled with reasonable accuracy. Thus, with a 100-mesh screen 8 in. in diameter, the maximum amount of material of the specific gravity of quartz (as cubes) should be

$$2 \times \frac{(0.0208 + 0.0147)}{2} \times \frac{\pi}{4} \times 8^2 \times 2.54^2 \times 2.65, \text{ or } 30 \text{ g.}$$

One of the great hindrances to perfect screening is the adherence of extremely fine particles to coarser particles, or to each other, through electrostatic action or because of the presence of

TABLE 8.—COMPARATIVE SIZING ANALYSES BY DRY SCREENING AND WET-AND-DRY SCREENING ON TWO WET-GROUND PRODUCTS DRIED PRIOR TO SCREENING

Size, mesh	Sample A		Sample B	
	Direct percentage weight		Direct percentage weight	
	Dry	Wet-and-dry	Dry	Wet-and-dry
+48	13.3	12.4	0.2	0.0
48/65	9.0	6.1	1.0	0.2
65/100	13.0	9.6	3.2	0.7
100/150	19.0	11.5	3.6	1.3
150/200	10.9	9.9	7.0	2.4
−200	34.8	50.5	85.0	95.4

minute amounts of moisture or because of dried salts. The best way to take care of the dust and "slime" problems is to conduct a wet-and-dry screening operation. Table 8 presents comparative data obtained by dry screening and wet-and-dry screening with a Ro-tap on two sets of split samples.(16)

Wet-and-dry screening is conducted as follows.(16) The sample to be screened is pulped with water and dispersed, with a minute quantity of sodium silicate dissolved in water if necessary; then it is stirred in a pan and decanted over the finest screen to be used in the screening operation. This washing and decanting is repeated until the supernatant liquor becomes practically clear in a short time (about half a minute). Additional water is then allowed to play on the material retained on the screen until the liquor passing through the screen is practically clear. The undersize is flocculated by suitable chemicals, *e.g.*, lime, copper sulphate, or calcium chloride, the water is decanted off, and the flocculated sediment is dried. The oversize is dried (preferably in a low-temperature oven to avoid decomposition and decrepitation), and the screen itself is dried, preferably by a draft of air from a fan. The dry, granular oversize is then screened dry on all the screens, including the screen that was used in the wet-screening operation. The undersize obtained by dry screening from the finest screen is combined with the undersize obtained by wet screening. The products are weighed. Wet-and-dry screening is of course not applicable to materials that are altered by contact with water, such as cement. In that case, liquids other than water can be used in place of it.

Laboratory sizing by means of screens is generally conducted to 200-mesh (0.074 mm.), but recent improvements in screen manufacture have made it possible to extend this limit to one-half that size, or 0.037 mm.

Classification Technique. The second most important method of sizing is by classification, also known as elutriation and sedimentation.(8) Although these three terms are not strictly equivalent, they all three imply use of the principle of sizing material in accordance with differences in settling velocity in a fluid medium. The fluid medium is either a gas (air) or a liquid, generally water.

Particles finer than 0.05 to 0.2 mm. (depending upon the specific gravity of the solid and viscosity of the fluid) settle at a velocity proportional to the square of the particle size, proportional to the apparent specific gravity of the settling particle, and inversely proportional to the viscosity of the fluid medium. This relationship is known as *Stokes' law of sedimentation by viscous resistance* (see Chapter VIII for derivation). Particles coarser than the limit stated above settle according to a different relationship, but since they are more readily sized by screening, there is no need to consider at this stage their settling behavior. Particles finer than 0.1 micron in diameter settle somewhat faster than would be indicated by Stokes' law. But particles so fine cannot be sized by elutriation or sedimentation techniques because it is impossible to regulate the procedure accurately enough for such slow-settling particles (convection currents, flocculation, evaporation, time). It follows therefore that within the scope of laboratory elutriation or sedimentation the only settling law is the law of Stokes. For practical purposes, it may be sufficient to remember that 200-mesh particles of quartz at room temperature settle 4 in. in water in 15 sec. and to derive other settling times from this datum.

Sedimentation. In sizing particles by *sedimentation*, the sample is first suspended in the fluid medium by suitable agitation, then allowed to settle for the proper length of time. The settling operation is stopped by withdrawing the fluid: the material remaining in suspension is finished material in so far as the size at which the split is being made is concerned, but the sediment contains many particles which, having settled but a fraction of the total depth of the settling vessel, are smaller than the size at which a split is being attempted. Accordingly, it is necessary to submit the sediment to a repetition of the sedimenting operation. This must be done as many times as necessary, sometimes 20 or more. Strictly speaking, there is no end point to a sedimentation operation since a portion of the material that should remain in suspension can always manage to settle. But here again, as in the case of screening, a satisfactory practical end point can be reached.

Suitable vessels in which to conduct sedimentation operations are large beakers (800-, 1,000-, or 2,000-cc. capacity) or enamelware buckets. Withdrawal of the suspended material is usually effected by a siphon of large bore made of glass tubing and rubber, and provided with an upturned intake.

Elutriation.[3,6,10] In elutriation, the material to be separated is again suspended in the fluid medium but this time the fluid medium instead of being at rest is caused to ascend at a rate exactly equal to the chosen settling velocity. Here again, there is no practical end point to the operation because those particles whose size is very close to the size at which the split is being made will remain in apparent suspension a long time, and because the ascending velocity of the fluid is somewhat larger in the center of the elutriator than along its walls.

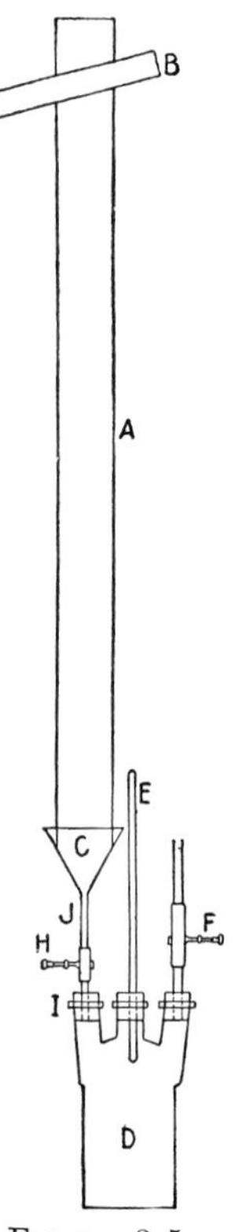

FIG. 35.—Sketch of elutriator. Parts are as follows: (*A*) elutriator column, (*B*) overflow launder, (*C*) teter chamber, (*D*) bottle, (*E*) thermometer, (*F*) and (*H*) clamps.

Elutriators have been designed of many shapes. Among the more satisfactory are long cylindrical glass tubes with a short conical section at the bottom for feeding fluid and an overflow launder at the top to guide the effluent to a receiving tank (Fig. 35).

If the material that is being sedimented is handled in a liquid, repetition of the sedimentation from ten to twenty times usually yields a satisfactory separation when a split is being made to remove the finest particles. Elutriators are used when greater accuracy is desired; then a displacement of fluid equal to twenty to forty times the volume of the elutriator is generally adequate. In making splits at sizes coarser than the finest split, a smaller number of sedimentation operations is sufficient and a smaller volume of fluid can be put through an elutriator.

Flocculation and Other Limitations. One of the greatest difficulties inhibiting the wider use of sedimentation and elutriation is the fact that it is difficult to eliminate completely the *flocculation* of the fine particles. Fine particles, when adhering to form relatively large floccules, settle more rapidly than if not flocculated. Generally, flocculation can be overcome either by the addition of suitable salts or colloids (sodium carbonate, sodium silicate, gelatin, gum arabic) or through the use of a suitable fluid medium in which flocculation happens not to occur. Thus, sulphide flotation concentrates are generally dispersed in acetone or amyl alcohol provided they do not contain a large proportion of fine silicate or carbonate slimes. If they contain a large proportion of the latter, the nonsulphide slimes can be dispersed in water and removed that way, and then the sediment from the water sedimentation or elutriation can be sedimented or elutriated in a suitable organic liquid.

In conducting either a sedimentation or elutriation operation, it is preferable to make the size separation at the finest size first and at the coarsest size last. In that way, the material rehandled from one step to another is of small bulk instead of being unwieldy.

Attention must be called to the fact that sizing by elutriation or sedimentation cannot be conducted without involving the specific gravity of the minerals being separated. Therefore, *if the sample contains many different mineral species, each size class will contain a different range of actual sizes for each mineral.* As a result, the total size range in each size class will be considerably larger than if one mineral species only were involved. There is unfortunately no remedy for this difficulty.

Sizing by elutriation in liquids is conveniently conducted at a divisional size as fine as 5 to 10 microns (for ordinary silicate minerals). Sizing by sedimentation in beakers or buckets can be conducted at a divisional size as low as 2 or 3 microns, although the operation becomes increasingly inaccurate and tedious with increasing fineness of the divisional size.

Centrifuging. Recently, experiments have been conducted to ascertain whether the *centrifuge* can be used to size material finer than that which can be handled by elutriation or sedimentation. The preliminary experiments concluded to date show that the centrifuge can be used in the range of sizes extending from 2 microns to 0.05 micron. These limits can probably be stretched somewhat by further search. The greatest difficulty that besets sizing by sedimentation in a centrifuge is that it is practically impossible to check on the operation by any visual method: Most of the particles are too fine to be measured accurately or even to be seen with the most high-powered microscopes; also, flocculation is more obnoxious than ever.

Sizing by Measurement under the Microscope. Sizing by measurement under the microscope[20] differs in no essential way from sizing by measurement with a yardstick. In practice, the material is sprinkled on a glass slide or is mounted in some way on a slide or a briquette; the diameter of each particle encountered in moving the slide under the objective by means of a mechanical stage is measured, this measurement being taken in a direction chosen at random without relation to the large and intermediate dimensions of the particle. If a and b are the two largest dimensions of the particle (situated in the plane of the stage of the microscope), the measurement yields a value that in the long run averages close to $\sqrt[2]{ab}$. As has already been indicated, this technique of measurement of particles under the microscope neglects the third dimension c. If desired, c can be determined by focusing first on top of the particle and then on the glass slide and determining the distance traveled by the objective. This is ascertained by reading the scale that accompanies the fine adjustment of the microscope for the two end positions of the objective. The neglect of the c dimension can also be overcome by briquetting, polishing, and measurement; but that introduces other difficulties.

One of the limitations of measurement of particles with the microscope arises from the fact that the method is not suitable when there are particles of widely different sizes (1) because some of the large particles cover fine particles which are thus not seen, (2) because it is not possible to have all the particles in focus at once, and (3) because fine particles stick to coarser particles and to each other, and occasionally form utterly unresolvable floccules. In using the microscope for determining the size of fine material, it is therefore necessary first to wash and to size the material.

Another objection to the microscope method of measuring particle size is that after the particles have been measured they are still all left together, it being impossible to ascertain whether particles of a certain composition report in a certain size group in preference to others (unless a mineragraphic analysis is also made with the microscope).

Measurement of particles under the microscope can be conveniently carried out for particles as fine as 0.5 micron, or even somewhat finer, although the accuracy of the determination with particles of that size is not high. This is because the limit of visibility under the microscope and with the best oil-immersion objective is only about one-third of the wave length of the light used; with dry objectives, the limit of visibility cannot exceed one-half the wave length of the light used. In the case of yellow light, $\lambda = 0.5$ micron, it is not safe to consider detail as visible if it is finer than 0.2 to 0.25 micron.

Average Size. Reference is frequently made in connection with ore-dressing problems to the *average size* of particulate products. If the product is fairly closely sized, average size may be regarded as the arithmetic mean of the limiting sizes. Thus the average size of a $-65 + 100$-mesh product is

$$\frac{(0.0208 + 0.0147)}{2} = 0.0177 \text{ cm.}$$

But in dealing with a crudely sized or an unsized product, average size is a meaningless expression since many averages can be struck for various purposes, and with justification.[15,19] The most useful average is one such that the specific surface (square centimeters per cubic centimeter) of the particles of the average size equals the specific surface of the sample.[5,9]

In this connection, it is interesting to observe that the specific surface of particles of the same shape is inversely as the size: *e.g.*, a cube 1 cm. on edge has a surface of 6 sq. cm.; the same *weight* of material in the form of cubes 0.1 cm. on edge has a surface of 60 sq. cm., and in the form of cubes n cm. on edge it has a surface of $6/n$ sq. cm.

If $f_1, f_2, f_3, \ldots, f_p$ denote the volume abundance (on a percentage basis) of successive grades of size $n_1, n_2, n_3, \ldots, n_p$, the average size x is then such that

$$\frac{\Sigma f_1 + f_2 + f_3 + \cdots + f_p}{x} = \sum \frac{f_1}{n_1} + \frac{f_2}{n_2} + \cdots + \frac{f_p}{n_p}.$$

Since $\Sigma f_1 + f_2 + \cdots + f_p = 100$ (by definition), average size can be defined by the relation:

$$x = \frac{100}{\sum \frac{f}{n}}. \qquad \text{[III.1]}$$

This relationship is useful in connection with crushing and grinding efficiency, at which place it is discussed further.

Example. The following tabulation shows a simple way of calculating average size on the basis of Eq. [III.1].

Size, mesh	Average size of grade, cm. (n)	Volume abundance, per cent (f)	$\frac{f}{n}$
28/35	0.0503	2	39.8
35/48	0.0356	8	224.5
48/65	0.0251	18	717.5
65/100	0.0178	30	1685.0
100/150	0.0125	20	1600.0
150/200	0.0089	12	1350.0
200/270	0.0063	7	1110.0
270/400	0.0044	3	681.0
			$\sum \frac{f}{n} = 7407.8$

$$x = \frac{100}{\sum \frac{f}{n}} = \frac{100}{7407.8} = 0.0135 \text{ cm.}$$

Literature Cited

1. ANON.: Standard Hand Method for Screen Testing of Ores, *Mining and Met.*, **13**, 447–449 (1932).
2. ANON.: Screen Testing Ores by Hand, *Eng. Mining J.*, **134**, 149–150 (1933).
3. CLEMMER, J. BRUCE, and WILL H. COGHILL: Improved Laboratory Elutriator and Its Application to Ores, *Eng. Mining J.*, **129**, 551–554 (1930).
4. DISBRO, G. A.: Screen-scale Sieves Made to a Fixed Ratio, *Proc. Am. Soc. Testing Materials*, **13**, 1053–1068 (1913).
5. GAUDIN, A. M.: An Investigation of Crushing Phenomena, *Trans. Am. Inst. Mining Met. Engrs.*, **73**, 253–316 (1926).
6. GAUDIN, A. M., J. O. GROH, and H. B. HENDERSON: Sizing by Elutriation of Fine Ore-dressing Products, *Ind. Eng. Chem.*, **22**, 1363–1366 (1930).
7. GEER, M. R., and H. F. YANCEY: Expression and Interpretation of the Size Composition of Coal, *Am. Inst. Mining Met. Engrs., Tech. Pub.* 948 (1938).
8. GESSNER, HERMANN: "Die Schlämmanalyse," Akademische Verlagsgesellschaft, Leipzig (1931).
9. GREEN, HENRY: A Photomicrographic Method for the Determination of Particle Size of Paint and Rubber Pigments, *J. Franklin Inst.*, **192**, 637–666 (1921).

9a. GROSS, JOHN: A Proposed Standard Sizing Test, *mimeographed report, Committee on Milling Methods, Am. Inst. Mining Met. Engrs.* (1925).

10. GROSS, JOHN, S. R. ZIMMERLEY, and ALAN PROBERT: A Method for the Sizing of Ore by Elutriation, *U. S. Bur. Mines, Rept. Investigations* 2951 (1929).

11. HATCH, T., and S. P. CHOATE: Statistical Description of Size Properties of Non-uniform Particulate Substances, *J. Franklin Institute*, **207**, 369–387 (1929); *Harvard Eng. School Publ.* 35, 369–387 (1928–1929).

12. HERSAM, ERNEST A.: Clastic Form in Size Measurement, *Eng. Mining J.*, **131**, 403–404 (1931).

13. HOOVER, THEODORE J.: A Standard Series of Screens for Laboratory Testing, *Trans. Inst. Mining Met.*, **19**, 486–508 (1910), with large bibliography; *Eng. Mining J.*, **90**, 27 (1910).

14. MARTIN, G., C. E. BLYTH, and H. TONGUE: Researches on the Theory of Fine Grinding. I. Law Governing Connection between Number of Particles and Their Diameters in Grinding Crushed Sand, *Trans. Ceram. Soc.*, **23**, 61–120 (1923–1924).

15. PERROTT, G. ST. J., and S. P. KINNEY: The Meaning and Microscopic Measurement of Average Particle Size, *J. Am. Ceram. Soc.*, **6**, 417–439 (1923).

16. PROBERT, ALAN: Standardized Sieving Methods, *Eng. Mining J.*, **131**, 311–312 (1931).

17. ROSIN, P., and E. RAMMLER: The Laws Governing the Fineness of Powdered Coal, *J. Inst. Fuel*, **7**, 29–36 (1933).

18. WADELL, HAKON: Sphericity and Roundness of Rock Particles, *J. Geol.*, **41**, 310–331 (1933), with large bibliography.

19. WEIGEL, W. M.: Size and Character of Grains of Non-metallic Mineral Fillers, *U. S. Bur. Mines, Tech. Paper* 296 (1924).

20. WORK, LINCOLN T.: "The Graphical Analysis of Fineness Distribution Curves for Pulverized Materials," dissertation, Columbia University, New York (1928).

21. ANDREASEN, A. H. M.: The Fineness of Solids and the Technological Importance of Fineness, *Ingeniorvidenskabelige Skrifter*, **3**, Akad. Tekn. Videnskaber Dansk. Ing., Copenhagen (1939).

CHAPTER IV

LIBERATION

One of the objects of comminution is to liberate dissimilar minerals from attachment to each other. In fact this is the principal goal of comminution whenever a separation is sought that has chemical rather than physical objectives.

It has commonly been asserted that mineral dressing consists of two steps, liberation and separation, and that the second step is impracticable if the first has not been successfully accomplished. This is only partly true, since some degree of separation is possible between locked particles of various kinds, but it should be evident that production of a reasonable degree of liberation is prerequisite to making a fair separation.

Particles of ore can, of course, consist of a single mineral in which case they are termed *free particles*, or they may consist of two or more minerals in which case they are *locked particles*. Locked particles are binary, ternary, quaternary, etc., as they contain two, three, four, or more minerals.

The *degree of liberation* f of a certain mineral or phase is the percentage of that mineral or phase occurring as free particles in relation to the total of that mineral occurring in the free and locked forms. Conversely, the degree of locking of a mineral is the percentage occurring in locked particles in relation to the total occurring in the free and locked forms.

In the following analysis, *grain* and *grain size* pertain to the uncrushed rock and *particle* and *particle size* to the crushed or ground rock.

The very fact that a large lump is reduced in size to small particles does not result in rupturing the bond between adjacent dissimilar minerals; but it restricts the occurrence of locking to a relatively small portion of the original lump, and this produces a result practically equivalent to freeing. However, if the physical properties of the adjacent minerals are sufficiently dissimilar, or if the bond between them is notably weaker than either of them, fracture may take place preferentially at the boundary. In that case, comminution results in true freeing of minerals in addition to a restriction in the occurrence of locking. It can, therefore, be said that liberation is increased by comminution by two means, *liberation by size reduction* and *liberation by detachment.*[14]

LIBERATION BY SIZE REDUCTION

To illustrate liberation by size reduction consider, first, a cube of rock 10 cm. in diameter consisting of cubic grains 10 mm. in diameter of two equally abundant minerals. Assume that the crushing can be conducted to yield grains all of the same size (this simplification does not represent actual conditions, but

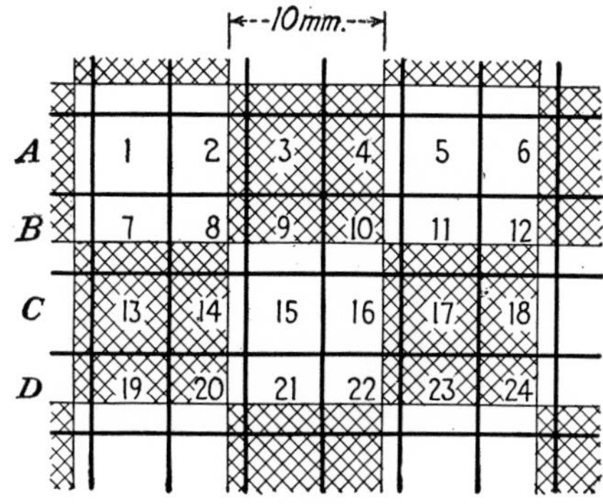

Fig. 36.—Superposition of a crushing lattice (heavy lines) on a grain lattice (faint lines).

it does not vitiate the value of the example), and assume further that the crystals in the rock are thoroughly fast to each other. Until the rock is crushed to particles 10 mm. in diameter, all the grains are locked. If the rock is crushed to particles 5 mm. in diameter, some of the grains will be free, others locked. It is not correct to assume that all the particles are free just because 5 mm. is smaller than 10 mm. That the particles are not all free is readily concluded from reasoning with two superimposed cubic lattices, one having a parameter of 10 mm. and the other of 5 mm.

Minerals of Equal Abundance. This may be visualized by reference to Fig. 36 in which two lattices of parameters 10 and 5 mm. have been superimposed with axes parallel. It will appear that in the plane of the section there are no free particles in each alternate row, *A* and *C* alone having free-looking particles. Within these rows, every other particle appears free. That is, the free particles seem to be Nos. 1, 3, 5, 13, 15 and 17. But there is an even chance that these are locked as may be realized by considering the relationship of the lattices at right angles to the plane of the paper. So the free particles for the two minerals put together would be 6 out of 48 or $\frac{1}{8}$, not 6 out of 24. The degree of liberation of each mineral is the same, 3 out of 24 or $\frac{1}{8}$. If the lattices were not superimposed coaxially, the proportion

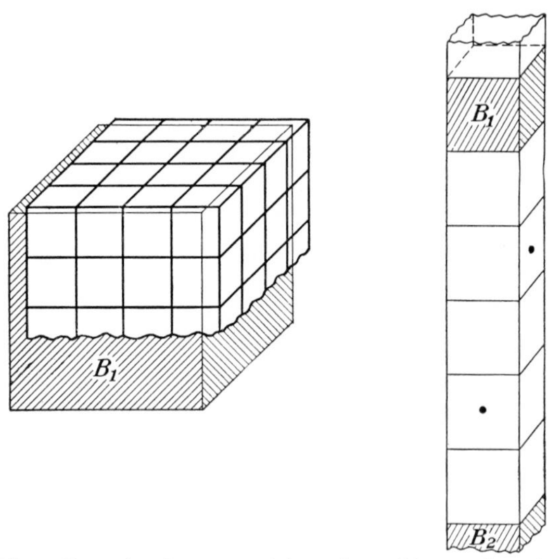

Fig. 37.—Three-dimensional superposition of crushing and grain lattices in unit prism.

of free particles would be even smaller. The only way in which greater freeing could occur would be by superposing every other plane of the crushing lattice on a plane of the grain lattice. But the chance of this happening is nil (as it is the ratio cubed of the width of a line to a definite length).

This calculation shows that if the particles are one-half as large as the grains the liberation is only 12.5 per cent (if the two constituents are equally abundant). This may seem a somewhat surprising result.

Minerals of Unequal Abundance. Calculation for the more general case in which the two constituents are not equally abundant shows that the results for the scarce constituent are as obtained in the case of minerals of equal abundance, but that the freeing of the abundant constituent is greater. The derivation of the basic formulas is as follows (Fig. 37).

Consider an aggregate of cubes of phases A and B of the same size α in which A is n times as abundant as B, and in which the various cubes are arranged so that the various B's are on the average as far apart as possible. Within such a lattice, consider further a prism of cross section α and of height $(n + 1)\alpha$, consisting of n A's and one B.

If a fracturing lattice of parameter β such that $\alpha = \beta k$ is superimposed upon the grain lattice and parallel to it, the unit prism will contain *the equivalent* of $(n + 1)k^3$ particles. Each layer will contain k^2 particles, and there are $(n + 1)k$ layers. Of these layers, $(k - 1)$ will cut B only, $(nk - 1)$ will cut only A, and two will cut both A's and B's.

In the layers cutting B's only, the sets of rows on two of the four sides of the prism will cut also cubes outside the prism, and these will be cubes of A. Hence the particles wholly within B are $(k - 1)(k - 1)(k - 1)$, and the liberation of B will be

$$f_B = \left(\frac{k - 1}{k}\right)^3. \qquad \text{[IV.1]}$$

In the layers cutting A's only, numbering $(nk - 1)$, the sets of rows on two of the four sides of the prism will cut also cubes outside the prism. These outside cubes will be made of A except one per prism face (marked by a dot in Fig. 37). Each of these two loci of locking causes a reduction of $k(k + 1)$ in the number of free particles of A. Generally, then, the particles of free A number, per unit prism,

$$(nk - 1)k^2 - 2(k + 1)k, \qquad \text{or} \qquad (nk^2 - 3k - 2)k.$$

The liberation of A is then

$$f_A = \frac{nk^2 - 3k - 2}{nk^2}. \qquad \text{[IV-2]}$$

This is correct provided $n \geqslant 5$, as there are geometric limitations if $n < 5$. Actually

$$\left.\begin{aligned}
&\text{For } n = 1 \qquad f_A = f_B = \left(\frac{k - 1}{k}\right)^3 \\
&\text{For } n = 2 \qquad f_A = \frac{(k - 1)^2 (2k + 1)}{2k^3} \\
&\text{For } n = 3 \qquad f_A = \frac{k - 1}{k} \\
&\text{For } n = 4 \qquad f_A = \frac{(k - 1)(4k^2 + k - 1)}{4k^3}.
\end{aligned}\right\} \qquad \text{[IV.2}a\text{]}$$

Quantitative theoretical results are presented in Table 9 for both the less abundant constituent (phase B) and the more abundant constituent (phase A). This table relates the percentage liberation f with the ratio k of the grain size in the uncrushed rock to the particle size in the crushed rock. This relation is given for various relative abundances n of the more abundant phase A to the less abundant phase B. The grain

TABLE 9.—RELATIONSHIP OF DEGREE OF LIBERATION f TO RATIO k OF GRAIN SIZE TO PARTICLE SIZE, AND TO RELATIVE VOLUMETRIC ABUNDANCE n OF THE TWO PHASES ($A > B$)
(Based on assumptions of particle lattice and grain lattice detailed in the text)

k	f_B, per cent	f_A, per cent						
		$n = 1$	$n = 2$	$n = 4$	$n = 10$	$n = 25$	$n = 100$	$n = 1{,}000$
0.10	0	0	0	0	0	0	0	77.0
0.25	0	0	0	0	0	0	56.0	95.5
0.5	0	0	0	0	0	44.0	86.0	98.6
1.0	0	0	0	0	50.0	80.0	95.0	99.5
2.0	12.5	12.5	31.3	53.1	80.0	92.0	97.9	99.8
4.0	42.2	42.2	63.3	78.6	91.2	96.5	99.1	100.0
8.0	66.9	66.9	81.3	89.9	95.9	98.5	99.6	100.0
16.0	82.4	82.4	90.6	95.3	97.9	99.4	99.9	100.0
32	90.9	90.9	95.4	97.6	99.1	99.9	100.0	100.0
64	95.5	95.5	97.6	99.0	99.8	100.0	100.0	100.0

size is, of course, the size of the grains of B. Table 9 shows the following:

1. The less abundant phase is not freed at all unless the particles are finer than the grain size.
2. The particles must be much finer than the grain size if the less abundant phase is to be freed materially.
3. If n is large, the more abundant phase may be appreciably freed even if the particles are coarser than the grain size.
4. The more abundant phase is always freer than the less abundant phase. Figure 38 presents in graphical form some of the data of Table 9.

Mineral Particles of Many Sizes. So far in this analysis, it has been assumed that all particles that are produced by comminution have the same size. This assumption can be dis-

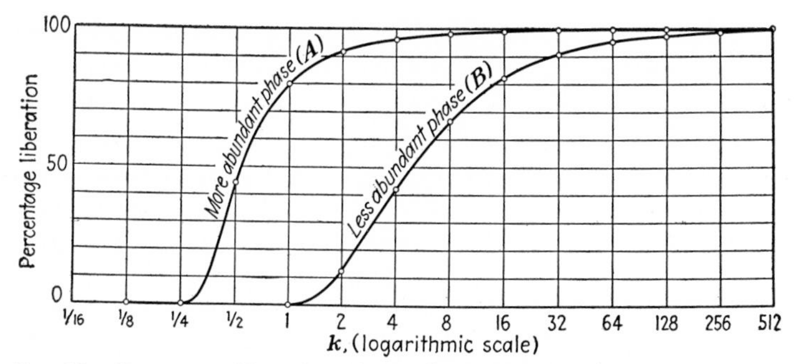

FIG. 38.—Percentage liberation of two phases A and B ($A > B$), in function of ratio k of grain size in uncrushed rock to particle size in crushed rock, on the assumptions that all particles are of one size, and that A is volumetrically 25 times as abundant as B.

pensed with since crushed or ground products have some characteristics of well-nigh universal occurrence (see Chapter VI). Table 10 presents the sizing analysis of a rod-mill product. This can be regarded as typical of many crushed or ground products.

TABLE 10.—TYPICAL SIZING ANALYSIS OF A CRUSHED PRODUCT (ROD-MILL PRODUCT)

Size Range, Microns	Weight, Per Cent
Plus 4,699	0.1
4,699/3,327	0.6
3,327/2,362	2.9
2,362/1,651	12.8
1,651/1,168*	17.8
1,168/833	15.9
833/589	12.7
589/417	9.8
417/295	7.0
295/208	5.3
208/147	4.0
147/104	2.9
104/74	2.0
74/52	1.6
52/37	1.3
37/26	1.0
26/18	0.7
18/13	0.5
Minus 13	1.1

* This includes the size of greatest weight frequency γ. Here

$$\gamma = \frac{(1{,}651 + 1{,}168)}{2} = 1{,}410 \text{ microns, approximately.}$$

Broadly speaking, it can be converted to represent the sizing analysis of some other crushed product by merely sliding one of the columns of figures with respect to the other. (Table 7, Chapter III, presents similar data for a product crushed by rolls.) The size of greatest weight frequency γ will be used as reference size.

Multiplication of the percentage weight of each size grade (Table 10, column 2) with the appropriate percentage liberation of that size (Table 9), summation of these products, and division by 100 yields the average percentage liberation of the crushed product containing all sizes of particles. This has been done for

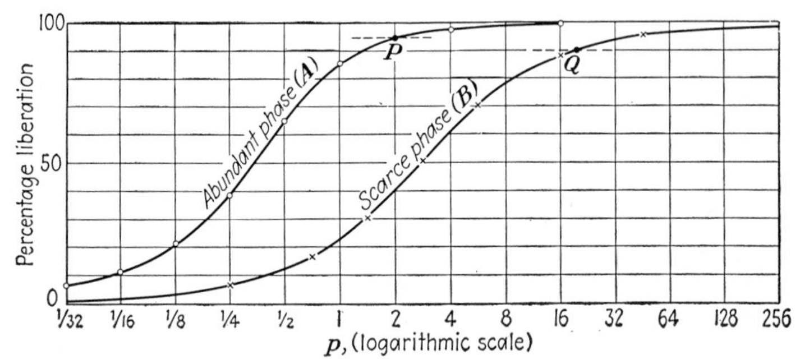

Fig. 39.—Percentage liberation of two phases A and B ($A > B$), in function of the ratio p of the grain size in the uncrushed rock to the size of the particles of greatest weight frequency γ, on the assumptions that the crushed product is typified by rod-mill grinding, and that A is volumetrically 25 times as abundant as B.

the special case in which the gangue is twenty-five times as bulky as the valuable mineral ($n = 25$)—a condition typical of base-metal sulphide ores. Calculations were carried out for values of p, the ratio of the grain size to the size of greatest weight frequency γ, ranging from $\frac{1}{32}$ to 128. The results are presented in Fig. 39.

Because of the limitative assumptions made in preparing Fig. 39, its data are not strictly applicable in practice. Broadly, however, it seems that under the most favorable conditions an increase in liberation of some 20 per cent can be expected from doubling the reduction in size. It is also clear that some liberation is obtained in the scarcer phase, even if p is less than one, and that considerable liberation is obtained in the more abundant phase, even if p is considerably less than one.

Example. As a practical application of Fig. 39, consider the following case: A zinc ore consists of sphalerite in limestone occurring to the extent of $\frac{1}{25}$ by volume as grains about 2 mm. in diameter; to what fineness must the ore be ground (1) in order to make possible the rejection of 95 per cent of the gangue as free particles, (2) in order to achieve a liberation of about 90 per cent of the sphalerite?

Figure 39 (point P) shows that condition (1) is fulfilled by crushing so that ratio p of the grain size to the size of maximum weight frequency is 2, *i.e.*, if the size of maximum frequency γ is 1 mm.

Condition (2) is fulfilled by crushing so that the ratio p of the grain size to the size of maximum frequency γ is 20; *i.e.*, so that the size of maximum weight frequency is about 0.10 mm. (point Q). In other words, rejection of some 95 per cent of the gangue should be possible by tabling, but production of a high-grade concentrate would necessitate the use of flotation.

LIBERATION BY DETACHMENT

If the rock to be crushed is made of grains bonded loosely, fracturing to the grain size of the rock should result in complete liberation and no further liberation should be obtainable by additional comminution. Although such a condition is theoretically possible, practically it is no more than partly realized.

Outstanding practical examples of mineral products liberable by detachment are partly indurated gravels, land-pebble phosphate rock, weathered rocks, etc.

At the other extreme, may be mentioned intimate intergrowths of sulphide minerals of similar properties, as marmatite and pyrrhotite or marmatite and chalcopyrite. Freeing of these minerals is almost entirely a case of reducing the locking by size reduction.

Most ores, however, belong to neither of the extremes just mentioned but to an intermediate class in which more or less preferential fracturing occurs at grain boundaries. This fracturing can be traced to one of the following causes:

1. Macrostructural weaknesses, as bedding planes in coals.
2. Microstructural weaknesses, as the schistosity of some ores.
3. Microstructural differences in physical properties of adjacent minerals, as in hardness, brittleness, cleavability. Examples are pyrite and chalcocite, galena and quartz.

Existence of liberation by detachment is shown by a sizing analysis of the crushed product obtained from a uniformly sized product fed to a crusher: if there is preferential fracturing, there will exist an excessive amount of particles of the size of the grains that are involved in the liberation by detachment. This may be seen at a glance from Fig. 40 in which are plotted the sizing analyses of products crushed similarly from homogeneous and heterogeneous rocks. (The adjectives *homogeneous* and *heterogeneous* are used to describe rocks lacking the property of liberation by detachment, and distinguished by it, respectively.) Figure 40 also includes the grain-size frequency curve as deter-

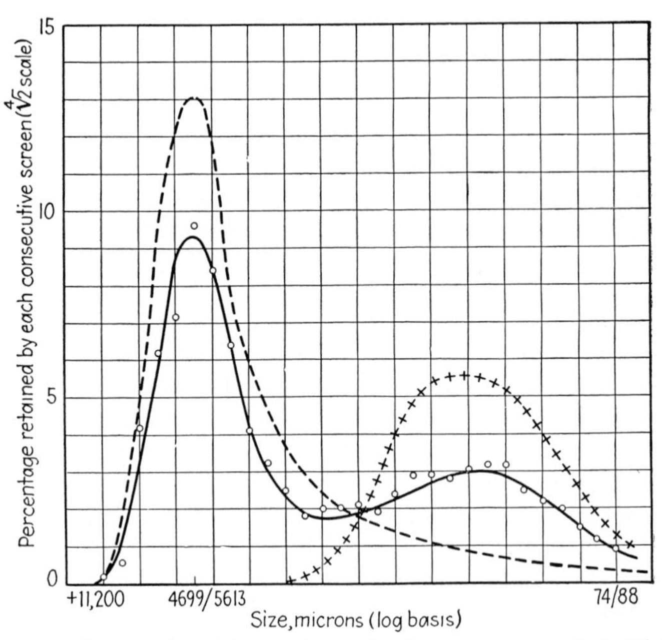

FIG. 40.—Comparative sizing analyses of a heterogeneous rock (solid line) and of a homogeneous rock (broken line) after crushing to the same size in rolls; and grain-size frequency curve (cross-dot line) as determined microscopically on the uncrushed rock.

mined microscopically on the uncrushed ore. The accord between the curves concerning the heterogeneous rock is striking.

BEHAVIOR OF LOCKED PARTICLES

It has already been indicated that it is possible to effect substantial concentration of a mineral product without first completing its liberation. This becomes understandable when inquiry is made into the behavior of locked particles in separating devices. Thus, a locked particle consisting of one-half galena and one-half quartz by volume has an average specific gravity somewhat in excess of five; this should make it possible to separate it readily by gravity concentration from either free quartz or free galena whose respective specific gravities are 2.65 and 7.50. Likewise, a particle consisting of quartz and magnetite behaves in a manner intermediate between that in which pure quartz and pure magnetite behave when subjected to the influence of magnetic forces.

Although locked particles behave, broadly speaking, in a manner intermediate between that of their constituent minerals, and although separation of the locked particles from free particles of the same constituent minerals is possible, the very fact that locked particles grade from those consisting mostly of gangue with only a few specks of mineral to those consisting mostly of mineral with only a few specks of gangue shows that hard-and-fast separation cannot be anticipated.

Quantitative Make-up of Locked Particles. Locked, coarse particles of sphalerite and jasper containing one-tenth of sphalerite by weight are well-worth separating from free gangue particles (with the ultimate aim of freeing their constituents by grinding, and of separating them by flotation); yet such a separation can hardly be envisaged by the usual jigging practices since the spread in specific gravity between locked particles of this type and free gangue particles is but 0.09. Thus the quantitative make-up of locked particles is of significance in their response to separating operations: no one would expect the same response of particles containing mostly mineral and of particles containing mostly gangue by any method of concentration.

This quantitative make-up of locked particles is of paramount importance in gravity concentration. For in that process, all minerals respond to gravity—to different extents, it is true—but still to extents in the same order of magnitude.

The quantitative make-up of locked particles is of less importance in dealing with magnetic separation, since in this instance the relative susceptibilities of minerals to magnetic forces differ enormously more than do the specific gravities of any minerals. The spread in response to magnetic force between a locked particle containing one-tenth of magnetite with nine-tenths quartz on the one hand and free quartz on the other can be made large enough to permit adequate separation.

In flotation, the quantitative make-up of locked particles becomes of no real importance; whatever importance this factor may appear to command is entirely ascribable to the correlation between it and the quantitative make-up of the surface of the particles (Figs. 41 and 42). It is to this surface make-up of locked particles that attention must now be directed.

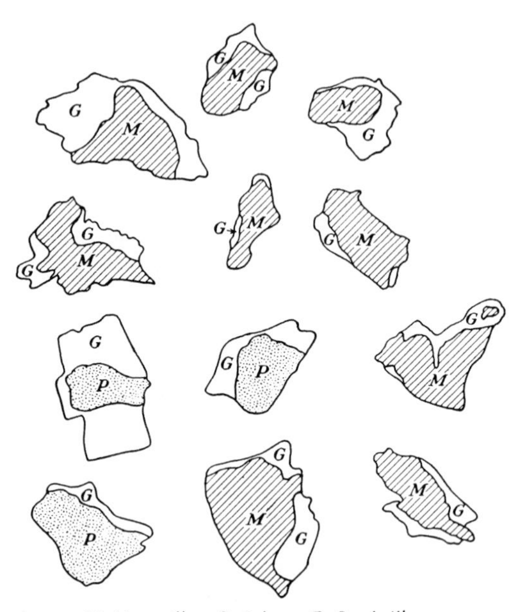

FIG. 41.—Typical locked particles in a lead concentrate (Sullivan Mill, Kimberley, B. C.) Note that galena is more abundant on the outside of the particles.

Qualitative Make-up of Locked Particles. Examination of many thousands of locked particles has shown that they may be grouped into several types. For simplicity's sake, consideration is here limited to binary locked particles, but the same remarks apply, of course, to more complex particles.

In one type, the two phases adjoin each other with rectilinear or gently curving boundaries, suggesting that the particles of this type have arisen from a rock in which the grain size is considerably larger than the particles in question (Fig. 43). In this type of particle, the quantitative occurrence of the phases is in substantial accord with their surface occurrence. The flotation behavior of particles of this type agrees with their behavior by other methods of concentration when due allowance is made for characteristics of flotation and of other means of concentration. Locked particles of this type can be freed readily by regrinding.

A second type of locked particle is the vein type in which one phase appears as a veinlet in another phase (Fig. 44). This type is much rarer than type I. It responds in flotation much

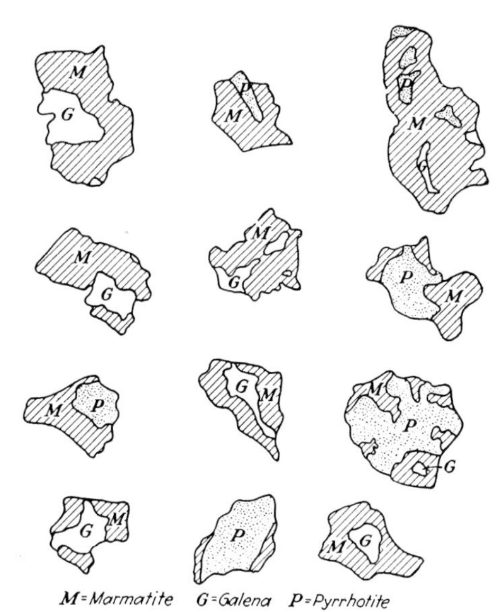

Fig. 42.—Typical locked particles in a zinc concentrate (Sullivan Mill, Kimberley, B. C.). Note that marmatite is more abundant on the outside of the particles.

as might be expected from the relative abundance of the two phases. It is naturally more difficult to free than type I.

A third type consists of shells of one phase surrounding another phase. In its perfect development (Fig. 45) it is, of course, not common, but particles showing a part of a shell on a core are frequent (Fig. 45*a*). Particles of this type do not respond to flotation as might be expected from the relative abundance of the phases, but rather in proportion to the surface area of the two phases (*i.e.*, in proportion to the perimeters of the two phases, when the particles are viewed in polished section). Particles of this type are very common in ores containing one hard phase in juxtaposition with soft phases. They are typified by pyrite cores with chalcocite, bornite, or galena shells.

Particles of type III result in the dilution of copper or lead flotation concentrates with much worthless pyrite. Suitable liberation and reconcentration of rougher concentrates made up largely of locked particles of type III is one of the fundamental yet difficult tasks in some concentrators. The regrinding of type III particles usually leads to the production of very fine particles of the rim substance together with coarser particles of locked character, and with still coarser particles of the core substance. This fortunate feature permits some degree of reconcentration but, by and large, type III locked particles are difficult to re-treat.

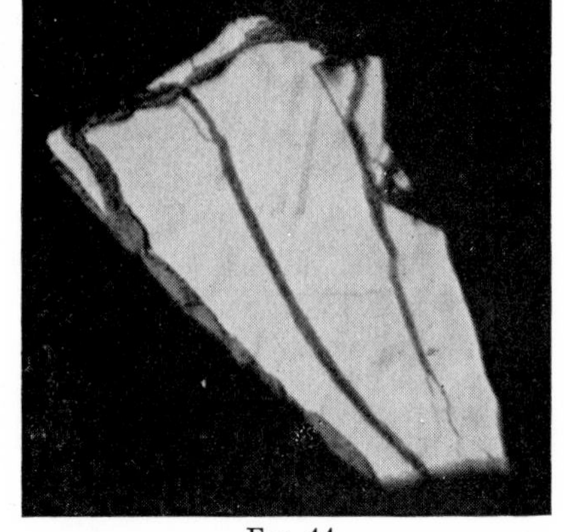

Fig. 43. Fig. 44.

Fig. 43.—Locked particle, type I. Coarse association of galena and sphalerite, with gently-curving boundaries. 150/200-mesh, Midvale zinc concentrate.

Fig. 44.—Locked particle, type II. Vein-like occurrence of chalcocite in pyrite. 100/150-mesh Anaconda copper concentrate.

Fig. 45.—Locked particle, type III. Chalcocite shell on pyrite. 150/200-mesh. Cananea copper concentrate.

The fourth type of locked particles consists of occlusions of one phase in another,[32] in the form of fine blebs (Fig. 46). This type of locked particles accounts for much of the loss of metal in current flotation tailings. It accounts also for much of the inclusion of copper in zinc concentrates[2,39] made by flotation from complex ores. Locked particles of this type respond to flotation and to other means of concentration almost like the unadulterated host phase, largely because the inclusions form but a small proportion of the total, usually less than one-twentieth. Common examples are inclu-

FIG. 45*a*.—Locked particle, type III*a*. Galena shell on pyrite. Midvale lead concentrate. 200/280-mesh

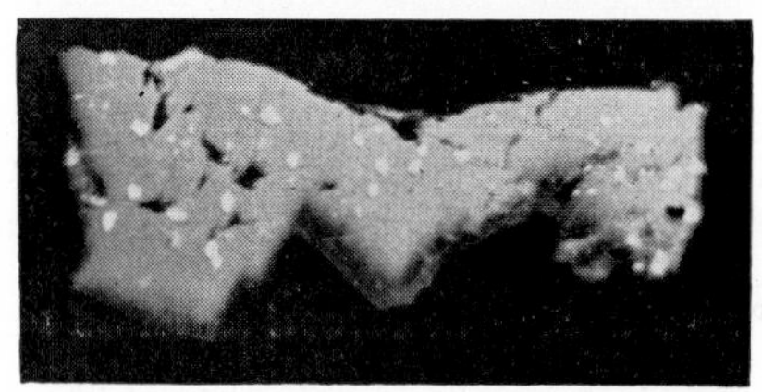

FIG. 46.—Locked particle, type IV. Inclusions of chalcopyrite blebs in sphalerite. 200/280-mesh. Cananea middling.

FIG. 47.—Locked particle, type IV. Intergrowth of bornite and chalcopyrite. 200/280-mesh. Cananea concentrate.

sions of chalcopyrite or tetrahedrite in silicates, of gold in silicates or pyrite, of chalcopyrite in marmatite, of pentlandite in pyrrhotite. In many instances, the blebs appear to average not over 5 microns in diameter. Freeing of such particles is entirely out of question at present. In the case of Fig. 46, even so modest a liberation as two-thirds would require grinding to 1 micron (size of greatest weight frequency)—a performance that so exceeds the possibilities of current industrial grinding as to stagger the imagination, and a stunt that would be worthless since separation of paintlike particles is almost impossible.

Figure 47 presents a fairly common variant of type IV, in which both constituents of the intergrowth are desired in the same concentrate. Although mineralogically a locked particle, this can be regarded as free, from a practical standpoint.

DETERMINATION OF DEGREE OF LIBERATION

Ascertainment of the most advantageous degree of liberation can be achieved empirically by making a large number of separating tests, each on a product crushed or ground to a different degree. By weighing the extra cost of increased comminution against the metallurgical improvement, the conditions most advantageous from an economic viewpoint may be selected. This is the time-honored technique, and it is perhaps the most satisfactory in the long run whenever gravity concentration is involved because there is no way of tampering with the force of gravity. But if flotation is to be used, it is not possible to ascertain by chemical analysis whether failure of a certain proposed operation should be ascribed to inadequate liberation or to erroneous selection of the flotation agents. It is in this connection that quantitative determination of the degree of liberation is of practical value to the operator or testing engineer.(38)

Quantitative determination of degree of locking is attained in one of two ways, depending upon whether the important minerals are transparent or opaque.

If the important minerals are transparent, they are viewed with a polarizing microscope as loose grains sprinkled on a slide in an oil of suitable index of refraction. If the minerals are opaque, they are viewed with a metallographic microscope equipped to use polarized light;(19) to that end, a sample is briquetted with Bakelite and polished.

Identification of transparent grains[13,29,30,36,43] with the polarizing microscope involves the use of three criteria: color, index of refraction, and birefringence, all of which can be ascertained by inspection, by movement of the fine adjustment of the microscope, and by crossing the Nicols. If the birefringence is to be ascertained, even approximately, the particles must be sized prior to inspection. That is also required for quantitative purposes.

Identification of opaque grains[1,7,12,34,37,40,42] involves the use of three criteria also: color, relief, and behavior under polarized light.[8,9,44,45] These criteria can be ascertained by inspection and by crossing the Nicols. In addition, the task of identification may be facilitated by first filming iridescently and selectively some of the minerals by a technique recently developed and standardized.[16,17,18] Selectively staining transparent minerals is also useful.[10,21,22]

A number of studies of mineral liberation are available in the technical literature.[4,5,6,15,23,24,25,26,31,41]

Microscope Counting Technique. In making a count of a mineral product that has been briquetted,[3,11,20,27] it is essential to keep the records in a systematic way. Figure 48 is a facsimile of a record sheet.

Mineral particles are classified into free particles, binary particles, and ternary particles. Under the heading *free particles* are recorded the particles which appear free; under the heading *binary particles* are recorded those which appear to consist of two mineral species only, and under the heading *ternary particles,* those which appear to be composed of three or more mineral species. Each free particle is represented by a stroke, each binary particle by one number, and each ternary particle by three or more numbers, depending upon the number of mineral species that make it up. The number describing a binary particle records the parts per 20 parts of particle surface occupied by the first constituent of the binary particle. The numbers describing each ternary particle record likewise the parts per 20 occupied by each constituent. These numbers are arrayed in a column, each number appearing in a row set aside for one particular mineral species.

In making a count, it is preferable to operate in crews of two, one person observing the material under the microscope and the

Free		
Galena	𝍸 𝍸	
	𝍸 𝍸	
	𝍸 𝍸	
	𝍸 𝍸 𝍸 𝍸 𝍸 𝍸 𝍸 𝍸 𝍸 𝍸 𝍸 𝍸 𝍸 𝍸 𝍸 𝍸 𝍸 //	432
Marmatite	𝍸 𝍸 𝍸 𝍸 𝍸 𝍸 //	32
Pyrite	𝍸 𝍸 𝍸 𝍸 ////	24
Chalcopyrite	//	2
Silicate	𝍸 𝍸 𝍸	15
Binary		
Marmatite-galena	8-4-6-3-7-5-0-2-1-1-6-15-7-3-11-10-8-2-1-	
	1-3-4-5-5-0-1-16-1-2-4-4-3-2-12-10-8-2-	
	1-0-10-18-4-3-5-1-2-7-6-9-13-2-5-6-0-	
	1-2-0-8-15-14-1-17-2-0-14-6-8-9-11-	69
Pyrite-galena	12-6-14-5-7-3-2-5-4-1-8-12-10-4-2-1-	
	0-8-7-16-1-2-3-0-1-6-6-4-10-0-2-13-	
	17-1-3-2-9-11-	38
Chalcopyrite-galena	3-12-5-18-2-	5
Silicate-galena	5-12-8-7-3-5-2-1-15-13-14-8-7-3-13-	
	16-5-11-19-19-9-3-2-1-0-6-10-2-8-3-2-	
	6-5-14-12-3-17-5-6-12-14-13-1-0-6-	
	7-2-1-8-9-5-4-3-	53
Pyrite-marmatite	3-16-14-7-2	5
Chalcopyrite-marmatite	2-4-3-5-7-9-2-1-0-0-16-2-0-1	14
Silicate-marmatite	8-14-3-5-1-7-	6
Silicate-pyrite	3-12-	2

Ternary																			
Galena	8	12	7	2	6	13	10	8	3	4	2		10						
Marmatite	6	3	4	8	4	2	5	2	1	4	16	3							
Pyrite	6		9		10		4	10	16			8	6						
Chalcopyrite						5	1				2								
Silicate		5		10						12		9	4						13
Total																			710

FIG. 48.—Facsimile of record sheet used for particle counting.

other making the record from the first operator's spoken observations. Alternation in observing and recording minimizes eye fatigue and ensures greater speed and accuracy in the results. From the results obtained and recorded, it is possible to calculate not only an ultimate chemical analysis, but also, and this is the important point, the degree of association of minerals to form locked particles and the extent to which each one of the minerals contributes to each particular type of locked particle.

Locking Factor. In this connection, it should be pointed out that although a free particle cannot be cut so as to appear locked, a locked particle may be cut so as to appear free. This is appreciated if reference is made to Fig. 49 which shows what appears to be a free silicate until the glint of sulphide attached to the bottom of the particle is properly valued. It follows that the observations made under the microscope yield a result that is overstating the liberation of the material under consideration. To correct this error inherent in the method of examining polished surfaces of briquetted mill products, it is necessary to increase the number of particles actually observed to be locked by a locking factor.

FIG. 49.—Silicate particle in polished section (with sulphide attached to it inside of the briquette); 150/200-mesh.

Accurate determination of the locking factor is a difficult matter, but some estimate of its magnitude can be made by a mathematical analysis of the problem.[33,35] Figure 50 is a chart giving the magnitude of the locking factor in terms of the relative abundance of the constituents of a binary particle. For practical purposes, a ternary particle may be considered as being a binary particle composed of mineral and gangue only and the data of Fig. 50 be applied to it. Thus in the data reproduced in Fig. 48, there appear 69 galena-marmatite particles containing $^{403}/_{20}$, or 20.15, grain equivalents of marmatite. On the average, those grains appear to be 20.15/69, or 29 per cent, marmatite. According to Fig. 50, the appropriate locking factor is 1.30.

As a first approximation, adoption of a single locking factor, perhaps 1.4, is reasonable. This means that locked particles are some 40 per cent more abundant than they appear to be.

Application. To one who is not especially familiar with microscope technique and mineralogy, it may appear that it is rather difficult to apply the technique of quantitative mineragraphy to ore-dressing problems. This, however, is not commonly the case because the number of mineral species

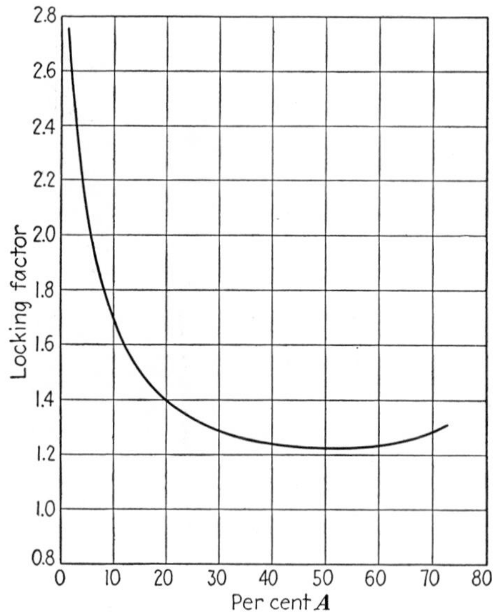

FIG. 50.—Relation of locking factor to percentage of phase A in the total mineral surface.

to be observed in a given ore is generally not large. In lead-zinc ores from the Coeur d'Alene district, for example, the following minerals are usually found: galena, marmatite, pyrite, chalcopyrite, and gangue. This makes five recordable minerals. In the ore from the Sullivan mine of British Columbia, there are six important minerals, pyrrhotite being added to the minerals in the list given for the Coeur d'Alene ores. In most sulphide copper ores, the number of minerals is rarely over eight; and in the handling of coal, the number of recognizable species rarely exceeds five. Once, therefore, the essential minerals have been recognized by an expert on matters relating to mineragraphy, it is an easy task for even an untrained man to make the necessary records.

The securing of the data may be very tedious, if high accuracy is desired, but they need not consume much time if a lesser accuracy is acceptable. Actually, it may be estimated from the Gaussian "least-square" method that the accuracy of a determination varies as the square root of the number of particles that are identified: a sixteenfold increase in the number of observations from 400 to 6,400 yields, for example, but a fourfold increase in precision. A count of some 500 grains may be ample for concentrates, but several thousand may have to be counted in tailings.

The liberation of the minerals at the sizes at which separation is generally attempted is quite variable. In the concentration of many sulphide ores treated by flotation, the average liberation of all minerals is frequently of the order of 75 per cent, at the beginning of flotation. Because of regrinding operations, the final liberation as gauged from concentrates and tailing frequently exceeds 90 per cent, the tailing being as a rule appreciably freer (in accordance with Fig. 39). On the other hand, in the treatment of most coals, the liberation often does not reach 10 to 20 per cent even when the separating operation is completed. In other words, the problem of separating minerals in the treatment of sulphide ores may be regarded primarily as a problem in the separation of free mineral particles from each other, and the problems involved in cleaning coals (and iron ores) may be regarded as primarily problems in the separation of high-grade locked particles from low-grade locked particles.

Literature Cited

1. ANON.: "The Role of the Microscope in Ore Dressing," American Cyanamid Company, New York (1935).
2. BIBOLINI, ALDO: Il ferro ed il rame nella fluttuazione della blenda, *Atti accad. sci. Torino*, **68**, 213–225 (1933).
3. BIRD, PAUL H.: Mounting Concentrates and Tailings for Microscopic Study, *Eng. Mining J.*, **136**, 233–234 (1935).
4. COGHILL, WILL H.: Degree of Liberation of Minerals in the Alabama Low-grade Red Iron Ores after Grinding, *Trans. Am. Inst. Mining Met. Engrs.*, **75**, 147–165 (1927).
5. COGHILL, WILL H., WARREN HOWES, and S. R. B. COOKE: Mineragraphic Aid in the Concentration of Manganiferous Iron Ores, *Eng. Mining J.*, **131**, 361–365 (1931).
6. COOKE, S. R. B.: Microscopic Structure and Concentratability of the Important Iron Ores of the United States, *U. S. Bur. Mines, Bull.* 391 (1936).

7. DAVY, W. M., and C. M. FARNHAM: "Microscopic Examination of the Ore Minerals," McGraw-Hill Book Company, Inc., New York (1920).
8. DAYTON, R. W.: "Theory and Use of the Metallurgical Polarization Microscope," Rensselaer Polytechnic Institute, Troy, N. Y. (1935).
9. DAYTON, R. W.: Theory and Use of the Metallurgical Polarization Microscope, *Am. Inst. Mining Met. Engrs., Tech. Pub.* 593 (1935).
10. ENGEL, A. L.: Staining Method for the Differentiation between Feldspar and Quartz, *U. S. Bur. Mines, Rept. Investigations* 3370 (1938).
11. ERDMAN, E. A.: A Transparent Mounting for Microsections, *Metals and Alloys*, **8,** 27 (1937).
12. FARNHAM, C. M.: "Determination of the Opaque Minerals," McGraw-Hill Book Company, Inc., New York (1931).
13. FORD, WILLIAM E.: "Dana's Textbook of Mineralogy," 4th ed., John Wiley & Sons, Inc., New York (1932), pp. 236–239.
14. GAUDIN, A. M.: An Investigation of Crushing Phenomena, *Trans. Am. Inst. Mining Met. Engrs.*, **73,** 253–316 (1926).
15. GAUDIN, A. M.: Unusual Minerals in Flotation Products at Cananea Mill Studied Quantitatively by Microscope, *Eng. Mining J.*, **134,** 523–527 (1933).
16. GAUDIN, A. M.: Staining Minerals for Easier Identification in Quantitative Mineragraphic Problems, *Econ. Geol.*, **30,** 552–562 (1935).
17. GAUDIN, A. M.: The Identification of Opaque Solids by Selective Iridescent Filming, Part I, Optics, *J. Phys. Chem.*, **41,** 811–859 (1937).
18. GAUDIN, A. M., and DONALD W. MCGLASHAN: Sulfide Silver Minerals. A Contribution to Their Pyrosynthesis and to Their Identification by Selective Iridescent Filming, *Econ. Geol.*, **33,** 143–193 (1938).
19. GRATON, L. C.: Technique in Mineralography at Harvard, *Am. Mineral.*, **22,** 491–516 (1937).
20. HEAD, R. E.: Mounting Flotation Products for the Microscope, *Eng. Mining J.*, **137,** 336–337 (1936).
21. HEAD, R. E., and A. L. CRAWFORD: A Staining Method for Distinguishing Cerussite and Anglesite, *U. S. Bur. Mines, Rept. Investigations* 2932 (1929).
22. HEAD, R. E., and A. L. CRAWFORD: Utilizing Staining Methods in the Identification of Minerals, *Eng. Mining J.*, **127,** 877 (1929).
23. HEAD, R. E., ARTHUR L. CRAWFORD, F. E. THACKWELL, and GLEN BURGENER: Statistical Microscopic Examination of Mill Products of the Copper Queen Concentrator of the Phelps Dodge Corporation, Bisbee, Ariz., *U. S. Bur. Mines, Tech. Paper* 533 (1932).
24. HEAD, R. E., ARTHUR L. CRAWFORD, F. E. THACKWELL, and GLEN BURGENER: Detailed Statistical Microscopic Analyses of the Ore and Mill Products of the Silver King Flotation Concentrator, Park City, Utah, *U. S. Bur. Mines, Rept. Investigations* 3236 (1934).
25. HEAD, R. E., ARTHUR L. CRAWFORD, F. E. THACKWELL, and GLEN BURGENER: Detailed Statistical Microscopic Analyses of Ore and Mill Products of the Utah Copper Co., *U. S. Bur. Mines, Rept. Investigations* 3288 (1935).
26. HEAD, R. E., ARTHUR L. CRAWFORD, F. E. THACKWELL, and A. LEE CHRISTENSEN: Statistical Microscopic Study of Ores and Mill Prod-

ucts from the Anyox Plant of the Granby Consolidated Mining, Smelting and Power Co., Ltd., Anyox, B. C., *U. S. Bur. Mines, Rept. Investigations* 3290 (1935).

27. Head, R. E., and Morris Slavin: A New Development in the Preparation of Briquetted Mineral Grains, *Utah Eng. Exp. Sta., Tech Paper* 10 (1930).
28. Hecht, Arthur S.: Microscopical Determination of Ore-treatment Methods, *Eng. Mining J.*, **134,** 14–16 (1933).
29. Johannsen, Albert: "Manual of Petrographic Methods," 2d ed., McGraw-Hill Book Company, Inc., New York (1918), pp. 247–265.
30. Loewe, F., and G. Gerth: Die Bestimmung des Quarzgehaltes in Aufbereitungsprodukten mit Hilfe der Tetralinmethode, *Metall u. Erz*, **32,** 481–485 (1935).
31. Martin, H. S.: Microscopic Studies of Mill Products as an Aid to Operation at the Utah Copper Mills, *Am. Inst. Mining Met. Engrs., Tech. Pub.* 255 (1929).
32. McLachlan, C. G.: Increasing Gold Recovery from Noranda's Milling Ore, *Trans. Am. Inst. Mining Met. Engrs.*, **112,** 570–596 (1934).
33. Moncrief, R. L.: "A Microscope Analysis of the Flotation Products from the Retreatment of the Middling from the Anaconda Concentrator," Montana School of Mines (1934).
34. Murdoch, J.: "Microscopical Determination of the Opaque Minerals," John Wiley & Sons, Inc., New York (1916).
35. Ravitz, S. Frederick, and Harold A. Steane: An Experimental and Theoretical Investigation of Quantitative Microscopic Mineralogical Analysis, *Am. Inst. Mining Met. Engrs., Contribution* 76 (1935).
36. Rogers, Austin F., and Paul F. Kerr: "Thin-section Mineralogy," McGraw-Hill Book Company, Inc., New York (1933), pp. 55–60.
37. Schneiderhöhn, H., and Paul Ramdohr: "Lehrbuch der Erzmikroskopie," Part II, Gebrüder Bornträger, Berlin (1931).
38. Schwartz, G. M.: Review of the Application of Microscopic Study to Metallurgical Problems, *Econ. Geol.*, **33,** 440–453 (1938).
39. Shenon, Philip J.: Chalcopyrite and Pyrrhotite Inclusions in Sphalerite, *Am. Mineral.*, **17,** 514–518 (1932).
40. Short, M. N.: Microscopic Determination of the Ore Minerals, *U. S. Geol. Survey, Bull.* 825 (1931).
41. The Staff (A. L. Blomfield, H. S. Rood, B. S. Crocker, C. L. Williamson): Milling Investigations into the Ore As Occurring at the Lake Shore Mine, *Trans. Can. Inst. Min. Met.*, **39,** 279–434 (1936).
42. Van der Veen, R. W.: "Minerography and Ore Deposition," G. Naeff, The Hague (1925).
43. Winchell, N. H., and Alexander N. Winchell: "Elements of Optical Mineralogy," 2d ed., Part I, John Wiley & Sons, Inc., New York (1922), pp. 73–78.
44. Wright, Fred E.: Examination of Ores and Metals in Polarized Light, *Trans. Am. Inst. Mining Met. Engrs.*, **63,** 370–381 (1920).
45. Wright, Fred E.: Polarized Light in the Study of Ores and Metals, *Proc. Am. Phil. Soc.*, **58,** 401–447 (1919).

CHAPTER V

GRINDING

There is little to differentiate grinding from crushing aside from the fact that crushing bears on relatively coarse material and grinding on relatively fine material. At first it was believed that a real difference existed, to wit, that in crushing rupture of the particles was effected through the application of compressive stresses, whereas in grinding rupture was effected through the application of shearing stresses. Thus, crushing has been considered as resulting from impact and grinding from rubbing. Later developments in the field of grinding machinery have brought out machines in which shearing stresses are not the only ones used, and perhaps not the most important. It is, therefore, patent that the subdivision of comminution into crushing and grinding is now rather artificial. It may be said that grinding is the breaking down to the ultimate fineness to which comminution is conducted of the relatively coarse particles made by crushing.

Originally "grinding" was used to refer to that method of comminution in which a heavy surface is caused to slide or roll over another surface, the material to be ground being caught between the two moving surfaces. Buhr stones, which have been used in the flour-milling industry from time immemorial, are typical. In the laboratory, mortars and pestles have been used as a means of reducing to fine size solid materials of all kinds.

Arrastres, Chilean mills, and *Huntington mills* were used in dressing plants for many years, but they have been replaced almost entirely by ball mills, rod mills, or tube mills.[3,22] Arrastres were primitive machines, used largely in Mexico. Chilean mills were used extensively as recently as 1910 for secondary size reduction in gravity-concentrating plants. A Chilean mill consists of an iron tub with screened window in which two or three runners roll around from a central drive shaft

Fig. 51.—Chilean mill. (*Fried. Krupp A. G. Grusonwerk.*)

Fig. 52.—Huntington mill. (*Allis-Chalmers Mfg. Co.*)

(Fig. 51). Huntington mills differ from Chilean mills in that the grinding is done, not by the runners pressing down on the ore, but by the loosely suspended roller-shod hammers pressing by centrifugal force on an annular tire (Fig. 52). Chilean mills and Huntington mills were designed for wet grinding. The modern *Bethlehem coal pulverizer* is essentially a dry-grinding Chilean mill. Another modern variant of the Huntington mill is the *Raymond bowl mill* (Fig. 53), in which the roller-tired hammers

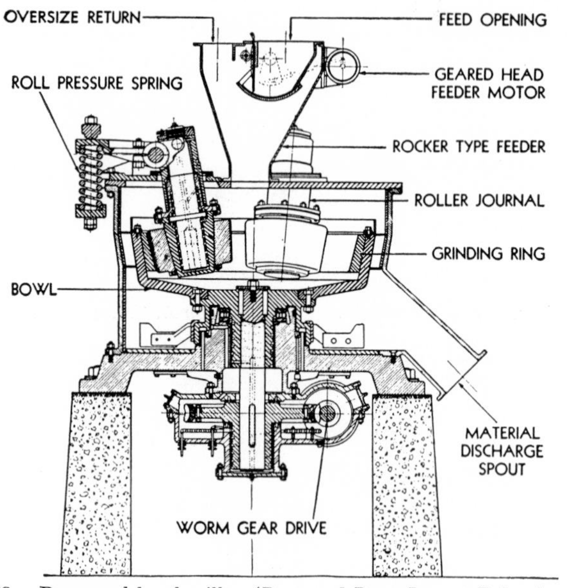

FIG. 53.—Raymond bowl mill. (*Raymond Bros. Impact Pulverizer Co.*)

are stationary and the tub or bowl is rotating. This is used for grinding soft materials, such as coal, limestone, and dolomite. The *dry pan* for clay grinding bears the same relation to the Chilean mill that the bowl mill bears to the Huntington mill.

New machines all ape the grinding action of fast streams in that they consist of a load of grinding medium that is made to tumble with the ore within a rotating cylindrical or cylindro-conical shell. To this group of devices belong the ball mill, the rod mill, the tube mill, and others of recent design. These grinding mills are operated either wet or dry. It is of interest to note that ball mills date back to 1876 at which time a mill was invented by William Brückner, which may be regarded as the forerunner of the modern mills.[1] Extensive adoption of ball mills, however, is much more recent.

BALL MILLS

Ball mills are characterized by the use of steel or iron balls as the grinding medium. The ball mills used in practice are continuous machines, crushed rock being taken in at one end through a feeder and ground rock being discharged at the other end or through the periphery. Laboratory machines are either intermittent or continuous.

Ball mills can be classified according to the shape of the mill, the method of discharging the ground ore, and whether the grinding is conducted dry or wet.

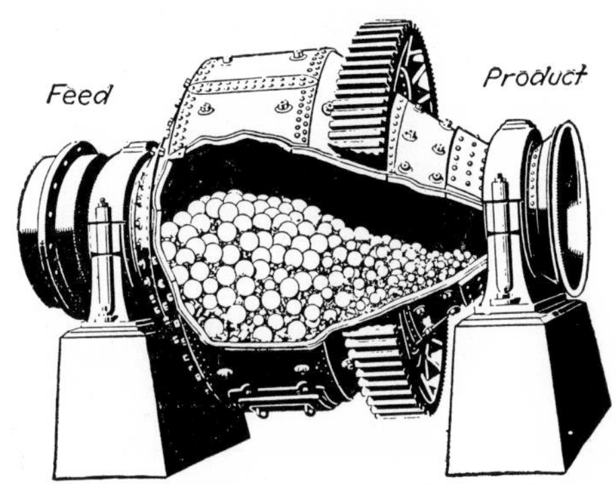

Fig. 54.—Hardinge ball mill (shell cut away to show inside arrangement of balls). (*Hardinge Co.*)

Cylindroconical Mills. The only significant representative of this class of mills is the Hardinge mill,[31,32,51,52] which consists of two conical sections connected by a short cylindrical section (Fig. 54) supported by end bearings on which hollow trunnions revolve. Feed is taken in through a feeder located beyond one of the trunnions, and a continuous stream of thick pulp pours from a discharge lip located axially beyond the other trunnion. The feeder is of the spiral-scoop type (Fig. 55). The mill is gear-driven from a countershaft which may in turn be driven by a belt (V-rope or flat) or through a speed reducer.

The conical section toward the feed end is obtuse, and that toward the discharge end is acute. It is said that the conical shape compels coarse particles and large balls to seek the cylin-

FIG. 55.—Standard ball-mill scoop feeder. (*Allis-Chalmers Mfg. Co.*)

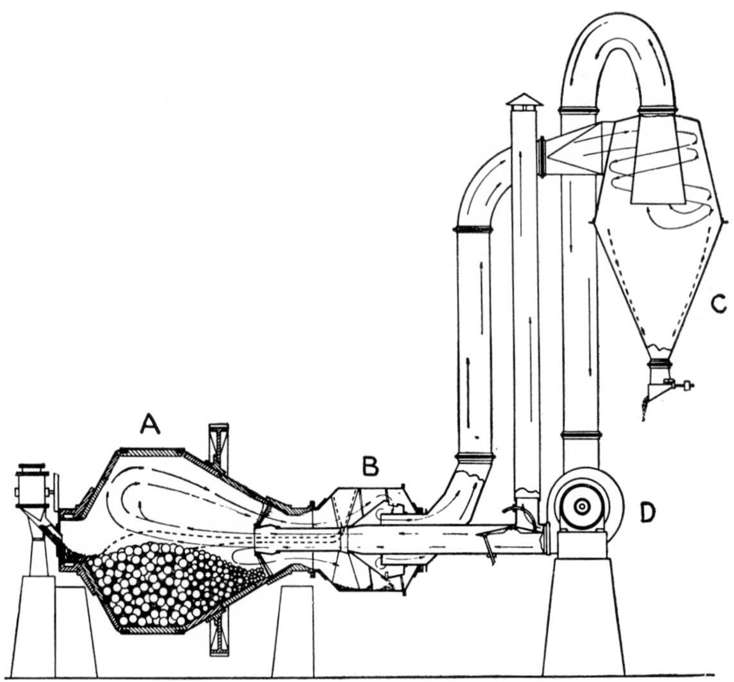

FIG. 56.—Principle of operation of Hardinge cylindroconical mill with rotary air classifier. (*Hardinge Co.*)

drical section of the mill, whereas fine particles and small balls are found in the conical sections, especially that near the discharge end of the mill, thus making possible the preferential grinding of coarse particles by large balls, and of fine particles by small balls.

Hardinge mills are very widely used. In the metallurgical field, they are usually adapted for wet grinding, but they are also used for dry grinding in conjunction with pneumatic classifiers. The principle of operation of dry-grinding mills is shown by Fig.

Fig. 57.—Marcy grate mill. (*Mine and Smelter Supply Co.*)

56 in which the arrows indicate the directions of air flow; *A* is the ball mill, *B* is the pneumatic classifier, *C* is the collector of ground product, and *D* is the fan. Attention is drawn to the closed-circuit character of the air flow. Dry-grinding Hardinge mills are used for pulverizing coal, clay, limestone, and cement clinker.

Cylindrical Mills. Cylindrical mills are made by a number of different manufacturers, the various mills being differentiated principally by the mode of discharge of the ground product. Three discharge methods may be recognized:

1. Through screens along the cylindrical shell (peripheral discharge mill).
2. By free overflow from the axis of the mill (overflow mill).

3. Through a grate extending as a diaphragm across the full section of the mill near the discharge end (grate mills).

Peripheral-discharge mills[36] are typified by the old Krupp mill. Mills of this type are uncommon in the United States because of the expensive screen wear. They are still used here and there for grinding nonabrasive industrial products, either wet or dry.

Overflow mills are typified by the Traylor mill. They have been used extensively, but recent trend in mill selection appears

FIG. 58.—Discharge grate for Marcy mill. (*Mine and Smelter Supply Co.*)

to favor grate mills. Overflow mills are the simplest of all in design and construction.

Grate mills (Fig. 57) are made principally by Allis-Chalmers Manufacturing Company and by Mine and Smelter Supply Company (Marcy mill). They are wet-crushing mills with the discharge end fitted with grates (Fig. 58). Between the grate and the end of the mill are lifters which raise the product so as to discharge it through the discharge trunnion of the mill. Grate mills feature a low pulp line. This is said to result in a high grinding capacity without overgrinding of finished material inasmuch as there is a rapid efflux of material from the grinding to the discharging compartment.[53] In the Allis-Chalmers grate mill, the level of the pulp can be regulated by plugging or unplugging the discharge holes at the end of the mill, outside of the grate diaphragm (Fig. 59).

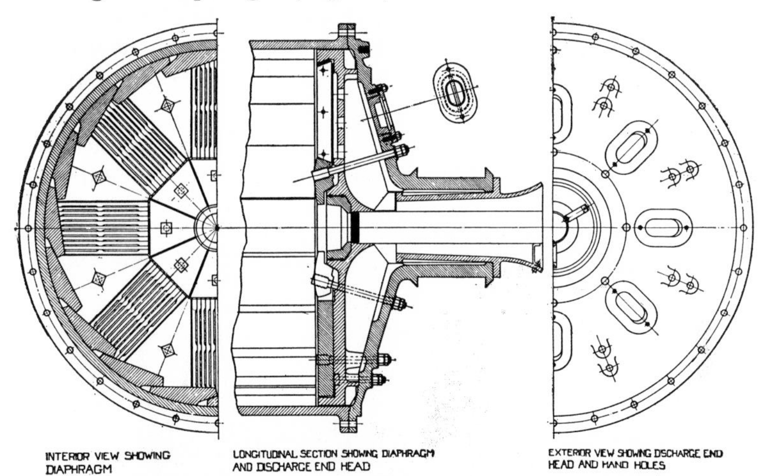

FIG. 59.—Variable-discharge diaphragm for Allis-Chalmers grate mill. (*Allis-Chalmers Mfg. Co.*)

In addition to the types of mill that have been described, mention should be made of the Allis-Chalmers Compeb mill. It consists of two, three, or four compartments in series for

FIG. 60.—Three-compartment Compeb mill. (*Allis-Chalmers Mfg. Co.*)

successively finer grinding (Fig. 60). It is used in the cement industry.

The Williamson mill, which has come into use more recently, is furnished with either grate or overflow discharge. The purpose of its design (Fig. 61) is to prevent slippage and increase ball action.

In spite of differences in construction and operation, the various types of ball mills have many points in common: the mill is carried by two trunnions located, respectively, between the feeder and the mill itself and between the mill itself and the discharge lip; the trunnions are carried on bearings fastened directly to concrete piers; the feeder is usually a spiral scoop sometimes concealed within a circular shell.

Fig. 61.—Williamson mill. (*Williamson Co.*)

Balls and Liners. The interior of ball mills is lined by replaceable liners usually made of alloy steel but sometimes of rubber.(4) Liners may be of several types, smooth, shiplap, wave, wedge bar, etc. (Fig. 62). Liners other than smooth are designed to help lift the ball load as the mill is revolved and to minimize the slip between layers of balls.

Balls range in size from 1 to 6 in. They are made of cast iron, forged steel, cast steel, or alloy steel. The largest balls are used for coarser grinding, especially in mills of large diameter. The usual size is 3 to 4 in.

A considerable item of expense in grinding is the wear of the liners and balls. According to DeVaney and Coghill, ball wear is closely related to power input.[14] Ball consumption usually ranges from 1 to 3 lb. per ton of ore ground (Table 11). In large-scale operations where scrap is available at the dressing plant, it is generally economic to make the balls at the plant even

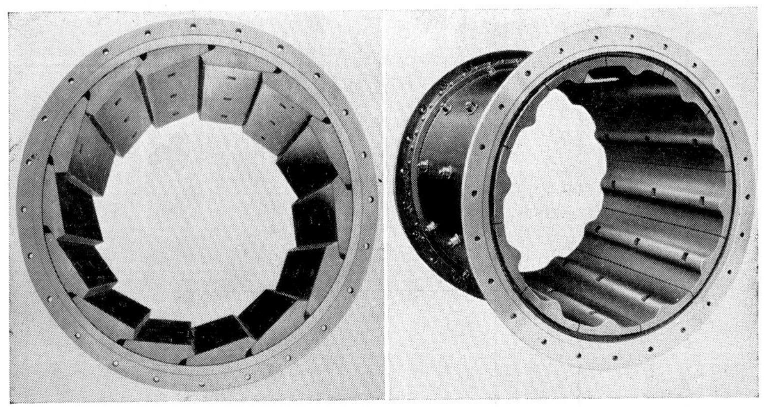

FIG. 62a and b.—Shiplap liner and wave liner. (*Allis-Chalmers Mfg. Co.*)

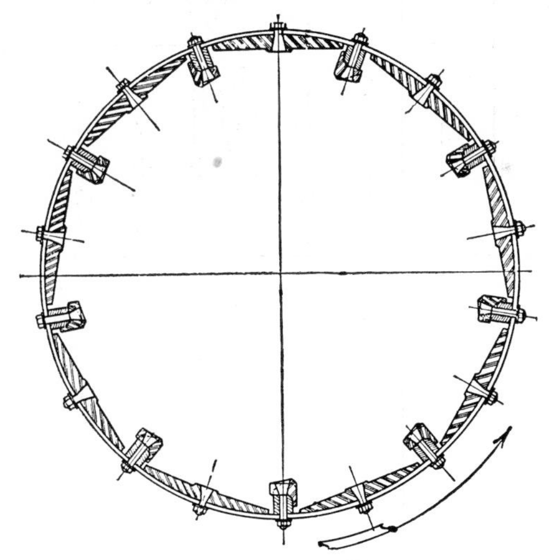

FIG. 62c.—"Komata" liner.

though they be of poor quality, rather than to buy them. On the contrary, for small operations, the local manufacture of balls is inadvisable. The purchase of alloy-steel balls of the highest quality, usually of chrome, chrome-nickel, or chrome-molyb-

TABLE 11.—CONSUMPTION OF BALLS AND LINERS IN BALL MILLS
(According to data published by the U. S. Bureau of Mines)

Name of plant	Size of		Ball consumption, pounds per ton of ore	Metal used in balls	Ball size, inches	Liner consumption, pounds per ton
	Feed	Product				
Questa (Molybdenum Corporation of America).[6]	5% + 1½ in.	1% + 100-mesh	1.5	Forged steel	4	Not stated
Hayden (Nevada Consolidated Copper Company).[23]	1% + 10-mesh	2% + 65-mesh	0.99	Scrap steel and scrap iron	2	0.07
Verde Central (Verde Central Mines Inc.)[15]	Minus ⅝ in.	4% + 48-mesh	3.5	Cast iron	3	1ct./ton
Britannia (Britannia Mining and Smelting Company).[43]	3.5% + 4-mesh	3% + 35-mesh	2.5	Scrap steel	3½	Not stated
Superior (Engels Copper Mining Company.)[44]	1.8% + 1-in.	1% + 48-mesh	1.03	Cast steel	5	0.23
Walkermine (Walker Mining Company)[42]	5.4% + 1-in.	2% + 30-mesh	2.07	Forged steel	4	0.56
Matahambre (Minas de Matahambre)[41].	1.5% + ¾ in.	1% + 28-mesh	0.69	Steel	4½	0.54
Morning (Federal Mining and Smelting Company).	3.5% + 1½ in.	2.5% +100-mesh	2.19	Cast iron	3 and 4	0.4
Pecos (American Metal Company).........	0.6% + 1½ in.	2.2% + 65-mesh	1.8	Forged steel	4½	0.3
Miami (Miami Copper Company)[40]......	3.8% + 3-mesh	4.6% + 48-mesh	1.86	Steel	Not stated	0.23

denum steel is advisable for inaccessible operations, but cheap balls may be preferable if freight costs are inconsequential.

In the case of large-scale operations, the cost of making balls from scrap averages from ½ to 1 ct. per pound of balls (in addition to the cost or value of the scrap). Purchased steel balls cost about 5 cts. per pound. In comparative tests, it has been shown that cast-iron balls wear several times as fast as alloy-steel balls and that high-carbon steel balls wear up to twice as fast as alloy-steel balls.[24]

Liner wear ranges from 0.1 to 0.5 lb. per ton of ore ground. Liners are more expensive per pound than balls because of their shape. In addition, their installation involves a shutdown expense. Liner wear is more variable than ball wear perhaps because it is affected to a greater extent by mill size and design. Overflow mills (cylindrical or cylindroconical) and mills of largest volume, other things being equal, have the least linear wear. In order to reduce the cost of liner replacement, some large plants use liners made of cut scrap rail set on end in a matrix of cement. The first cost of such a lining is much smaller than that of fabricated linings, and even though the wear may be greater, the final cost is substantially less.

Ball Load. The ball load should be such that when the mill is stopped it is slightly more than half full (even in the absence of ore and water; the ore and water distend the ball charge very little as they naturally occupy the interstices between the balls). To compensate for the wear on balls, it is customary to add to each mill, perhaps once per day or per shift, one or more balls of the largest diameter. It is not customary to make up a new charge with balls all of the same size as that would leave too much "pore" space; rather it is customary to start with a ball load composed of balls of various sizes so as to make possible the introduction into the mill of the greatest crushing load possible. The daily additions, however, are of the largest balls only; this compensates for size reduction by wear.

Theory of Ball-mill Operation. The path of balls[10] is compounded of a circular section (as the balls are lifted) and of a parabolic or near-parabolic section as the balls are dropped (Fig. 63). A ball at a distance r from the center of the mill revolving at n r.p.m. abandons the circular path for a parabolic path when the centripetal component of gravity exceeds the centrifugal component of angular acceleration, *i.e.*, according to Davis (Fig. 64), when

$$\frac{mv^2}{r} = mg \cos \alpha, \qquad [V.1]$$

in this relation m is the mass of the ball, v its linear velocity, r the

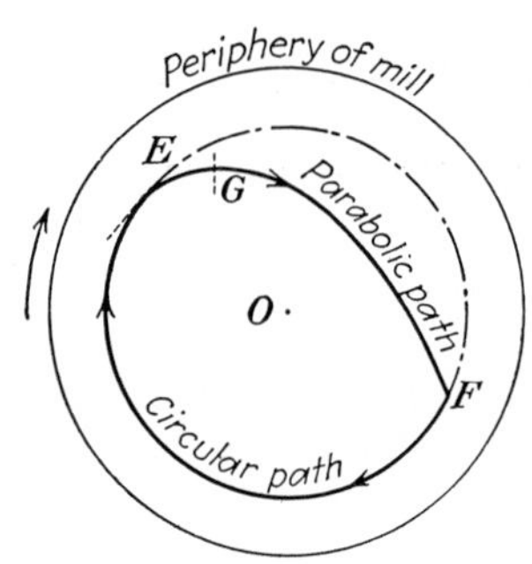

FIG. 63.—Path of a typical ball in a ball mill. (*According to Davis.*)

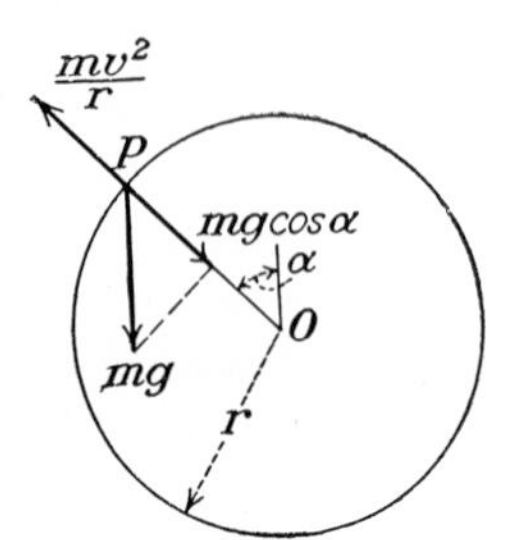

FIG. 64.—Forces acting on a ball at distance r from center of mill.

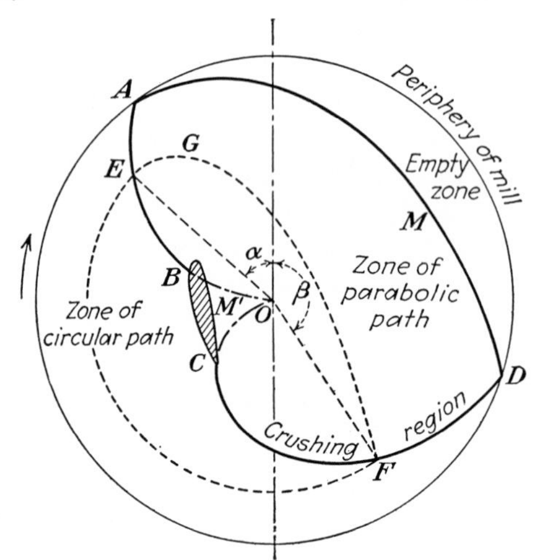

FIG. 65.—Zones in a ball mill. (*After Davis.*)

radius of the mill, g the acceleration of gravity, all expressed in consonant units, *e.g.*, in c.g.s. units. Hence, since $v = 2\pi rn$,

$$\cos \alpha = \frac{4\pi^2}{g} n^2 r. \qquad [V.2]$$

The locus of the points E that delineate the beginning of the parabolic path (Fig. 63) for various ball positions from center of mill to periphery is the curve OBA of Fig. 65. This curve can be determined graphically from calculated values of α in function of r.

Centrifuging of the outermost layer of balls takes place as $\cos \alpha$, as determined by Eq. [V.2], exceeds unity.

It can be shown, according to Davis,[10] that (see Fig. 65)

$$\beta = 3\alpha. \qquad [V.3]$$

From this relationship, the locus of the points F that delineate the end of the parabolic path can be drawn in the same manner as the locus of the points E. This gives curve DCO of Fig. 65.

Davis also shows that the arcs CO and BO correspond to unstable equilibrium so that the zone $BM'C$ is a dead zone in which there is no effective motion.

The inside of a ball mill consists then of four zones, an empty zone, a dead zone, a zone of circular path in which balls adjoin each other, and a zone of parabolic path in which the balls are spread out, as shown in Fig. 65.

It is implicit in the philosophy of Davis that the crushing action takes place along CD, and nowhere else.

Experiment verifies the existence of the Davis zones (see the photographs and moving pictures by Davis, also by Haultain and Dyer), but the speed required to attain them is greater than that predicted by the theory, according to Haultain and Dyer.[33,34] This is explained, and has been experimentally proved by Haultain and Dyer, as due to slippage between layers of balls. This slippage represents dissipation of kinetic energy, probably associated in part with grinding. It seems, then, that the Davis philosophy must be modified to include the existence of a substantial crushing action within the zone of circular ball path arising from the rolling of balls on each other and the slippage between ball layers. Slippage is certainly a function of the many variables that enter into the operation of a ball mill, *e.g.*, pulp dilution, size and kind of mineral particles, pore space in ball charge, speed of mill. But the effect of these variables has not yet been quantified.

Still another correction seems in order according to Gow, Campbell, and Coghill[25] who point out that the path of each ball is not compounded of two segments, one circular and one parabolic, but of three segments, one circular, one parabolic, and one near-parabolic. The last corresponds to the segment *EG* in Fig. 63. Its consideration is required, because balls slow down on a rising parabolic path, something that they cannot do since they are in juxtaposition at the end of the circular segment of their path. *EG* is of course tangent to the two other segments of the path. Mathematical derivation of the equation for curve *EG* agrees with the facts, as may be seen from Fig. 65*a*. Broadly

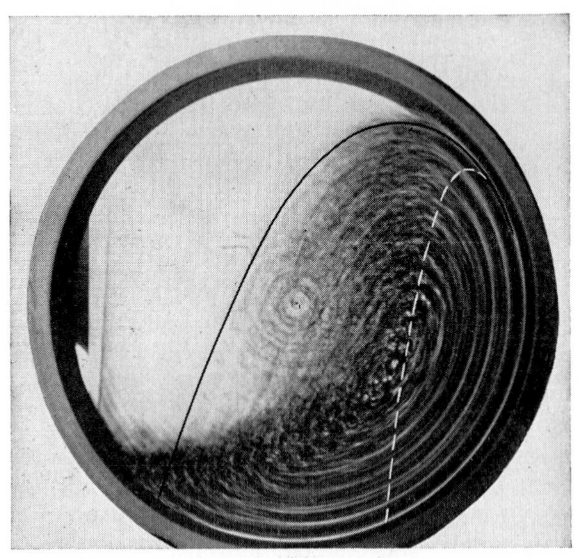

Fig. 65*a*.—Comparison of actual path of outside layer of balls with their theoretical paths as determined by the Davis formula (white line) and the Gow, Campbell, and Coghill formula (black line). (*After Gow, Campbell, and Coghill.*)

speaking, the path of the balls is wider than had been thought according to earlier work. Davis' mathematical analysis may represent a convenient first approximation and that of Gow, Campbell, and Coghill a second approximation. It may be noted that some criticism has been leveled at the new theory.[35]

Operating Details. The *speed* of ball mills should be as high as possible without centrifuging the charge. As the speed of the mill is increased, work input increases at first in proportion to the speed. Then, as slippage increases, the work input increases more slowly than the speed; it increases until a critical speed is reached beyond which the power input (aside from that wasted as friction in the drive and bearings[39]) decreases rapidly to the vanishing point (Fig. 66). Of course, when that condition is reached, the solids are centrifuged on the mill shell and there is no

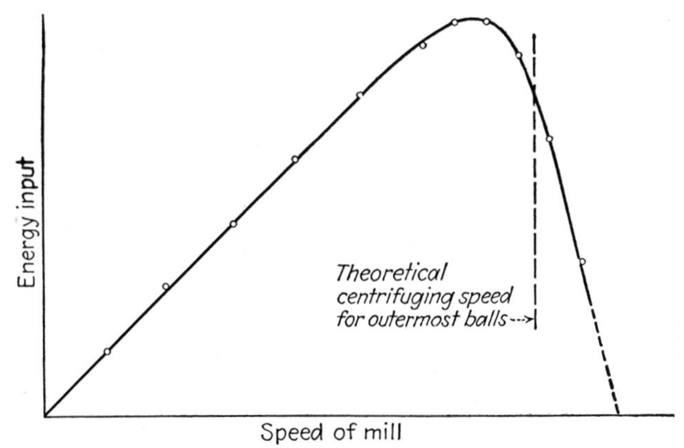

FIG. 66.—Relationship of energy input in ball mill to speed of mill.

work done by the mill. Fahrenwald[18,20] gives for the critical speed

$$N = \frac{54.2}{\sqrt{S - s}}. \qquad [V.4]$$

In this relationship, S is the radius of the mill and s the radius of the balls, both expressed in feet.

The *energy* that the mill can be made to consume[26,28] is a function not only of the speed, but also of the ball load, the specific gravity of the ore, and the dilution of the pulp (in case wet grinding is employed). If the ball load is increased in a closed mill, the power that the mill can consume is increased at first, but not in proportion to the load, until a maximum is reached; beyond this stage, the power input decreases gradually to practically nothing (Fig. 67). This is due to the fact that as the ball load is increased the center of gravity of the load comes nearer and nearer the axis of rotation of the mill; the energy input being proportional to the product of the load and leverarm, it is clear that the energy input will increase or decrease depending upon whether the load increases more or less rapidly than the leverarm decreases, as the charge is increased.

It is interesting to observe that the pulp dilution has a considerable effect on the possible energy input. For a certain critical range in pulp dilution, the energy input is maximum. This usually is obtained in a pulp consisting of 60 to 75 per cent solids—the range depending upon the chemical composition, particle size, and specific gravity of the ore.

Nature of Grinding Action. It is a question of considerable debate whether the work done by the mill in grinding the material is accomplished at the point of impact of the tumbling balls with the "restive load" of balls and ore, or whether it takes place within the restive load by the slippage of one layer of balls on another. It might be contended that the rubbing action is one

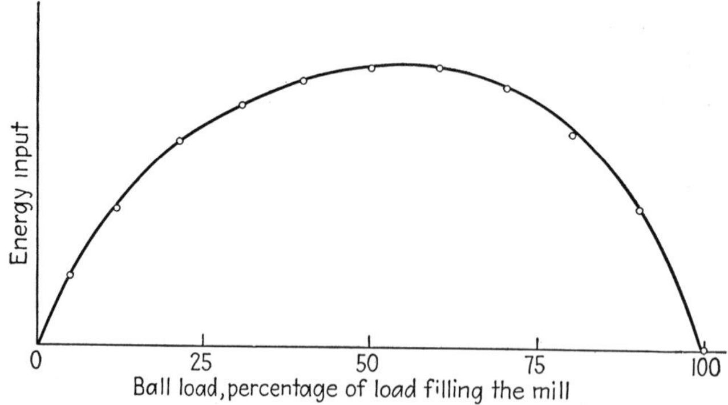

Fig. 67.—Relationship of energy input in ball mill to ball load.

that abrades the particles. Yet the weight of the evidence seems to favor the view that fracturing is by impact, whether the impact is caused by the blow of falling balls or by the rolling of one layer of balls on another. If fracturing is by impact in the rolling of layers of balls, it may resemble that obtained in crushing rolls, even though in this case balls do not revolve toward each other, as roll shells do, but in the opposite way. Where attrition is the mode of size reduction, the size of the attritus is usually very small as compared with the size of the feed. This is shown by Fig. 68, curve *A*, which presents the sizing analysis of the product obtained by allowing quartz grains about an inch in diameter to tumble in a ball mill emptied of balls. It is of interest to compare this with Fig. 68, curve *B*, which presents the sizing analysis of a typical product obtained by crushing quartz in a closed ball mill and to further compare these two figures with Fig. 33, Chapter III, which presents the sizing analysis of the crushed product obtained by crushing quartz in rolls.

To further substantiate the contention that very little actual grinding in a ball mill results from attrition, it is sufficient to grind in successive batches (in a laboratory batch mill) coarser

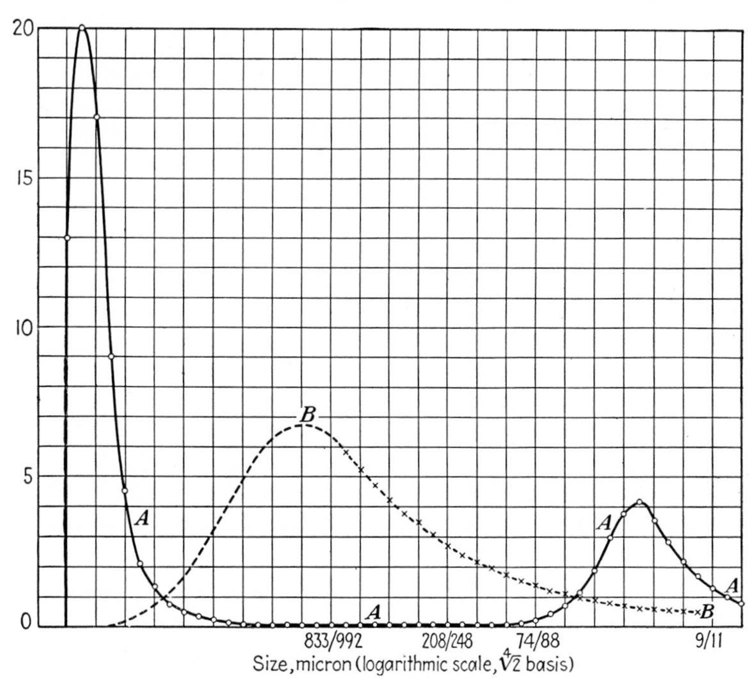

FIG. 68.—Sizing analysis of a typical ball-mill product (*B*) compared to sizing analysis of an abraded product (*A*).

and coarser material with the same balls. It is found that finer feeds are ground normally, the ground products having sizing analyses like that shown in Fig. 68*B*, but that coarser feeds yield products resembling that represented in Fig. 68*A*. Gradations, also, can be observed between products like 68*A* and products like 68*B*, depending upon the size of the balls and of the ore particles.

The practical conclusion which can be drawn from these experiments is that balls must be of a size commensurate with the material fed to the mill, or else nipping of the particles will not occur, and the particles will merely become rounded, superficially abraded, but not ground.

Ball mills do not exert a *sizing action* on their product in contrast to other crushing or grinding machines. This is because, regardless of its size, the particle that finds itself in the path of a falling ball will be broken unless indeed it happens to be too large to be nipped between the spherical surfaces of the balls in apposition, in which case it merely has a corner nicked off and eventually assumes a rounded outline. As a result, ball mills cannot be expected to regulate their performance without the help of a sizing device of some kind.

In peripheral-discharge mills, screens were inserted at the periphery of the mill to yield a sized product. But screening is not so effective at a fine size as classification, and peripheral-discharge mills have become obsolete for fine grinding.

Ball-mill-classifier Circuits. The sizing action required to regulate the performance of ball mills is obtained by operating the mill in closed circuit with a classifier. The mill discharges into a classifier which produces a finished product as overflow and an unfinished sand requiring further grinding; this sand is returned to the ball mill for further grinding. In the case of large installations where a number of units are employed, it is sometimes advantageous to send the classifier sand to a ball mill other than the one in which this sand was made so that circuits of somewhat different type are obtained.

Figure 69 shows a flow sheet suggested by Dorr and Marriott[16] as a composite picture of recent developments. This thoughtfully planned flow sheet visualizes an arrangement that is increasingly adopted at large flotation plants in its philosophy rather than its detail, with substantial improvement in over-all economy over simpler flow sheets. It is, of course, impractical for small plants.

If a product all finer than a certain critical size is required, the capacity of the ball mill is increased considerably by using it in closed circuit with a classifier and this increase is made still greater by increasing the circulating load[11] between the ball mill and the classifier (Fig. 70). In practice, circulating loads[5,54] between ball mills and classifiers are rarely less than 200 per cent and frequently exceed 700 per cent. It may appear strange at first sight that the circulating load between a

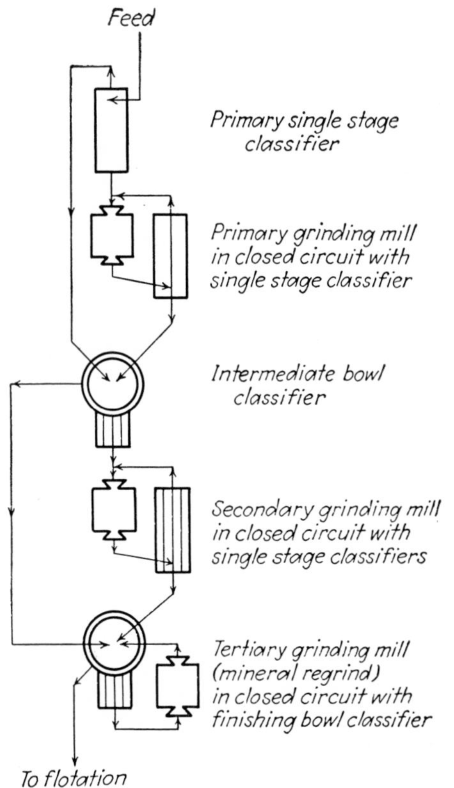

FIG. 69.—Modern grinding flow sheet. (*After Dorr and Marriott.*)

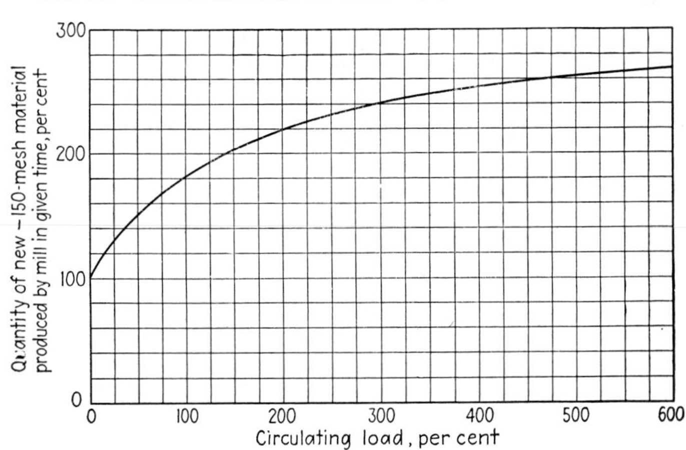

FIG. 70.—Relationship between circulating load and mill capacity.

mill and classifier producing, for example, 100 tons of finished product per day can be as much as 400 tons or more, but this will be made clear by Fig. 71. The trend to the use of higher and higher circulating loads in ball-mill-classifier circuits has resulted in certain mechanical requirements in ball mills, classifiers, and auxiliary equipments. These are summarized as follows by Dorr and Anable(17):

1. Rotary mills should be relatively great in diameter and short in length and provided with extra large feed and discharge trunnions to accommodate circulating sand loads at least ten times the new feed.

2. Spiral feed scoops should be of extra heavy construction, armed with abrasive resisting tips and renewable internal scoop liners.

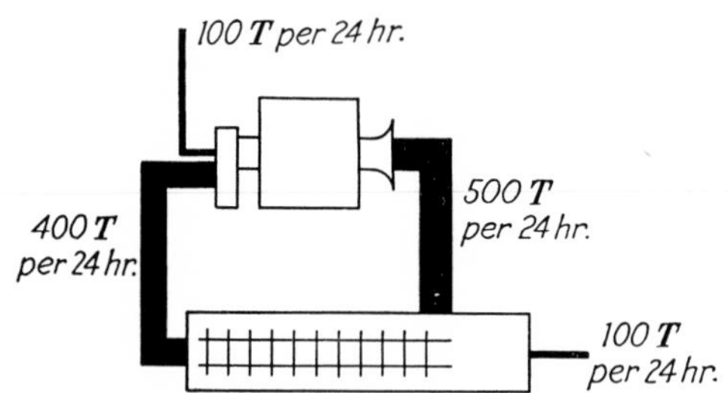

FIG. 71.—Ball-mill-classifier circulating load of 400 per cent.

3. Provision should be made for measuring the loss of grinding media and making up deficiencies at frequent intervals.

4. Wattmeters should be installed to serve as guide to the correct loading of the mill.

5. The classifier should be of rugged simple construction, large enough to carry an adequate circulating load with ease.

6. Suitable meters should be installed to show power consumption, which broadly is a function of the sand tonnage handled.

7. A constant head tank for dilution water should be available, so that the rate of addition of water may be regulated to a nicety and then held constant. The dilution of the overflow should be checked constantly, either manually or by some type of hydrometer connected with the tank.

8. Classifiers may generally be installed at such an elevation and at such a distance from the rotary mill that an all-gravity closed circuit is secured; *i.e.*, mill discharge flows by gravity to the feed launder of the classifier, and classifier sands similarly gravitate to the feed box of the mill. Where this proves impossible, it is preferable to locate the classifier at such an elevation that sands return to the mill by gravity, provid-

ing auxiliary apparatus for raising the more dilute and less abrasive mill discharge to the classifier feed launder. For the latter purpose, a rotary pipe launder and spiral pickup scoop have been used to advantage in several cases.

Reduction Ratio. The reduction ratio that can be secured by the use of ball mills is large compared with the reduction ratios obtainable with coarse or intermediate crushing devices. Instead of ranging from 5 to 8, the reduction ratio range is as large as 50 to 100 for ball-mill-classifier circuits. However, where such large reduction ratios are used, and if the capacity of the plant is large, it is usually more economical to arrange the ball mills so that they are not in parallel but in series, one ball mill doing the

TABLE 12.—CAPACITY OF BALL MILLS

Size of mill	Capacity range, tons per 24 hr.	Number of examples
A. Cylindroconical Mills*		
3 × 8	20 to 30	2
4½ × 16	20 to 50	3
5 × 16	50 to 70	2
6 × 16	100 to 350	7
6 × 22	120 to 300	7
8 × 22	150 to 450	10
8 × 30	160 to 500	8
8 × 36	150 to 700	12
8 × 48	500 to 1,000	7
10 × 48	700 to 1,200	3
B. Cylindrical Mills, Grate Type†		
4 × 4	40 to 60	2
5 × 4	50 to 75	2
5 × 6	90 to 100	4
6 × 4½	130 to 250	5
6 × 6	180 to 275	5
6 × 12	275	1
7 × 6	275 to 600	3
8 × 5	800	1
8 × 6	350 to 750	8

* Outside diameter of cylindrical portion, in feet, by length of cylindrical portion, in inches.
† Outside diameter by length, both in feet.

coarse grinding and another or several others doing the fine grinding (Fig. 69). If such an arrangement is followed, the reduction ratio may be 20 for the ball mill doing the coarse grinding and only 5 for those doing the fine grinding.

Capacity. The capacity of ball mills depends upon the size of the mills, the hardness of the ore, the reduction in size attempted, and the efficiency of the operation. Accordingly, average-capacity figures are practically meaningless. Table 12 presents some data from practice and is merely intended to give some idea of the tonnage that can be handled by mills of various sizes. Considerable experimentation is required to permit judicious selection of the proper mill for a given duty, particularly experimentation to determine the friability of the ore.

Cost. The cost of grinding is divisible essentially into three classes, power, supplies, and labor. Except in plants where the tonnage handled is small, the labor item is less important than the other two. Relative importance of power and supply costs is dependent upon the cost of unit power, hardness of the ore,

TABLE 13.—OPERATING GRINDING COST (BALL MILLS),
(As reported by the U. S. Bureau of Mines, 1934)

Mill	Energy consumed, kw.-hr. per ton	Cents per ton					Capacity of plant, tons per day
		Labor	Power	Supplies	Miscellaneous	Total	
Hayden	3.7	1.47	3.46	2.24	0.14	7.31	12,000
Verde Central	.....	7.2	15.5	11.1	....	33.8	400
Britannia	7.8	1.04	1.48	2.08	....	4.60	7,000
Superior	.....	4.67	8.95	11.04	....	24.66	1,200
Morning	7.35	2.31	8.33	8.66	....	19.30	1,200
Page (Federal Mining and Smelting Company)	.....	7.0	13.1	21.0	....	41.1	300
Pecos (American Metal Company)	10.96	5.00	6.70	15.46	....	27.16	600
Montana Mill (Eagle-Picher Lead Company)	13.35	2.33	12.58	10.29	....	25.20	250
Miami	6.06	1.63	3.42	3.75	0.33	9.13	16,000

and unit cost of balls and liners. Table 13 presents grinding costs from practice.

Dry Grinding. Ball mills are usually employed to grind ore in wet pulp. But for some purposes,[2] especially in the chemical industries, ball mills are employed to grind dry. This is true, for instance, of grinding coal for the preparation of pulverized fuel, and of certain chemicals. Instead of being connected to a classifier using water as the classifying medium, dry-grinding ball mills are connected in closed circuit with pneumatic classifiers (Fig. 56). Dry-grinding mills are often employed to produce an extremely fine product. This arises from the high settling speed of solids suspended in air as compared with solids suspended in water.

ROD MILLS

Rod mills are similar to ball mills in appearance and in general principle. They may be defined as rotating cylindrical shells loaded with rods that grind the ore by tumbling within the shell. Rod mills are made of a length greater than their diameter in order to avoid jamming of the rods in the mill. Ball mills, on the other hand, generally have a diameter commensurate with their length, since that type of construction yields the greatest capacity for a given weight of machine. Cylindroconical mills can be filled with rods instead of balls, provided the cylindrical section is relatively long and the diameter small.

Rod mills differ from ball mills in their grinding action in that the rods are kept apart by the coarsest particles. The grinding action is exerted preferentially on the coarsest particles. This peculiar property of rod mills recommends their selection where mills are used in connection with concentrating processes that fail on fine particles. Conversely, wherever flotation is to be used, ball mills are preferable. This is not so much because "slimes" are preferable to granular particles for flotation as because ball mills are generally cheaper to operate than rod mills. An objection to rod mills is that as rods wear down they become thin and occasionally warp, causing an entangling of the rod load, unless removed at intervals. To avoid this difficulty, periodic opening and inspection of rod mills are in order. It is also possible to minimize this difficulty by using rods of high-carbon steel, which is brittle rather than malleable. The worn rods are then discharged as small bits of steel.

Figure 72 shows a Marcy rod mill. This mill is characterized by the fact that the discharge end is open, the lid (appearing in the right of the picture) being so designed as merely to prevent the rods from falling out. The Marcy rod mill, consequently, has a very low pulp level.

FIG. 72.—Marcy rod mill, viewed from discharge end, showing open-end construction. (*Mine and Smelter Supply Co.*)

Marcy rod mills differ from ball mills also in respect to the method of supporting the mill. Instead of two trunnions, Marcy mills may be carried on one trunnion and two rollers or on four rollers.

Allis-Chalmers and Hardinge rod mills have also been used extensively. They are like the corresponding ball mills in general form, except for their proportionately greater length and for the provision in the Hardinge rod mill of a low-level pulp discharge.

Figure 73 presents sizing analyses of products obtained from the same feed with a ball mill and rod mill, both used without closed-circuit classifier. It is clear that the rod mill produces a more closely sized product than the ball mill. The sizing action exerted by a rod mill makes it unnecessary to use a classifier in closed circuit. Recent plant experiments by E. H. Rose, mill superintendent of the International Nickel Company, have shown no practical difference in the grind obtained with ball and rod mills, if the mills are in closed circuit with sizing devices.

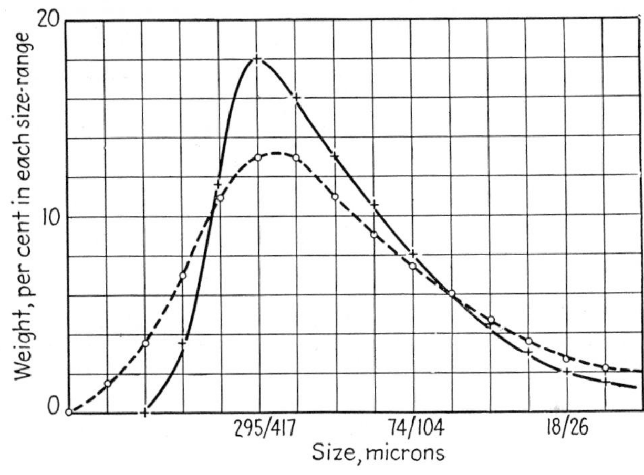

FIG. 73.—Comparative sizing analyses of a rod-mill product (full line) and of a ball-mill product (broken line), both mills being operated in open circuit.

Consumption of rods and liners is approximately the same as in the case of a ball mill employed for a similar duty, using steel balls.

TUBE MILLS

Tube mills are merely long ball mills of relatively small diameter in which flint pebbles are used instead of steel balls. Tube mills 5 ft. 6 in. in diameter and 22 ft. long are standard. This shape is necessary to secure an acceptable capacity; it follows from two facts: (1) Mills of large diameter are not possible in view of the fragility of the pebbles and (2) a greater bulk of grinding medium has to be used for equivalent duty as compared with ball mills, because of the relatively low specific gravity of the pebbles. In the United States, tube mills have been displaced by ball mills except where abraded iron is definitely objectionable, as in some ceramic operations. In South Africa, many of the mills grinding gold ore preparatory for cyanidation are still equipped with flint pebbles.

The pebbles used in tube mills come mostly from northwestern Europe, especially from the British Isles, France, and Denmark where they are secured from the disintegrated materials of the chalk cliffs of the area. They have occasionally been replaced by local siliceous pebbles or chosen ore lumps.[9]

Several different types of liners are used, one of the best known being the "El Oro" liner. It consists of grooved plates bolted to the shell of the mill, in which the pebbles wedge. In this fashion, the consumption of metal is reduced, most of the wear being at the expense of the pebbles.

Tube mills are used in open circuit, or in closed circuit with a classifier; they are adapted to either wet or dry grinding.

Ball and rod mills have been equipped at times with grinding mediums of special shapes, such as rods with a hexagonal section, double-conical or cubic blocks. These various grinding mediums have had their advocates, but generally speaking they have not shown enough advantage (if any) over balls or rods to justify their higher cost.[30,48]

HADSEL MILL

The Hadsel mill[29] (Fig. 74) consists of a revolving short cylindrical shell of very large diameter mounted on rollers. The ore (up to 6 in. in size) acts both as grinding medium and as medium to be ground; it is elevated by pockets recessed in the shell, much as in a sand elevator, dropped at the top of its path, and falls on striking plates already covered with ore. Grinding is supposed to take place only between ore and ore, and wear is expected to be low. The shell contains water to the pouring point. The overflow is either a finished product or one ready for classification. The sediment that falls on the elevator pockets is automatically re-treated. The machine is attractive in principle as it represents a substantial simplification over standard practice, but it is beset by the difficult requirement that there must be just the right proportion of coarse to fine particles in the ore if there is to be neither lack nor excess of grinding medium.

A recent variant is a dry-grinding Hadsel mill operated with a pneumatic classifier and a spraying device. This variant is

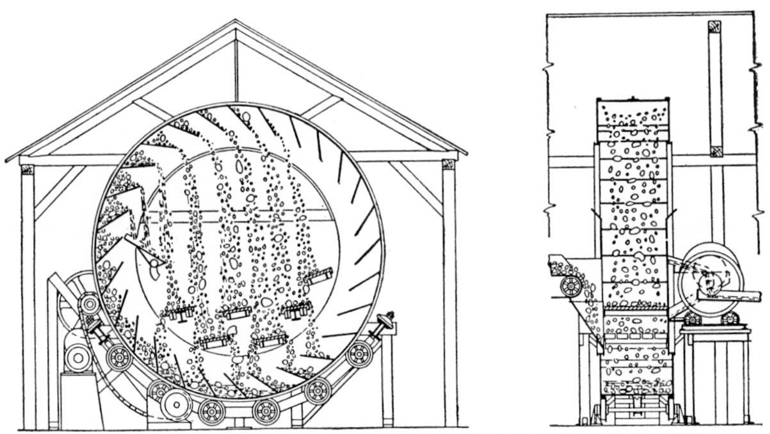

FIG. 74.—Principle of operation of Hardinge-Hadsel mill. (*After Hall.*)

meant to deliver a pulp of the right consistency for flotation. Both types of Hadsel mill are still on trial.

DISINTEGRATION BY INTERNAL FORCES

Fire setting has been used as a means of breaking ore in mines since antiquity; it is rapidly becoming a lost art, however, as explosives are almost universally employed. Fire setting consists in building a fire near a face so as to heat the surface of the rock, then in quenching the face. The sudden cooling and contraction of the rock result in fracturing to an extent sufficient to permit relatively easy picking.

A somewhat similar procedure can be used, according to Holman,[38,49] to granulate quartz or quartzose ores. It seems that several factors enter into play, including at least (1) contraction on sudden cooling, (2) differential contraction on cooling of outside and inside of particles, (3) differential expansion of outside and inside, and (4) passage of silica from one allotropic form to another (alpha and beta quartz, tridymite, cristoballite).

The physical properties involved in this process may then include the coefficient of expansion, the heat conductivity, the brittleness (or its converse, malleability), and the existence of transformations in the solid state. Clearly, since these properties are specific to each solid, each solid may behave in its own way, and in different ways in different size and temperature ranges. If the process is applied to a heterogeneous rock, it is bound to favor fracturing at boundaries of physical discontinuity, *i.e.*, precisely where it is desired to cause fracturing if liberation is to be most effectively accomplished. This feature, alone, would seem to justify further work with the process.

A related procedure is the explosive shattering devised by Dean and Gross of the Bureau of Mines.(12,13) It consists in soaking water into the pores, cracks, and fissures of a rock, applying superheated steam for a length of time, and suddenly releasing the pressure.

Literature Cited

1. Anon.: A Forerunner of the Modern Ball Mill, *Eng. Mining J.*, **129,** 85 (1930).
2. Anon.: Feldspar Industry Adopts Modern Grinding Methods, *Eng. Mining J.*, **126,** 746–747 (1928).
3. Blickensderfer, F. C.: A Comparative Test of the Marathon, Chilean and Hardinge Mills, *Trans. Am. Inst. Mining Met. Engrs.*, **55,** 678–706 (1916).
4. Bennett, W. E.: Rubber Linings for Rotary Grinding Mills, *Mining and Met.*, **14,** 399–400 (1933).
5. Bond, Fred C.: Determination of the Circulating Load in a Wet Closed-circuit Grinding System, *Mining and Met.*, **18,** 507 (1937).
6. Carman, J. B.: Milling Methods at the Questa Concentrator of the Molybdenum Corporation of America, Questa, New Mexico, *U. S. Bur. Mines, Information Circ.* 6551 (1932).
7. Coghill, W. H.: Evaluating Grinding Efficiency by Graphical Methods, *Eng. Mining J.*, **126,** 934–938 (1928).
8. Coghill, Will H., and R. G. O'Meara: Milling Methods and Costs at a Flat River (Mo.) Mill, *U. S. Bur. Mines, Information Circ.* 6658 (1932).
9. Davis, Carl R., J. L. Willey, and S. E. T. Ewing: Recent Developments in the Fine Grinding and Treatment of Witwatersrand Ores, *Trans. Am. Inst. Mining Met. Engrs.*, **71,** 983–1017 (1925).
10. Davis, E. W.: Fine Crushing in Ball Mills, *Trans. Am. Inst. Mining Met. Engrs.*, **61,** 250–296 (1919).
11. Davis, Edward W.: Ball-mill Crushing in Closed Circuit with Screens, *Minn. Sch. Mines Exp. Sta., Bull.* 10 (1925).
12. Dean, R. S., and John Gross: Explosive Shattering of Minerals as a Substitute for Crushing Preparatory to Ore Dressing, *U. S. Bur. Mines, Rept. Investigations* 3118 (1932).

13. Dean, R. S., John Gross, and Carl E. Wood: Preparing Ore by Explosive Shattering, *Eng. Mining J.*, **136,** 281–283 (1935).
14. De Vaney, Fred D., and Will H. Coghill: The Relation of Ball Wear to Power in Grinding, *Eng. Mining J.*, **138,** 337–340 (1937).
15. Dickson, R. H., and E. M. Smith: Milling Methods and Costs at the Verde Central Concentrator, Jerome, Ariz., *U. S. Bur. Mines, Information Circ.* 6489 (1931).
16. Dorr, J. V. N., and A. D. Marriott: Importance of Classification in Fine Grinding, *Trans. Am. Inst. Mining Met. Engrs.*, **87,** 109–154 (1930).
17. Dorr, J. V. N., and Anthony Anable: Fine Grinding and Classification, *Trans. Am. Inst. Mining Met. Engrs.*, **112,** 161–177 (1934).
18. Fahrenwald, A. W., and H. E. Lee: Ball-mill Studies, *Am. Inst. Mining Met. Engrs., Tech. Pub.* 375 (1931).
19. Fahrenwald, A. W., G. W. Hammar, H. E. Lee, and W. W. Staley: Ball-mill Studies. II. Thermal Determinations of Ball-mill Efficiency, *Am. Inst. Mining Met. Engrs., Tech. Pub.* 416 (1931).
20. Fahrenwald, A. W.: Some Fine-grinding Fundamentals, *Trans. Am. Inst. Mining Met. Engrs.*, **112,** 88–115 (1934).
21. Fischer, H.: Der Arbeitsvorgang in Kugelmühlen, insbesondere in Rohrmühlen, *Z. ver. deut. Ing.*, **48,** 437–441 (1904).
22. Franke, Robert: Hardinge Mills vs. Chilean Mills, *Trans. Am. Inst. Mining Engrs.*, **47,** 50–64 (1914).
23. Garms, W. I.: Concentrator Methods and Costs at the Hayden Plant of the Nevada Consolidated Copper Co., Ariz., *U. S. Bur. Mines, Information Circ.* 6241 (1930).
24. General Discussion: Composition of Mill Balls and Determination of Wearing Qualities, *Trans. Am. Inst. Mining Met. Engrs.*, **79,** 206–215 (1928).
25. Gow, A. M., A. B. Campbell, and W. H. Coghill: A Laboratory Investigation of Ball Milling, *Trans. Am. Inst. Mining Met. Engrs.*, **87,** 51–81 (1930).
26. Gow, A. M., and M. Guggenheim: Dead-load Ball-mill Power Consumption, *Eng. Mining J.*, **133,** 632 (1932).
27. Gow, Alexander M., Morris Guggenheim, and Will H. Coghill: Review of Fine Grinding in Ore Concentrators, *U. S. Bur. Mines, Information Circ.* 6757 (1934).
28. Gow, Alexander M., M. Guggenheim, A. B. Campbell, and Will H. Coghill: Ball Milling, *Trans. Am. Inst. Mining Met. Engrs.*, **112,** 24–78 (1934).
29. Hall, R. G.: The Hadsel Mill, *Trans. Am. Inst. Mining Met. Engrs.*, **112,** 15–24 (1934).
30. Hardinge, Harlowe: Theory and Practice in Selecting Grinding Media, *Eng. Mining J.*, **124,** 695–698 (1927).
31. Hardinge, H. W.: The Hardinge Conical Mill, *Trans. Am. Inst. Mining Engrs.*, **39,** 336–341 (1908).
32. Hardinge, H. W.: The Hardinge Conical Mill, *Trans. Am. Inst. Mining Engrs.*, **45,** 194–209 (1913).

33. HAULTAIN, H. E. T., and F. C. DYER: Ball Paths in Tube Mills, *Faculty Applied Sci. and Eng. Sch. Eng. Research, Univ. of Toronto, Bull.* 3 (1922).
34. HAULTAIN, H. E. T., and F. C. DYER: Ball Paths in Tube Mills and Rock Crushing in Rolls, *Trans. Am. Inst. Mining Met. Engrs.*, **69**, 198–207 (1923). Moving-picture films (Am. Inst. Mining Met. Engrs.).
35. HAULTAIN, H. E. T., and F. C. DYER: The Study of Ball Paths in Tube Mills, *Mining and Met.*, **12**, 108 (1931).
36. HINES, PIERRE R.: Notes on Theory and Practice of Ball Milling, particularly Peripheral-discharge Mills, *Trans. Am. Inst. Mining Engrs.*, **59**, 249–262 (1918).
37. HODGES, FRED: Milling Methods at the Hurley Plant of the Nevada Consolidated Copper Company, Hurley, New Mexico, *U. S. Bur. Mines, Information Circ.* 6394 (1931).
38. HOLMAN, BERNARD W.: Heat-treatment as an Agent in (Quartz) Rock Breaking, *Trans. Instn. Mining Met.*, **36**, 219–262 (1927).
39. HORNE, HARLAN F.: A Grinding Mill with Micarta Bearings, *Eng. Mining J.*, **138** [No. 10] 43–44 (1937).
40. HUNT, H. D.: Milling Methods and Costs at the Concentrator of the Miami Copper Company, Miami, Arizona, *U. S. Bur. Mines, Information Circ.* 6573 (1932).
41. KIRCHNER, A. R., J. V. GALLOWAY, and W. P. SCHODER: Milling Methods and Costs of the Minas de Matahambre. S. A., Concentrator, *U. S. Bur. Mines, Information Circ.* 6544 (1931).
42. MCKENZIE, M. R., and H. K. LANCASTER: Milling Methods at the Concentrator of the Walker Mining Co., Walkermine, California, *U. S. Bur. Mines, Information Circ.* 6555 (1932).
43. MUNRO, A. C., and H. A. PEARSE: Milling Methods and Costs at the Concentrator of the Britannia Mining and Smelting Co., Ltd., Britannia Beach, B. C., *U. S. Bur. Mines, Information Circ.* 6619 (1932).
44. NELSON, W. I.: Milling Methods and Costs at the Superior Concentrator of the Engels Copper Mining Co., Plumas County, Calif., *U. S. Bur. Mines, Information Circ.* 6550 (1932).
45. PEARSE, H. A.: The Concentration of Britannia Ores, *Trans. Can. Inst. Mining Met.*, **30**, 915–928 (1927).
46. PEARSE, H. A.: Three-product Flotation at the Britannia, B. C., Mill, *Mining and Met.*, **15**, 379–383 (1934).
47. PEARSE, H. A.: Three-product Flotation at Britannia. Separation of Copper-zinc-iron from Low-grade Ore, *Trans. Can. Inst. Mining Met.* (1934). (In *Can. Mining Met., Bull.* 267, pp. 341–350.)
48. ROSE, E. H.: Hexagonal Grinding Rods Tested in Patiño Mill, *Eng. Mining J.*, **123**, 49–50 (1927).
49. STŎCES, BOHUSLAV: Anwendung der Feuermethode in modernen Bergbau, Cited by Prof. Henry Louis in discussion of paper by Bernard W. Holman, *Trans. Instn. Mining Met.*, **36**, 258 (1927).
50. STRACHAN, C. B.: Milling Methods of the American Zinc Company of Tennessee, Mascot, Tennessee, *U. S. Bur. Mines, Information Circ.* 6379 (1930).

51. TAGGART, ARTHUR F.: Tests on the Hardinge Conical Mill, *Trans. Am. Inst. Mining Engrs.*, **58**, 126–177 (1918).
52. TAGGART, ARTHUR F., and R. W. YOUNG: Grinding Brass Ashes in the Conical Ball Mill, *Trans. Am. Inst. Mining Engrs.*, **54**, 26–33 (1916).
53. VAN WINKLE, C. T.: Recent Tests of Ball-mill Crushing, *Trans. Am. Inst. Mining Engrs.*, **59**, 227–248 (1918).
54. WEINIG, A. J.: A Chemical Method of Determining Tonnages in Mill Circuits, *Mining and Met.*, **14**, 505–506 (1933).
55. WHITE H. A.: The Theory of Tube Milling, *J. Chem. Met. Mining Soc. S. Africa*, **15**, 176–193 (1915).

CHAPTER VI

ATTRIBUTES OF COMMINUTION

Some attributes of comminution deserve a rather detailed examination from a cost or an utilization standpoint. They pertain to the comminuting operation, *e.g.*, crushing and grinding efficiency, or to the comminuted product, *e.g.*, the shape and size of the particles in the product.

Shape of Comminuted Particles. A widespread belief exists that whereas particles visible to the eye may be angular, fine particles are more or less rounded, until a fineness is reached beyond which the particles are definitely spherical. Nothing could be farther removed from the truth. Actually, the shape of crushed or ground particles varies but little with size, and the finest particles that can be examined under the microscope are as angular as any. In some instances, the finest particles are more angular than the coarsest particles, as will be explained presently, but in no case is there truth in the reverse.

In the first place, it is physically absurd to expect sphericity in extremely fine particles of crystalline solids inasmuch as these particles come closer to a unit crystal than coarse particles. Although it is conceivable that a large grain may be broken (or more exactly, worn) to appear round, a very small particle whose diameter may be but that of a few hundred unit crystals is bound to display angularity since matter is not infinitely divisible, but only divisible in definite *quanta*. In the second place, if fine particles of rounded outlines were to be formed by comminution, even finer particles should also have formed that would have been fitted between the rounded particles in the unbroken grains. In other words, the finest particles produced by comminution cannot be rounded but must have sharp corners.

Among extremely fine particles are particles which approximate the finest detail that can be resolved microscopically. It might be recalled at this point that, according to the classical theory, this ultimately resolvable detail is $\lambda/2n$, in which λ is the wave length of the light used and n the "numerical aperture" of the objective of the microscope. In the best lenses, $n = 1.4$, and with blue light $\lambda = 0.45$ micron, so that the ultimately resolvable detail is about 0.15 micron. A sharp corner in a submicron particle will appear rounded simply because that detail is too small in reference to the light wave used to examine it: light goes around it just as sound goes around a building.

Particles somewhat coarser, yet still very fine (micron size and coarser), are all angular as may be verified easily with a microscope.

Although angular, the coarsest particles in comminuted products are frequently shaped somewhat differently than fine particles: their shape is more regular, in the sense that their three dimensions of length, breadth, and thickness are more nearly equal. This seems to be due to the sizing action of most grinding devices, which imposes a slight preferential fracturing of long or flat particles in the coarsest sizes.

If wear has been applied (as in a ball mill that is made to grind a feed too coarse for the balls), truly rounded coarse particles are obtained. But this is an ineffective means of comminuting and should be regarded as an exception.

Effect of Cleavage and Crystalline Habit. Angularity, of course, is modified by cleavage and general fracturing habit of each mineral species.

Quartz, which has a conchoidal habit and practically no cleavage, produces the sharpest of splinters.

Galena and sphalerite, which display good and fair cleavage, respectively, yield particles with points and edges, although generally not regarded as sharply angular.

Mica, molybdenite, and graphite yield plates and are not usually regarded as angular.

Stibnite and other crystals of acicular habit yield needlelike particles.

Chalcocite, bornite, and pyrite yield equidimensional angular particles.

Sizes Present in Comminuted Products. It is important to observe that, irrespective of whether the feed to a crusher or grinder is sized or not, the product is not fully sized, *i.e.*, it contains particles of all sizes from the coarsest to the finest. The coarsest size in the product of a comminuting operation is generally determined by the setting of the comminuting device in the case of a jaw crusher, a gyratory crusher, a cone crusher, or rolls; by the size of the screen in a stamp mill; and by the rate of rock passage through a tube, rod, or ball mill. But there is no way of establishing a lower size limit for crushed or ground particles: there are always produced some particles finer than is desirable. The practical issue, then, is to limit the overgrinding to the strict minimum.(1) Overgrinding costs money and causes increased losses by rendering material otherwise subject to concentration unfit for that purpose. This desire to avoid overgrinding is responsible for the wide utilization of rolls, rod mills, and other crushers and grinders that have the reputation, more or less justified, of minimizing overgrinding. This desire to avoid overgrinding is also at the root of elaborate flow sheets featuring alternations of comminuting with concentrating steps—as, for example, the concentration by jigging at a size at which liberation is far from complete, regrinding of the tailing or of a middling product in which locked particles have been concentrated, and further concentration of the reground product by tabling or flotation.

It has also been sought to minimize overgrinding by the use of comminuting methods that do not utilize a force applied externally to the particle. Thus, it has been proposed to soak the particles to be ground in water at a temperature well above the boiling point and under considerable pressure, then to suddenly release the pressure so as to form steam under pressure in the pores and cracks within particles, and thus explode them with an internal force selectively applied at boundaries of heterogeneity. The idea is very attractive and if practically utilizable may prove revolutionary.

It is of interest to inquire whether any lower size limit at all exists among the particles in a comminuted product. From an experimental point of view, all that may be said is that there exist particles finer than any size at which observation can be made. Whether the limit is the finest screen (37 microns), elutriation (about 5 microns), the microscope (about 0.25 micron), or the centrifuge (about 0.05 micron), material finer than the limiting size does occur. From a theoretical point of view, on the other hand, a particle finer than a molecule is unthinkable. The question even arises whether a molecule, *e.g.*, of SiO_2, consisting of one atom of silicon and two of oxygen, can be anything but a gas, and whether, if it is produced by grinding, it will not immediately condense on a neighboring particle. It would seem to be safer to think of the smallest possible solid particle as the unit crystal, *i.e.*, the smallest physical assemblage of atoms or molecules possessing the characteristics of crystallinity. In the case of silica, such a unit crystal consists of three SiO_2 molecules, in the case of common salt, of four atoms of sodium and four of chlorine, etc. The unit crystal of sodium chloride is a cube 5.6Å. on edge; the unit crystal of quartz has a volume equivalent to a cube 4.8Å. on edge; that of galena is a cube 5.9Å. on edge. Generally speaking, it is practically certain that the size of unit crystal is rarely smaller than 0.0005 micron, and rarely larger than 0.001 micron. The finest particles produced by comminution, then, are of the order of magnitude of 0.001 micron (see Table 6, Chapter III).

Relative Abundance of Particles of Various Sizes. It has already been explained in Chapter III that a geometric sequence of divisional sizes for a sizing scale is preferable to any other. It was also stated that the customary sizing scale is one in which each size is finer than that preceding it in the ratio of $\sqrt[2]{2}$.

When the sizing analysis of a crushed product is charted, a certain regularity in relative abundance of the various sizes suggests itself. This regularity becomes more obvious when allowance is made for the manufacturing imperfection in screens. For example, the amount of material retained by a 200-mesh screen and passing a 150-mesh screen covers the size range from 104 to 74 microns. It is clear that even a slight irregularity in the manufacturing process may cause some openings to be larger than they should be and others smaller. A 200-mesh screen, then, instead of retaining material ranging from 104 to 74 microns (*i.e.*, across a 30-micron range) may retain material only from 102 to 76 microns, or across a 26-micron range. Such a minor defect as deviation by only 2 microns in the size of the screen openings may well throw off the sizing analysis by 10 per cent or more.

If due allowance is made for the imperfection of screens, the regularity in relative abundance of the various sizes in a comminuted product can be expressed in definite mathematical form.[5,16,18,34,36,37] This is suggested by the characteristic shape of sizing analyses, which in direct-plot charts remind one of probability curves, and in cumulative-plot charts of probability-integral curves.

Many attempts have been made, especially in recent years, to express analytically sizing curves. Considerable progress has

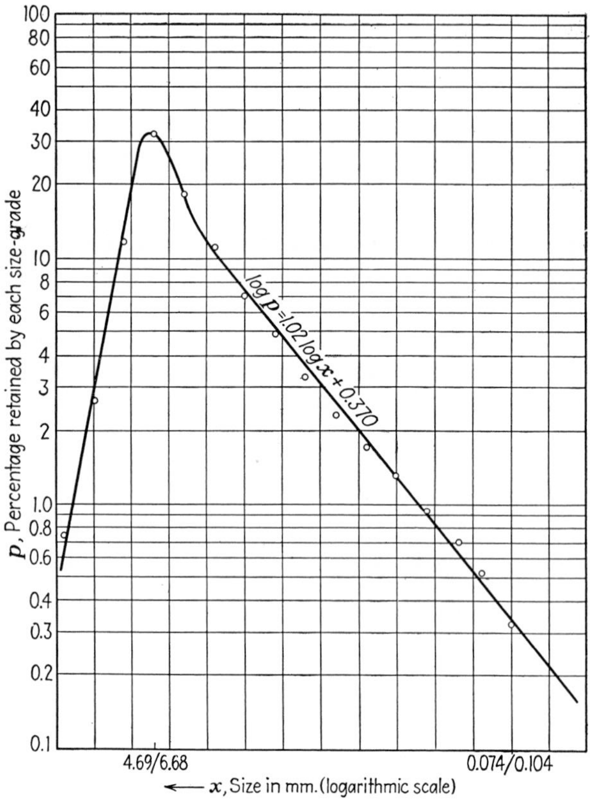

FIG. 75.—Logarithmic plot of the sizing analysis stated in Table 7 and charted in Figs. 30 to 33.

already been made, especially in dealing with the fine end of sizing curves.

One of the simplest relations involves the use of logarithmic paper,[16] as may be seen from Fig. 75, which presents the same data as Fig. 33, Chapter III. The only change has been the use as ordinate of the logarithm of the weight retained by each screen.

Generally, if p is the percentage retained in the size fraction averaging about size x,

$$\log p = k \log x + \log A, \qquad \text{[VI.1]}$$

in which k and $\log A$ are constants for a given product. This relationship is valid only in the fine-size range. In the particular instance of Fig. 75, the equation takes the numerical form

$$\log p = 1.02 \log x + 0.370, \qquad \text{[VI.1}a\text{]}$$

in which p is expressed in percentage of material per $\sqrt[2]{2}$ increment in size, x is the arithmetic average size of the particles in one size grade, in millimeters, and in which both logarithms are to the base 10.

In this expression, $k = 1.02$ is the most significant criterion that characterizes the product under consideration. The higher the value of k, the less abundant are the fine particles in terms of coarse particles, and the less the overgrinding. The relationship between the numerical value of k and overgrinding, however, is exponential and not direct, since Eq. [VI.1] can be put in the form

$$p = Ax^k \qquad \text{[VI.2]}$$

so that a small change in k may make a very great change in p.

In Eq. [VI.1a], $\log A = 0.370$ is also a significant criterion that characterizes the product. The higher the value of $\log A$, the finer the product. Here the relationship is not exponential.

In comparing various crushed or ground products, k becomes a means of ascertaining which machine causes the most or the least overgrinding. Table 14 gives average values of this criterion of relative size abundance for various types of comminuting operations. This table brings out that k is largest for coarse crushing, for least reduction ratio, and for machines featuring positive control of particle size.

CRUSHING AND GRINDING EFFICIENCY

In view of the large fraction of crushing and grinding costs that is represented by energy, it is clear that crushing and grinding efficiencies are among the most important of the subjects of interest to mineral-dressing engineers. It is difficult to quantify the efficiency of a crushing operation, especially since no universally acceptable definition can be offered for that entity.

TABLE 14.—EXPERIMENTAL VALUES OF CRITERION OF RELATIVE SIZE ABUNDANCE k

Material	Comminuting device	Reduction ratio	Time of grinding	k
Quartz	Rolls	2.8	...	0.92
Quartz	Rolls	5.6	...	0.87
Quartz	Rolls	8.0	...	0.80
Galena	Rolls	1.5	...	0.92
Galena	Rolls	2.0	...	0.84
Quartz	Batch Rod mill	...	3 min.	0.95
Quartz	Batch Rod mill	...	7 min.	0.90
Quartz	Batch Rod mill	...	15 min.	0.84
Quartz	Batch Rod mill	...	30 min.	0.76
Quartz	Batch Rod mill	...	1 hr.	0.70
Quartz	Batch Rod mill	...	2 hr.	0.61
Quartz	Batch Rod mill	...	4 hr.	0.56
Quartz	Continuous Ball mill	...	Fed rapidly	0.85
Quartz	Continuous Ball mill	...	Fed slowly	0.65
Quartz	Continuous Ball mill	...	Fed very slowly	0.5
Montana gold ore*	Gyratory	...	...	1.02
Hard, yet clayey ore*	Gyratory	...	...	0.85

* According to Bond and Maxson.(5)

Thermodynamic Definition. Obviously, energy is required to effect comminution; in other words, comminution may be regarded as a process of conversion of energy from one form to another. The energy expended is in the form of kinetic energy. The energy recovered is in the form of potential energy (surface energy), heat, and sound. Of these, only the potential energy is desired. Hence the efficiency of a comminuting operation is the ratio of the surface energy produced to the kinetic energy expended.

Defined in this fashion, crushing or grinding efficiency becomes an acceptable thermodynamic criterion of comminution, in the same fashion as the ratio of output of an electric motor as kinetic energy to input as electric energy is a thermodynamic criterion of conversion of one certain form of energy to another, useful, form of energy.

Although this theoretical definition of crushing or grinding efficiency is simple, practical measurement is at present extremely difficult and inaccurate.

Determination of Crushing or Grinding Efficiency.[20] The quantities that must be determined are

1. Total energy input.
2. Energy lost in transmission from point of measurement of energy input to point of application in comminuting device.
3. Total new surface produced.
4. Specific surface energy.

The accuracy with which measurements are possible decreases in the order 1, 2, 3, 4. The total energy input can be measured accurately, mechanically, or electrically. The transmission loss is more difficult to measure. It might be considered that a determination of energy consumption with the crushing or grinding device idling constitutes a measure of the transmission loss, but this involves an assumption that is not thoroughly justifiable. So much for factors dealing with energy input. After all, with due allowance for the unjustifiability of the assumption to which attention has just been drawn, the measurement of net energy input can be made to a precision of a few per cent.

Determination of New Surface Produced. Determination of new surface produced can be made with high accuracy in those instances in which a precise chemical reaction is available, or in those instances in which a precise physical property is available that is proportional to surface. These methods can be used only if the crushed or ground product is chemically homogeneous; even then they are available now for only a few substances, typically for quartz[21,22,23,34] and for magnetite.[8,19] In such favorable instances, the precision of the measurement can be made better than 10 per cent; it may even equal the precision of measurement of the net energy input.

In most cases, however, the crushed rock is chemically heterogeneous so that chemical or magnetic surface determination is not available. The only method consists in estimating the surface from a sizing analysis together with a factor designed to allow for the irregularity in shape of the crushed particles.

Fundamentally, this method is based on the fact that the surface of a crushed product is inversely as the particle size, other things being equal.*

If all particles in a comminuted product were of one regular shape, as cubic, tetrahedral, or octahedral, a close measurement of the surface presented by any sized fraction could be made. Since, however, the shape of the particles is not regular, the only possible way of making such a measurement is to use a shape factor to allow for irregularity with reference to a regular shape, *e.g.*, the cube. A convenient figure for this factor is 1.75, *i.e.*, the irregularly shaped particles present 1.75 times as much surface as cubes of the same screen size. This factor may be in error on the average by 10 to 20 per cent, but the extent to which it may be in error cannot be ascertained. If the reaction rate or magnetic method may be used, a check is available on the factor. This check indicates that the shape factor ranges, for fine particles, from 1.3 to 2.0, with the average near 1.75.

The great difficulty in determining surface produced by comminution lies not in the uncertainty due to particle shape, but in the uncertainty with which the size of the finest size grade can be estimated. Until recently, estimations were currently based on the assumption that the material finer than the finest screen, or the finest elutriator fraction had as much surface as an equal weight of material having for size one-half the limiting size of the "minus" material. But it is now widely appreciated that such an estimate is ten to thirty fold too small. Accurate determination can be made by suitable adaptation of Eq. [VI.2], as follows:(16)

Consider 100 cc. of comminuted material containing $V_{x_1}^{x_2}$ cc. of material ranging from unit-crystal size x_1 to size x_2, and such that the particle distribution follows the relationship described by Eq. [VI.2].

Let dv be the volume of particles ranging in size from x to $x + dx$. The surface area of these particles is ds.

* Chapter III. p. 67.

$$ds = \frac{6h}{x}\,dv, \qquad \text{[VI.3]}$$

h being the shape factor.

The volume p ranging in size from x to tx ($t = \sqrt[2]{2}$ in the case of Fig. 75 and Eq. [VI.1a]) is

$$p = Ax^k,$$

and the volume ranging from x to $x + dx$ is $dx/(tx - x)$ times as large or,

$$dv = \frac{dx}{x(t-1)}Ax^k = \frac{A}{t-1}\,x^{k-1}dx. \qquad \text{[VI.4]}$$

Hence,

$$V_{x_1}^{x_2} = \frac{A}{t-1}\int_{x_1}^{x_2} x^{k-1}\,dx = \frac{A}{k(t-1)}(x_2{}^k - x_1{}^k) \qquad \text{[VI.5]}$$

and

$$S_{x_1}^{x_2} = \frac{6hA}{t-1}\int_{x_1}^{x_2} x^{k-2}\,dx = \frac{6hA}{(t-1)(k-1)}(x_2{}^{k-1} - x_1{}^{k-1}). \qquad \text{[VI.6]}$$

The specific surface of the minus x_2 material, or surface per unit weight is then

$$\Omega = \frac{S_{x_1}^{x_2}}{\Delta V_{x_1}^{x_2}} = \frac{6hk}{\Delta(k-1)}\,\frac{x_2{}^{k-1} - x_1{}^{k-1}}{x_2{}^k - x_1{}^k}, \qquad \text{[VI.7]}$$

Δ being the specific gravity of the material.

Application. What is the specific surface of the minus 200-mesh quartz in the case of Fig. 75?

Using Eq. [VI.7],

$$\Omega = \frac{6 \times 1.75 \times 1.02}{2.65 \times 0.02} \times \frac{0.0074^{0.02} - 0.00000005^{0.02}}{0.0074^{1.02} - 0.00000005^{1.02}},$$

or 5,780 sq. cm. per g.

This corresponds to an average particle size of 6.9 microns.

By this method, the average size of minus 200-mesh material can be estimated. It is found to range from 2 to 8 microns. The error in that determination depends upon the exactitude with which k has been determined, the validity or nonvalidity of Eq. [VI.2] in the extrapolated (colloidal) size range, and the correctness of the size assumed for the unit crystal. Precise measurements by screening and sedimentation give an average size that may be within 50 per cent of the true average size.

Since much of the surface in a crushed or ground product is in the finest size fraction (even in a roll product, more than half of the surface may be concentrated in the few per cent of the weight that are minus 200-mesh), the possible error arising from improper surface determination of the minus 200-mesh size fraction is *the* great source of error. Clearly, an accurate determination of total new surface can hardly be expected: a probable error smaller than 30 to 50 per cent should not be anticipated.

Specific Surface Energy of Solids. The specific surface energy of liquids can be measured with precision, as it is numerically the same as the surface tension. Water, for example, has at 20°C. a surface energy of 72.8 ergs per sq. cm. and a surface tension of 72.8 dynes per cm. In the case of solids, indirect methods only are available for the determination of surface energy (since surface tension cannot be said to exist in measurable form.[38] These methods are based on mathematical utilization of physicochemical quantities; the accuracy of the answers is necessarily in doubt by at least the extent to which assumptions are made. The surface energy of solids as obtained by these means is large compared with that of liquids, and enormously divergent values are reported by different chemists.[2,4,6,14,26,27,28,31,41,42,43,44] One of the most reliable estimates gives 920 ergs per sq. cm. for the surface energy of quartz.[10] Practically nothing is known of the surface energy of the common minerals.

In addition to these difficulties concerning specific surface energy, it is difficult to know what allowance to make for preferential fracturing at grain boundaries (see Chapter IV) and for the existence in the feed of cracks, visible or invisible. If it is assumed that the specific surface energy of quartz can be applied to the average siliceous ore, and if furthermore no allowance is made for preferential fracturing or for cracks, it is likely that the value obtained for the surface energy produced by crushing will be overestimated.

Order of Magnitude of Grinding Efficiency. The data now available indicate that crushing and grinding efficiencies range from a small fraction of 1 per cent to a very few per cent. The remainder of the energy input is almost wholly converted to heat and wasted. In this connection, it is of interest to note that the pulp flowing out of grinding mills is noticeably warmer than the inflowing water, sometimes by as much as 10 to 20°C.

It is not known why comminuting efficiencies of apparently effective machines should be so low. The current grinding principles are all wrong, or our data on specific surface energy are wide of the mark, or else there is a definite physical reason why grinding efficiencies cannot be appreciably larger, much as there is a definite thermodynamic reason why steam-engine efficiencies are limited in the range below about 30 per cent (the Carnot cycle). In any event, this field of study promises to be one of the most fruitful, theoretically and practically, in the whole of mineral dressing.

Thermal Determination of Grinding Efficiency. To avoid the difficulties inherent in surface-energy determinations, it was proposed by Fahrenwald[12,13] to calculate the useful energy produced in comminution by difference between the energy input and the heat evolved. This method may have the seed of a real solution to the problem of comminuting efficiency, but it is in need of refinement and will call for energy input, heat, and chemical measurements of the highest precision.

Crushing and Grinding Law. Considerable interest has been attached to so-called crushing laws which meant to relate the work done to the work input.[25] Two so-called laws have been debated with great zeal by their proponents. With the disproof[17] of the so-called Kick law,* there has remained the Rittinger law.

As originally stated by P. R. von Rittinger, in Lehrbuch der Aufbereitungskunde, p. 20, Berlin (1867), the law read, "Die zum Zerkleinern erforderliche Kraft steht mit dem Oberflächenzuwachse in geradem Verhältnisse," or that the energy necessary for reduction of particle size is directly proportional to the increase of surface.†

* The Kick "law" was stated by Stadler[39] and restated by Taggart.[40] Its proponents always argued its plausibility, never demonstrated its correctness by experiment. At best, it should therefore have been regarded as a postulate.

According to Stadler, Kick postulates that "The energy required for producing analogous changes of configuration in geometrically similar bodies of equal technological state varies as the volumes or weights of these bodies."

† Stadler translated Rittinger's statement "the increase of the surfaces exposed is directly proportional to the force required." This is an error in translation as may be observed from the context; Rittinger gave to "Kraft" the older meaning of *energy*, not the newer meaning of *force*. Another

Viewed from the surface-energy concept, the Rittinger law states the proportionality between work input and work output. Viewed from the standpoint of surface energy and of the definition of thermodynamic comminution efficiency, the Rittinger law merely states that crushing and grinding efficiency are constant regardless of the size to which comminution is conducted, and regardless of the device used for comminution.

Gross and Zimmerley[21,22,23] have experimentally proved with a drop-weight apparatus that the new surface produced in crushing is directly proportional to the energy consumed in the

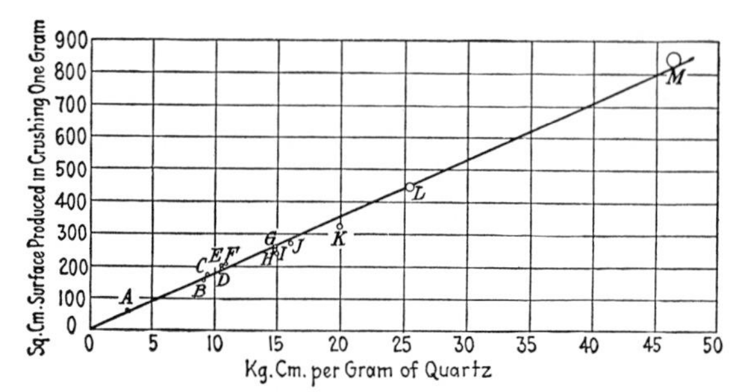

FIG. 76.—Relation of work input to surface produced in crushing quartz. (*After Gross and Zimmerley.*)

process. This may be seen from Fig. 76 which is reproduced from their original paper.

Practical Measures of Grinding Efficiency. The operator's measure of grinding efficiency can hardly be rigorous in view of what has been said above concerning the imprecision of efficiency measurements based on energy output/energy input. Actually, the practical test of grinding efficiency can be summed up in the query: how many tons can be put through a mill, and ground to a given size, per day, and at what energy expense in kilowatt-hours per ton? In other words, to the practical operator an efficient grinding operation connotes one efficient not only in terms of low power cost, but also one efficient in the sense that it has made full use of the possible capacity of the mill. By leaving out of consideration the "capacity" measure of grinding efficiency, there is substituted for the energy/energy

sentence in Rittinger's text is perhaps less ambiguous: "In zwei Zerkleinerungsfällen wird dann die angewendete Arbeit mit den Oberflächenzunahmen proportional sein."

factor of rigorous efficiency a ratio of weight ground to energy input. This would be regarded as an abominable "efficiency" by a physicist, but from an engineering standpoint it is a measure of the effectiveness of the operation and of the operator.(7)

Rated on the basis of tons ground to minus 200-mesh per kw.-hr., the efficiency of grinding mills shows a striking increase with increase in circulating load (Chapter V, Fig. 70). This does not denote necessarily an increase in absolute efficiency; rather it indicates a change in the size distribution of the minus 200-mesh material in the direction of increased proportion of the coarser particles and decreased overgrinding.

Crushing Resistance of Minerals. Because of differences in specific gravity, brittleness, specific surface energy, etc., different minerals display various resistances to crushing.(30,35,45) Zimmerley and Gross's data are presented in Table 15.

TABLE 15.—CRUSHING RESISTANCE OF MINERALS IN FALLING-BALL APPARATUS
(*After Zimmerley and Gross*)

Mineral	Surface per unit work, sq. cm. per kg. cm.	Relative to quartz		
		Surface per unit work	Work per unit surface	Work per unit weight
Quartz	8.9	1.00	1.000	1.000
Pyrite	11.3	1.27	0.788	0.418
Sphalerite	28.4	3.19	0.313	0.215
Calcite	38.5	4.32	0.231	0.225
Galena (average of three)	46.9	5.27	0.190	0.067

Literature Cited

1. ANON.: Efficient Concentration, *Mining and Met.*, **9**, 386–387 (1928).
2. BERDENNIKOV, V. P.: Measurement of the Surface Tension of Solid Substances, *Physik. Z. Sowjetunion*, **4**, 397–419 (1933).
3. BELL, JOHN W.: An Investigation of Rock Crushing Made at McGill University, *Trans. Am. Inst. Mining Engrs.*, **57**, 133–142 (1917).
4. BIEMÜLLER, J.: The Surface Tension of Alkali Halides, *Z. Physik*, **38**, 759–771 (1926).
5. BOND, FRED C., and WALTER L. MAXSON: Crushing and Grinding Characteristics as Determined from Screen Analyses, *Trans. Am. Inst. Mining Met. Engrs.*, **112**, 146–160 (1934).

6. Bružs, B.: The Surface Energy of Solids. I. Surface Energy of Barium Sulfate, *J. Phys. Chem.*, **34**, 621–626 (1930).

7. Coghill, Will H.: Evaluating Grinding Efficiency by Graphical Methods, *Eng. Mining J.*, **126**, 934–938 (1928).

8. Dean, R. S.: Magnetite as a Standard Material for Measuring Grinding Efficiency, *Am. Inst. Mining Met. Engrs., Tech. Pub.* 660 (1936).

9. Dean, R. S., John Gross, and C. E. Wood: Preparing Ore by Explosive Shattering, *Eng. Mining J.*, **136**, 281–283 (1935).

10. Edser, Edwin: The Concentration of Minerals by Flotation, *Fourth Rept. Colloid Chem. and Its General and Industrial Applications, Brit. Assoc. Adv. Sci.*, p. 281 (1922).

11. Fahrenwald, A. W., and H. E. Lee: Ball-mill Studies, *Am. Inst. Mining Met. Engrs., Tech. Pub.* 375 (1931).

12. Fahrenwald, A. W., G. W. Hammar, Harold E. Lee, and W. W. Staley: Ball-mill Studies. II. Thermal Determinations of Ball-mill Efficiency, *Am. Inst. Mining Met. Engrs., Tech. Pub.* 416 (1931).

13. Fahrenwald, A. W.: Some Fine-grinding Fundamentals, *Trans. Am. Inst. Mining Met. Engrs.*, **112**, 88–115 (1934).

14. Fricke, R., R. Schnabel, and K. Beck: Surface and Heat Content of Crystallized Magnesium Hydroxide, *Z. Elektrochem.*, **42**, 881–889 (1936).

15. Gates, Arthur O.: Kick *vs.* Rittinger: An Experimental Investigation in Rock Crushing, Performed at Purdue University, *Trans. Am. Inst. Mining Met. Engrs.*, **52**, 875–909 (1915).

16. Gaudin, A. M.: An Investigation of Crushing Phenomena, *Trans. Am. Inst. Mining Met. Engrs.*, **73**, 253–316 (1926).

17. Gaudin, A. M., John Gross, and S. R. Zimmerley: The So-called Kick Law Applied to Fine Grinding, *Mining and Met.*, **10**, 447–448 (1929).

18. Geer, M. R., and H. F. Yancey: Expression and Interpretation of the Size Composition of Coal, *Trans. Am. Inst. Mining Met. Engrs.*, **130**, 250–269 (1938).

19. Gottschalk, V. H.: Coercive Force of Magnetite Powders, *U. S. Bur. Mines, Rept. Investigations* 3268, pp. 83–90 (1935).

20. Gross, John: Summary of Investigation on Work in Crushing, *Trans. Am. Inst. Mining Met. Engrs.*, **112**, 116–129 (1934)—128 references.

21. Gross, John, and S. R. Zimmerley: Crushing and Grinding. I. Surface Measurement of Quartz Particles, *Trans. Am. Inst. Mining Met. Engrs.*, **87**, 7–26 (1930).

22. Gross, John, and S. R. Zimmerley: Crushing and Grinding. II. Relation of Measured Surface of Crushed Quartz to Sieve Sizes, *Trans. Am. Inst. Mining Met. Engrs.*, **87**, 27–34 (1930).

23. Gross, John, and S. R. Zimmerley: Crushing and Grinding. III. Relation of Work Input to Surface Produced in Crushing Quartz, *Trans. Am. Inst. Mining Met. Engrs.*, **87**, 35–50 (1930).

24. Gross, John, and C. E. Wood: Explosive Shattering as a Possible Economical Method of Ore Preparation, *U. S. Bur. Mines, Rept. Investigations* 3268, pp. 11–19 (1935).

25. HAULTAIN, H. E. T.: A Contribution to the Kick *vs.* Rittinger Dispute, *Univ. Toronto Press Bull.* 4, *Faculty Applied Sci. and Eng. Res.* (1924); *Trans. Am. Inst. Mining Met. Engrs.*, **69,** 183–197 (1923).
26. KOSSEL, W.: The Energy of Surface Processes, *Ann. Physik*, **21,** 457–480 (1934).
27. KRUSTINSONS, J.: Dependence of the Dissociation Temperature of Solids on the Size of Crystal Grains, *Acta Univ. Latviensis Kim. Fakultat*, **1,** 273–277 (1929).
28. KRUSTINSONS, J.: Influence of the Size of the Grain upon the Dissociation Pressure of Solids. Red Mercuric Oxide. Iceland Spar, *Z. physik. Chem., Abt. A.*, **150,** 310–316 (1930).
29. LAZAREV, V. P.: Determination of the Work of Separation and the Surface Energy of Mica, and the Effect of Water on Them, *J. Phys. Chem.* (*U.S.S.R.*), **7,** 320–327 (1936).
30. LENNOX, LUTHER W.: Grinding Resistance of Various Ores, *Trans. Am. Inst. Mining Met. Engrs.*, **61,** 237–249 (1919).
31. LIPSETT, S. G., F. M. G. JOHNSON, and O. MAASS: The Surface Energy and the Heat of Solution of Solid Sodium Chloride. I. *J. Am. Chem. Soc.*, **49,** 925–943 (1927).
32. LIPSETT, S. G., F. M. G. JOHNSON, and O. MAASS: A New Type of Rotating Adiabatic Calorimeter. The Surface Energy and Heat of Solution of Sodium Chloride. II. *J. Am. Chem. Soc.*, **49,** (1927) 1940–1949.
33. LIPSETT, S. G., F. M. G. JOHNSON, and O. MAASS: The Surface Energy of Solid Sodium Chloride. III. The Heat of Solution of Finely Ground Sodium Chloride, *J. Am. Chem. Soc.*, **50,** 2701–2703 (1928).
34. MARTIN, GEOFFREY: Recent Research in the Science of Fine Grinding Done by the British Portland Cement Research Association between 1923 and 1925, *J. Soc. Chem. Ind.*, **45,** 160T–163T (1926).
35. MAXSON, W. L., F. CADENA, and F. C. BOND: Grindability of Various Ores, *Trans. Am. Inst. Mining Met. Engrs.*, **112,** 130–145 (1934).
36. MELDAU, R., and E. STACH: The Fine Structure of Powders in Bulk with Special Reference to Pulverized Coal, *J. Inst. Fuel*, **7,** 336–354 (1934).
37. ROSIN, P., and E. RAMMLER: The Laws Governing the Fineness of Powdered Coal, *J. Inst. Fuel*, **7,** 29–36, 109–112 (1933).
38. SAAL, R. N. J., and J. F. T. BLOTT: On the Surface Tension of Solid Substances, *Physica*, **3,** 1099–1110 (1936).
39. STADLER, H.: Grading Analyses and Their Application, *Trans. Inst. Mining and Met.*, **19,** 471–485, 509–537 (1910); **20,** 420–438 (1911).
40. TAGGART, ARTHUR F.: The Work of Crushing, *Trans. Am. Inst. Mining Engrs.*, **48,** 153–179 (1914).
41. TAMMANN, G., and W. BÖHME: Surface Tension of Gold Foils, *Ann. Physik*, **12,** 820–826 (1932).
42. TZENTNERSHVER, M., and J. KRUSTINSONS: The Influence of Particle Size upon the Dissociation Pressure of Solid Substances, *Z. physik. Chem.*, **130,** 187–192 (1927).

43. Williams, A. R., F. M. G. Johnson, and O. Maass: The Heats of Solution and Specific Heats of Rhombic Sulfur in Carbon Disulfide: The Surface Energy of Solid Rhombic Sulfur, *Can. J. Research*, **13B,** 280–288 (1935).
44. Yap, Chu-Phay: The Surface Energy of Iron Carbide, *Trans. Am. Soc. Steel Treating*, **20,** 289–312 (1932).
45. Zimmerley, S. R., and John Gross: Crushing Resistance of Minerals, *U. S. Bur. Mines, Rept. Investigations* 2948 (1929).
46. Carman, P. C.: The Size and Surface of Fine Powders, *J. Chem. Met. Min. Soc. South Africa*, **39,** 266–281 (1939); **39,** 338–342 (1939); **40,** 142–146 (1939).

CHAPTER VII

INDUSTRIAL SCREENING

There are two methods of industrial sizing corresponding to two of the three methods of laboratory sizing discussed in Chapter III, *viz.*, screening and classification.

Screening can be practiced from a dividing size of several inches to a dividing size as fine as about 0.1 mm. Likewise, classification can be practiced from a dividing size as coarse as 2 or 3 mm. to as fine a dividing size as 0.02 to 0.03 mm. It is often stated as a mineral-dressing rule that for sizing coarser than 20 mesh screens are preferable, and for sizing finer than 35 mesh classifiers are preferable. Although this is a good general rule, it must be applied with the realization that there are many exceptions. For example, products that must be dry, like talcum powder, aluminum powder, cement, or foundry sand, are screened dry, down to 200-mesh. Conversely, if it is desired to introduce the effect of specific gravity in the sizing, or if the material is in suspension in water, classification is preferred.

In studying industrial screening, consideration must be given to the type of screening surfaces and to the type of machine that is employed with these various screening surfaces.

SCREENING SURFACES

Screening surfaces are of three varieties, parallel rods, punched plates, and woven wire. Generally speaking, parallel rods are used for the coarsest sizing and woven wire for the finest.

Parallel-rod screening surfaces are usually made of steel bars, steel rails, cast iron, or wood. The trend is to a wider use of secondhand steel rails locally assembled to form the screening surfaces.

Punched plates are made of sheet steel punched by dies in various patterns. The openings are circular, square, or slotlike. If the screening surface is to make a separation at a coarse size, circular openings are the rule; if the separation is to be at a fine size, slotted openings are used. This preference arises from the lesser tendency to blind observed with slotted than with circular openings. Slotted openings make a separation at a size equivalent to that obtained with round openings whose diameter is slightly larger than the width of the slot.

Woven-wire screening surfaces are woven of carefully gauged wire, generally made of steel, but sometimes of copper, bronze, monel metal, or some other alloy. One or both wires are crimped

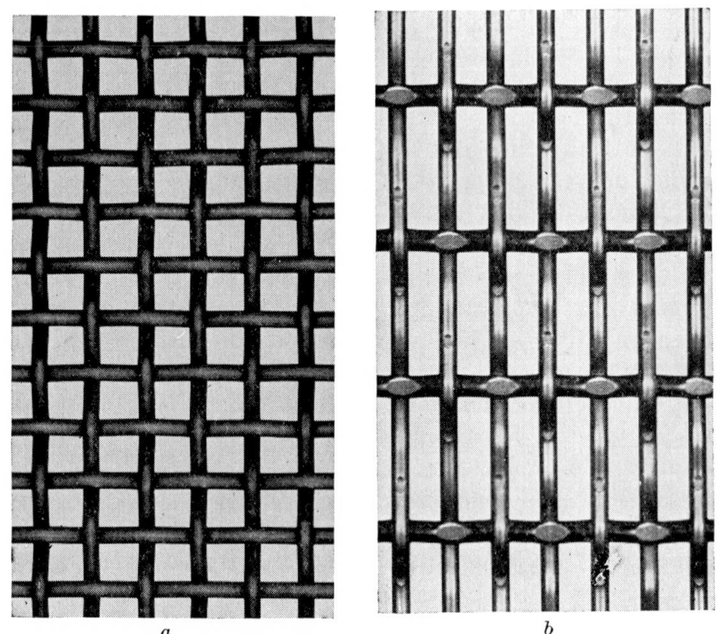

a *b*

FIG. 77.—Woven-wire screen cloth. *a*, square-opening; *b*, rectangular-opening (ton-cap). (*W. S. Tyler Co.*)

to prevent distortion. The weaving may be such as to produce square or rectangular openings (Fig. 77). Rectangular openings present the advantage of greater percentage of openings in ratio to the area of the screen; square openings make a more rigid cloth. Consequently, for very coarse screening a square-opening woven-wire is preferable whereas for intermediate and fine screening, rectangular openings are advantageous. Table 16 presents some data as to percentage of openings in square-opening and rectangular-opening cloth. The examples included in this table, from data by the W. S. Tyler Company, show an

TABLE 16.—PERCENTAGE OF OPENINGS TO TOTAL SCREEN-CLOTH AREA FOR SQUARE-OPENING AND RECTANGULAR-OPENING WOVEN-WIRE CLOTHS
(*W. S. Tyler Company*)

Comparison	Opening, inches	Heavy cloth		Light cloth	
		Square	Rectangular	Square	Rectangular
A	0.781	....	60.6		
	0.773	....	.....	....	66.9
	0.750	49.8			
	0.750	....	....	63.4	
B	0.256	....	49.5		
	0.262	....	....	....	62.1
	0.250	39.4			
	0.250	....	....	49.6	
C	0.126	....	....	....	57.3
	0.1225	....	43.2		
	0.125	29.5			
	0.125	....	....	48.7	
D	0.052	....	40.6		
	0.047	....	....	....	49.5
	0.053	28.1			
	0.046	....	....	41.5	
E	0.024	....	35.0		
	0.023	....	...	....	47.6
	0.023	24.5			
	0.0245	....	....	37.3	

average increase in openings of 28 per cent for the rectangular-opening cloth over the square-opening cloth; this advantage is more marked (36 per cent) for heavy cloth than for light cloth (19 per cent). Open area in punched plates is much less than in corresponding woven-wire screens.

MECHANISM OF PASSING THROUGH A SCREENING SURFACE

Many factors affect the rate at which undersize material may pass through a screening surface. The most important are the following:

1. The absolute size of the openings.

2. The relative size of the particle to that of the opening it must penetrate.
3. The percentage of openings to total surface in the screening surface.
4. The angle at which the particle strikes the screening surface.
5. The speed with which the particle strikes the screening surface.
6. The moisture content of the material that is being screened.
7. The opportunity for stratification in size layers afforded to the material which is being screened.

Clearly, the principal aim in designing a screen is to produce a device that permits passage of as great a proportion of undersize as possible at the highest possible rate, per unit of surface.

Effect of Size of Screen Openings. Since particles of various sizes have similar configurations, passage of a given weight of undersize particles (averaging, say, one-half the size of the openings in screens) will involve passage of a number of particles inversely proportional to the cube of the screen aperture. At the same time, the number of openings per unit of surface of the screen varies inversely as the square of the screen aperture. The number of passages to be effected per opening will then vary inversely as the size of the apertures in the screen. This leads to the fundamental conclusion that other things being equal *the capacity of a screen is directly proportional to the screen aperture.*

This fundamental conclusion is the reason why screens are impractical for fine sizing although displaying a tremendous capacity for coarse sizing. It is also the reason why screen capacities are stated in terms of tons per square foot per 24 hr. *per millimeter screen aperture* rather than in terms of tons per square foot per 24 hr.

Effect of Relative Particle Size. The relative size of the particle and of the opening controls passage or nonpassage of the particle. To illustrate how important this factor is, let a be the screen aperture (assumed square) and d be the particle diameter (assumed spherical). If the particle is striking the screen normally, it will not go through the opening undeflected, unless the center of the particle (Fig. 78) is more central with reference to the opening than the two particle positions shown in the figure. For wires of very small section compared with the openings (Fig. 78a), the probability of passage p is

$$p = \left(\frac{a - d}{a}\right)^2 \qquad \text{[VII.1]}$$

since two dimensions must be considered.

For wires of substantial diameter (Fig. 78*b*), the chance of passage without bumping against the wires is likewise

$$p = \left(\frac{a - d}{a + b}\right)^2. \qquad \text{[VII.2]}$$

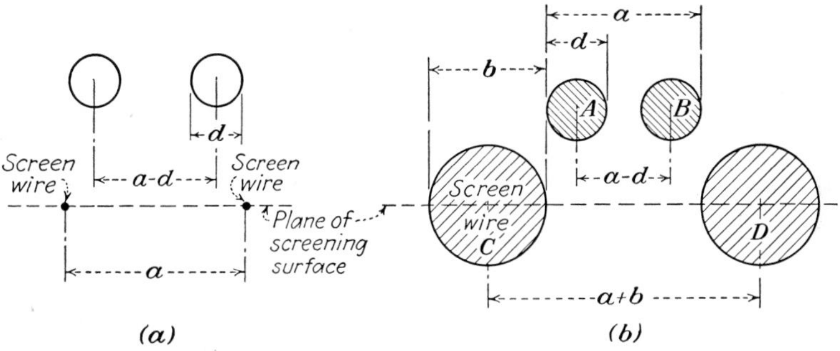

FIG. 78.—Factors involved in passage of particles through a wire screen (normal incidence).

This relationship takes into account the percentage of openings to the total surface $\left(\frac{a}{a + b}\right)^2$ since

$$\left(\frac{a - d}{a + b}\right)^2 = \left(\frac{a - d}{a}\right)^2 \cdot \left(\frac{a}{a + b}\right)^2.$$

Effect of Reflection of Particles on Wires; Normal Incidence. The ratio $\left(\frac{a - d}{a + b}\right)^2$ does not reflect accurately the chance of passage; it is a lower limit. This is because passage may well occur after the particle strikes a wire and is reflected into the opening. This is shown, *e.g.*, by consideration of Fig. 78*b*. Particle A could be shifted appreciably to the left so as to strike wire C and yet so as to go through the opening CD. Actually, the limiting position that A could assume would be approximately one such that the path of A after reflection on C would strike D equatorially, as shown in Fig. 79. What is true of the particle in the position A is of course true also if it is in the position B (Fig. 78*b*). Accordingly, the probability of passage through the screen opening is not $\left(\frac{a-d}{a+b}\right)^2$, but larger. This probability can be expressed by Eq. [VII.3].

$$\left.\begin{aligned} m &= \frac{2a+b-d}{b+d} && (a) \\ \cos\alpha &= \frac{1+\sqrt{1+8m^2}}{4m} && (b) \\ p &= \left[\frac{(a+b)-(b+d)\cos\alpha}{a+b}\right]^2. && (c) \end{aligned}\right\} \quad [\text{VII.3}]$$

In these equations, p is the probability, and $\cos\alpha$ and m are parameters expressible in terms of wire size b, opening size a,

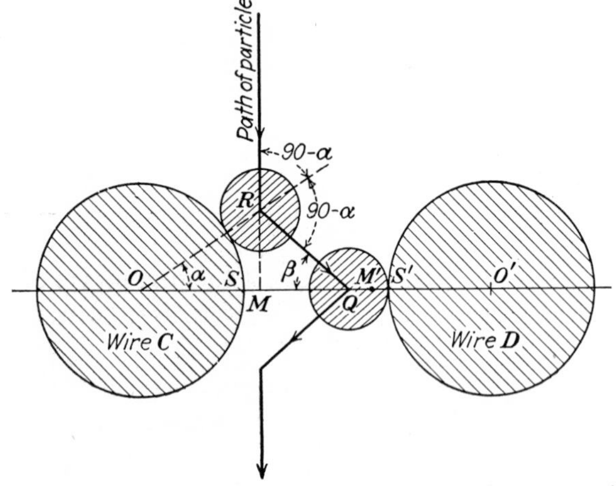

FIG. 79.—Limiting position for passage of a particle through a woven-wire screen, if reflection on the wires is considered (normal incidence).

and particle size d. These equations can be derived as follows.

Referring to Fig. 79,

$$OQ = OM + MQ = OM + MR \cot\beta.$$
$$\beta = 2(90-\alpha) - 90 = 90 - 2\alpha$$
$$\cot\beta = \cot(90-2\alpha) = \tan 2\alpha = \frac{2\tan\alpha}{1-\tan^2\alpha}$$

Also,

$$OQ = OS + SS' - S'Q = \frac{b}{2} + a - \frac{d}{2} = a + \frac{b-d}{2}$$

$$OM = OR\cos\alpha = \left(\frac{b+d}{2}\right)\cos\alpha$$

$$MR = OR \sin \alpha = \left(\frac{b + d}{2}\right) \sin \alpha$$

Hence,

$$a + \left(\frac{b - d}{2}\right) = \left(\frac{b + d}{2}\right) \cos \alpha + \left(\frac{b + d}{2}\right) \sin \alpha \cdot \frac{2 \tan \alpha}{1 - \tan^2 \alpha}.$$

Dividing each term by $\frac{b + d}{2}$, and putting

$$m = \frac{2a + b - d}{b + d},$$

$$m = \cos \alpha + \sin \alpha \frac{2 \tan \alpha}{1 - \tan^2 \alpha}. \qquad \text{[VII.3}a\text{]}$$

Transposing $\cos \alpha$ to the first term, and multiplying by $(1 - \tan^2 \alpha) \cos^2 \alpha$,

$$(m - \cos \alpha)(2 \cos^2 \alpha - 1) = 2 \cos \alpha (1 - \cos^2 \alpha)$$

which after expansion and cancellation of like terms becomes

$$2m \cos^2 \alpha - \cos \alpha - m = 0$$

whose positive root $(0 < \alpha < 90°)$ is

$$\cos \alpha = \frac{1 + \sqrt{1 + 8m^2}}{4m}. \qquad \text{[VII.3}b\text{]}$$

The probability of passage of the particle is

$$\left(\frac{MM'}{OO'}\right)^2 \text{ or } \left(\frac{OO' - 2OM}{OO'}\right)^2,$$

or

$$p = \left[\frac{(a + b) - (b + d) \cos \alpha}{a + b}\right]^2. \qquad \text{[VII.3}c\text{]}$$

Table 17 gives values of p according to Eqs. [VII.3], and also according to Eq. [VII.2], for various values of d/a ranging from 0 to 1, and for the two cases of the wire diameter equal to the opening and equal to one-quarter of the opening. Table 17 shows that the chance of passage varies enormously with the relative size of particle and opening and that the variation is particularly notable for particles whose size approaches the size of the openings of the screen.

Table 18 gives values of the probability of passage for particles of various relative sizes of 10, 100 and 1,000 opportunities for passage are afforded. The cases of 10 chances are representative of poor industrial screening, those of 1,000 chances, of conditions

TABLE 17.—PROBABILITY OF SPHERICAL PARTICLES PASSING AT NORMAL INCIDENCE THROUGH A WOVEN-WIRE SCREEN HAVING SQUARE OPENINGS

Ratio of particle diameter to opening, d/a	Wire diameter equal to opening		Wire diameter equals one-quarter of the opening	
	Probability of passage, per cent		Probability of passage, per cent	
	According to Eq. [VII.2]	According to Eq. [VII.3]	According to Eq. [VII.2]	According to Eq. [VII.3]
0.0	25.0	36.4	64.0	72.7
0.1	20.2	30.8	51.9	62.4
0.2	16.0	25.6	40.9	52.4
0.4	9.0	16.7	23.0	34.1
0.6	4.0	8.1	10.2	18.0
0.8	1.0	2.3	2.56	5.47
0.9	0.25	0.65	0.64	1.55
0.95	0.062	0.185	0.16	0.42
0.99	0.0025	0.0075	0.006	0.018
1.00	0.0000	0.0000	0.0000	0.0000

TABLE 18.—PROBABILITY OF SPHERICAL PARTICLES PASSING THROUGH A WOVEN-WIRE SCREEN HAVING SQUARE OPENINGS IF THE WIRE DIAMETER IS EQUAL TO THE OPENING (CASE A) OR EQUAL TO ONE-QUARTER OF THE OPENING (CASE B) AND IF 10, 100, AND 1,000 CHANCES OF PASSAGE AT NORMAL INCIDENCE ARE AFFORDED TO THE PARTICLES

Ratio of particle diameter to opening, a/d	Probability of passage, per cent					
	10 chances		100 chances		1,000 chances	
	A	B	A	B	A	B
0.0	99.0	100.0	100.0	100.0	100.0	100.0
0.1	97.5	100.0	100.0	100.0	100.0	100.0
0.2	94.8	99.9	100.0	100.0	100.0	100.0
0.4	83.9	98.8	100.0	100.0	100.0	100.0
0.6	57.0	86.2	99.9	100.0	100.0	100.0
0.8	20.8	43.0	90.2	99.4	100.0	100.0
0.9	6.3	14.5	47.9	79.0	99.8	100.0
0.95	1.8	4.1	16.9	34.4	84.3	98.5
0.99	0.1	0.2	0.7	1.8	7.2	16.5
1.00	0	0	0	0	0	0

in test screening. It is seen that even test screening does not offer an opportunity for perfect sizing, and that industrial screening does not permit appreciable passage of particles much coarser than 80 to 90 per cent of the screen aperture.

If screens having a rectangular opening instead of a square opening are used, the probability of passing is considerably increased as hindrance to passage is unidirectional instead of bidirectional.

Equation [VII.4], similar to Eq. [VII.2], gives the probability of passage through rectangular openings if reflection on wires is left out of account.

$$p = \frac{a - d}{a + d} \cdot \frac{A - d}{A + d}. \qquad \text{[VII.4]}$$

In this relationship, a and A are the dimensions of the opening. For example, if $b = a$, $d/a = 0.6$, the probability of passage is 4 per cent on a square-mesh screen for one chance, but on a rectangular-mesh screen in which $A = 4a$, the probability becomes 13.6 per cent, *i.e.*, it shows a much greater increase than is suggested by the relative opening percentages of 25 and 40.

A more accurate relationship is likewise provided by relationship [VII.5], if it is desired to allow for reflection of particles on wires.

$$\left.\begin{aligned} p &= \left[\frac{(a + b) - (b + d)\cos\alpha}{(a + b)}\right] \cdot \left[\frac{(A + b) - (b + d)\cos\alpha'}{A + b}\right] \\ \cos\alpha &= \frac{1 + \sqrt{1 + 8m^2}}{4m}; \qquad \cos\alpha' = \frac{1 + \sqrt{1 + 8M^2}}{4M} \\ m &= \frac{2a + b - d}{b + d} \qquad M = \frac{2A + b - d}{b + d}. \end{aligned}\right\} \qquad \text{[VII.5]}$$

Applying relationship [VII.5] for $A = 4a$, $b = a$, $d/a = 0.6$, p becomes 21.6 per cent as compared with 8.1 per cent in the case of the corresponding square-mesh screen. The probability of passage if 10 chances are afforded becomes 91.2 per cent for the rectangular-mesh screen as compared with 57.0 per cent for the square-mesh screen and 86.2 per cent for a square-mesh screen of very light weight. Clearly, rectangular-mesh screens have a decided industrial advantage.

Effect of Reflection of Particles on Wires; Oblique Incidence. The angle at which the path of the particle strikes the screening surface is of very great importance, but a rigorous analysis is tedious. Derivation of relationships similar to [VII.3] and

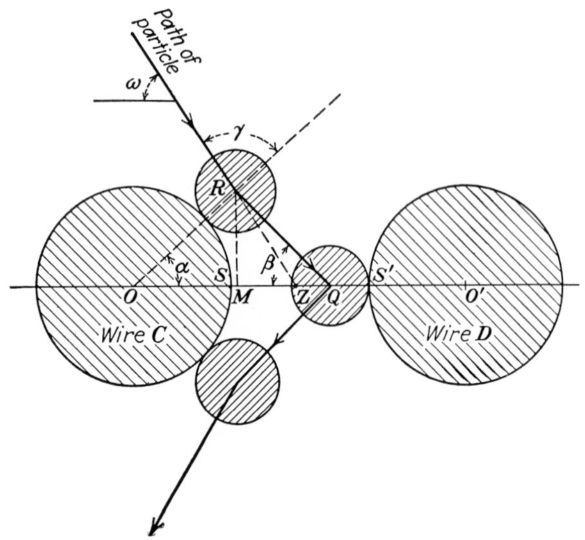

Fig. 80.—Limiting position for passage of a particle through a woven-wire screen if reflection on the wire is considered (oblique incidence).

[VII.5] lead to equations in $\cos \alpha$ of the fourth degree; these equations can be solved, but not readily.

Trigonometric analysis of the problem (similar in its broad lines to that used for normal incidence) leads to the following relationships (see also Fig. 80):

$$k = \tan \omega \qquad (a)$$

$$m = \frac{2a + b - d}{b + d} \qquad (b)$$

$$4m^2(k^2 + 1)\cos^4 \alpha - 4m(k^2 + 1)\cos^3 \alpha - (4m^2 - 1)(k^2 + 1)\cos^2 \alpha + 2m(k^2 + 2)\cos \alpha + (k^2m^2 - 1) = 0. \qquad (c)$$

$$p = \left[\frac{(a + b) - (b + d)\cos \alpha_0}{(a + b)}\right] \cdot \left[\frac{(a + b) - \left(\frac{b + d}{2}\right)\left(\frac{\sin(\omega + \alpha_1) + \sin(\omega - \alpha_2)}{\sin \omega}\right)}{(a + b)}\right] \qquad (d)$$

[VII.6]

In these relationships, $\cos \alpha_0$ is the value of $\cos \alpha$ for normal incidence, Eq. [VII.3], and α_1 and α_2 are angles corresponding to positive roots of Eq. [VII.6c], so chosen that $\alpha_1 > \alpha_2$.

Angle ω is the angle of the particle path to the surface of the screen when particle and screen meet.

Numerical values of the probability of passage as a function of the angle of incidence (for the case $b = a$) is presented in

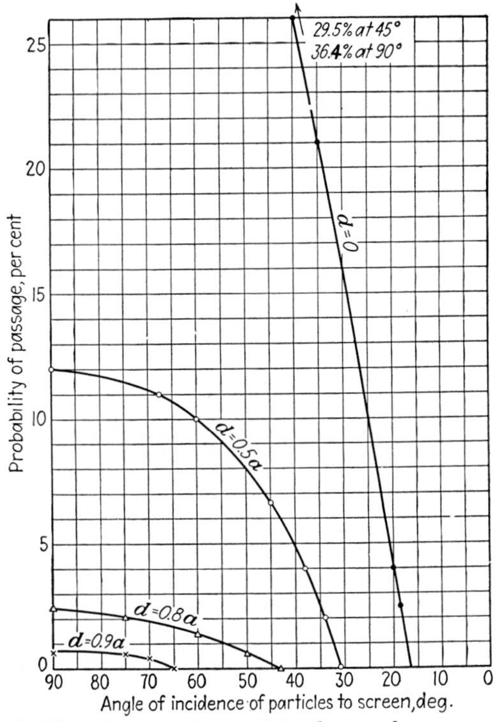

Fig. 81.—Probability of passage for particles that reach a screen at an angle (see text for details).

Fig. 81. It appears from this figure that with increasing deviation from the normal there is at first a slight impairment in the chance of passage followed, at increasing obliquity, with a rapid reduction to the vanishing point for the chance of passage.

Effect of Moisture. It is impossible to overemphasize the importance of *moisture* content of screen feed. Either bone-dry or wet pulp is relatively easy to screen, but even a small percentage of water in a dry feed increases the difficulty of screening out of all proportion to its abundance.

TYPES OF SCREENS

Screens can be classified into stationary screens and moving screens.

Stationary Screens. Stationary screens are principally of the type known as *grizzlies* (Fig. 82) consisting of parallel rods set at an angle to the horizontal and employed in conjunction with coarse crushing.

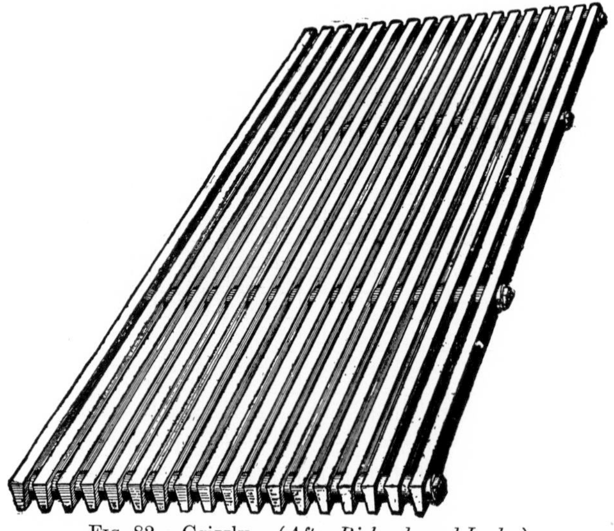

Fig. 82.—Grizzly. (*After Richards and Locke.*)

Grizzlies may be made of bars having a rectangular cross section with or without taper from feed to discharge end, or they may be of special sectional shape, as rails, I beams, diamond headed. The bars are held together by depressed cross rods, or by sleeves around long through bolts. The bars are now usually made of tough steel, but steel-capped wooden bars have been used.

The slope of grizzlies is generally such that the material fed to it will just flow; however, cases are not unknown where the grizzly slope is deliberately made slightly too low for free flow so that the attendant to the crusher is compelled to feed the crusher by pulling with a hoe or bar on the grizzly oversize; in this way, the attendant is placed in a position permitting him to remove objectionable objects (dynamite sticks, sledges, tramp iron, wood) and thus keep them out of the crusher.

Types of stationary screens other than grizzlies have been used in the past and are still in use in some antiquated mills. These, however, are the exception.

Moving Screens. Moving screens include moving grizzlies, trommels, vibrating screens, and shaking screens.

Moving grizzlies are of various types, designed to save headroom, act as feeders, and effect better screening. As the sizing that grizzlies are required to do is usually merely preliminary and

Fig. 83.—Plain cylindrical trommel arranged for delivering four products. (*Allis-Chalmers Mfg. Co.*)

essentially designed to facilitate the task of coarse crushers, there is little necessity for the refinement implied by the introduction of moving in place of stationary grizzlies.

In moving-bar grizzlies, alternate bars alternately rise and subside, so that the material moves forward gently with considerable turning over. Chain, traveling-bar, and live-roll grizzlies are other types of moving grizzlies.

Trommels (Fig. 83) consist of rotating cylindrical, prismatic, conical, or pyramidal shells of punched plate, commonly 3 to 4 ft. in diameter and 5 to 10 ft. long into which the material to be screened is made to flow. These shells are driven from a central shaft fastened to them by four- or six-armed spiders. The central shaft in turn is driven from a pulley by means of a bevel gear. In so-called *revolving stone screens*, the drive is directly from a gear on the periphery of the screen or by tires and rollers.

Cylindrical and prismatic trommels are set at a slant. Conical and pyramidal trommels, on the other hand, are installed with a horizontal drive shaft but retain the conveying feature of cylindrical and prismatic trommels set at an angle. Prismatic and pyramidal trommels, as compared with the cylindrical and conical types, are said to provide better tumbling of the material, and easier repairs. Yet cylindrical trommels outnumber all other kinds.

In operation, the oversize is delivered at the lower end of the trommel, and the undersize, after passage through the openings,

FIG. 84.—Triple-compound trommel. (*Allis-Chalmers Mfg. Co.*)

falls into a trough in which it is carried away either by sliding in the dry state or by being flushed with water. For wet screening, a much flatter launder may be used.

Trommels can be arranged to deliver several oversizes as shown in Fig. 83. But such an arrangement requires passage of the coarse material over the finest screen first—a condition that is bound to increase upkeep costs. Trommels of a compound type (Fig. 84) are theoretically better, but they are open to the objection that the outer screen must be removed to repair the inner screens.

Vibrating Screens. Vibrating screens have been developed in the past 20 years to the point where they have made other types of screening devices obsolescent. They are characterized by the utilization of a reciprocating motion imparted to the screening surface in a direction normal or at a high angle to the screening surface.

The problems whose solution has been required are as follows:

1. Production of a suitable reciprocating motion.
2. Even transmission of this reciprocating motion to all of the screening surface.
3. Nontransmission of vibration to supporting structure.
4. Dustproofing of vital vibrating parts.
5. Ease of replacement of worn screen cloth.
6. Minimizing of loss of headroom on passage through the screens.

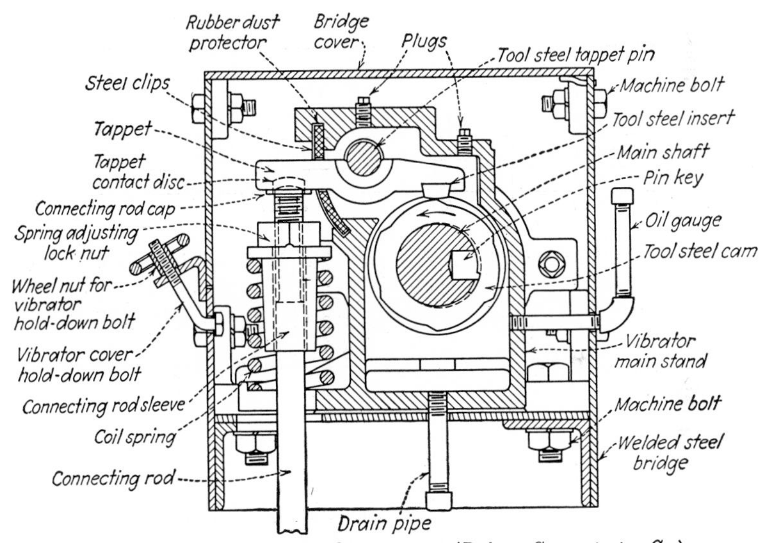

FIG. 85.—Vibrator for Leahy screen. (*Deister Concentrator Co.*)

These various aims have been achieved to a greater or lesser extent in several screens. These can be classified into mechanically vibrated and electrically vibrated screens. They can again be classified into screens providing symmetrical and asymmetrical vibration.

The *Leahy* screen consists of an inclined rectangular main frame fastened to a mechanical vibrator connected to the center of the screen frame by a rod. The vibrator (Fig. 85) is actuated by a multiple cam acting on a tool-steel button and working against a coil spring. The motion is of the asymmetrical type (Fig. 86*a*). The screen frame is sealed against the main frame by strips of live rubber. A part of the vibration is necessarily transmitted to the supporting structure.

The *Deister Plat-O* screen typifies mechanically vibrating screens using unbalanced flywheels to induce motion. Other

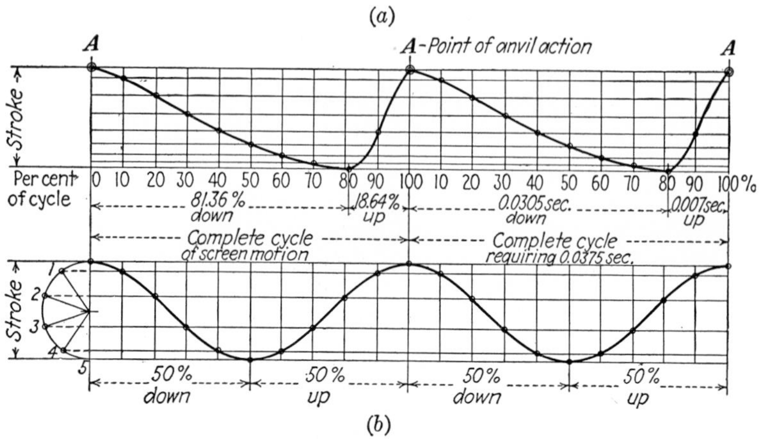

FIG. 86.—Vibrations for vibrating screens. (*a*) Asymmetrical vibration; (*b*) symmetrical vibration. (*Deister Concentrator Co.*)

screens of the same general type are the Link-Belt, Allis-Chalmers, and Kennedy-Van Saun screens.

The Deister Plat-O screen (Fig. 87) consists of a base frame carrying, through heavy coil springs, the screen frame proper. This screen frame is rigidly fastened to the vibrating mechanism. The vibrating mechanism consists of a shaft, pulley, two spherical roller bearings, and four unbalanced flywheels. One pair of flywheels is keyed rigidly to the shaft and, alone, would actuate the screen. The other pair of flywheels (smaller) is adjustable thus permitting control of the amplitude of vibration. The screen cloth is held in the screen frame by a sash-type mounting.

The Niagara, Tyler-Niagara, Telsmith, Robins-Gyrex, and Symons screens are mechanically vibrating screens actuated by a shaft concentric with the drive pulley but eccentric at the bearings to which the screen frame is fastened (Fig. 88).

In the *Symons screen* (Fig. 89), the vibration divides itself between the vibrating deck and the carrying deck inversely in

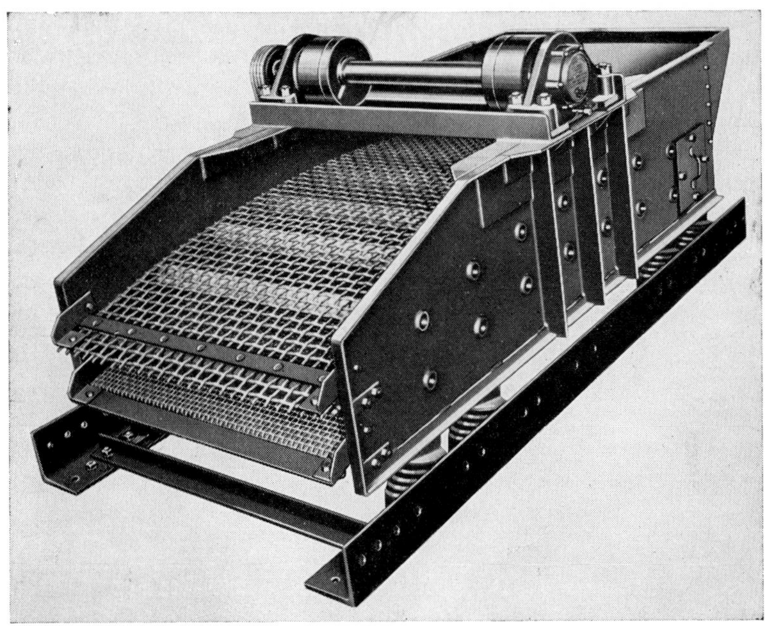

FIG. 87.—Triple-deck Plat-O vibrating screen. (*Deister Machine Co.*)

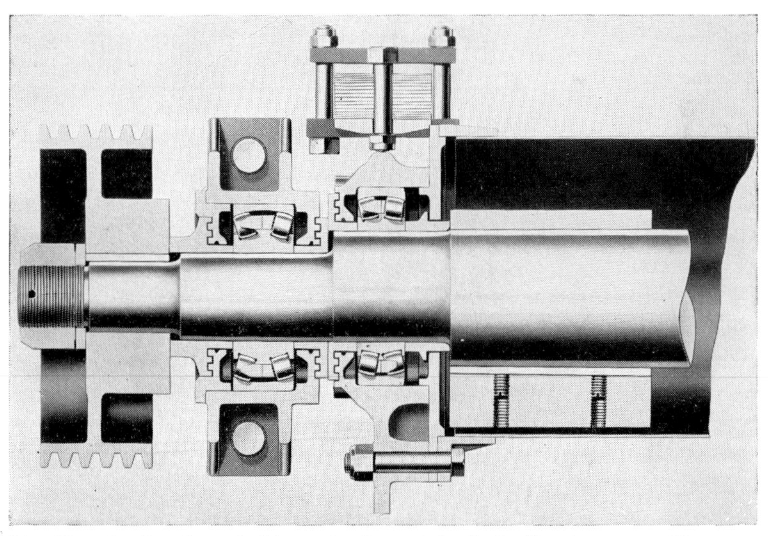

FIG. 88.—Section through drive unit of eccentric-shaft vibrating screen (Symons) showing eccentric shaft, bearings, and seals. (*Nordberg Mfg. Co.*)

proportion to their masses. The carrying deck thus counterbalances the vibrating deck, and since it is heavier its amplitude of vibration is smaller. The carrying deck in turn is carried on leaf springs set at an angle of 30° to the horizontal. This compels travel of the particles even if the screen is in the horizontal position, while still using a uniform vibrating motion (Fig. 86*b*).

The *Niagara*, *Tyler-Niagara* (Fig. 90), *Telsmith*, and *Robins-Gyrex* screens differ from the Symons screen in that counterbalance to the vibration of the deck is obtained by means of unbalanced flywheels. The screens are operated at a slope of 15 to 30°. Niagara-type screens are widely used.

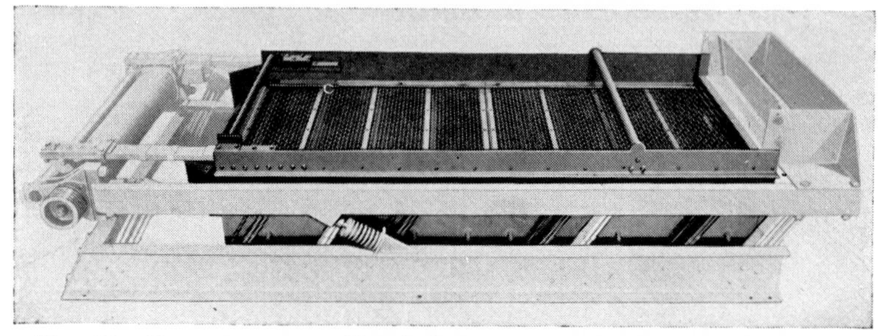

FIG. 89.—Symons screen. The dark portion shows that part of the screen which comprises the screening deck and its supports; the light portions, the parts that are in counterbalance, namely, the side bars, feed hopper, and power unit. (*Nordberg Mfg. Co.*)

The *Mitchell* screen should properly be regarded as a mechanically vibrated screen, although it is often described as an electrically vibrated screen. It consists of a rectangular frame carrying the vibrator from its center on a spherical bearing, and of a screen suspended by side plates from the vibrator.

The vibrator consists of an a. c. motor rotor with extra long shaft, with cages at either end in each of which one 1½- to 2-in. ball is driven in an appropriate race. The balls are set 180° apart, and by their unbalanced rotation cause the shaft to gyrate.

The *Hum-mer* vibrating screen is truly an electrically vibrated screen. Vibration is obtained by the action of an electromagnet on an armature (Fig. 91). The upstroke is suddenly interrupted when the armature hits the striking block, but the downstroke is

FIG. 90.—Two-surface Tyler-Niagara screen. (*W. S. Tyler Co.*)

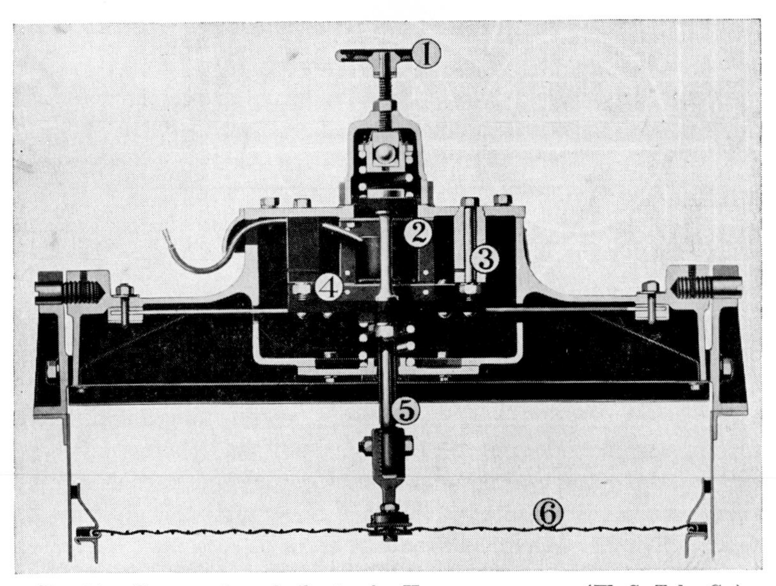

FIG. 91.—Cross section of vibrator for Hum-mer screen. (*W. S. Tyler Co.*)

not so interrupted. This differential motion results in constant unblinding of the screen; it furthermore simulates jigging and causes the particles to stratify, the coarsest particles riding in the uppermost layer and the finest next to the screen. Coarse particles are therefore discharged most rapidly, and fine particles are given a wider opportunity to pass through the screen. The screen cloth is held at drumhead tension by a simple stretching device which permits quick replacement of worn cloths. Figure 92, for example, shows the tensioning device in case a fine

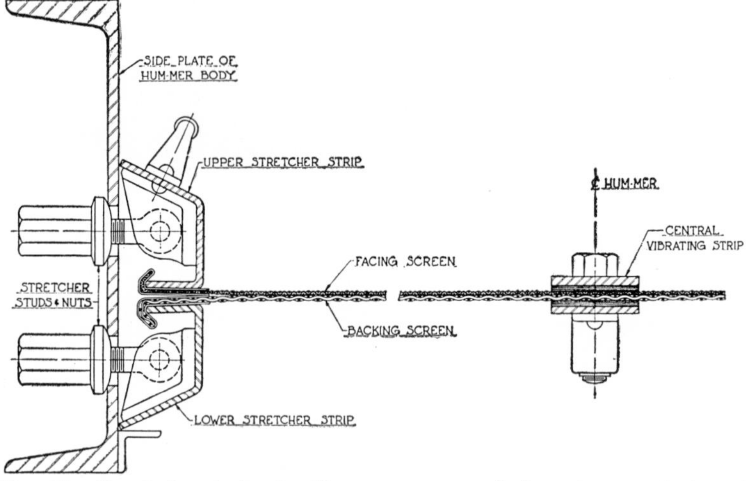

Fig. 92.—Tensioning device for Hum-mer screen. Independent tensioning of the facing screen and backing screen is shown. (*W. S. Tyler Co.*)

top screen and coarser backing screen are used. A special alternating current (15 cycle) is used; a motor generator set is therefore part of the necessary equipment.

By the use of rectified alternating current, the frequency of vibration can be doubled; this is useful in screening fine difficult products such as damp coal, cement slurry, and chemicals.

The type 400 Hum-mer has four vibrators, one at each corner, that vibrate in unison. This does away with localization of vibration under heavy load and permits of screening a greater tonnage.

The *Jeffrey-Traylor* vibrating screen is also an electric vibrating screen operated by the action of an oscillating armature on a stationary coil. In this case, however, there is no striking block to interrupt suddenly the motion of the screen. The direction of the motion is at a considerable angle from the normal to the screen in one type (Conveyanscreen, Fig. 93) but normal in another. All of the screening surface is vibrating. Nontransmission of the vibration to the supporting framework is obtained by suspension through vibration absorbers.

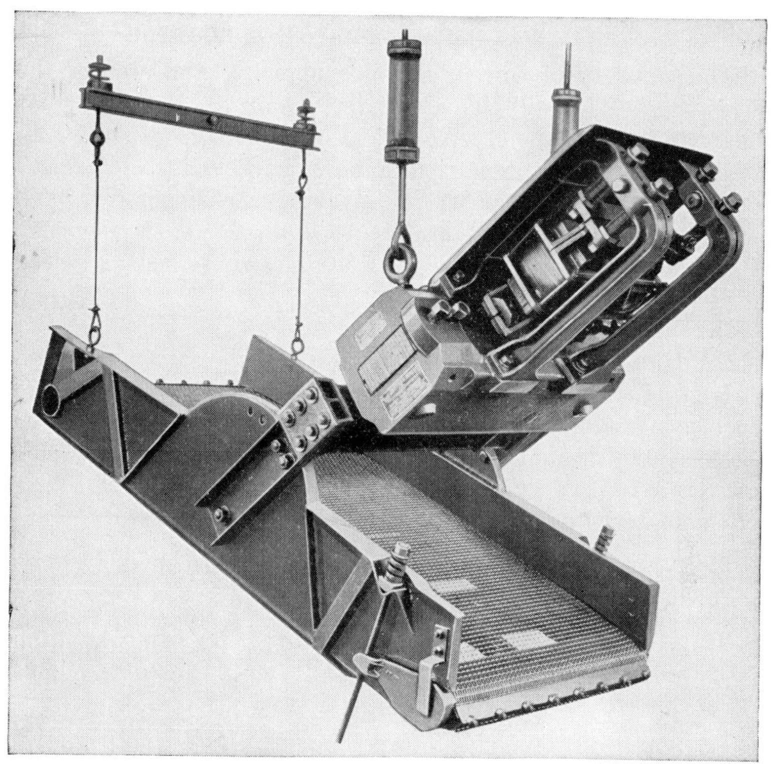

Fig. 93.—Jeffrey-Traylor Conveyanscreen. (*Jeffrey Manufacturing Co.*)

Shaking screens are used in coal preparation. They differ from vibrating screens in that the shocks imparted to the particles are much less frequent, of greater amplitude, and so directed as to convey as much as to screen. In shaking screens, the surface is horizontal or nearly so. Punched plate is used to avoid blinding, but this cuts the open area almost in half. Travel of the feed is obtained by a nonuniform harmonic motion of the deck. An example is the American Anti-gravity Coal screen in which the deck slopes upwards a few degrees to the discharge end.

Shaking screens have the advantage over most vibrating screens that they can be set almost flat, but they are prone to cause more rack and vibration to their supports, and repairs are more expensive.

OPERATING CHARACTERISTICS OF SCREENS

The capacity of screens depends upon (1) the area of the screening surface, (2) the size of the openings, (3) the peculiar characteristics of the ore such as specific gravity, moisture content, temperature, proportion of fines, particularly of "slime" or clay, etc., and (4) the type of screening mechanism used. To an extent, the capacity of the screen is a function of the efficiency that it is expected to attain, a sacrifice in capacity being required for an improvement in efficiency.

Because of the direct dependence of screening *capacity* upon the area of the screening surface and upon the screen aperture, it is customary to reckon capacity in terms of tons per square foot per millimeter screen aperture per 24 hr. Comparisons should be made judiciously, however, as such factors as specific gravity of the material, size distribution of the feed, and moisture of the material have a very marked effect on the capacity.

TABLE 19.—CAPACITY OF SCREENS OF VARIOUS TYPES

Type of Screens	Capacity Range (Ore), Tons per Sq. Ft. Area per Mm. Aperture per 24 Hr.
Grizzlies	1 to 5
Trommels	0.3 to 2
Shaking screens	2 to 8
Vibrating screens	5 to 20

Table 19 presents data for the capacity of screens of various types.

The *efficiency* of screens is difficult to quantify. Efficiency, in the field of mechanical engineering, is defined as the ratio of energy output to energy input. Strictly speaking then, what is currently termed screen efficiency is not an efficiency but a measure of the effectiveness of the screening operation as compared with a perfect screening operation. If a screen is regarded as expected to separate material finer than a certain size out of a mixed feed containing that fine material together with coarser material, the screen efficiency can be defined as the ratio of the quantity of this fine material that is taken out of the feed to the quantity of fine material in the feed. The percentage efficiency E can be expressed as

$$E = \frac{10{,}000U}{uF}, \qquad \text{[VII.7]}$$

in which U is the tonnage passing through the screen for each F tons of feed and u is the percentage of undersize in the feed as determined by test screening.

This formula, however convenient, fails to take into account that particles which are just finer than a given screen opening are difficult to screen out, whereas particles which are much finer than the screen opening are readily screened out, so that the same screen on different feeds might display vastly different efficiencies. To overcome this difficulty, it has been proposed to define screening efficiency as the ratio of the *difficult* grains that are taken out by the screen to the *difficult* grains in the feed. The objection to such a definition is, of course, that it is almost impossible to agree on what constitutes a difficult grain.

The operating cost of screens is small. In the case of stationary screens there is, of course, no power cost; the labor cost is low, as but a small part of a man's time has to be devoted to checking up on screens; and the cost of supplies is merely that of replacing screen cloth or other parts of the screen when they wear out. In the case of moving screens there is, in addition to these items, the power cost. However, this requirement is small; with vibrating screens it ranges from 0.1 to 0.005 ct. per ton.

Screens add to the cost of treatment of a mineral product mainly because of the first cost of the screens themselves, because a larger building is required to accommodate insertion of screens in the flow sheet, and because screens cause a fall in the ore flow and require compensating elevation of ore.

Literature Cited

1. Cerckel, H. O. H.: Screening and Dewatering, *Gas World*, **106** [No. 2752] Coking Sect., 61–66 (1937).
2. Ivers, E. J.: Progress in Screening Technique, *Metall u. Erz*, **27**, 209–215, (1930).
3. Taggart, Arthur F.: "Handbook of Ore Dressing," John Wiley & Sons, Inc., New York (1927), pp. 498–502.
4. Warner, R. K.: Efficiency of Screening, *Trans. Am. Inst. Mining Met. Engrs.* **70**, 631–640 (1924).

CHAPTER VIII

THE MOVEMENT OF SOLIDS IN FLUIDS

A number of mineral-dressing processes deal essentially with the movement of solids in fluids. This is true in particular of classification, thickening, filtration, and gravity concentration. Indeed, the movement of solids in fluids plays a part in all mineral-dressing processes. It is therefore a subject of extreme theoretical and practical importance.

Fluid Resistance and Terminal Velocity. It is obvious that if fluids did not exert a resistance to the motion of solids in them, the latter, under the influence of gravity or of other forces, would move with constant acceleration *ad infinitum.* Fluids that would behave in that fashion, however, exist only as a mathematical convenience. Even the most fluid liquid or the most tenuous gas exerts a resistance to the motion of a foreign body in its midst. Since the resistance is nil when the solid is at rest with respect to the fluid, but not nil when it is in motion, the resistance R is a function of velocity.

$$R = f(v). \qquad \text{[VIII.1]}$$

The exact determination of the form of this function for all possible cases is *the* great problem of hydrodynamics and aerodynamics. This equation enters as a fundamental factor in problems as varied as those of ballistics, of the flight of airplanes and airships, of the streamlining of motor cars and trains, of geologic processes, and of colloid behavior. In many ways, it may be regarded as the physical relationship of greatest significance to man.

After more than two centuries of search, both theoretical and practical, complete determination of the form of Eq. [VIII.1] is still wanting. For special cases, however, it is known with considerable accuracy.

If this force which we term resistance becomes equal in amplitude and opposite in direction to the resultant of all the other forces acting on a body in a fluid, the acceleration of the body will be nil and the velocity constant. This *terminal* or *maximum velocity* v_m is of special importance in mineral dressing.

It is necessary to evaluate v_m in terms of (1) the physical characteristics of the solid, such as specific gravity, size, shape; (2) the physical properties of the fluid such as viscosity, specific gravity, and extent; and (3) the forces that are called into play. It is further necessary to ascertain the character of the function relating the velocity to the time, and to estimate how much time is required for the terminal velocity to become established.

This study will first be directed to a single homogeneous sphere in a fluid extending in all directions to infinity in a uniform field of force. This is tantamount to elimination of variation in particle shape, in extent of the liquid, and in intensity of the field of force.

Settling of Fine Spheres. Under the limiting conditions stated in the preceding section, *Stokes* has shown[45] that if the velocity of motion is low enough to cause so-called *viscous* or *laminar* flow, the resistance to the motion of the sphere is

$$R = 6\pi\mu rv, \qquad \text{[VIII.2]}$$

in which μ is the viscosity of the fluid, r the radius of the sphere, and v the velocity, all expressed in c.g.s. units. This is true, of course, of a large or small sphere, provided v is small enough. The fact should be emphasized that Stokes' law, although holding for low velocities of *all* moving spheres, holds for the terminal velocity of small spheres but fails for the terminal velocity of large spheres.

Stokes' proof is entirely mathematical and difficult to follow for one not especially trained in mathematical physics. A somewhat simpler proof is to be found in a book by Page,[30a] but that proof requires a working knowledge of vector analysis which is also an advanced mathematical subject. Page calls attention to the fact that one-third of the resistance $2\pi\mu rv$ is due to the difference in pressure on the two sides of the sphere and the remaining two-thirds $4\pi\mu rv$ to the shearing stresses.

Stokes' law is accurately verified by experiment for the terminal velocity of small spheres, *e.g.*, of spheres of quartz less than 50 microns in diameter, falling in water.

Stokes' law is one of physics' best established and most famous relationships. Its fame is connected to the measurement of the mass of the electron by Millikan[30b]. In that experiment, Millikan modified the settling velocity of minute oil droplets (radius 0.00001 to 0.0001 cm.) by charging them with an integral number of electrons and placing them in an electric field. This was tantamount to changing the force acting on the drops by definite increments from the force of gravity for an uncharged drop to the force of gravity plus or minus an electric force in integral increments.

In connection with these experiments, Millikan indicated that, for extremely small spheres (r less than 0.1 micron), an appreciable correction is needed to the Stokes formula because the hydrodynamical assumption of Stokes that the fluid has a continuous structure (*i.e.* is indefinitely subdivisible) ceases to be valid for spheres so near molecular dimensions. In present-day mineral-dressing work, the Millikan correction can be overlooked, but it should be mentioned for the sake of completeness.

The second law of motion (mass $\times$ acceleration $= \Sigma$ forces), applied to a sphere falling in a fluid by laminar flow, then takes the form

$$m\frac{dv}{dt} = mg - m'g - R,$$

in which m stands for the mass of the sphere, m' for the mass of the displaced fluid, v for velocity, t for time, g for the acceleration of gravity, and R for resistance; or

$$\frac{4}{3}\pi r^3\Delta\frac{dv}{dt} = \frac{4}{3}\pi r^3(\Delta - \Delta')g - 6\pi\mu r v,$$

in which Δ and Δ' are the specific gravities of the solid and the fluid.

This equation can be simplified by dividing by $\frac{4}{3}\pi r^3\Delta$ and yields the fundamental differential equation

$$\frac{dv}{dt} = \frac{\Delta - \Delta'}{\Delta}g - \frac{9}{2}\frac{\mu v}{\Delta r^2} \qquad \text{[VIII.3]}$$

whose solution, developed on page 177, gives the relationship between time and velocity.

The maximum or terminal velocity can be obtained by putting $dv/dt = 0$, whence

$$v_m = \frac{2}{9}\frac{(\Delta - \Delta')r^2 g}{\mu}, \qquad \text{[VIII.4]}$$

in which form Stokes' law is usually formulated.

It is desirable to keep clearly in mind the limitations within which Stokes' law is applicable. Stokes himself has said[45]:

It has been shown experimentally by Coulomb (*Mémoires de l'Institut,* Tome III, p. 246) that in the case of very slow motions the resistance of a fluid depends partly on the square and partly on the first power of the velocity. The formula [VIII.2] determines in the particular case of a sphere that part of the whole resistance which depends on the first power of the velocity, even though the part which depends on the square of the velocity be not wholly insensible.

Quartz spheres falling in water follow accurately Stokes' law up to diameters of the order of 50 microns, and with small deviations up to 100 microns. Clearly then, thickening and the classification in water of fine flotation pulps are largely accountable on the basis of Stokes' law; but gravity concentration, which in the main deals with particles coarser than 100 microns, is not accountable on the basis of Stokes' law.

Settling of Coarse Spheres. Nearly two centuries before Stokes, *Newton*[27] calculated that the resistance to motion of a solid whose surface is at right angles to the direction of fluid flow equals the product of the density of the fluid by the square of the velocity and by the cross-sectional area.

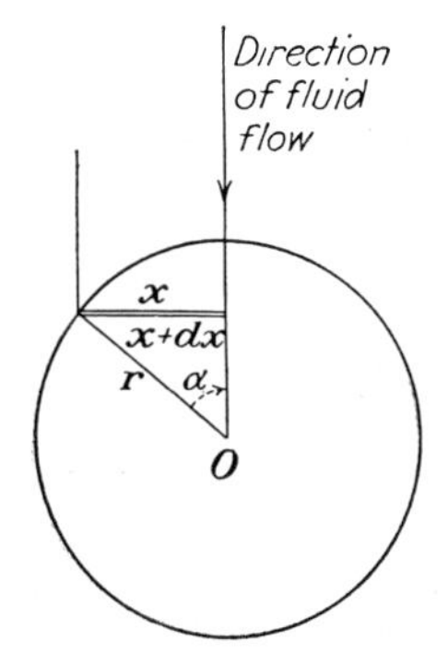

FIG. 94.—Newtonian resistance to the motion of a sphere.

The resistance of a circular disk of radius r is then $\pi r^2 \Delta' v^2$.

In the case of a sphere, however, the surface is at an angle to the direction of flow, the angle varying from 90° at the pole to 0° at the equator. Allowance for this condition can be made if the resistance which is offered by an element of surface dA set at an angle can be calculated, since proper integration then will give the total resistance[49] (Fig. 94).

The normal component of the velocity on the element dA is $v \cos \alpha$, hence the pressure exerted on the element dA is $\Delta' v^2 \cos^2 \alpha \, dA$. Of this pressure, the along-stream component* is $dR = \Delta' v^2 \cos^3 \alpha \, dA$.
But

$$\begin{aligned} dA &= 2\pi r \sin \alpha \cdot r \, d\alpha \\ &= 2\pi r^2 \, d(\cos \alpha); \end{aligned}$$

so

$$dR = 2\pi \Delta' r^2 v^2 \cos^3 \alpha \, d(\cos \alpha),$$

* The analysis presented here assumes collision of fluid particles with the sphere to be inelastic.

whence

$$R = 2\pi\Delta' r^2 v^2 \int_{\pi/2}^{0} \cos^3 \alpha \, d(\cos \alpha) = \tfrac{1}{2}\pi\Delta' r^2 v^2 \Big[\cos^4 \alpha \Big]_{\pi/2}^{0}$$

$$= \frac{\pi}{2}\Delta' r^2 v^2.$$

The Newtonian resistance to motion of a sphere in a fluid is then

$$R = \frac{\pi}{2}\Delta' r^2 v^2 \qquad \text{[VIII.5]}$$

notwithstanding the erroneous "correction" proposed by Finkey.(12)

Taken as a whole, the experimental verification of Newton's relationship is not satisfactory. By insertion in the formula of a coefficient Q known as the *coefficient of resistance*, the facts can be made to fit a relationship in which the resistance varies as the velocity squared. Equation [VIII.6] then accurately represents the facts, but Q cannot be regarded as even approximately constant except within narrow limits.

$$R = Q\frac{\pi}{2}\Delta' r^2 v^2. \qquad \text{[VIII.6]}$$

For spheres of minerals in water, Q is about 0.4 if r is larger than 0.2 cm.

Eddying vs. Laminar Flow. Equation [VIII.6], sometimes called the *Rittinger equation* after one of its chief protagonists,(40) deals with motion under conditions of *eddying* or *turbulent* resistance. One of the best brief discussions of *laminar vs. turbulent* flow is that given by Brindley,(4) who says:

Whenever there is transference of momentum in a fluid from faster to slower moving layers, stresses are set up. In cases of laminar flow . . . momentum is transferred by molecular collisions, and the viscous forces which arise are explained by the kinetic theory in terms of the mean free path of the molecules. In turbulent flow transference of momentum occurs between masses of liquid, that is to say, between large groups of molecules rather than between individual molecules. The main characteristics of turbulent flow are, firstly, that pressure differences are not proportional to the first power of the velocity, but increase more quickly; secondly, the flow is not stationary, since the pressure and the velocity do not remain constant at a particular point, but oscillate about a mean value . . . thirdly, that the stresses which arise are proportional to the square of the velocity, and therefore when a body moves through a liquid producing turbulence, the resistance to motion should be proportional to the square of its velocity.

Pictorial evidence of the difference between viscous and turbulent flow is presented in Figs. 95 and 96 which are photographs of a cylinder in viscous flow and in turbulent flow.[1,34]

Failure of Newton's Equation. It is somewhat disconcerting to find that experimental evidence is constantly accumulating discredit on Newton's relationship, which in its main aspects is sound. Yet the fact must be recognized that for speeds higher

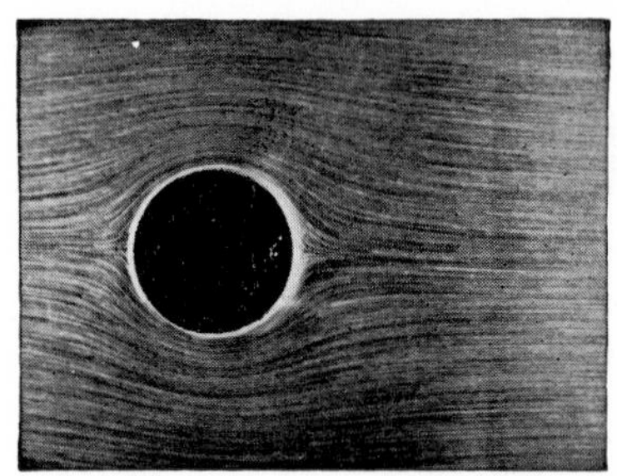

Fig. 95.—Fluid flowing by viscous flow past a cylinder. (*After Prandtl-Tietjens.*)

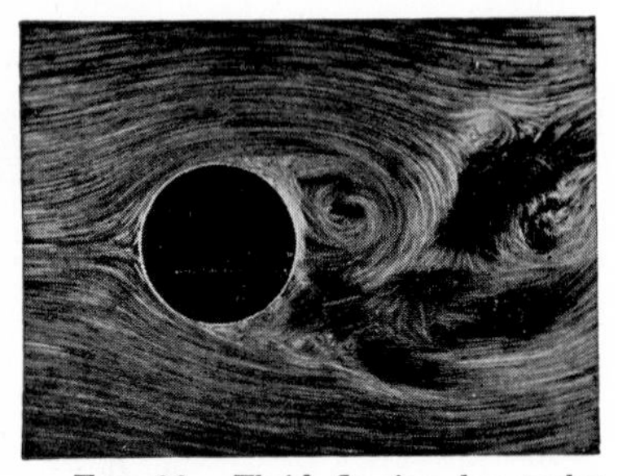

Fig. 96.—Fluid flowing by turbulent flow past a cylinder. (*After Prandtl-Tietjens.*)

than are usual in mineral dressing the coefficient Q not only does not remain stationary at 0.4, or at any other value, but suddenly drops to about 0.15.[7,8,42] At still higher speeds, the coefficient Q is again variable.

As usual, the reasons for the failure of the Newton formula is related to noncompliance with the conditions under which it is intended to apply. This is particularly well set forth by Zahm[49] who states the Newtonian premises as follows:

1. The discontinuous rare medium consists of indefinitely small equal particles freely disposed at equal, great distances from each other; they may be elastic or inelastic, at rest or in uniform motion.

2. The particles, say in uniform rectilinear motion, strike only the front part of a fixed obstacle, leaving a vacuum behind extending to infinity, bounded by parallel stream lines grazing the body.

3. The particles striking the hard smooth front retain their tangential velocity component and lose their normal component, if inelastic, or have it reversed if elastic. In all cases before and after rebound they

continue in straight lines without mutual interference: *i.e.*, with substantially infinite free paths.

The conditions encountered in the real problem of movement of a body through a fluid are so different from those set forth by Newton that it is amazing to find any agreement between theory and fact.

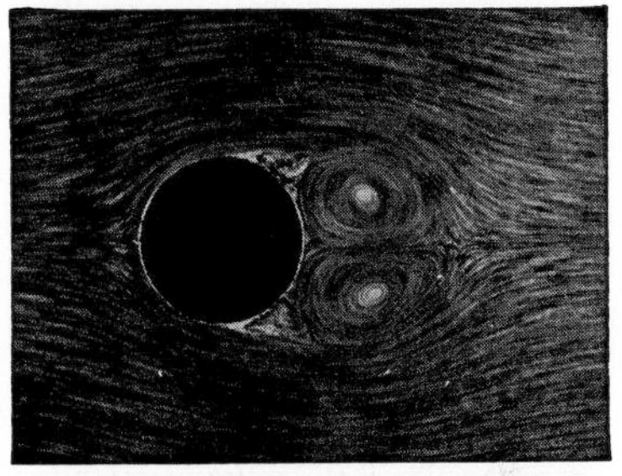

FIG. 97.—Stationary vortex in the wake of a cylinder (turbulent flow). The camera was at rest with respect to the cylinder. (*After Prandtl-Tietjens.*)

Vortices. Recent attempts have been made to build a theory of eddying resistance without the too elementary considerations of Newton. These theoretical attempts have yet to achieve their aim. But they have already indicated the existence of two types of vortices in the rear of moving spheres, one called *stationary vortices,* which occur at moderate speeds and move with the spheres, and the other called *drifting vortices,* which occur at high speeds and are left in the wake of the spheres, traveling

FIG. 98.—Drifting vortices in the wake of a cylinder (turbulent flow). The camera was at rest with respect to the moving fluid. (*After Prandtl-Tietjens.*)

in the same direction as the spheres, but slower. Figures 97 and 98 show stationary and drifting vortices photographed by Tietjens in Prandtl's laboratory at Göttingen, Germany.

Newton's consideration took stock only of the upstream shape of the moving body. The newer researches* have focused attention on the downstream shape of the moving body; it is somewhat belittling to human pride that men have failed for so long in imitating the streamlining of fishes and birds.

Terminal Velocity under Newtonian Resistance. If turbulent resistance prevails, the fundamental equation can be obtained from the second law of motion, as follows:

$$m\frac{dv}{dt} = mg - m'g - R$$

or

$$\frac{4}{3}\pi r^3\Delta\frac{dv}{dt} = \frac{4}{3}\pi r^3(\Delta - \Delta')g - Q\frac{\pi}{2}\Delta' r^2 v^2,$$

which on simplifying becomes

$$\frac{dv}{dt} = \frac{\Delta - \Delta'}{\Delta}g - \frac{3Q}{8}\frac{\Delta'}{\Delta r}v^2. \qquad \text{[VIII.7]}$$

The solution of this equation, page 178, gives the relationship between time and velocity.

The maximum or terminal velocity can be obtained most simply by putting $dv/dt = 0$, whence

$$v_m = \sqrt{\frac{8}{3Q} \cdot g\frac{\Delta - \Delta'}{\Delta'} \cdot r}. \qquad \text{[VIII.8]}$$

The important fact to remember about formulas [VIII.4] and [VIII.8] is that in viscous flow the terminal velocity varies as the square of the particle diameter, and in turbulent flow it varies as the square root of the particle diameter.

Settling of Spheres of Intermediate Size. It is of more than passing interest to note that the range in which neither the Stokes nor the Newton-Rittinger relation applies is precisely that range in which the bulk of gravity-concentrating operations is conducted. Many attempts have been made to devise a formula that fits the facts. One of the proposed formulas, that of Oseen,[(29)] has been widely used,[(14)] but it fails to represent the facts[(47)] even though it is said to have the best of theoretical backings.

* Especially those of Eiffel in France and Prandtl in Germany.

The Oseen formula, usually expressed

$$R = 6\pi\mu r v_m\left(1 + \frac{3}{8}\frac{\Delta' r v_m}{\mu}\right), \qquad \text{[VIII.9]}$$

actually implies that the resistance experienced by the sphere is the sum of two resistances, one Stokesian $6\pi\mu r v_m$ and the other Newtonian $\frac{9}{4}\pi\Delta r^2 v_m{}^2$.

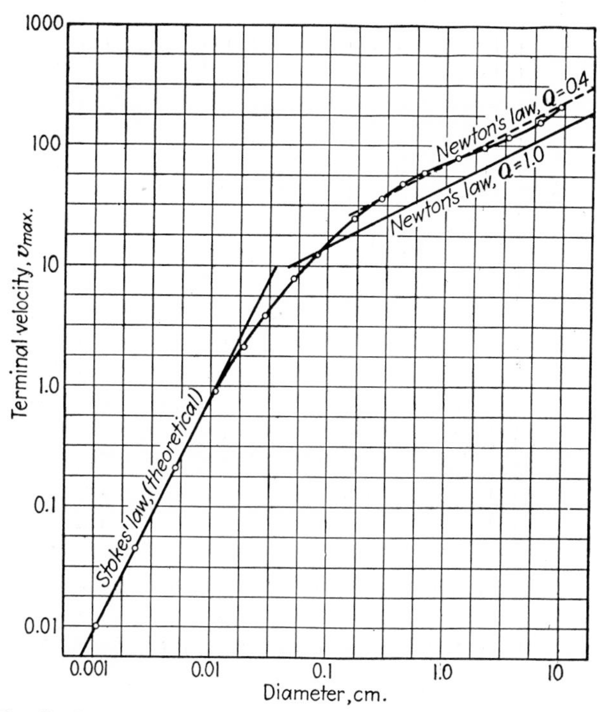

FIG. 99.—Settling velocities of quartz spheres in water, as determined experimentally compared with theoretical settling velocities according to Stokes' and Newton's laws.

The Oseen formula may be regarded as an approximation of a more general formula due to Goldstein.(15) The Goldstein formula comes closer to the facts than Oseen's, but again it fails at moderate and high speeds.

The idea of regarding the fluid resistance as consisting of two resistances, one due to viscous flow, and the other to eddying flow, is expressed very simply by Budryk.(5) Budryk's formula is

$$R = 6\pi\mu r v_m + Q\frac{\pi}{2}\Delta' r^2 v_m^2. \quad \text{[VIII.10]}$$

Unfortunately it also deviates from the facts.

Other formulas[46] that feature a fractional exponent of the velocity have a form that one would not expect to fit the facts.

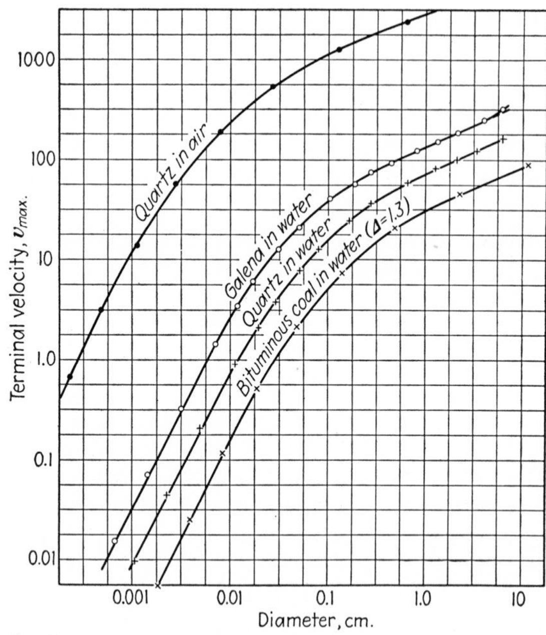

FIG. 100.—Settling velocities of galena, coal, and quartz spheres in water, and of quartz spheres in air.

By far the best-fitting formula published is that of Wadell.[47] It can be put in the form

$$R_{\text{Total}} = (R_{\text{Stokes}}) + (R_{\text{Newton}}) \quad \text{[VIII.11]}$$

which logic requires. Unfortunately, the formula contains a fractional exponent of the velocity and is wholly empirical.

Relationship of Terminal Velocity to Size. Figure 99 shows the relationship between v_m and r for spheres of quartz in water. This figure is based on the best data. It is drawn with log r and log v as coordinates rather than with r and v, since that kind of a plot permits the presentation of the whole range of sizes and velocities. The portions of the curve that fit exactly the Stokes equation and approximately the Newtonian equation are straight lines since in one case log v varies as 2 log r, and in the other as ½ log r. The Newton equation is drawn for $Q = 0.40$. The parallel line ($Q = 1.00$) below the factual curve is that which would be required if the Newton formula were strictly accurate.

Figure 100 shows the relationship between v_m and r for spheres of galena, quartz, and bituminous coal in water, and of quartz in air.

Reynolds Number. A more suitable way of presenting the same data is to make use of the coefficient of resistance Q and the *Reynolds number.* These quantities permit joint charting of data for *all* solids and *all* fluids, a result that it is not possible to achieve otherwise.

The coefficient of resistance Q and the Reynolds number P are defined by the relationships

$$\left.\begin{aligned} Q &= \frac{8r(\Delta - \Delta')g}{3v^2\Delta'} \qquad (a) \\ P &= \frac{2rv\Delta'}{\mu} \qquad (b) \end{aligned}\right\} \qquad [\text{VIII.12}]$$

in which the various symbols have the same meaning as heretofore. Q is of course the same coefficient introduced in the Newton-Rittinger equation to make it fit. In the charting that is to be proposed, the use of Q is extended to conditions under which the Newton equation fails completely.

The Reynolds number is an absolute number, lacking physical dimension. Dimensionally,*

$$P = \frac{2rv\Delta'}{\mu} \equiv \frac{L \cdot \frac{L}{T} \cdot \frac{M}{L^3}}{\left(M\frac{L}{T^2}\right) \cdot T \cdot \frac{1}{L^2}} = L^0M^0T^0$$

It was devised by Osborne Reynolds[35] and has since become an indispensable convenience in fluid dynamics.

Attention should be called to the fact that if P and Q are stated, and for a given pair of substances at a predetermined tempera-

* Radius is length, velocity is length divided by time, density is mass divided by length cubed, and viscosity is force multiplied by time divided by length squared (dyne-seconds per square centimeter).

ture, r and v can be calculated: Eqs. [VIII.12a] and [VIII.12b] are simultaneous equations in r and v from which r and v can be calculated if P, Q, Δ, Δ' and μ are known.

If Stokes' conditions prevail, Q should be proportional to r^{-3} (since v is proportional to r^2), and P should be proportional to r^{+3}, so that $P \cdot Q$ should be independent of r, and therefore also of v. A chart of log Q *vs.* log P then should be a diagonal line of slope -1 if the Stokes relationship prevails. Under conditions of ideal applicability of the Newton-Rittinger relation, Q is constant (and independent of P); the chart of log Q *vs.* log P is then a horizontal line parallel to the log P axis.

Figure 101 is such a diagram reproduced from the 1937 report of the Committee on Sedimentation of the National Research Council.[41] This diagram shows Stokes' law to be *accurately* valid up to $P = 0.6$ and the Newton-Rittinger relationship to be *approximately* valid from $P = 800$ to $P = 200{,}000$ with Q ranging from 0.35 to 0.48. From $P = 0.6$ to $P = 800$, no theoretical formula is valid, and from $P = 200{,}000$ to $P = 300{,}000$, there is an exceptional drop in Q from about 0.40 to about 0.15. This is the phenomenon to which allusion has already been made in connection with Eiffel's and Prandtl's aerodynamic investigations.

The values of P, Q, and v can be determined graphically from Fig. 101 in terms of r, Δ, Δ', μ and g as follows:
From Eq. [VIII.12], v can be eliminated,

$$\log Q = \log \frac{8r(\Delta - \Delta')g}{3\Delta'} - 2 \log v.$$

$$\log P = \log \frac{2r\Delta'}{\mu} + \log v.$$

Hence,

$$\log Q + 2 \log P = \log \frac{8r(\Delta - \Delta')g}{3\Delta'} + 2 \log \frac{2r\Delta'}{\mu}.$$

For given values of r, g, Δ and Δ', this takes the form

$$\log Q + 2 \log P = C,$$

in which C is a constant.

The locus of this line is a straight line of slope -2:

$$\log Q = C - 2 \log P$$

Intersection of this line with the log Q *vs.* log P curve defines P and Q, and hence v.

Figure 101 is of general applicability in dealing with the motions of any spherical solid in any fluid without exception. Minor modifications permit of its application to nonspherical bodies.

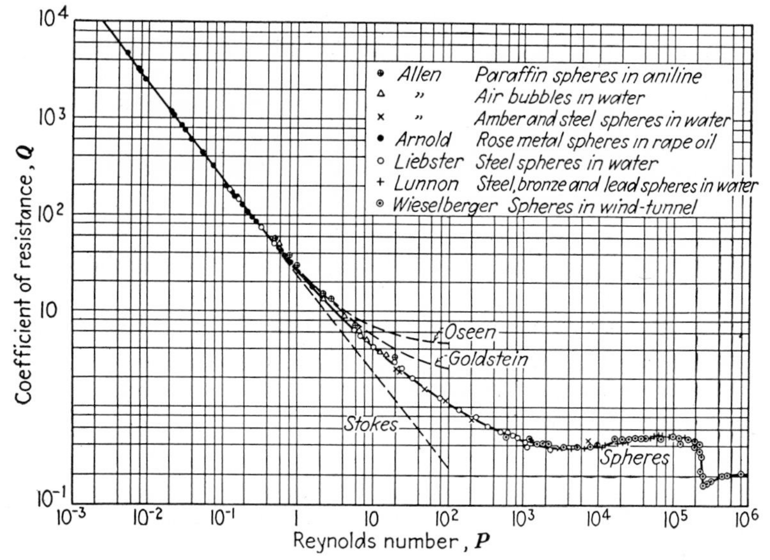

FIG. 101.—Relationship of coefficient of resistance to Reynolds number.

Relationship between Time and Velocity of Moving Solid. The differential equations relating time and velocity of moving solid have already been obtained for both viscous and turbulent resistance (Eqs. [VIII.3] and [VIII.7]) on the assumption that the particles under consideration are starting from rest. It is desired to solve these equations.

Equation [VIII.3] can be written

$$dt = \frac{\Delta}{\Delta - \Delta'} \cdot \frac{dv}{g - \dfrac{9}{2}\dfrac{\mu v}{(\Delta - \Delta')r^2}} = \frac{\Delta}{(\Delta - \Delta')g} \cdot \frac{dv}{1 - \dfrac{v}{v_m}}.$$

This equation is of the form

$$dt = k\frac{dv}{A + Bv}; \qquad \text{or} \qquad t = k\int_0^{v_0} \frac{dv}{A + Bv}$$

in which

$$k = \frac{\Delta}{(\Delta - \Delta')g}, \qquad A = 1, \qquad B = -\frac{1}{v_m},$$

whose general solution[31a] is

$$t = \frac{k}{B} \ln (A + Bv).$$

The definite solution, after all substitutions and satisfaction of boundary conditions, becomes

$$t_x = \frac{\Delta v_m}{(\Delta - \Delta')g} \ln \left(\frac{v_m}{v_m - v_x} \right).$$

If $v_x = xv_m$, t_x can be written more simply,

$$t_x = \frac{\Delta v_m}{(\Delta - \Delta')g} \ln \frac{1}{1 - x}. \qquad \text{[VIII.13]}$$

The Newtonian differential equation [VIII.7] can be solved in similar fashion. It can be written:

$$dt = \frac{\Delta}{(\Delta - \Delta')g} \cdot \frac{dv}{1 - \dfrac{3Q\Delta' v^2}{8(\Delta - \Delta')rg}}$$

$$= \frac{\Delta}{(\Delta - \Delta')g} \frac{dv}{1 - \dfrac{v^2}{v_m^2}} = \frac{\Delta v_m^2}{(\Delta - \Delta')g} \frac{dv}{v_m^2 - v^2}.$$

This equation is of the form:

$$dt = k' \cdot \frac{dv}{C^2 - v^2}$$

whose general solution is [31b]

$$t = \frac{k'}{2C} \ln \frac{C + v}{C - v}.$$

The definite solution, after all substitutions and satisfaction of boundary conditions, becomes

$$t_x = \frac{\Delta v_m}{2(\Delta - \Delta')g} \ln \frac{v_m + v_x}{v_m - v_x}.$$

If $v_x = xv_m$, t_x can be put in the form

$$t_x = \frac{\Delta v_m}{2(\Delta - \Delta')g} \ln \frac{1 + x}{1 - x}. \qquad \text{[VIII.14]}$$

In Eqs. [VIII.13] and [VIII.14], the quantity x is that fraction of the terminal velocity which is reached in the time t_x.

If the velocity is chosen to be some particular fraction of the maximum velocity, say one-half or one-quarter, the time for a given mineral and fluid depends only on r. That particular fraction of the maximum velocity is then reached in a time proportional to v_m, *i.e.*, to the square of the radius if the particle is small or to the square root of the radius if the particle is large. This in turn leads to the conclusion that extremely small particles reach a velocity almost equal to their ultimate velocity in an extremely short time and large particles require an appreciable length of time.

To visualize how the time varies with the velocity relative to the maximum velocity, *i.e.*, to visualize the t, x function in Eqs. [VIII.13] and [VIII.14], it is convenient to introduce as reference time, the time t_h at which the velocity is just one-half of the maximum velocity. This time at which the velocity is one-half of the maximum velocity can conveniently be termed the *half time*. Equation [VIII.13] then takes the form given by the equation pair [VIII.15], [VIII.16]; likewise Eq. [VIII.14] takes the form given by the equation pair [VIII.17], [VIII.18]. For convenience, these equations are expressed in terms of common logarithms instead of natural logarithms.

$$t_h \text{ (Stokes)} = 708 \times 10^{-6} \frac{\Delta}{\Delta - \Delta'} v_m \qquad \text{[VIII.15]}$$

$$\frac{t}{t_h} \text{ (Stokes)} = 3.322 \log \frac{1}{1 - x} \qquad \text{[VIII.16]}$$

$$t_h \text{ (Newton)} = 560 \times 10^{-6} \frac{\Delta}{\Delta - \Delta'} v_m \qquad \text{[VIII.17]}$$

$$\frac{t}{t_h} \text{ (Newton)} = 2.095 \log \frac{1 + x}{1 - x}. \qquad \text{[VIII.18]}$$

Figure 102 shows the curves relating t/t_h to v/v_m for viscous flow and for turbulent flow. The features brought out by this chart are (1) the remarkable similarity of the curves for viscous resistance and for turbulent resistance, (2) the relatively small increase in the time beyond the half time which is required to bring the velocity within reach of the maximum velocity. Thus, the velocity is already within 5 per cent of the maximum velocity if the time is $3.3t_h$ (viscous flow) or $4.4t_h$ (turbulent flow).

Relationship of Half Time to Size. Tables 20 and 21 show the calculated relationship between the half time and the radius for the case of galena, quartz, and bituminous coal settling in water and of quartz settling in air. The values for the half time

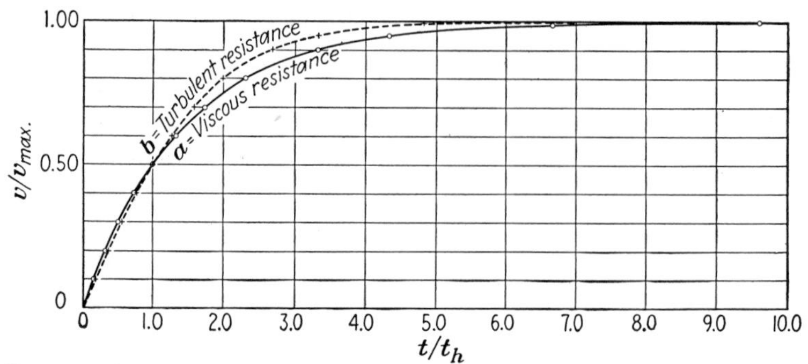

FIG. 102.—Relationship between relative velocity and relative time of settling for spherical particles of any substance and any size in any fluid.

were calculated from formulas [VIII.17] and [VIII.18], but for spheres of low Reynolds numbers the results were checked with formulas [VIII.15] and [VIII.16]. The agreement was good.

TABLE 20.—CALCULATED TIME IN SECONDS REQUIRED FOR SPHERICAL PARTICLES TO REACH ONE-HALF OF THEIR MAXIMUM SETTLING VELOCITY IN WATER AT 20°C.

Reynolds number, P	Coefficient of resistance, Q	Galena $\Delta = 7.50$		Quartz $\Delta = 2.65$		Coal $\Delta = 1.30$	
		Radius, cm. r	Time, t_h	Radius, cm. r	Time, t_h	Radius, cm. r	Time, t_h
100,000	0.32	1.679	0.1935	2.650	0.1698	4.68	0.2600
20,000	0.35	0.592	0.1099	0.934	0.0967	1.650	0.1470
5,000	0.40	0.246	0.0662	0.388	0.0583	0.685	0.0889
1,000	0.50	0.0972	0.0373	0.1534	0.0328	0.271	0.0498
200	0.80	0.0362	0.0180	0.0571	0.0157	0.1008	0.0242
50	1.55	0.0179	0.0091	0.0283	0.0080	0.0500	0.0117
10	4.3	0.0086	0.0038	0.0136	0.0034	0.0240	0.0052
2	15.0	0.0045	0.00146	0.0070	0.00128	0.0124	0.00195
0.5	52.	0.0027	0.00060	0.0042	0.00053	0.0075	0.00082
0.1	250.	0.0015	0.00024	0.0024	0.00018	0.0043	0.00030
0.01	2400.	0.0007	0.00005	0.0011	0.00003	0.0020	0.00004

TABLE 21.—CALCULATED TIME IN SECONDS REQUIRED FOR SPHERICAL PARTICLES OF QUARTZ TO REACH ONE-HALF OF THEIR MAXIMUM SETTLING VELOCITY IN AIR

Reynolds number, P	Coefficient of resistance, Q	Particle radius, centimeters r	Time, seconds t_h
100,000	0.32	1.050	2.67
20,000	0.35	0.370	1.52
5,000	0.40	0.153	0.91
1,000	0.50	0.056	0.50
200	0.80	0.0226	0.247
50	1.55	0.0112	0.125
10	4.3	0.0054	0.052
2	15.0	0.0028	0.020
0.5	52.	0.0017	0.0082
0.1	250.	0.0009	0.0029
0.01	2400.	0.0002	0.0004

Tables 20 and 21 show that the half time is extremely small for small bodies, but appreciable for large bodies. Comparison of the results for quartz in water and in air shows that the greater resistance exerted by the water reduces very considerably the half time. In highly viscous mediums, this reduction is enormously greater. A sphere of quartz 0.01 cm. in radius reaches half the maximum velocity in glucose of viscosity 720 c.g.s. units in about one thirty-millionth part of one second, but a boulder 10 in. across dropped from an airplane would require about 10 sec. to reach half its maximum velocity.

Relationship between Distance Traveled and Velocity. If s is the distance traveled by a moving particle, then

$$v = \frac{ds}{dt}$$

and the relationship between s and v can be obtained by suitable integration of equations [VIII.13] and [VIII.14] after substitution of ds/dt for v. This is useful in jigging. It is therefore carried out, but only for particles coarse enough to settle by eddying resistance.

Equation [VIII.14] is of the form

$$\frac{t}{A} = \ln \frac{1 + x}{1 - x} \qquad \text{or} \qquad e^{t/A} = \frac{1 + x}{1 - x}$$

in which

$$A = \frac{\Delta v_m}{2(\Delta - \Delta')g} \quad \text{and} \quad x = \frac{v}{v_m} = \frac{\frac{ds}{dt}}{v_m}.$$

Hence,

$$e^{t/A} = \frac{v_m \cdot dt + ds}{v_m \cdot dt - ds},$$

or, after assembling the terms in ds and dt,

$$ds = dt \cdot v_m \cdot \frac{e^{t/A} - 1}{e^{t/A} + 1}.$$

Changing the variable by putting $w = t/A$, hence $dt = A\, dw$, we get

$$ds = Av_m \frac{e^w - 1}{e^w + 1} dw.$$

Changing the variable again to $z = e^w$,

$$dz = d(e^w) = e^w\, dw \quad \text{or} \quad dw = \frac{dz}{z}$$

$$ds = Av_m \frac{z - 1}{z + 1} \frac{dz}{z} = Av_m \left[\frac{dz}{z + 1} - \frac{dz}{z(z + 1)} \right].$$

Integrating,

$$s = Av_m \left[\ln (z + 1) + \ln \frac{z + 1}{z} + C \right]$$

$$= Av_m \left[\ln \frac{(z + 1)^2}{z} + C \right] = Av_m \left[\ln \left(z + 2 + \frac{1}{z} \right) + C \right]$$

$$= Av_m [\ln (e^{t/A} + 2 + e^{-t/A}) + C].$$

The boundary conditions are satisfied if $s = 0$ when $t = 0$, hence $C = - \ln 4$, and

$$s = Av_m \ln \frac{2 + e^{t/A} + e^{-t/A}}{4}. \qquad \text{[VIII.19]}$$

Numerical results for quartz and galena spheres settling in water are presented in Fig. 103. In this figure, both scales are logarithmic.

Figure 103 shows that the distance traversed before the settling velocity approaches the maximum velocity increases very rapidly with particle size. This allows the following conclusions:

1. For practical purposes, very small particles settle at the rate of the terminal velocity from the start of their fall; the distance traversed before that velocity is assumed is wholly negligible.

2. This is not true of coarse particles; for very coarse particles it is wholly at variance with the facts.

Effect of Vessel Walls. It has been assumed until now that an endless fluid is concerned. But if a fluid is experimentally available, it must be in a container, and if the walls of the vessel are relatively near the moving solid, the velocity gradient across the fluid cannot be the same as if there were no walls, since the walls compel the velocity of the fluid to become nil at a near, finite distance instead of at infinity. Thus walls introduce an added resistance to the motion, and they impede particle speed.

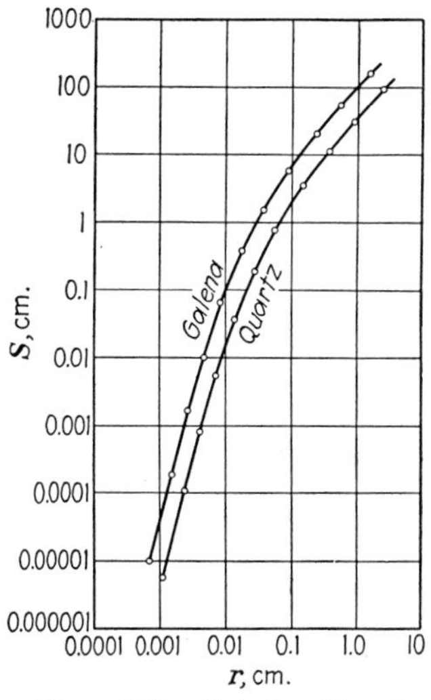

FIG. 103.—Relationship between radius of a settling sphere and distance traveled before the velocity reaches 95 per cent of the terminal velocity (galena and quartz in water).

Another way of viewing the same effect is to realize that it must be much more difficult for a sphere, say 0.9 cm. in diameter, to fall in a fluid through a tube 1 cm. in diameter than to fall through a tube 100 cm. in diameter since the fluid constantly displaced by the sphere must squeeze through an annulus 0.05 cm. wide instead of through an annulus 49.55 cm. wide.

Several corrections to the Stokesian and Newtonian formulas have been proposed to take care of the wall effect, the most widely used being the Ladenburg-Faxén correction.[11,20] This correction is usually formulated as a multiplier for the Stokesian resistance which becomes

$$R = 6\pi\mu r v\left(1 + 2.1\frac{r}{r'}\right). \qquad \text{[VIII.20]}$$

In this equation, r and r' are the radii of the sphere and enclosing cylinder. The other symbols have the same significance as in Eq. [VIII.2].

A correction that seems to fit the facts better in the case of viscous resistance is due to Francis.[13] Francis expresses wall impedance as a factor less than unity by which the viscosity of the fluid must be divided or by which the Stokesian settling velocity must be multiplied.

$$f = \left(1 - \frac{r}{r'}\right)^{2.25}; \qquad \text{[VIII.21]}$$

r and r' are the radii of the sphere and enclosing cylinder. The formula was experimentally verified with glucose and glycerol as fluids.

Table 22 presents a tabulation of the Ladenburg correction factor $f = \left(\frac{1}{1 + 2.1\frac{r}{r'}}\right)$, the experimental factors obtained by

TABLE 22.—LADENBURG AND FRANCIS WALL-CORRECTION FACTORS FOR VISCOUS FLOW

Ratio of radii of sphere and cylinder	Calculated Ladenburg factor	Experimental factor (Francis)		Calculated Francis factor
		For glucose, μ = 720 c.g.s.	For glycerol, μ = 4 c.g.s.	
0.0	1.0	1.0	1.0	1.0
0.1	0.81	0.79	0.79	0.79
0.2	0.67	0.62	0.62	0.61
0.3	0.58	0.44	0.44	0.45
0.4	0.51	0.32	0.30	0.32
0.5	0.46	0.21	0.19	0.21
0.6	0.42	0.14	0.10	0.12
0.7	0.37	0.07	0.045	0.07
0.8	0.33	0.04	0.02	0.03

Francis, and the calculated Francis factor according to Eq. [VIII.21].

The enormity of the disparity between the Ladenburg and Francis corrections at large r/r' ratios should be noted, as well as the excellence of the agreement between experimental results and the Francis equation.

A wall-correction factor was also proposed by Munroe[26] on the basis of experiments dealing with spheres suspended in a rising current of water within a glass tube. The Munroe correction is not strictly comparable with the Francis correction since it deals with conditions of turbulent flow rather than of viscous flow—or at least with conditions in the velocity range characteristic of the transition between viscous and turbulent flow.

Munroe's correction (Table 23) is in the shape of a multiplier $\left(\frac{1}{1-(r/r')^{3/2}}\right)^2$, which is applied to the Newtonian resistance given by Eq. [VIII.5] and which becomes

$$R = \frac{\pi}{2} r^2 v^2 \left(\frac{1}{1 - \left(\frac{r}{r'}\right)^{3/2}} \right)^2. \qquad \text{[VIII.22]}$$

In this relationship, r' denotes the radius of the cylinder. Consequently, the settling velocity becomes

$$v_m = \left[1 - \left(\frac{r}{r'}\right)^{3/2} \right] \sqrt{\frac{4}{3Q} g \frac{\Delta - \Delta'}{\Delta'} r}. \qquad \text{[VIII.23]}$$

The numerical value of the Munroe factor is presented in the following table.

Table 23.—Munroe Wall-correction Factor for Turbulent or Near-turbulent Flow

Ratio of Radii of Sphere and Cylinder	Munroe Factor
0.0	1.00
0.1	0.97
0.2	0.91
0.3	0.84
0.4	0.75
0.5	0.65
0.6	0.54
0.7	0.41
0.8	0.28
0.9	0.15

In making his experiments, Munroe introduced not only a wall effect due to interference of the vessel wall with the velocity gradient across the flowing fluid, but also an additional effect due to the motion of the fluid against the wall. This circumstance raises some questions as to how the correction should be applied, as Munroe himself has pointed out.[26]

Equal-settling Particles. Particles are said to be *equal settling* if they have the same terminal velocities in the same fluid and in the same field of force. Spheres of one substance and of the same size are equal settling. Spheres of different substances may be equal settling if they bear to each other the proper size ratio. Thus spheres of specific gravities Δ_1 and Δ_2 are equal settling if

$$(\Delta_1 - \Delta')r_1^2 = (\Delta_2 - \Delta')r_2^2 \quad \text{under Stokes' law and}$$
$$(\Delta_1 - \Delta')r_1 = (\Delta_2 - \Delta')r_2 \quad \text{under Newton's law,}$$

and more generally if

$$r_1(\Delta_1 - \Delta')^{m_1} = r_2(\Delta_2 - \Delta')^{m_2}. \qquad \text{[VIII.24]}$$

In Eq. [VIII.24], the exponents range from ½ under conditions of viscous resistance to unity under conditions of eddying resistance.

TABLE 24.—CALCULATED FREE-SETTLING RATIOS OF SOME MINERAL PAIRS IN WATER

Mineral pair	Free-settling ratios for			
	$m = 1$ ($2r >$ 0.1 cm.)	$m = 0.85$ ($2r$ about 0.03 cm.)	$m = 0.65$ ($2r$ about 0.01 cm.)	$m = 0.5$ ($2r <$ 0.005 cm.)
Native Gold—quartz (17) (2.65)	9.70	6.91	4.38	3.11
Galena—quartz (7.5) (2.65)	3.94	3.20	2.44	1.98
Cassiterite—quartz (6.9) (2.65)	3.57	2.95	2.29	1.89
Hematite—quartz (5.2) (2.65)	2.55	2.22	1.84	1.60
Sphalerite—quartz (4.1) (2.65)	1.88	1.71	1.51	1.37
Rhodochrosite—quartz (3.5) (2.65)	1.51	1.42	1.30	1.23
Slate—anthracite (2.7) (1.7)	2.43	2.13	1.78	1.56
Slate—bituminous coal (2.5) (1.3)	5.00	3.94	2.85	2.24

The ratio $\rho_f = \frac{(\Delta_1 - \Delta')^{m_1}}{(\Delta_2 - \Delta')^{m_2}}$ is the *free-settling ratio* of spheres of specific gravity Δ_1 and Δ_2 in the fluid of specific gravity Δ'. Generally m_1 and m_2 are very close to each other so that the free-settling ratio can be expressed as

$$\rho_f = \left(\frac{\Delta_1 - \Delta'}{\Delta_2 - \Delta'}\right)^m$$

Table 24 gives the calculated free-settling ratios for a number of common mineral pairs.

Table 25 gives the free-settling ratios for some mineral pairs, as determined by Richards.[38] They differ somewhat from the

TABLE 25.—FREE-SETTLING AND HINDERED-SETTLING RATIOS OF VARIOUS MINERALS, WITH RESPECT TO QUARTZ
(*According to Richards*)

Mineral	Free-settling ratios for 228 mm. per sec. fastest grain	Hindered-settling ratios
Copper	3.75	8.60
Galena	3.75	5.84
Wolframite	3.26	5.15
Antimony	3.00	4.90
Cassiterite	3.12	4.70
Arsenopyrite	2.94	3.74
Chalcocite	2.17	3.11
Pyrrhotite	2.08	2.81
Sphalerite	1.56	2.13
Epidote	1.46	2.04
Anthracite	....	5.61*

* Anthracite to quartz.

calculated ratios, probably because of systematic differences in shape between minerals: native copper is usually rather rounded, galena cubic, quartz very irregular.

Table 26 summarizes from Table 87 in Richards' book[37] the range of free-settling ratios of galena to quartz for particles of various size ranges. It is clear that there is general rather than specific agreement between theory dealing with perfect spheres and practice dealing with particles of several shape habits.

TABLE 26.—FREE-SETTLING RATIOS OF GALENA TO QUARTZ AS EXPERIMENTALLY DETERMINED

(Condensed from Table 87 in R. H. Richards' "A Textbook of Ore Dressing")

Size Range of Galena, Millimeter	Free-settling Ratio Galena-quartz
1.3 and coarser	5.4
0.6 to 1.3	4.7
0.25 to 0.6	4.2
0.09 to 0.25	4.5
0.07 to 0.09	3.6
0.037 to 0.07	2.6
0.015 to 0.037	2.4

Effect of Simultaneous Movement of Many Particles. If a number of spheres of the same size are moving simultaneously against a fluid, conditions are different from those in which a single sphere is moving.

In the first place, and as in the case of a sphere moving in a cylinder of limited cross section, there must be an appreciable displacement of fluid in the direction opposite to the direction of motion of the spheres; in the case of spheres falling in "still" water, the water ceases to be still, even as a first approximation, if there are enough spheres. It is reasonable to consider with Munroe that each sphere moves in a cylinder of appropriate radius, and to then apply the suitable wall correction, *i.e.*, either the Munroe equation [VIII.23] or the Francis equation

$$v_m = \frac{2}{9}\left(1 - \frac{r}{r'}\right)^{2.25}\left(\frac{(\Delta - \Delta')r^2 g}{\mu}\right). \qquad \text{[VIII.25]}$$

The whole difficulty would seem to lie in determination of the appropriate radius, the difficulty arising from the fact that the "tube wall" is a real, stationary boundary in the case of the Francis and Munroe experiments, and an imaginary and moving boundary in the case of application of their results to the movement of particles en masse. This may be seen from Fig. 104 in which vectors are used to represent fluid movement at various points. With reference to Fig. 104*b*, the proper radius for sphere *A* is *AE*, not *AD* or *AF*. But although *AD* and *AF* can be calculated with fair accuracy from the solid content of the suspension, *AE* is difficult to evaluate exactly since determination of the curve *C'D'EF'* depends on too many factors to be easily ascertained. Furthermore, the resemblance of *C'D'E* in Fig.

104*a* to $C'D'E$ in Fig. 104*b* may not be so complete as shown in the diagram.

In the second place, the distance between particles varies all the time so that the actual average distance may not be as assumed from the foregoing observations.

Application of the Munroe and Francis corrections to the motion of suspensoids en masse is therefore nothing more than a first approximation of somewhat questionable validity.

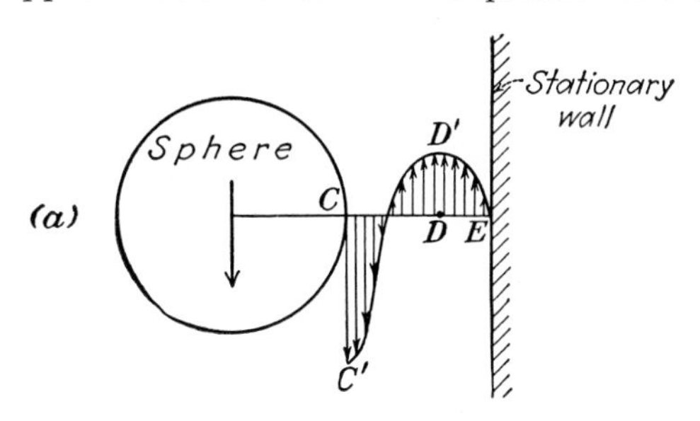

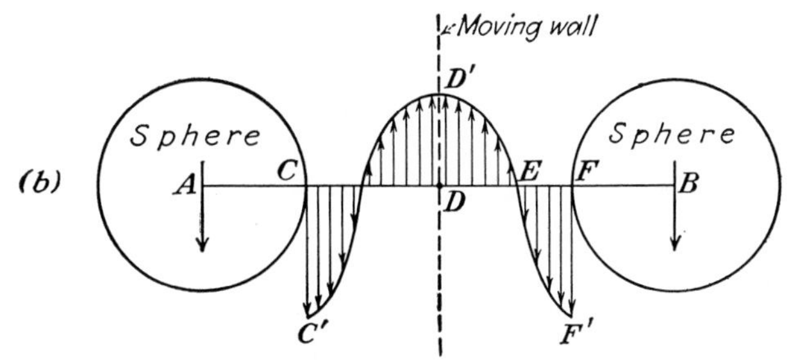

Fig. 104.—Comparison between the effects on a sphere of (*a*) a stationary wall, and (*b*) a moving wall such as is postulated to exist in the simultaneous sedimentation of many spheres.

Experimental work on the settling velocity of dispersoids is not abundant. No data other than Munroe's were found to cover the case of turbulent flow. In what concerns the case of viscous flow, the experiments of Einstein[9,10] Guth and Gold[16] and Kermack, McKendrick, and Ponder[18] are of interest.

Einstein, dealing with dilute suspensions, found that the change in settling velocity may be regarded as due to an increase in viscosity from μ to μ_s.

$$\mu_s = \mu(1 + 2.5\gamma), \qquad \text{[VIII.26]}$$

γ being the volumetric fraction of the suspension occupied by the solid.

Guth and Gold find that this equation fails for thicker suspensions and propose, instead,

$$\mu_s = \frac{\mu}{1 - 2.5\gamma - 14.1\gamma^2}. \qquad \text{[VIII.27]}$$

This equation, however, indicates infinite effective viscosity in a suspension less dilute than 20 per cent solids; and this is not in agreement with the facts.

Kermack, McKendrick, and Ponder express the settling velocity as

$$v_\gamma = v_o(1 - 7.1\gamma). \qquad \text{[VIII.28]}$$

This equation not only indicates lack of settling velocity in a suspension less dilute than about 15 per cent solids by volume, but also belies the careful data of its proponents (see Table 27).

An equation that fits the Kermack-McKendrick-Ponder data is Eq. [VIII.29]

$$v = \frac{2}{9}(1 - \gamma^{2/3})(1 - \gamma)(1 - 2.5\gamma)\frac{g(\Delta - \Delta')r^2}{\mu}. \qquad \text{[VIII.29]}$$

TABLE 27.—SETTLING VELOCITY CORRECTIONS IN TERMS OF VOLUMETRIC ABUNDANCE γ OF THE DISPERSED SOLID

γ	Kermack M'Kendrick and Ponder experiments, average	Kermack M'Kendrick and Ponder equation	Ladenburg equation for wall effect	Francis empirical equation for wall effect	Munroe correction	Equation [VIII.29]
0	100	100	100	100	100	100
0.0025	97	98	77	73	95	97
0.005	94	96	72	66	93	95
0.01	89	93	66	58	90	92
0.02	83	86	60	49	86	86
0.04	73	72	54	39	80	76
0.08	51	43	49	26	72	60
0.16	29	Negative value	45	14	60	35
0.32	13*	Negative value	38	6	43	8

* Results stated to be somewhat uncertain.

This equation, which will presently be discussed, is in better agreement with the facts than any other, as may be seen from Table 27.

In Eq. [VIII.29], the symbols have the same significance as heretofore. It shows three corrections for the Stokesian settling velocity. The first correction is intended to allow for the restriction in effective fluid cross section because of space occupied by solids. This term can be shown to be $(1 - \gamma^{2/3})$. The second correction is intended to allow for the fact that the average specific gravity of the fluid is not Δ' but greater, so that the average spread in specific gravity between solid particles and fluid *as a whole* is not $(\Delta - \Delta')$ but $(1 - \gamma)(\Delta - \Delta')$. It can be shown that the term $(1 - \gamma)(\Delta - \Delta')$ represents this spread in specific gravity, but it is not wholly satisfying from the standpoint of logic to insert the factor $(1 - \gamma)$ in Eq. [VIII.29]. The third correction is similar to the Einstein viscosity correction.

Although not wholly justifiable on logical grounds, Eq. [VIII.29] fits the best data more accurately than any of the other formulas that have been proposed. It is therefore adopted for further use in this text.

Settling of Large Spheres in a Suspension of Fine Spheres. Large spheres settle in a suspension of fine spheres as if the fine spheres were a part of the fluid. This is the principle used in determining the specific gravity of suspensions with a hydrometer; it is also the principle used in cleaning coal with a suspension of sand in water (see Chapter XI).

Thus a large sphere of quartz falling in a suspension of quartz sand having an average specific gravity of 1.7 falls at a speed determined by Eq. [VIII.30], not by Eq. [VIII.8].

$$v = \sqrt{\frac{8g}{3Q}\,\frac{\Delta - \Delta''}{\Delta''}\,r}. \qquad \text{[VIII.30]}$$

A large sphere of quartz in a suspension of fine galena in water of specific gravity 2.8 would rise, and one in a suspension of fine galena in water of specific gravity 2.65 would neither rise nor fall.

Suppose now particles of quartz of various sizes are placed in a suspension of very fine galena in water of specific gravity 2.8. Coarse pieces will rise, as already explained, because the fluid appears to them to have a specific gravity greater than their own. Particles of the same size as the galena also rise because they are slower settling than the denser mineral. It is likely that particles of all sizes will rise in such a suspension, but such an effect has yet to be experimentally established. Crude experiments of this nature are affirmative, but they cannot be regarded as conclusive.

Hindered-settling Ratio. By analogy, the hindered-settling ratio of two minerals may be defined as the ratio of the apparent specific gravities of the mineral *against the suspension* (not against the liquid) raised to a power between one-half and unity. It is given by the following relation:

$$\left.\begin{aligned} \rho_h &= \left(\frac{\Delta_1 - \Delta''}{\Delta_2 - \Delta''}\right)^m \\ &\text{(Newton) } 1 > m > \tfrac{1}{2} \text{ (Stokes).} \end{aligned}\right\} \qquad \text{[VIII.31]}$$

In this equation, Δ'' is the specific gravity of the suspension, not that of the liquid.

A typical example is afforded by a suspension of quartz and water, consisting of 40 per cent solids and 60 per cent water, by volume. Such a quicksand has an effective specific gravity of

$$\Delta'' = \gamma\Delta + (1 - \gamma)\Delta' = (2.65)(0.40) + (1.00)(0.60) = 1.66.$$

The settling ratio of galena to quartz instead of being 3.94 for $m = 1$ and 1.98 for $m = 0.5$ (Table 28) is

$$(7.5 - 1.66)/(2.65 - 1.66) = 5.90 \text{ for } m = 1$$

and 2.43 for $m = 0.5$.

The hindered-settling ratio defined in the fashion in which it is defined in this chapter becomes a powerful tool in the study of classification and of jigging.

Tables 28 and 29 give the calculated hindered-settling ratios for the mineral pairs galena-quartz and slate-bituminous coal for several volume percentages of solid, for several proportions of the minerals, and for several values of the exponent m appearing in Eqs. [VIII.24] and [VIII.31]. Table 25 has already presented experimental results by Richards.

Tables 28 and 29 show (1) a great variation in the magnitude of the hindered-settling ratio in function of changes in percentage solids by volume; (2) an even greater variation in the magnitude of the settling ratio with changes in the proportion of the heavy

TABLE 28.—CALCULATED HINDERED-SETTLING RATIOS*
GALENA–QUARTZ–WATER

Percentage heavy mineral out of total solids	Percentage solids by volume				
	0	10	20	30	40
0	3.94	4.27	4.65	5.2	5.9
	3.20	3.43	3.69	4.07	4.53
	2.44	2.57	2.72	2.92	3.17
	1.98	2.01	2.15	2.29	2.43
25	3.94	4.55	5.5	6.45	10.7
	3.20	3.63	4.26	4.93	7.5
	2.44	2.68	3.03	3.36	4.67
	1.98	2.13	2.35	2.55	3.27
50	3.94	4.9	6.8	12.3	∞
	3.20	3.86	5.1	8.5	↓
	2.44	2.82	3.40	5.1	↓
	1.98	2.22	2.61	3.51	↓
75	3.94	5.3	9.2	82.0	∞
	3.20	4.13	6.6	42.3	↓
	2.44	2.96	4.23	17.5	↓
	1.98	2.31	3.03	9.1	↓
100	3.94	5.85	14.9	∞	∞
	3.20	4.49	9.8	↓	↓
	2.44	3.15	5.8	↓	↓
	1.98	2.42	3.87	↓	↓

* The successive values in each block are for m = 1.0; 0.85; 0.65; and 0.50.

and light minerals; and (3) a number of instances in which the settling ratio is infinite, which means that the light mineral is excluded from participation in the suspension. Thus if a suspension of 40 per cent solids in water (by volume) is made, consisting of galena and quartz, such a suspension cannot persist if the galena is more than about one-third of the total solids. In other words, regardless of the size of the minerals, a suspension of 20 per cent galena and 20 per cent quartz in 60 per cent water (by volume) before it reaches dynamic equilibrium must segregate into two suspensions one of galena and water and the other of galena, quartz, and water. Preliminary experiments conducted

TABLE 29.—CALCULATED HINDERED-SETTLING RATIOS*
SLATE–BITUMINOUS COAL–WATER

Percentage heavy mineral out of total solids	Percentage solids by volume				
	0	10	20	30	40
0	5.00	5.45	6.00	6.72	7.67
	3.94	4.23	4.59	5.05	5.65
	2.85	3.00	3.21	3.45	3.76
	2.24	2.34	2.45	2.59	2.77
25	5.00	6.00	7.67	11.00	21.00
	3.94	4.59	5.65	7.70	13.30
	2.85	3.21	3.76	4.76	7.25
	2.24	2.45	2.77	3.31	4.59
50	5.00	6.72	11.00	41.00	∞
	3.94	5.05	7.70	23.50	
	2.85	3.45	4.76	11.20	
	2.24	2.59	3.31	6.40	↓
75	5.00	7.67	21.00	∞	∞
	3.94	5.65	13.30		
	2.85	3.76	7.25		
	2.24	2.77	4.59	↓	↓
100	5.00	9.00	∞	∞	∞
	3.94	6.48			
	2.85	4.17			
	2.24	3.00	↓	↓	↓

* The successive values in each block are for $m = 1.0$; 0.85; 0.65; and 0.50

in the author's laboratory indicate the correctness of this conclusion, provided the solid content at which the effect is expected is low enough so that the suspension has no plastic properties.

Comparison of Table 29 and Table 30 shows that an enormous difference exists in attainable hindered-settling ratios as between settling in water and in air for the mineral pair coal-slate.

Effect of Particle Shape. The effect of particle shape has been disregarded until now, not because it lacks importance, but because in many ways it is one of the most difficult to evaluate. Natural particles are not only not spherical, but they have irregular shapes with reentrant angles, which seem to defy quantifica-

TABLE 30.—CALCULATED HINDERED-SETTLING RATIOS*
SLATE–BITUMINOUS COAL–AIR

Percentage heavy mineral out of total solids	Percentage solids by volume				
	0	10	20	30	40
0	1.92	2.02	2.16	2.32	2.54
25	1.92	2.05	2.23	2.47	2.82
50	1.92	2.08	2.31	2.61	3.22
75	1.92	2.11	2.40	2.87	3.86
100	1.92	2.14	2.50	3.18	5.00

* For $m = 1.0$.

tion. A fairly successful attempt, however, has already been made by Wadell[47,48] who defines the degree of sphericity ψ of a particle as the ratio of the surface of sphere s which has the same volume as the particle to the actual surface S of the particle.

$$\psi = \frac{s}{S}. \quad \text{[VIII.32]}$$

Before considering Wadell's empirical results, it is of interest to examine the theoretical findings to date. They may be summarized as follows:

1. The resistance of flat plates of infinitesimal thickness at low velocities is $R = 16\mu rv$ if the plate is normal to the fluid flow and $R = 10.67\mu rv$ if the plate is parallel to the fluid flow, as compared with $R = 6\pi\mu rv$ for spheres.[28]

2. At velocities low enough so that there is no eddying, particles have a slight tendency to be orientated parallel to the fluid flow, but this tendency is so slight that it has been asserted that particles retain the same orientation, whether that orientation sets up a large resistance or a small one.

3. At velocities high enough so that eddying resistance is dominant, the particles tend to orientate themselves crosswise to the fluid flow, *i.e.*, in such a way as to cause the greatest possible resistance. This condition is particularly well met if the particle has an axis of symmetry passing through its center of gravity and at right angles to that plane which if placed crosswise to the lines of fluid flow will produce the greatest resistance. Lack of such

an axis of symmetry (the general case) results in greater or lesser vibration or rotation or wobbling, as noted by Richards.[36]

To get an idea of the relative settling velocity of thin disks and a sphere (all of the same substance) moving by viscous resistance, calculations were made by Oberbeck's formula[28] of the relative diameters of disks and spheres that are equal settling in function of the ratio n of the disk thickness to its diameter. The results are presented in Table 31, motion of the disks being considered first in facing orientation, then in edgewise orientation, and finally in average orientation.

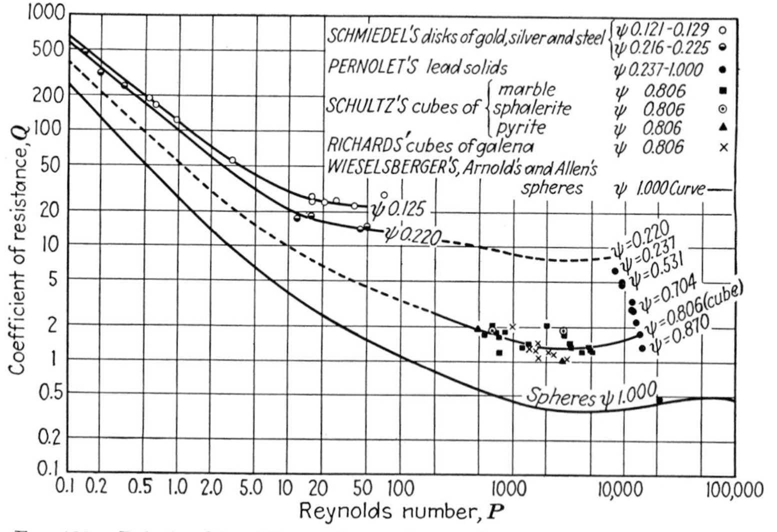

Fig. 105.—Relationship of Reynolds number to the coefficient of resistance for nonspherical bodies. (*After Wadell.*)

Table 31 shows that disks that are equal settling with spheres may appear much larger if viewed under the microscope in a flat-lying position. This effect was experimentally observed in a qualitative way. From a volumetric standpoint, however, the disparity in size between disks and spheres is surprisingly small.

The effect of shape on settling velocity is shown in Fig. 105 in which the coefficient of resistance Q is plotted against the Reynolds number P for values of the latter ranging from 10^{-1} to 10^5. This figure presents for reference the $P - Q$ curve for spheres already included in Fig. 101, and $P - Q$ curves for cubes ($\psi = 0.806$) and for thin plates ($\psi = 0.125$ and $\psi = 0.220$)

TABLE 31.—RELATIVE DIAMETERS $2r/2r'$ OF EQUAL-SETTLING, THIN DISKS AND SPHERES MOVING BY VISCOUS RESISTANCE, IN FUNCTION OF THE RATIO $n = h/2r$ OF THE DISK THICKNESS TO THE DISK DIAMETER

$n = h/2r$	$2r/2r'$ (facing)	$2r/2r'$ (edgewise)	$2r/2r'$ (average)
½*	(1.06)	(0.87)	(0.97)
¼*	(1.50)	(1.23)	(1.37)
1/10	2.38	1.94	2.16
1/20	3.36	2.75	3.05
1/30	4.12	3.37	3.75
1/40	4.75	3.87	4.31

* Such a stout disk does not approach Oberbeck's assumptions, and his equations probably do not do more than give a rough approximation.

experimentally obtained. It also includes a number of points charted from the pioneer data by Pernolet.(32)

Values of v and r can be secured from this figure in the same way as for spheres, using P', Q' values of nonspherical bodies with Eqs. [VIII.12].

To give an idea of the values of ψ for a few common solids, Table 32 has been prepared, and to give some values of the rela-

TABLE 32.—SPHERICITY OF VARIOUS GEOMETRIC BODIES

Body	Sphericity
Sphere	1.000
Cube	0.806
Prism $a \times a \times 2a$	0.767
Prism $a \times 2a \times 2a$	0.761
Prism $a \times 2a \times 3a$	0.725
Disk $h = r$	0.827
Disk $h = \frac{1}{3}r$	0.594
Disk $h = \frac{1}{10}r$	0.323
Disk $h = \frac{1}{15}r$	0.220
Cylinder $h = 3r$	0.860
Cylinder $h = 10r$	0.691
Cylinder $h = 20r$	0.580

tive sizes of equal-settling bodies in function of ψ, Table 33 was prepared.

Table 33 shows that shape affects the settling velocity of large particles to a great degree; on the contrary, its effect is relatively small among fine particles. If the ratio of the disk radius to sphere radius, or *shape-settling ratio*, is regarded as a settling ratio

TABLE 33.—RELATIVE SIZES OF SPHERES, THIN DISKS (ψ = 0.22; r/h = 15) AND CUBES (ψ = 0.806) OF QUARTZ WHICH ARE EQUAL SETTLING IN WATER
(Calculated from Fig. 105)

Sphere, r, cm.	Disk, r, cm.	Cube, ½a, cm.	Radius of disk / Radius of sphere	Face diagonal of cube / Diameter of sphere
0.615	50.4	2.28	82.0	5.24
0.138	9.02	0.378	65.0	3.87
0.040	0.894	0.064	22.4	2.28
0.0134	0.117	0.0183	8.7	1.94
0.0050	0.033	0.0062	6.6	1.72
0.0024	0.0112	0.0025	4.7	1.48
0.00112	0.00396	0.00108	3.5	1.35

somewhat analogous to the free-settling ratio, it is seen that this ratio changes from 82 to 3.5 for very flat disks and spheres of the same specific gravity, as the size is reduced from gravel size to silt size. Even if cubes are compared with spheres, the shape-settling ratio changes from 5.25 to 1.35 as the size of the spheres is reduced from gravel to silt.

The particles currently encountered in mineral-dressing problems have sphericities of the order of magnitude of 0.5 to 0.7. Minerals with strong cleavages may give rise to particles with characteristically larger sphericities (galena) or smaller sphericities (micas).

Literature Cited

1. AHLBORN, F.: Die Theorie der diskontinuierlichen Flüssigkeitsbewegungen und die Wirklichkeit, *Physik. Z.*, **29**, 34–41 (1928).
2. ARNOLD, H. D.: Limitations Imposed by Slip and Inertia Terms upon Stokes' Law for the Motion of Spheres through Liquids, *Phil. Mag.* [6], **22**, 755–775 (1911).
3. BINGHAM, EUGENE C.: "Fluidity and Plasticity," McGraw-Hill Book Company, Inc., New York (1922), p. 55.
4. BRINDLEY, G. W.: Turbulent and Streamline Damping of a Pendulum in Viscous Media, *Phil. Mag.* [7], **12**, 522–534 (1931).
5. BUDRYK, WITOLD: Contribution à la théorie du lavage, *Rev. Ind. Minér.*, **16**, 740–750 (1936).
6. DORR, J. V. N., and E. J. ROBERTS: Sedimentation. The Velocity of Fall of Individual Spheres in Liquids, *Trans. Am. Inst. Chem. Engrs.*, **33**, 106–115 (1937).

7. EIFFEL, G.: Sur la résistance des sphères dans l'air en mouvement, *Compt. Rend.*, **155**, 1597–1599 (1912).
8. EIFFEL, G.: "Nouvelles recherches sur la résistance de l'air et l'aviation," H. Dunod and E. Pinat, Paris (1914).
9. EINSTEIN, ALBERT: Eine neue Bestimmung der Moleküldimensionen, *Ann. Physik* [4], **19**, 289–306 (1906).
10. EINSTEIN, ALBERT: Bemerkung zu der Abhandlung von W. R. HESS "Beitrag zur Theorie der Viskosität heterogener Systeme," *Kolloid Z.*, **27**, 137 (1920).
11. FAXÉN, H.: "Einwirkung der Gefässwände auf den Widerstand gegen die Bewegung einer kleinen Kugel in einer zähen Flüssigkeit," Uppsala (1921), p. 44; also in SCHILLER, ref. 42, p. 343, and in BARR, G. "A monograph of Viscometry," Oxford University Press, London (1931).
12. FINKEY, JOSEF: The Scientific Fundamentals of Gravity Concentration, translated by C. O. Anderson and M. H. Griffitts, *Univ. Missouri, School of Mines and Met. Bull.* 1, *Tech. Series* **11** (1930).
13. FRANCIS, ALFRED W.: Wall Effect in the Falling-ball Method for Viscosity, *Physics*, **4**, 403–406 (1933).
14. GESSNER, HERMANN: "Die Schlämmanalyse," Akademische Verlagsgesellschaft, Leipzig (1931).
15. GOLDSTEIN, S.: The Steady Flow of a Viscous Fluid Past a Fixed Spherical Obstacle at Small Reynolds Numbers, *Proc. Roy. Soc. London* [A], **123**, 225–235 (1929).
16. GUTH, EUGENE, and OTTO GOLD: On the Hydrodynamical Theory of the Viscosity of Suspensions, Preprint for *Physics* (1938) presented at Dec. 1937 meeting American Physical Society. Abstract in *Phys. Rev.*, **53**, 322 (1938).
17. HANCOCK, R. T.: The Teter Condition, *Mining Mag.*, **55**, 90–94 (1936).
18. KERMACK, WM. O., ANDERSON G. M'KENDRICK, and ERIC PONDER: The Stability of Suspensions. III. The Velocities of Sedimentation and of Cataphoresis of Suspensions in a Viscous Fluid, *Proc. Roy. Soc. Edinburgh*, **49**, 170–197 (1929).
19. LADENBURG, RUDOLF: Über die innere Reibung zäher Flüssigkeiten und ihre Abhängigkeit vom Druck, *Ann. Physik* [4], **22**, 287–309 (1907).
20. LADENBURG, RUDOLF: Über den Einfluss von Wänden auf die Bewegung einer Kugel in einer reibenden Flüssigkeit, *Ann. Physik* [4], **23**, 447–458 (1907).
21. LEMIN, C. E.: Motion of a Sphere through a Viscous Liquid, *Phil. Mag.* [7], **12**, 589–596 (1931).
22. LUNNON, R. G.: The Resistance of Air to Falling Spheres, *Phil. Mag.* [6], **47**, 173–182 (1924).
23. LUNNON, R. G.: Fluid Resistance to Moving Spheres, *Proc. Roy. Soc. London* [A], **110**, 302–326 (1926).
24. LUNNON, R. G.: Fluid Resistance to Moving Spheres, *Proc. Roy. Soc. London* [A], **118**, 680–694 (1928).
25. MASON, MAX, and WEAVER WARREN: The Settling of Small Particles in a Fluid, *Phys. Rev.*, **21**, 212 (1923).

26. MUNROE, H. S.: The English *vs.* the Continental System of Jigging--Is Close Sizing Advantageous?, *Trans. Am. Inst. Mining Engrs.*, **17,** 637–659 (1888–1889).
26a. MUNROE, H. S. *Ibid.*, footnote to p. 641.
27. NEWTON, ISAAC: "Mathematical Principles of Natural Philosophy," Book II. Trans. into English in 1729.
28. OBERBECK, A.: Über stationäre Flüssigkeitsbewegungen mit Berücksichtigung der inneren Reibung, *Crelles J.*, **81,** 62–80 (1876).
29. OSEEN, C. W.: Über den Gültigkeitsbereich der Stokesschen Widerstandsformel, *Arkiv. Mat. Astron. Fysik,* **9** [No. 16], 1–15 (1913).
30. PAGE, LEIGH: "Introduction to Theoretical Physics," D. Van Nostrand Company, Inc., New York (1928); (*a*) pp. 239–246; (*b*) pp. 247–248.
31. PEIRCE, B. O.: "A Short Table of Integrals," Ginn and Company Boston (1910); (*a*) formula 26, p. 5; (*b*) formula 48, p. 8.
32. PERNOLET, V.: A l'étude des préparations mécaniques des minerais, ou expériences propres à établir la théorie des différents systèmes usités ou possibles, *Ann. Mines* [IV], **20,** 379–425, 535–596 (1851).
33. PRANDTL, L., and O. G. TIETJENS: "Fundamentals of Hydro- and Aeromechanics," McGraw-Hill Book Company, Inc., New York (1934).
34. PRANDTL, L., and O. G. TIETJENS: "Applied Hydro- and Aeromechanics," McGraw-Hill Book Company, Inc., New York (1934).
35. REYNOLDS, OSBORNE: An Experimental Investigation of the Circumstances Which Determine Whether the Motion of Water Shall be Direct or Sinuous, and of the Law of Resistance in Parallel Channels, *Phil. Trans. Roy. Soc. London,* **174,** 935–982 (1883).
36. RICHARDS, ROBERT H.: Velocity of Galena and Quartz Falling in Water, *Trans. Am. Inst. Mining Engrs.*, **38,** 210–235 (1907).
37. RICHARDS, ROBERT H.: "A Textbook of Ore Dressing," McGraw-Hill Book Company, Inc., New York (1909), p. 268.
38. RICHARDS, ROBERT H.: Development of Hindered-settling Apparatus, *Trans. Am. Inst. Mining Engrs.*, **41,** 396–453 (1910).
39. RICHARDS, ROBERT H., and BOYD DUDLEY, JR.: Experiments on the Flow of Sand and Water through Spigots, *Trans. Am. Inst. Mining Engrs.*, **51,** 398–404 (1915).
40. RITTINGER, P. RITTER VON: "Lehrbuch der Aufbereitungskunde," Ernst und Korn, Berlin (1867).
41. ROUSE, HUNTER: "Nomogram for the Settling Velocity of Spheres, Report of the Committee on Sedimentation, 1936–1937, National Research Council" (1937), pp. 57–64.
42. SCHILLER, LUDWIG: "Handbuch der Experimentalphysik," Vol. IV, Part II, "Hydro und Aerodynamik," Akademische Verlagsanstalt, Leipzig, (1932).
43. SCHMIEDEL, J.: Experimentelle Untersuchungen über die Fallbewegung von Kugeln und Scheiben in reibenden Flüssigkeiten, *Physik. Z.*, **29,** 593–610 (1928).

44. Stewart, R. F., and E. J. Roberts: The Sedimentation of Fine Particles in Liquids. A Survey of Theory and Practice, *Trans. Inst. Chem. Engrs. (London)*, **11,** 124–141 (1933).
45. Stokes, G. G.: "Mathematical and Physical Papers" (1901); also in *Trans. Cambridge Phil. Soc.*, **9,** Part II, pp. 51 *et seq.* (1851).
46. Taggart, Arthur F.: "Handbook of Ore Dressing," John Wiley & Sons, Inc., New York (1927), p. 552.
47. Wadell, Hakon: Some New Sedimentation Formulas, *Physics*, **5,** 281–291 (1934).
48. Wadell, Hakon: The Coefficient of Resistance as a Function of Reynolds' Number for Solids of Various Shapes, *J. Franklin Inst.*, **217,** 459–490 (1934).
49. Zahm, A. F.: Superaerodynamics, *J. Franklin Inst.*, **217,** 153–166 (1934).

CHAPTER IX

CLASSIFICATION

In classification, particles of various sizes, shapes, and specific gravities are separated by being allowed to settle in a fluid. The coarser, heavier, and rounder grains settle faster than the finer, lighter, and more angular grains. The fluid is in motion, carrying away the slow-settling grains while a sediment of fast-settling grains is removed simultaneously from the classifier.

Water or air is used as fluid, and the size at which a separation is made ranges currently from 20- to 300-mesh. Occasionally sizing at the equivalent of 400- to 600-mesh, or even finer, is attained, but that is rare. Sizing coarser than 20-mesh was common in the days when gravity concentration was more widely used than now. But it is now rare, largely because of the disuse of gravity concentration.

Classification may be regarded as based principally on Stokes' law of sedimentation. The agreement, in fact, is better than appears to be the case inasmuch as the suspensions have an appreciable solid content, and therefore an increased specific gravity and an increased viscosity. Both of these effects result in a decreased settling velocity and decreased importance of eddying resistance.

Settling Velocity in a Classifier. It will be remembered that fine particles settle at a velocity directly proportional to the square of their diameter, directly proportional to their apparent specific gravity, and inversely proportional to the viscosity of the fluid. It will be recalled, also, that since classification is carried out in a relatively thick suspension the settling velocity is reduced in proportion to a correcting factor f_1, which can be expressed as (see Eq. [VIII.29])

$$f_1 = (1 - \gamma^{2/3})(1 - \gamma)(1 - 2.5\gamma), \qquad [IX.1]$$

in which γ is the fraction of the volume of the suspension occupied by the solid.

Another factor is required to take care of the irregularity in the shape of the particles. A relationship exists between shape and settling velocity (see Chapter VIII, page 194). It can be expressed as the ratio of the "size" (defined as the diameter of a sphere of equal volume) of equal-settling particles, one of which is irregular in shape, and the other spherical. This ratio is 1.00 for spheres, 1.19 for cubes compared with spheres, and 1.28 for very thin disks compared with spheres.* It is clear then that in the case of particles fine enough to settle by viscous resistance little variation occurs in the shape-correction factor. A ratio of 1.24 can be selected as a good average for irregular particles whose degree of roundness, or sphericity (page 197), is intermediate between that of cubes and of very thin disks. Since the velocity depends upon the size squared, the correction factor f_2 is approximately

$$f_2 = \frac{1}{(1.24)^2} = 0.65. \qquad \text{[IX.2]}$$

TABLE 34.—NUMERICAL FACTOR FOR CALCULATING SETTLING VELOCITY IN A CLASSIFIER AT VARIOUS PULP DILUTIONS AND TEMPERATURES

$F = f_1 \cdot f_2 \cdot f_3$; $f_1 = (1 - \gamma^{2/3})(1 - \gamma)(1 - 2.5\gamma)$, in which γ is the fraction of the suspension volume occupied by solids, $f_2 = 0.65$; $f_3 = \frac{2}{9} 980\mu^{-1}$, in which μ is the viscosity of water

Temperature, °C......	5°	10°	15°	20°	25°	30°	35°
Viscosity of water.....	0.0152	0.0131	0.0114	0.0100	0.0089	0.0080	0.0072
γ, %							
0	9,310	10,800	12,420	14,150	15,900	17,700	19.650
0.5	8,810	10,300	11,850	13,500	15,150	16,900	18,750
1	8,560	9,940	11,400	13,020	14,630	16,300	18,050
2	8,020	9,310	10,720	12,160	13,700	15,250	16,950
3	7,540	8,750	10,050	11,460	12,880	14,340	15,900
4	7,100	8,240	9,480	10,800	12,130	13,500	15,000
6	6,240	7,240	8,320	9,480	10,650	11,850	13,150
8	5,580	6,470	7,440	8,480	9,530	10,600	11,770
10	4,920	5,720	6,580	7,480	8,420	9,370	10,390
12	4,340	5,030	5,790	6,600	7,410	8,240	9,160
14	3,800	4,410	5,070	5,780	6,490	7,220	8,020
16	3,310	3,830	4,410	5,020	5,640	6,280	6,980
18	2,860	3,320	3,820	4,340	4,880	5,440	6,030
20	2,470	2,860	3,290	3,750	4,210	4,690	5,210

* These results are calculated from Table 33, Chapter VIII.

If, furthermore, the viscosity of the liquid and the acceleration of gravity are viewed as constituting a third factor f_3, the velocity (in centimeters per second) can be expressed as

$$v = F(\Delta - \Delta')r^2 \qquad \text{[IX.3]}$$

in which

$$F = f_1 \cdot f_2 \cdot f_3. \qquad \text{[IX.4]}$$

Table 34 gives the value of F for various suspensions at various temperatures. This table brings out that changes in pulp density are more significant in respect to settling velocity than changes in viscosity. The well-known fact that the size of overflow in a Dorr classifier is increased by speeding up the rakes—an action that results in maintaining a denser suspension—is directly attributable to an increase in γ.

The factors presented in Table 34 do not give settling velocities that agree exactly with those based on screen sizing, but that is because particles carefully screen sized by woven-wire screens are coarser than would be judged from the edges of the square *apertures.** Agreement with Table 34 is obtained if screen-sized particles are measured under the microscope.

* The following experiment was carried out: Deslimed pyrite was screened for 10 min. in a Ro-tap machine using a nest of Tyler sieves. Grains $-150 + 200$-mesh were spread on a glass slide and measured in their *a*- and *b*-orientations with a calibrated micrometer eyepiece. The *c*-dimension was also measured by focusing on the top and bottom of each grain. Average determinations on 25 particles were as follows:

a—length	169 microns
b—breadth	121 microns
c—thickness	76 microns
$\sqrt[3]{abc}$—average size	116 microns

It is interesting to note that the openings of the 150- and 200-mesh screens have projections in the plane of the screen of 74 and 104 microns, indicating an expectable average size of 89 microns. The dimension of the projection in the plane of the screen of the diagonal of the opening is, respectively, 104 and 147 microns, average 126 microns.

Explanation of the discrepancy is partly to be sought in the irregular shape of the particles, which can then work themselves around a corner, in the fashion in which a large table may be made to pass through a small door, and partly in that the opening in a screen has a skewed shape imposed on it by the double crimping of the wire. Indeed, it may be calculated that the opening of the 200-mesh screen at a slant is nearer 84 than 74 microns at the edge of the hole.

Desirable Classification Conditions. The most desirable conditions for classification depend on the use that is to be made of the classifier products.

If the classifier products are to be subsequently concentrated by tabling, it is advantageous to emphasize the effect of differences in specific gravity of the minerals. In such a case, classification is not strictly a sizing operation but rather that step in a concentrating operation which is called *sorting*. To emphasize the effect of specific gravity, it is desired to use hindered-settling conditions as much as possible, *i.e.*, to use as dense a suspension as possible.

If, on the other hand, the classifier is operated merely as a sizing device to assist a grinding mill, the effect of differences in specific gravity should be minimized as much as possible. This is accomplished by approximating free-settling conditions, *i.e.*, by using dilute suspensions.

If sizing is at a very fine size, dilute suspensions offer the additional advantages of increased settling rate and of decreased tendency to flocculate (Chapter XIV). In that case, very cold aqueous suspensions should be avoided, although it is not economic to heat pulps artificially.

Although dilution is metallurgically advantageous for true sizing, it is costly, as it connotes use of a large quantity of water for each unit of solid handled and a proportionately larger plant all around.

In practice, sizing classifiers are operated at dilutions ranging from a solid content of 3 to 4 per cent by weight (γ = 1 to 2 per cent) if sizing is at the extreme fine end of the practical range of classifiers, up to 30 or 35 per cent by weight (γ = 12 to 18 per cent) if sizing is at the extreme coarse end. Sorting classifiers require a solid content as high as possible. It may range from 40 to 70 per cent by weight, depending upon the specific gravity and size of the solids.

CLASSIFIERS

Classifiers fall into three broad classes:

1. Sorting classifiers using a relatively dense aqueous suspension as the fluid medium.

2. Sizing classifiers using a relatively dilute aqueous suspension as the fluid medium.

3. Sizing classifiers using air as the fluid medium.

Sorting Classifiers. In sorting classifiers, the settling conditions are more or less hindered. The separation achieved is

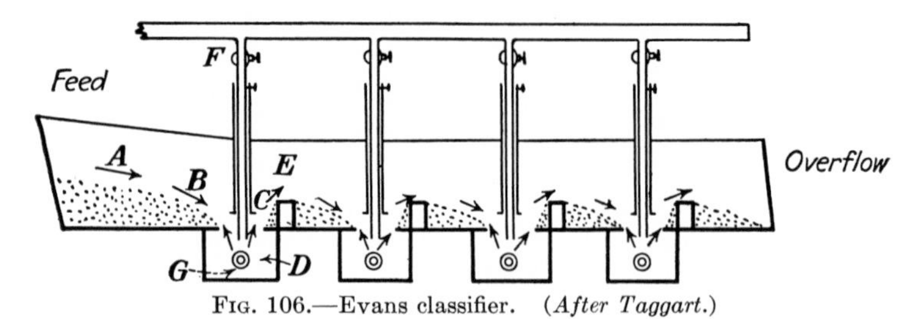

FIG. 106.—Evans classifier. (*After Taggart.*)

sorting, *i.e.*, sizing modified by specific gravity and shape; it is usually applied to relatively coarse products.

A simple type is the launder classifier, an example of which is the *Evans classifier* (Fig. 106). This consists of a sloping launder *A* to which are attached several rectangular boxes opening into

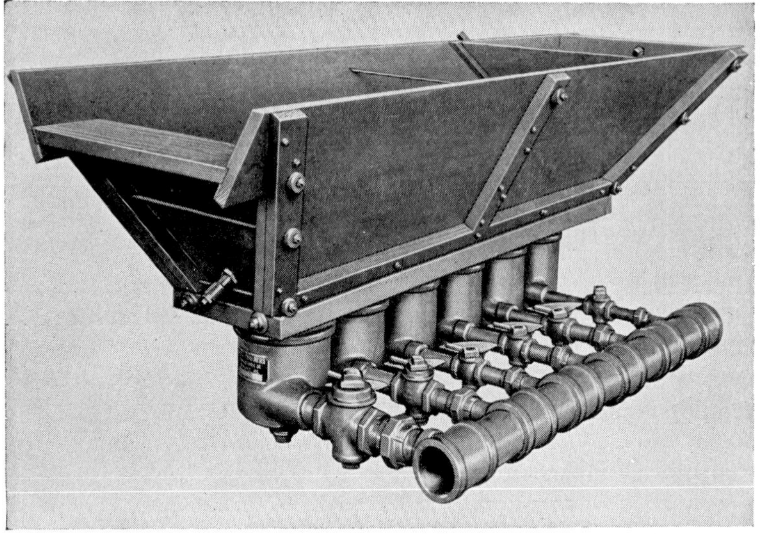

FIG. 107.—Richards hindered-settling classifier. (*Allis-Chalmers Mfg. Co.*)

the launder at *BC* and also capable of discharging through spigots *G*. Water is introduced through pipes which are controlled by valves at *F*. The faster settling particles discharge through the spigot *G*, and the slower settling particles overflow at E back into the launder. The water introduced through F is redundantly known as *hydraulic water*.

More elaborate types include the *Anaconda classifier*[1,10,18] and the *Richards hindered-settling classifier* (Fig. 107). In the Richards classifier, cylindrical sorting columns replace the rectangular boxes of the Evans classifier; the hydraulic water is fed from below into the cylindrical sorting column, and from there into an inner conical column through tangential and radial ports.

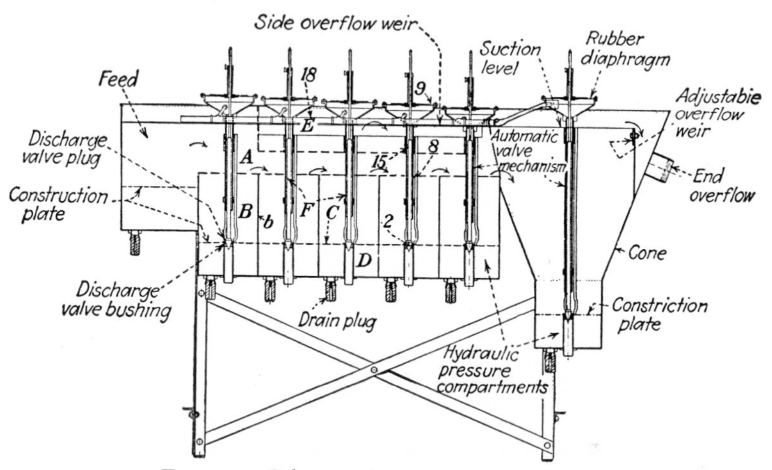

FIG. 108.—Fahrenwald sizer. (*Dorr Company.*)

A relatively recent development along the same lines, but with improved means for spigot discharge, is the *Fahrenwald sizer* (Fig. 108). The Fahrenwald sizer consists of a trapezoidal tank A, fitted with five rectangular classifying pockets B and one cylindrical pocket. Each rectangular pocket and the cylindrical pocket yield a spigot product whose size decreases gradually from the first rectangular pocket to the cylindrical pocket. The spigot is allowed to discharge whenever the head exerted by the tetering column exceeds a certain predetermined value. The water column in the standpipe ⑧ varies with the density of the tetering column, thus pushing up and down the rubber diaphragm ⑨ and the plug ② which is fastened to the diaphragm. Adjustment of the critical teter column density is possible by screwing the collar ⑮ up or down. Hydraulic water is supplied from a constant-head tank through a header ⑱ into the hydraulic-pressure compartments D, and from them into the teter column B, through the perforated constriction plate C. A trickle of water is also allowed to flow along each standpipe F to keep it clear of grit. Glass windows make observation of the teter columns possible.

The *Richards pulsator classifier* is characterized by the use of an intermittent or pulsating upward current of water designed to make settling as completely hindered as possible.

The *Richards-Janney* classifier, extensively used in the porphyry-copper mills, when gravity concentration was used in

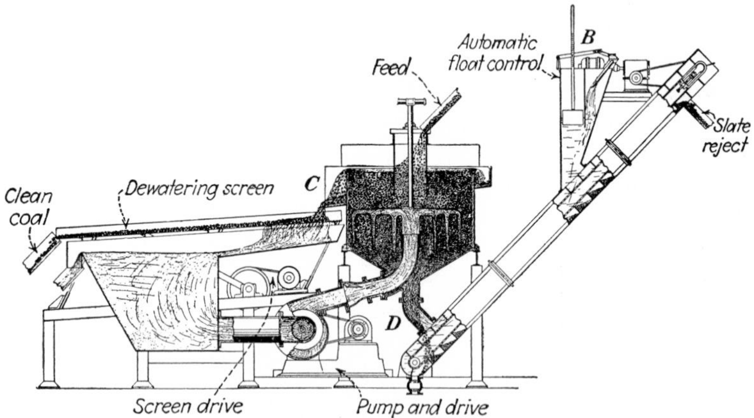

Fig. 109.—Hydrotator classifier used for cleaning coal. (*Wilmot Engineering Co.*)

those plants, does away with one of the annoying drawbacks of some of the other sorting classifiers, *viz.*, the accidental packing of the sorting column. This effect is avoided in the Richards-Janney classifier by the use of a positive-drive raking and stirring mechanism which stirs each teter column one or two times per minute.

In the *Hydrotator classifier*, a hindered-settling zone at the bottom of the classifier supplements a free-settling zone at the top. These zones are not brought about by a constriction of the bottom of the classifier, as in the preceding examples, but by an increase in velocity of flow in the bottom zone without increase in the top zone. Figure 109 shows a Hydrotator arrangement of recent design which includes a pulp-density-control mechanism

B and auxiliary devices for removal of slime particles from overflow *C* and underflow *D*. The type of Hydrotator used in this illustration is intended for coal cleaning, but other modifications are also available. The cardinal feature of the Hydrotator is the continual circulation through the machine of a relatively large quantity of fine mineral particles by use of an outside pump discharging through an arrangement reminiscent of a self-actuating revolving lawn sprinkler. In older machines, this revolving part was self-actuated, but as it frequently became stuck later models have been equipped with a positive drive. Discharge of the sediment takes place whenever the pulp density exceeds a predetermined figure. Control is by means of a float, which actuates a valve. Removal of very fine particles from the overflow is obtained by wet screening.

Sizing Classifiers. Sizing classifiers do not require additional water besides that present in the suspension which is being treated. They utilize free-settling conditions to effect sizing as much as possible unaffected by specific gravity and shape.

Sizing classifiers may be subdivided into settling cones which have no moving parts and mechanical classifiers which have moving parts.

Settling cones are conical sheet-metal shells with the apex at the bottom and a peripheral overflow launder. Feed intake is at the top (center), inside of a small cylindrical or cylindroconical bottomless shell whose function is to prevent by-passing of the feed to the overflow. Spigot discharge is through the bottom. A gooseneck pipe of adjustable height is provided to guide the sediment away from the tank.

In the *Allen Cone* (Fig. 110), the discharge is automatic and involves, of course, some moving parts. The float *F* situated within an inner cylindroconical shell *C* surrounding the feed shell *A* and the baffle *B* is working against a spring to keep the spigot *J* closed, but when the level of the sediment *E* rises sufficiently to prevent passage of pulp from the feed shell *C* to the body of the classifier, or when the density of the suspension is raised sufficiently, the float is raised and opens the spigot, only to close it again as the level or density is lowered. Regulation of the density of the spigot product is obtained by adjustment of the position of the weight *K*.

Mechanical Classifiers. Mechanical classifiers have gained importance with the increasing use of flotation as a dressing method.

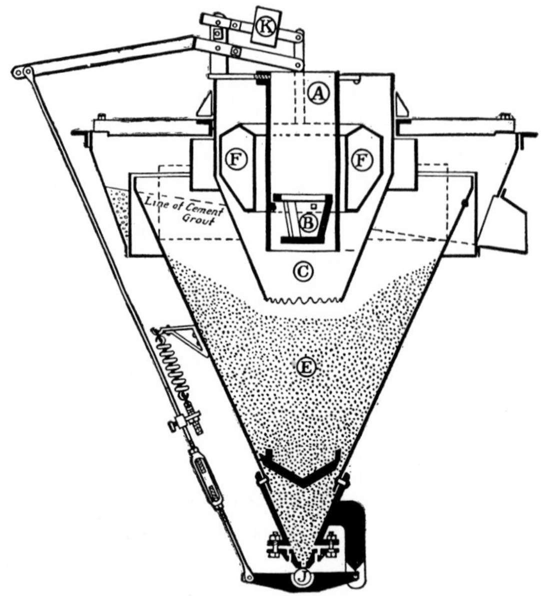

FIG. 110.—Allen sand cone. (*Allen Cone and Manufacturing Co.*)

The *Esperanza classifier* consists of a long sloping trough on the bottom of which the coarser and heavier grains settle and the finer

FIG. 111.—Side view of a quadruplex Dorr F classifier. (*Dorr Company.*)

and lighter grains overflow a terminal weir. In the trough, there revolves an endless belt fitted with scraping flights. This belt drives the settled material up the slope of the trough at the upper end of which it is discharged continuously.

The *Dorr classifier* produces the same type of separation, *viz.*, overflow of the fine particles and upslope dragging of the coarse particles along the bottom of a sloping trough (Fig. 111), but the dragging of the sand is intermittent instead of continuous.

The classifier tank, usually constructed of steel but sometimes of wood, concrete, or special metals, is an inclined, rectangular settling box. The upper or sand-discharge end is open, and the lower or slime-overflow end is partially closed by a tail board and overflow lip. In various installations, the inclination of the bottom of the tank is 1½ in. or more per foot, depending to a certain degree on the size at which a separation is to be made. Feed enters through either end of a transverse trough near the overflow which is equipped with splitter vanes for assuring even distribution across the full width of the tank.

The raking mechanism, usually of iron and steel, but sometimes of wood or special metals, consists of one or more mechanically operated rakes, swung from two hangers—one at the overflow- and the other at the sand-discharge end—and actuated by the head motion. The head motion, situated at the sand-discharge end of the classifier, actuates the raking mechanism through a system of heavy gears, pinions, cranks, and eccentrics with the necessary connecting links and hangers.

This head motion imparts to the rakes a motion that traces in the vertical plane a rectangular "indicator diagram" the upper corners of which are rounded. Just before the raking stroke begins, the rake blades are dropped vertically into the sand which they then convey forward toward the discharge end of the classifier. At the termination of the raking stroke, the blades are lifted and returned to the point where they are again dropped into the sand, preparatory to the next raking stroke.

In Duplex, Triplex, and other multiple-rake units, the sequence of operation of the various rakes is staggered so that they operate out of phase to reduce vibration and improve dynamic balance.

A variant of the Dorr rake classifier is the *Dorr bowl classifier* (Fig. 112) in which a larger settling tank is provided. The settling tank consists of a shallow cylindrical upper part and a lower part in the shape of a flat cone. A scraping member revolves gently in the settling tank, bringing the sand to the center where it falls into the trough of the classifier, whence it is scraped up to the discharge lip by the usual Dorr mechanism. The slime overflows from the periphery of the bowl. The Dorr bowl classifier with its much larger settling area is designed for separation at a finer size than the Dorr rake classifier; the long overflow weir and relatively large settling tank reduce the amplitude of the pulp surges (resulting from immersion and emergence of the rakes) and yield a more accurately sized overflow. Both the Dorr rake and Dorr bowl classifiers are very widely used. Water is sometimes added in the drag section to keep slimes out of the rake product.

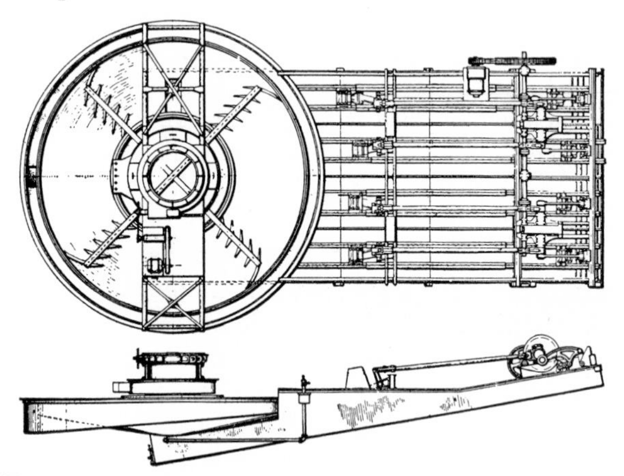

FIG. 112.—The Dorr turret-bowl classifier. (*Dorr Company.*)

In the new *Dorr multizone classifier*[5,8] (Fig. 113), two settling zones are utilized, a hindered-settling, or velocity, zone near the rakes and a freer settling, or quiescent, sorting zone in annular spaces above the hindered-settling zone. The lower end of the usual classifier is decked over and is surmounted by one or more circular overflow columns. Feed is introduced near the middle of the machine and passes downward along the underside of the deck to the base of the overflow columns. Interchangeable cylinders in the overflow column regulate the size of the coarsest particles in the overflow by increasing or reducing the ascending fluid velocity. This classifier is especially suitable for relatively coarse sorting.

A somewhat different type of mechanical desliming classifier is what might be called the spiral-ribbon classifier, typified by the

Akins classifier (Fig. 114) and the *Hardinge classifier*.[6] In the Akins classifier, there is the usual sloping trough of mechanical classifiers in which the pulp is kept gently agitated by one or more spiral ribbons mounted on a shaft. This spiral ribbon acts much as do the rakes in the Dorr classifier and the scraping flights in the case of the Esperanza classifier, not only to remove the settled material from the bottom of the tank, but also as an elevator for that material. In the Akins classifier, the settled solids are turned over and over before they are finally discharged; thus a good opportunity is presented for complete desliming of the sands. Metallurgical results are variously stated to be somewhat

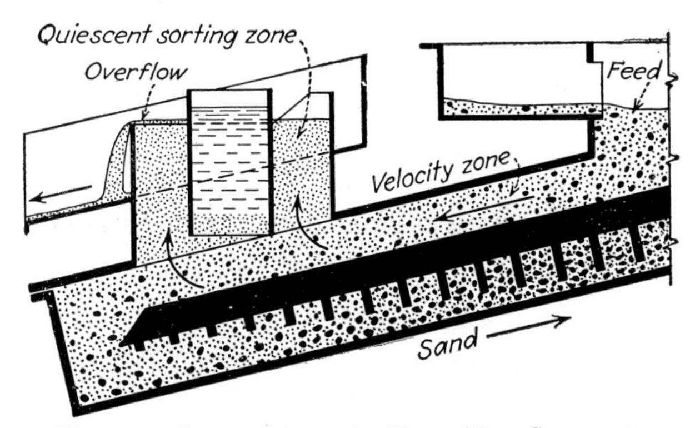

Fig. 113.—Dorr multizone classifier. (*Dorr Company*.)

inferior to somewhat superior to those attained with the Dorr classifiers.

The *Dorrco sand washer* (Fig. 115) is designed primarily for the desliming of common sand. Its principle is similar to that of the various mechanical classifiers and of the *sand wheel*.[21] In this case, however, the scraping mechanism consists of curved blades or buckets arranged in a circle about an inclined shaft. Motion of the scraper is continuous.

For very fine sizing in water, mechanical classifiers of the rectangular-trough type are not suitable, as they cause too much agitation. Suitable classifiers are of the tank type and include (1) a scraping mechanism at the bottom for the removal of the sediment, (2) overflow of the slime at the top periphery, and (3) feeding at the top center. Examples are the Allen slime classifier, Hydrotator slime classifier, Dorr bowl classifier, and

Dorr hydroseparator. The *Dorr hydroseparator* is substantially an undersized Dorr thickener (Chapter XX). It is said by the

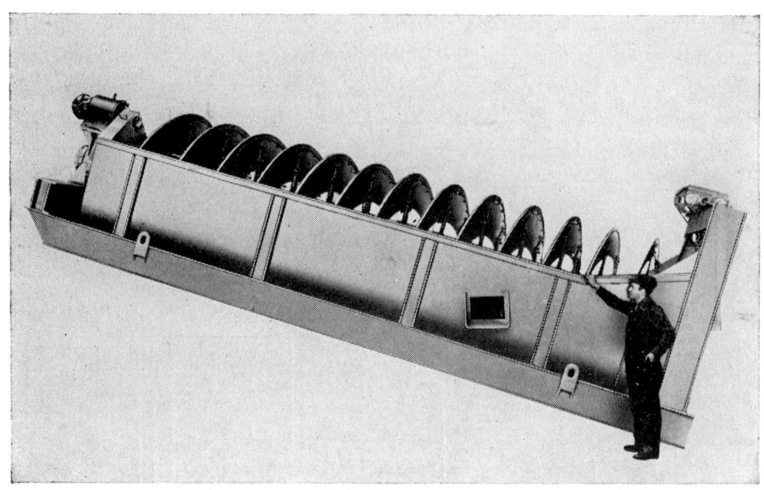

Fig. 114.—Akins classifier. (*Colorado Iron Works Co.*)

Fig. 115.—The Dorrco sand washer. (*Dorr Company.*)

makers to be capable of making separations at a size as fine as 5 microns (3,000-mesh).

One of the reasons for the wide use of mechanical classifiers of the Dorr and the Akins types is that they act not only as classifiers but also as elevators for the sands. Both of these machines are designed so that the sand is discharged higher than the pulp is received, thus compensating for the loss of head in going from the grinding mill to the classifier and from the classifier to the mill. This makes it possible to use mechanical classifiers in

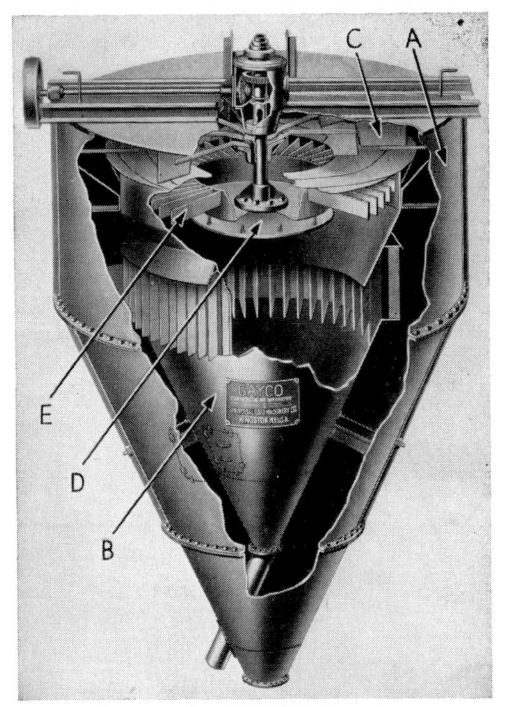

FIG. 116.—Sectional view of Gayco pneumatic classifier. (*Universal Road Machinery Corp.*)

closed circuit with grinding mills without elevating. Such a satisfactory arrangement is not possible with hydraulic classifiers.

Pneumatic Classifiers.[15,19] The medium used in penumatic classifiers is fifty to one hundred times less viscous than water. In addition, the apparent specific gravity of any pneumatic suspension is appreciably less than that of an aqueous suspension of equivalent volumetric composition. Both of these factors make for much higher settling velocities in air than in water. As a generalization, it might well be said that the settling velocity in pneumatic classifiers is roughly one hundred times greater than in water classifiers. Consequently, higher fluid speeds are the rule, and the suspensions are volumetrically more dilute.

Pneumatic sorting classifiers have not been designed, and the sizing classifiers are applied to fine particles only. Pneumatic sizing classifiers are used in conjunction with dust collectors (the equivalent of thickeners and filters) since some disposal must be made of the undersize.

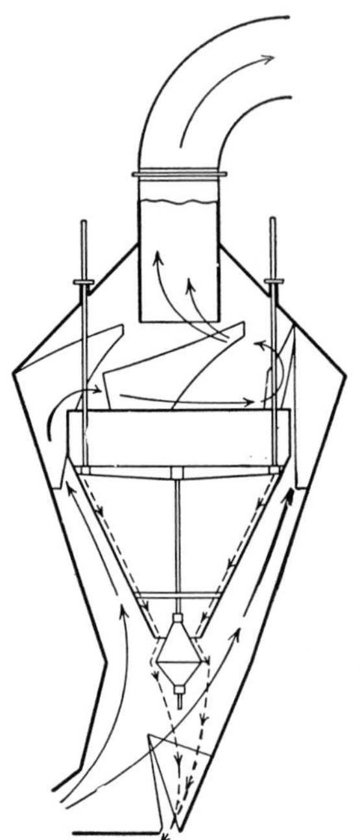

FIG. 117.—Hardinge pneumatic classifier. (*Hardinge Co.*)

The *Gayco air separator* (Fig. 116) consists of two concentric cylindroconical shells *A* and *B*, a circulating fan *C*, and a rotating feeding apron *D*. The outer shell is closed; the inner shell is open at the top and middle so that an ascending closed-circuit current of air can pass up the annular space between the inner shell and the feeding apron; the air current is of course produced by the fan.

In operation, the powdered solid is introduced as a constant stream through a hollow shaft on the revolving feeding apron. The centrifugal action of the apron throws the material radially across the ascending air stream. A centrifugal motion is imparted to the air stream by a rotating fan *E*. The net effect of the various forces acting on the suspension is to allow sedimentation of the coarser grains in the inner shell *B*, while the finer grains settle in the outer shell *A*. Centrifugal force is said to be an important factor in accomplishing the latter action.

The *Raymond Whizzer* separator is similar to the Gayco.

In the *Hardinge pneumatic classifier* (Fig. 117), a dilute suspension of solids in air is made to ascend in the annular space between an inner conical shell and an outer conical shell. At the top of the inner shell and fastened to the outer shell are stationary vanes partly radial and partly curved. Coarse particles fall within the inner shell, and they pass out across the annular flow of air at the bottom of the classifier. There is considerable recirculation of material of intermediate size. Regulation is obtained by raising or lowering the inner cone. Raising the inner cone increases the cross section of the annulus, hence reduces the velocity of the rising current. At the same time, it brings the suspension into the curved part of the guiding vanes, so that the centrifugal effect is magnified. Finally, the distance that the sediment has to cross the annular air flow is increased. All of these changes result in a reduction in maximum particle size in the undersize product. Conversely, lowering of the inner cone makes for a coarser undersize.

PERFORMANCE OF CLASSIFIERS

Capacity. The capacity of a classifier is directly proportional to (1) the cross-sectional area of the sorting column, (2) the rising velocity of the fluid (water or air) in the sorting column, (3) the solid content in the classifier intake.

$$C = aAv\gamma\Delta, \qquad \text{[IX.5]}$$

in which C is the tons of solid per hour, A the cross-sectional area in square feet, v the velocity in feet per minute, γ the percentage of solids by volume, Δ the specific gravity of the solids, and a the constant $(62.5 \times 60)/2{,}000 = 1.875$ required to secure C in terms of tons of solid per hour.

Thus, if a classifier has a cross-sectional area of 10 sq. ft., if the rising velocity of the fluid in the sorting column is 1.5 ft. per min. (equivalent to the settling velocity of 100-mesh quartz at 20°C. in a suspension containing 8 per cent solids by volume), the capacity is

$$C = \frac{(10 \times 1.5 \times 0.08 \times 2.65 \times 62.5 \times 60)}{2{,}000} = 5.96$$

tons of solids per hour.

Practical capacity in machines of the tank type such as hydroseparators is in agreement with the preceding formula.

In machines of the trough type such as the Dorr or Akins classifiers, the practical capacity is appreciably smaller if the full cross-sectional area of the settling trough is considered. Perhaps a part of the discrepancy arises from the fact that the solid content of the suspension is variable, but in all cases appreciably larger than the solid content of the inflowing pulp.

The capacity of trough-type classifiers is limited, from a practical standpoint, not only in accordance with the rate of overflow, but also in accordance with the sand load. A machine of this type may be considered as a combined classifier and conveyor, with the result that the ultimate capacity is controlled by either the classifier part or the conveyor part, depending upon circumstances. For detailed data, the reader is referred to catalogues of the classifier manufacturers and to handbooks.

Efficiency.[14,22] The efficiency of classifiers is difficult to quantify. The usual method consists in screen sizing the classifier overflow and underflow, and of calculating an efficiency from those data. It should be clear that such a practice is unfair since classifiers do not size material in the same way as screens. It would be fairer to judge classifier efficiencies from careful laboratory classifications of the industrial classifier's overflow and sediment. But laboratory classifiers are too temperamental to make this method easy. By keeping in mind, then, that the efficiency of classifiers is automatically penalized by the fact that it is usually based on screen sizings of the classifier products, the next question is to ascertain what formula to use to calculate the efficiency.

It is customary to use the same formula that is employed to calculate the recovery in a concentrating operation (Chapter X), *viz.*,

$$E = 100\frac{c(f - t)}{f(c - t)} \qquad [IX.6]$$

in which E is the efficiency expressed as a percentage, c, f, and t are the content of minus x-mesh material in the overflow, feed, and underflow, x being any size such that neither c, f, nor t is nil.

But all metallurgists do not agree that this formula should be used, the contention being made, for example, that if some feed is by-passed into the overflow the efficiency as given by Eq. [IX.6] is increased, which clearly should not be the case. To dispose of this objection, it has been proposed to use

$$E = \frac{10{,}000(c - f)(f - t)}{f(100 - f)(c - t)} \qquad [IX.7]$$

This formula expresses efficiency as the ratio, on a percentage basis, of the classified material in the overflow to classifiable material in the feed. It gives lower results than formula [IX.6].

The efficiency of classifiers is usually in the range of 50 to 80 per cent, but it would be appreciably greater if gauged by sedimentation analyses rather than by screen analyses.

One of the principal deterrents to very high efficiencies is the partly flocculated condition in which pulps are customarily classified. The subsequent treatment method, be it cyanidation or flotation, requires the use of some chemicals that influence or even control the dispersion or flocculation of the pulp.

Cost. The cost of classification is strikingly small, except when the size at which the classification is carried out is so fine that relatively large machines and extremely dilute pulps must be used. This is made clear by the following example: Classification at 300-mesh requires four times the settling space needed for classification at 150-mesh and about twelve times the settling space required by classification at 65-mesh. Considerably more water, also, is required for classification at a fine size.

The costs for supplies, electric current, repairs, and labor chargeable to classification are very small. First cost, and consequent capital charges, on the other hand, are comparatively large, especially if classification is charged with the cost of the enlarged building required to house it. In large plants, the total cost of classification may be of the order of magnitude of 1 ct. per ton, but this depends largely on bookkeeping, since much of that cost is overhead and capital.

CLASSIFICATION AS A MEANS OF CONCENTRATION

Although classification is generally merely a sizing operation or a sorting operation adjunct to gravity concentration, it can become a means of concentration, even the only means of concentration. This results in two general types of cases:

1. If the valuable constituents in the broken ore are in one range of sizes and the waste in another range of sizes.
2. If settling can be crowded enough for stratification to result (in accordance with the theory discussed in Chapter VIII), the lower stratum consisting of the heavy material and the upper stratum consisting of the light material.

Separation of Fine Valuable Mineral from Coarse Waste. This is illustrated by the dressing of clays. Clays consist of silicate minerals in an extremely fine state of subdivision and more or less contaminated by coarse impurities, principally quartz, feldspars, micas, and pyrite.

Deflocculation of the raw clay, classification into a sedimented residue and suspended washed clay, flocculation of the washed clay, thickening, and filtration are usually all the dressing that is required. In this case, classification results in concentration. Free settling is used.

Although this method of concentrating clays is very simple, much mystery has surrounded it. No doubt this is related to the use of rule-of-thumb methods for control of flocculation and dispersion. The old practice of settling in long troughs, with intermittent removal of the quartz, is being replaced by the more modern use of hydroseparators for continuous classification at a very fine size.[3,17]

Separation of Coarse Valuable Mineral from Fine Waste. This type of separation is the opposite of the preceding separation, in that a classifier sediment is the concentrate and a classifier overflow is the tailing. Where there is no difference in specific gravity between the mineral and gangue, free-settling classifiers may be used; in the more frequent cases where the coarse, valuable material is also of higher specific gravity, hindered settling is preferable.

Examples are found in iron-ore beneficiation, phosphate-rock beneficiation, and gravel washing.

1. In iron ores, siliceous or clayey impurities generally segregate in the finer sizes of the crushed ore. Classification into a sedimented washed ore and a suspended waste is often the principal means of concentration.

2. In phosphate rock of the land-pebble type found in Florida, phosphatic nodules or pebbles, bones, teeth, etc., occur loosely cemented in a matrix of fine phosphatic grains, quartz sand, and clay. There is practically no difference in specific gravity between the various constituents, but screens combined with classifiers permit segregation into (*a*) coarse phosphatic nodules, (*b*) sand plus fine phosphate grains, and (*c*) clay more or less mixed with colloidal phosphatic particles. The coarse nodules, until recently, have been the only concentrate.[23]

3. In gravel cleaning, the objectionable clay is removed by scrubbing and classification.[4,24]

Scrubbing Classifiers. Standard classifiers can be used for these various operations in which classification means concentration. Since, however, there is usually some tenuous bonding of dissimilar grains requiring rupture, a more or less gentle mixing, turning over, or tumbling of the ore is desirable. This can be accomplished in a trommel outside of the classifier, or in a trommel inside the classifier, as in the Dorr washer (Fig. 118),

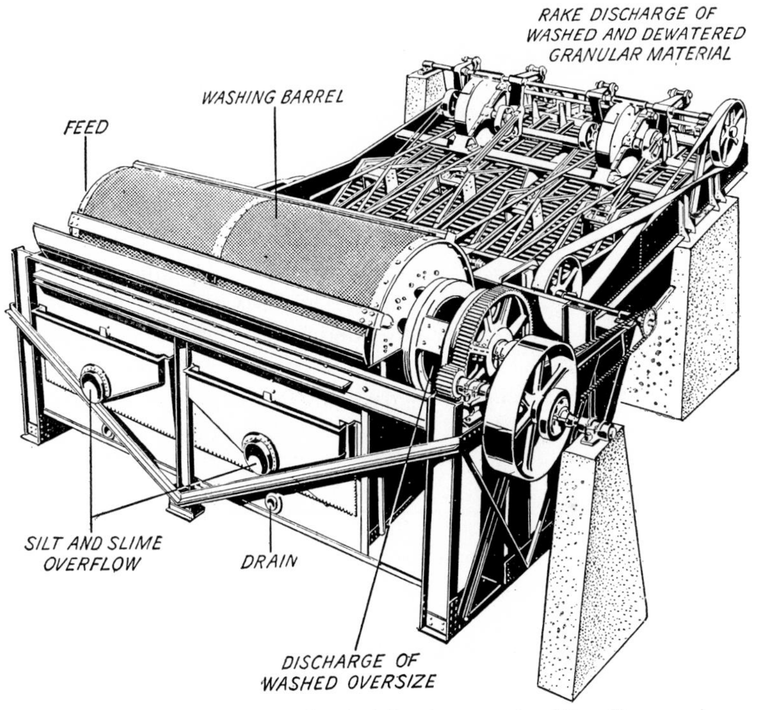

FIG. 118.—Dorr washer, with splash hood removed. (*Dorr Company.*)

in a log washer,[2,20] or in a combination of mill without grinding medium and a classifier.[34]

Log washers usually consist of one or more logs (Fig. 119) carrying metal blades (Fig. 120) at the surface and placed at a slope of 1 to 1½ in. per ft. in an inclined trough. The blades are inclined to the shaft, in the manner of airplane propeller blades, and they are located in a spiral along each log. The log rotates and thus conveys the ore lumps to the upper end of the trough while breaking them up and turning them over. The overflow carries the fines as usual.

Hydraulic gold traps represent a widespread use of classification to effect concentration.[7]

Concentration by Launder Stratification. Another instance of using classification as a means of concentration is in the sluice

FIG. 119.—Double-screw log washer for stone and gravel. (*Allis-Chalmers Mfg. Co.*)

box of gold-mining fame (Figs. 121 and 122) and the Rheolaveur coal washer (Fig. 123). Both consist essentially of a trough in which a suspension of coarse solids is flowing. The solids stratify according to size and specific gravity, mostly or even exclusively according to the latter if the spread in specific gravities between the various minerals is large, and if the locally prevailing proportions of minerals are suitable. This phenomenon seems to be an instance of stratification by extremely hindered settling (see Chapter VIII, page 193).

In sluice boxes, irregularities known as riffles act as traps for gold and other heavy minerals. The material accumulating in sluice boxes is usually known as *black sand* because of its color which is due to the preponderance of magnetite and ilmenite. Many different types of riffles have been devised as may be seen from Fig. 122. Some riffles consist of nailed wooden blocks, quarter-round moulding or peeled raw timber, or of a layer of selected rocks.

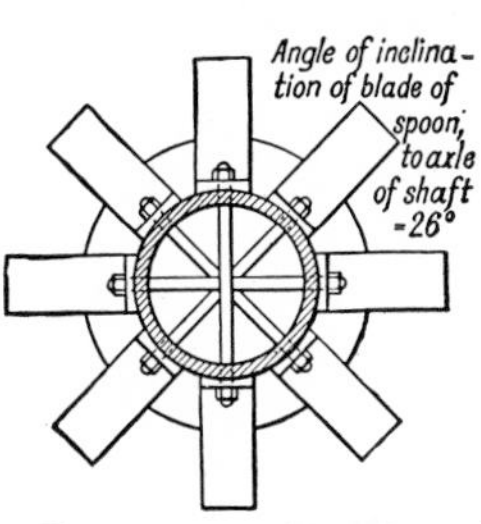

Fig. 120.—Detail of cast-iron log for a log washer. (*After Taggart.*)

In the *Rheolaveur* coal-washing process,(11,12,33) the slate, "bone," and coal stratify in that order from bottom to top of the launder. "Slate traps" occur at intervals along the launder. The slate falls in these traps, against a rising stream of "hydraulic" water, from which it is removed intermittently or continuously. Intermittent removal (Fig. 124) is used for coarse slate and continuous removal (Fig. 125) for fine slate. These are also known as "sealed" and "free" discharge *Rheo* boxes, respectively. Regulation of the discharge of the boxes is obtained by varying the rising current of water or by changing the gate settings.

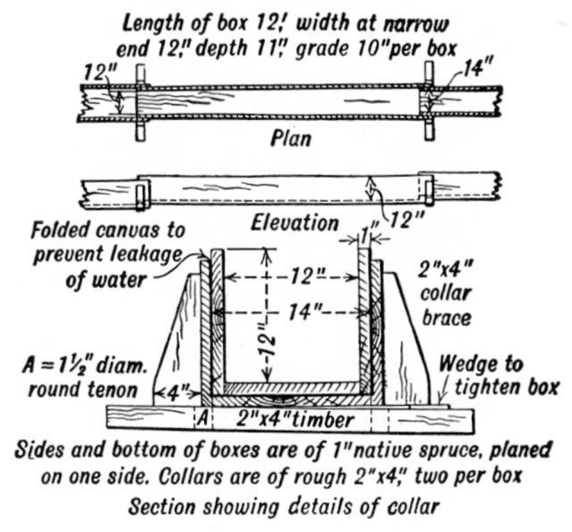

Fig. 121.—Typical sluice box. (*After Ellis, Peele and Taggart.*)

Recently there has been developed an automatic Rheo box for discharging coarse sediment. This box is controlled by the solid load pressing on a lever, one control opening the trap when the load reaches a certain level and the other closing the trap when

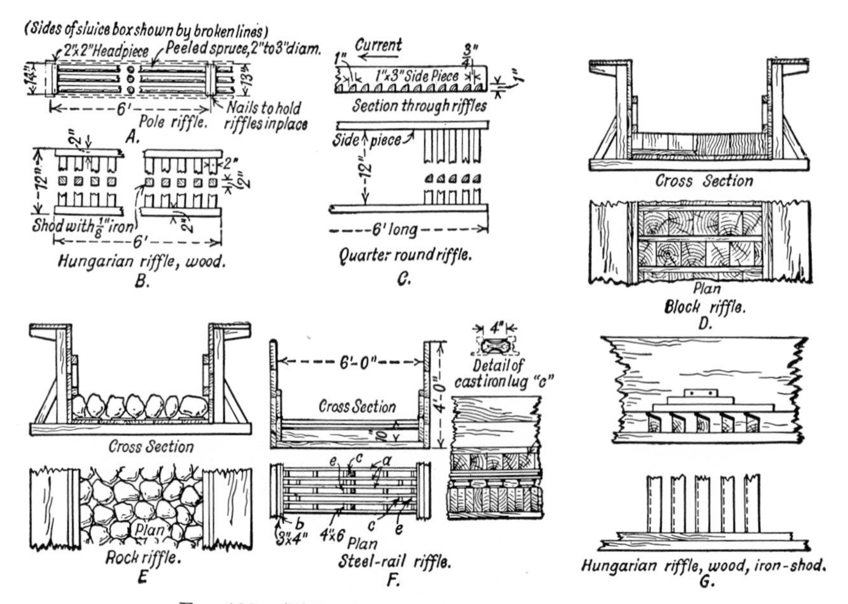

Fig. 122.—Riffles for sluice boxes. (*After Taggart.*)

Fig. 123.—Rheolaveur trough washer for coal. (*Koppers-Rheclaveur Co.*)

the load reaches a certain, other level. The box may then remain shut, open, or be changing from open to shut, or vice versa, as necessary. The control is electrically transmitted, use being made of a suitable electric-relay mechanism.[11] This device is said to have increased considerably the effectiveness of Rheolaveurs.

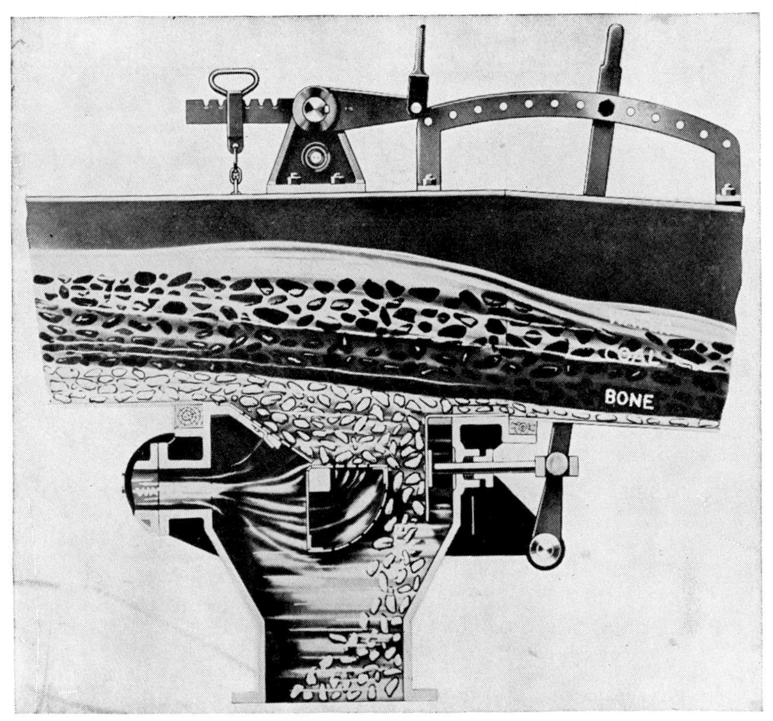

FIG. 124.—Sealed-discharge Rheo box for Rheolaveur trough washer. (*Koppers-Rheolaveur Co.*)

In the operation of this simple concentrating device, a large circulating load of middling is maintained in order to make possible the joint production of clean coal and clean waste. This middling is of higher slate or shale content than the raw coal and supplies the quantity of waste that is required to form a bed of sufficient thickness so that clean coal cannot be drawn in the Rheo boxes that make final waste. Considerable clean coal, however, is drawn through the Rheo boxes that make middling.

The tremendous importance of the Rheolaveur coal-washing process is hardly appreciated by those not specializing in coal washing. According to Berthelot, the amount treated in 1936 by this process, the world over, amounted to 115 million tons; roughly then, one-third of all the cleaned coal was cleaned by that process.

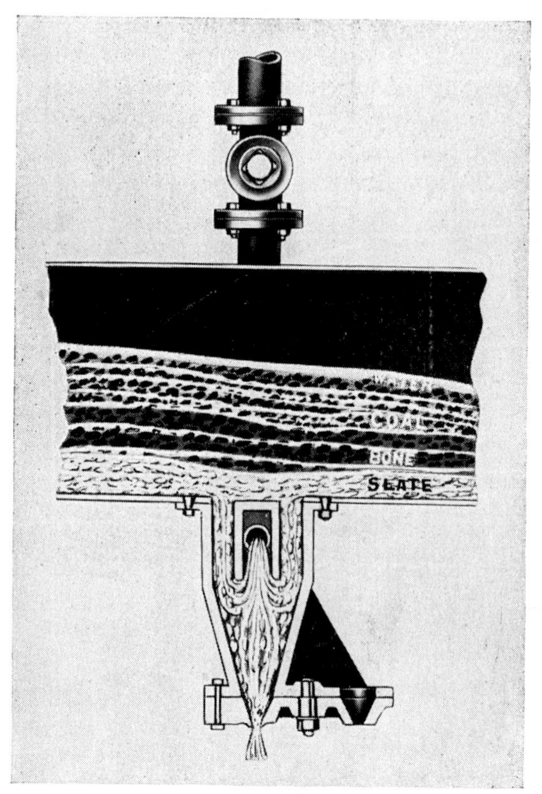

FIG. 125.—Free-discharge Rheo box for Rheolaveur trough washer. (*Koppers-Rheolaveur Co.*)

Recent investigations at the Battelle Institute, using glass-sided experimental launders, has pointed the way to an improved launder trap.[29] It is said by Richardson that the *Battelle launder* will permit cleaning coal with less middling circulation and in a smaller space than older launder classifiers.

The *Hydrotator* hindered-settling *classifier*[25,26] is another classifier adapted to operate as a concentrator (Fig. 109).

Concentration by Sizing and Sorting. Attention has already been drawn to the fundamental difference between screen sizing and classifier sizing, to wit, that specific gravity of the solid is not involved in screen sizing but is involved in classifier sizing.

This difference can be made the basis for a method of concentration: if an ore is first classified into several spigot products, then if each spigot product is screen sized, a separation is obtained first into equal-settling particles which are of somewhat different

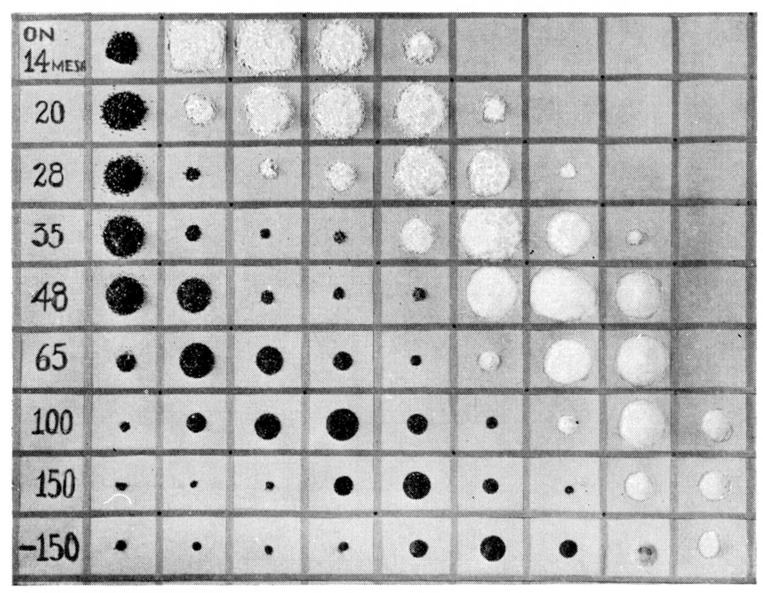

FIG. 126.—Separation of galena from quartz by sorting and sizing, in accordance with the procedure of Richards. Each column represents a spigot product; successive spigot products are arrayed from left to right.

sizes (if the minerals differ in specific gravity), then into particles of different sizes. If seven classified products are first obtained and if each is later divided into eight screened grades, there are altogether 7 × 8, or 56, groups of particles. Since the separations have different bases, the particles in the various groups differ in size or composition. By gathering together into one product those particle groups which have substantially the same chemical composition, and by gathering together in another product those particle groups which have some other chemical composition, a concentrate and a tailing are obtained.

The procedure can of course be reversed into screening followed by classification. Figure 126, made according to Richards,[27,28] is a photograph of an array of particle groups, each column representing one sorted product, and each row one screen size. This figure shows beautifully that an excellent separation is possible if the minerals to be separated are as different in specific gravity as galena and quartz and if no locked particles are present. Poorer results are obtained under less-favored circumstances. Experiments such as that pictorially reproduced in Fig. 126 were made by Richards in his famous experimental determinations of free-settling and hindered-settling ratios: from the size distribution of the various minerals in each classifier spigot product, the average size of each mineral was determined in each spigot product, and hence the settling ratio in each spigot product. Broadly speaking, a good agreement was obtained with the theoretical figures presented in Chapter VIII.

Asbestos Concentration. Pneumatic classification[30,31,32] is utilized to separate the relatively fluffy asbestos fiber from the granular gangue associated with it. To that end, the raw rock is first disintegrated in such a way as to open the asbestos fiber as little as possible. The comminuted product is then screened in such a way that the fiber remains parallel to the screening surface while the granular waste passes through. The dedusted product is then concentrated by pneumatic classification, the asbestos being sucked up while the gangue remains.

Centrifugal Classification. To save fine particles of heavy minerals in the days prior to the widespread use of flotation, there were designed many so-called centrifugal concentrators.[13,35] Most of these devices were in fact centrifugal classifiers and not concentrators in the sense in which jigs or tables are concentrators. Even in those very favorable cases where the spread in specific gravity between mineral and gangue is large, centrifugal classifiers cannot be expected to do more than yield a partial enrichment in the sediment. When it is recalled that a centrifugal device is bound to be expensive to build and operate, it becomes clear why centrifugal classifiers have remained a practical failure. This does not mean that a useful centrifugal concentrator could not be designed, but it does mean that such a concentrator would have to depart from the principles of simple classification.

Literature Cited

1. Ammon, Robert: The Anaconda Classifier, *Trans. Am. Inst. Mining Engrs.*, **46,** 277–325 (1913).
2. Amos, A. R., Jr., and S. B. Patterson: Log Washers in the Aggregate and Flux-stone Industries, *Trans. Am. Inst. Mining Met. Engrs.*, **129,** 145–155 (1938).
3. Anable, A.: Modernized Treatment Improves Results in a Non-metallic Industry, *Eng. Mining J.*, **127,** 566–568, 605–608 (1929).
4. Anable, Anthony: Preparation of High-specification Sand at the Grand Coulee Dam, *Trans. Am. Inst. Mining Met. Engrs.*, **129,** 156–169 (1938).
5. Anon.: An Important Improvement in Classifiers, *Mining and Met.*, **18,** 247 (1937).
6. Anon.: Countercurrent Classifier, *Eng. Mining J.*, **135,** 189–190 (1934).
7. Anon.: Hydraulic Gold Trap Improves Recovery, *Eng. Mining J.*, **137,** 290 (1936).
8. Anon.: New Dorr Multizone Classifier, *Eng. Mining J.*, **138,** 271 (1937).
9. Antisell, Toner: Mica Mining and Milling Methods, *Eng. Mining J.*, **122,** 894–896 (1926).
10. Bardwell, Earl S.: Application of Hindered Settling to Hydraulic Classifiers, *Trans. Am. Inst. Mining Engrs.*, **46,** 266–276 (1913).
11. Berthelot, Ch.: "Épuration, séchage, agglomération et broyage du charbon," Ch. Dunod, Paris (1938).
12. Crawford, J. T., C. P. Proctor, and M. J. Williams: Launder Washing of Coarse Coal, *Trans. Am. Inst. Mining Met. Engrs.*, **130,** 172–189 (1938).
13. Doerner, H. A.: Centrifugal Concentration, *U. S. Bur. Mines, Tech. Paper* 457 (1929).
14. Fahrenwald, A. W.: Classifier Efficiency, *Am. Inst. Mining Met. Engrs., Tech. Pub.* 275, 14 pp. (1930).
15. Gay, Rubert M.: Air Separation Effective in the Non-metallic Mineral Industries, *Eng. Mining J.*, **129,** 65–68 (1930).
16. Geismer, H. S.: The Preparation of Brown Iron Ores, *Trans. Am. Inst. Mining Engrs.*, **42,** 169–180 (1911).
17. Grout, J. R., Jr.: Better China Clay from Improved Beneficiation, *Eng. Mining J.*, **138,** 341, 352 (1937).
18. Hayden, Ralph: Concentration of Slimes at Anaconda, Mont., *Trans. Am. Inst. Mining Engrs.*, **46,** 239–265 (1913).
19. Hebley, Henry F.: The Dedusting of Coal, *Trans. Am. Inst. Mining Met. Engrs.*, **108,** 88–127 (1934).
20. Johnson, Guy R.: Ore Washer at Longdale, Va., *Trans. Am. Inst. Mining Engrs.*, **24,** 34–40 (1894).
21. Johnson, J. E., Jr.: An Apparatus for the Removal of Sand from Waste Water of Ore Washers, *Trans. Am. Inst. Mining Engrs.*, **28,** 225–235 (1898).
22. Newton, Harry W., and W. H. Newton: A Study of Classification Calculations, *Rock Prod.* [16], **35,** 26–30 (1932).

23. O'Meara, R. G.: Added Recovery by Hydraulic Sizing of Fine Material in the Land-pebble Phosphate District of Florida, *U. S. Bur. Mines, Rept. Investigations* 3139 (1931).
24. Price, Thomas M.: Aggregate Production at Hoover Dam, *Trans. Am. Inst. Mining Met. Engrs.*, **109**, 397–417 (1934).
25. Remick, W. L.: Fine-coal Cleaning by the Hydrotator Process, *Trans. Am. Inst. Mining Met. Engrs.*, **75**, 569–582 (1927).
26. Remick, W. L., and George B. Jones: Hydrotator Coal-cleaning Process, *Am. Inst. Mining Met. Engrs., Tech. Pub.* 219 (1929).
27. Richards, Robert H.: Development of Hindered-settling Apparatus, *Trans. Am. Inst. Mining Engrs.*, **41**, 396–453 (1910).
28. Richards, Robert H.: "A Textbook of Ore Dressing," McGraw-Hill Book Company, Inc., New York (1909), p. 276.
29. Richardson, A. C.: Mechanism of Launder Separations, *Trans. Am. Inst. Mining Met. Engrs.*, **130**, 156–171 (1938).
30. Ross, J. G.: Processing Canadian Asbestos, *Eng. Mining J.*, **135**, 563–565 (1934).
31. RuKeyser, W. A.: Asbestos Milling in Quebec, *Eng. Mining J.*, **133**, 102–106 (1932).
32. RuKeyser, Walter A.: Mechanical Cobbing of Chrysotile Asbestos, *Eng. Mining J.*, **134**, 235–237 (1933).
33. Schweitzer, Edgar: Operation of Rheolaveur Plant at Dorrance Colliery, Lehigh Valley Coal Co., *Trans. Am. Inst. Mining Met. Engrs.*, **94**, 311–323 (1931).
34. Towers, J. K.: Improvements in Mineral Scrubbing, *Eng. Mining J.*, **133**, 174 (1932).
35. Vivian, Godfrey T.: Centrifugal Machines for Ore Grading and Ore Concentrating, *Trans. Am. Inst. Mining Engrs.*, **44**, 676–683 (1912).

CHAPTER X

QUANTIFYING CONCENTRATING OPERATIONS

For the control of mill operations, as for the control of tests, a measure of the effectiveness of concentrating operations is needed. No measure has been universally accepted, but several are widely employed.

Direct Statement. The simplest way to express metallurgical results is by a statement of the weight of the various products derived from a given weight of feed, together with their percentage content of the various metals, minerals, or gangues whose separation is desired. For example, the results of concentrating a copper ore by flotation may be stated as in Table 35.

TABLE 35.—DIRECT PRESENTATION OF A CONCENTRATING OPERATION

Product	Weight, per cent	Metal content, per cent copper	Gangue content, per cent insoluble
Concentrate	9.3	18.52	12.6
Tailing	90.7	0.15	62.7
Feed	(100)	1.86	57.9

Presentation of the results in the form of Table 35, although direct, lacks in interpretative convenience. In order to weigh the advantages of the operation, eight numbers are to be kept in mind (five only if the gangue content is deemed of secondary importance).

To reduce the numbers required to quantify a concentrating operation, its results are often expressed in terms of the ratio of concentration and of the recovery of the valuable constituents (or of the recoveries of the valuable constituents).

Ratio of Concentration. The relative weight of feed to that of concentrate is a measure of the concentration in weight that has been effected. This ratio is known as the *ratio of concentration.*

Expressed analytically,

$$K = \frac{F}{C}, \qquad [X.1]$$

in which K is the ratio of concentration, F the weight of feed, C the weight of concentrate.

If it is difficult to ascertain the weights of concentrate and tailing, formula [X.2] can be used, in which K is expressed in terms of assays of feed f, concentrate c, and tailing t.

$$K = \frac{c - t}{f - t}. \qquad [X.2]$$

Equation [X.2] is obtained from the basic equations

$$F = C + T \qquad (a)$$
$$Ff = Cc + Tt. \qquad (b)$$

The first of these basic relations states that the weight of feed equals the sum of the weights of the products; the second states that the weight of one constituent in the feed equals the sum of the weights of that constituent in the products.

If (a) is multiplied by t, the terms in Tt in (a')

$$Ft = Ct + Tt \qquad (a')$$

and in (b) are the same.

Subtraction of the two equations, term by term, yields

$$F(f - t) = C(c - t)$$

or

$$\frac{F}{C} = \frac{c - t}{f - t}.$$

Hence

$$K = \frac{c - t}{f - t}.$$

The ratio of concentration, taken by itself, is almost useless as it discloses nothing concerning the quality of the concentrate and tailing. Also, it loses significance if more than one concentrate is produced. But it is very useful in conjunction with recovery.

Recovery. From the relative weight of the concentrate and tailing and their assays, that fraction of a certain metal or gangue contained in the feed which is recovered in the concentrate can be figured. This fraction, when expressed as a percentage, is known as *recovery*. Thus, in the case of Table 35 the copper recovery in the concentrate is 92.7 per cent and the "insoluble" recovery is 2.0 per cent.

With the same notation as above, the recovery expressed as a percentage is

$$R = 100\frac{Cc}{Ff}. \qquad [X.3]$$

If it is difficult to ascertain weights, R can be expressed in terms of assays, just as the ratio of concentration can be expressed in terms of assays.

$$R = 100\frac{c(f - t)}{f(c - t)}. \qquad [X.4]$$

Derivation of [X.4] flows from [X.2] and [X.3]:

$$R = 100\frac{c}{f} \cdot \frac{C}{F} = 100\frac{c}{f} \cdot \frac{1}{K} = 100\frac{c(f - t)}{f(c - t)}.$$

The conception of recovery is equally useful if more than one concentrate or if one or more middlings are made.

Thus if a feed is broken up into several concentrates of weights C_1, C_2, C_3, . . . and grades c_1, c_2, c_3, . . . the fundamental equation stating the recovery in any product is

$$R_n = \frac{100C_nc_n}{F \cdot f}$$

and that stating the ratio of concentration is

$$K_n = \frac{F}{C_n}.$$

In certain cases, R_n and K_n can be expressed in terms of assays only, excluding weights, but the equations are rather complicated. Furthermore, the results of the calculations may be wide of the mark, even to the extent of giving negative answers, if either the sampling or the analytical work is faulty. Derivation and use of the equations are set forth by Taggart.[6]

Rejection, Losses. By an extension of the definition, the percentage loss of metal in a discarded product, or else the rejection of gangue, may be regarded as recoveries. At any rate, such losses and rejections are mathematically symmetrical to recovery in the concentrate. In the instance of Table 35, the copper recovery in the tailing (copper loss) is 7.3 per cent and the insoluble recovery in the tailing (insoluble rejection) is 98.0 per cent.

Use of Recovery and Ratio of Concentration. Joint consideration of recovery and ratio of concentration is the traditional method of expressing metallurgical results, at least in connection with metal mining. In the nonferrous metal-mining field, recovery of metal is regarded as the most important criterion with grade of concentrate next and ratio of concentration third; in iron beneficiation, ratio of concentration is often most important with grade of concentrate next and recovery third. In coal washing, ratio of concentration is so much more important than "recovery" as to have usurped the term: to a coal operator the washery recovery is the ratio of weight of washed coal to weight of raw coal, or the reciprocal of the ratio of concentration, expressed as a percentage. This quantity is more properly termed the *yield*.

Efficiency of a Concentrating Operation. In comparing various operations or tests, it is difficult to make a selection of the best operation or test if recovery, ratio of concentration, and grade of concentrate must be weighed simultaneously.

To that end, several indices have been proposed. One of the simplest and most suitable is the metallurgical efficiency proposed by R. W. Diamond.[2] This is defined as the arithmetical average of the recoveries of the principal constituent of each product (tailing included). Thus in a separation of galena, sphalerite, and gangue, if the lead concentrate contains 90 per cent of the lead, the zinc concentrate 88 per cent of the zinc, and the tailing 95 per cent of the gangue, the metallurgical efficiency is $(90 + 88 + 95)/3 = 91\%$. Generally,

$$E = \frac{\Sigma R_{nN}}{n}, \qquad \text{[X.5]}$$

in which R_{nN} denotes the recovery of constituent n in the N product. Unfortunately, the metallurgical efficiency is not nil when there is no separation. This is demonstrated readily by considering the separation made by a sampling splitter as yielding a "concentrate" and a "tailing." In view of the construction of the splitter, one-half the mineral is in the concentrate and one-half the waste is in the tailing so that the metallurgical efficiency is 50 per cent.

A convenient yet abstract measure of the perfection of a two-way separation is the selectivity index.[4]

Selectivity Index. Selectivity index is the geometrical mean of the relative recoveries and relative rejections of two minerals, metals, or groups of minerals or metals.

If R_a is the recovery of a in A and J_a is the rejection of a in B, and R_b the recovery of b in A and J_b its rejection in B, the relative recovery of a to b is R_a/R_b, and the relative rejection of b to a is J_b/J_a. By definition:

$$\text{S.I.} = \sqrt{\frac{R_a}{R_b} \cdot \frac{J_b}{J_a}}. \qquad [\text{X.6}]$$

But

$$J_a = 100 - R_a; \qquad R_b = 100 - J_B.$$

Hence,

$$\text{S.I.} = \sqrt{\frac{R_a \cdot J_b}{(100 - R_a)(100 - J_b)}}. \qquad [\text{X.7}]$$

For example, if lead recovery in a lead concentrate is 95 per cent and gangue rejection in deleaded pulp is 96 per cent,

$$\text{S.I.} = \sqrt{\frac{95 \times 96}{5 \times 4}} = 21.3.$$

In some cases, it is easier to determine selectivity indices from grades than from recoveries. Thus if assays for substances a and b are M and N in the concentrate and m and n in the tailing, it can be shown that the selectivity index is

$$\text{S.I.} = \sqrt{\frac{M}{m} \times \frac{n}{N}}. \qquad [\text{X.8}]$$

In the case of Table 35, for instance, the selectivity index is

$$\text{S.I.} = \sqrt{\frac{18.5}{0.15} \times \frac{62.7}{12.6}} = 24.8.$$

If the grade of concentrate and tailing is the same (no separation), the selectivity index is unity. If, on the other hand, the concentrate is absolutely free of waste constituents and the tailing is absolutely free of mineral, the selectivity index is infinite. Such a happening, unfortunately, is wholly in the realm of theory. In practice, the index ranges from 4 to 40, exceptionally good and poor results falling outside this range.

Selectivity index gives an accurate scientific measure of the effectiveness of a separation. It seems, in fact, to be *the* most accurate measure that could be devised. But it fails to measure in proper economic fashion the value of successive improvements. Thus doubling the index does not mean that the returns from the sale of the concentrate are doubled. Although in some cases the returns may be more than doubled, in other cases the increase is slight. And what effect a doubling of the selectivity index may have on the operating margin is of course wholly unpredictable.

Economic Recovery or Efficiency. From an economic standpoint, neither recoveries nor ratios of concentration nor selectivity indices afford a comprehensive appreciation of the economic effectiveness of a concentrating operation. It has been proposed[5] to use to this effect a new quantity called *economic recovery* or *efficiency*. This is the percentage ratio of the actual value of the concentrate obtained per ton of ore to the value of that weight of concentrate theoretically obtainable in mineralogically pure form from a ton of ore.

Economic recovery may be illustrated by the following example: 0.15 ton of lead concentrate worth $50 per ton can be secured in practice from an ore that theoretically should yield 0.14 ton of mineralogically pure galena concentrate worth $75 per ton. Even though the actual lead recovery may be as high as 95 per cent, the economic recovery is only

$$\frac{0.15 \times 50}{0.14 \times 75} = 71.4 \text{ per cent.}$$

At first sight, economic recovery seems extremely attractive. However, it is open to several minor objections, some of which are technical and others economic. The principal objections are the following:

1. Since ores are variable (even those treated at any one mill) and frequently contain more than one valuable mineral, the composition of the mineralogically pure concentrate will vary from day to day.

2. The value of the mineralogically pure concentrate will vary from day to day, even though it were to have a uniform composition, because of fluctuations in metal prices. This is partly, but not wholly, compensated by variations in the value of the actual concentrates.

Allowance for the ore variations can be made by calculating in each instance the composition of the mineralogically pure concentrate.

Coal-washing Efficiency.[1,3] In dealing with coal and in general with nonmetallics, there is no such standard as smelter schedules to guide the determination of concentrating efficiency. There is used instead a *washing efficiency.*

Washing efficiency is the ratio of the weight of coal of a certain ash content actually obtained by washing to the weight obtainable according to float-and-sink tests (see Chapter XI). This measure, of course, is such that efficiencies of over 100 per cent can be obtained (theoretically at least) if the concentrating process uses some other physical property than gravity (*e.g.*, magnetic susceptibility or floatability).

The recovery penalizes the concentrating operation to the extent that locked particles are present, unless all these particles are gathered in the concentrate (in which case, the grade of the concentrate is lowered). The coal-washing efficiency, on the other hand, does not penalize the concentrating operation if locked particles are present. It is a less rigorous measure than those used in ore concentrating. Thus in coal washing, recovery of 90 per cent of the fuel with rejection of 50 per cent of the waste may represent a washing efficiency of 96 to 97 per cent.

Screen and Classifier Efficiencies. Screen or classifier efficiencies, it will be noted, are obtained by formulas [VII.7] and [IX.6] which are identical with the recovery formula [X.4]. In other words, the efficiency of a screen or classifier is regarded as the ratio of the attained to the attainable in a size separation; likewise the recovery in a concentrating operation is the ratio of the attained to the attainable in a composition separation.

Literature Cited

1. Coe, G. D.: An Explanation of the Washability Curves for Interpretation of Float-and-sink Data on Coal, *U. S. Bur. Mines, Information Circ.* 7045 (1938).

2. DIAMOND, R. W.: Ore Concentration Practice of the Consolidated Mining and Smelting Co. of Canada, Ltd., *Trans. Am. Inst. Mining Met. Engrs.*, **79**, 95–106 (1928).
3. FRASER, THOMAS, and H. F. YANCEY: Interpretation of Results of Coal-washing Tests, *Trans. Am. Inst. Mining Met. Engrs.*, **69**, 447–482 (1923).
4. GAUDIN, A. M.: Selectivity Index: A Yardstick of the Segregation Accomplished by Concentrating Operations, *Trans. Am. Inst. Mining Met. Engrs.*, **87**, 483–487 (1930).
5. HANDY, R. S.: Milling Methods and Costs at the Northern Idaho Mills of the Bunker Hill and Sullivan Mining and Concentrating Co., *U. S. Bur. Mines, Information Circ.* 6314 (1930).
6. TAGGART, A. F.: "Handbook of Ore Dressing," John Wiley & Sons, Inc., New York, (1927), pp. 1235–1242, 1250–1252.

CHAPTER XI

HEAVY-FLUID SEPARATION

If a fluid is available whose specific gravity is intermediate between that of two solids which it is desired to separate, no simpler process could be desired than to suspend the mixed solids in the fluid, allow one to rise and the other to sink, and draw off separate products from top and bottom of the separating vessel.[13] A typical example is the separation of wood chips from gravel or sand, using water as the medium.

Since all minerals are heavier than water, water is not suitable for the practice of "float-and-sink" separation. Some aqueous solutions are available, however, whose specific gravity is sufficient to permit coal to "cream" while associated impurities sink. Organic liquids having a specific gravity well above 2.75 but under 3.5 are available. They can be used to reject, as a light layer, the common gangue minerals quartz and calcite,[21] but they are relatively expensive. Liquids having a specific gravity over 3.5 are few and very expensive even for laboratory and research purposes.

Heavy pseudo liquids can be made by suspending solids in water, and these fluids can be used almost like true liquids, provided the particles to be separated are coarse in comparison to the size of the "medium" particles, provided the "medium" is thin enough not to acquire plastic properties, and provided the "medium" is agitated enough not to settle.

Pseudo liquids are very much cheaper than organic liquids of high specific gravity, so the practical disadvantage of fluid loss is not nearly so significant. On the other hand, the use of pseudo liquids is not so simple as that of high-specific gravity liquids.

Laboratory Use of Heavy Liquids. In the laboratory, heavy liquids are very useful for assessing the optimum separation obtainable by gravity concentration: by the use of a series of fluids of graduated specific gravities, a crushed solid can be separated into fractions whose specific gravity lies within narrow limits, as 1.40 and 1.45, or 2.75 and 2.85. In this way, locked particles are segregated from free particles and locked particles of different compositions are separated from each other.

The procedure has not been applied to extremely fine particles, but can readily be used for all sieve sizes and the coarsest subsieve sizes. For example, 200-mesh quartz settles approximately 4 in. in water in 15 sec. In this case, the viscosity is about 0.01 poise, and the apparent specific gravity is 1.65. If separation to 0.01 unit in specific gravity is desired and if the viscosity of the fluid is ten times as large as that of water (corresponding to the viscosity of a free-flowing oil), the settling time should be some 1,650 times as large, or about 7 hr. for 4 in. Because of its reduction in settling time, centrifuging permits an extension of the float-and-sink procedure to subsieve sizes.

One of the most useful heavy fluids is acetylene tetrabromide (or tetrabromethane) whose specific gravity is 2.96. This fluid can be diluted with carbon tetrachloride and give solutions of lower specific gravity down to 1.59, the specific gravity of carbon tetrachloride.

Another useful group of fluids of low specific gravity (for coal analysis) is aqueous solutions of zinc chloride and of calcium chloride.

For specific gravities higher than 2.96, the student is referred to the review by Sullivan[21] and to other special works.[12,16,18]

Figure 127 represents the results obtained by float-and-sink on one sample of coal.[6] This figure is typical of the studies currently made on the washability of coal. Determinations were made of the ash content of specific gravity fractions averaging, respectively, 1.28, 1.30, 1.38, 1.50, 1.70, 1.90, and 2.20. These fractions were obtained by the use of heavy fluids; their relative weights were determined. From the specific gravities and cumulative weights, curve I (specific gravity *vs.* cumulative weight) was drawn.

From the ash content calculated on a cumulative basis and the cumulative weight percentage, curve II was drawn. Curve III records also the elementary or actual ash content of each fraction against cumulative weight.

Thus if a separation is made by float-and-sink testing at specific gravity 1.40, the raw coal (containing 16.0 per cent ash) yields clean coal weighing 78 per cent of the raw coal (point

A on curve I) and containing 8.6 per cent ash (point *B* on curve II). If gravity methods are used, such a separation is the most that can be expected of that coal without further comminution.

Curves IV and IV*a* are designed to give a measure of the difficulty in separating the raw coal into cleaned coal and waste

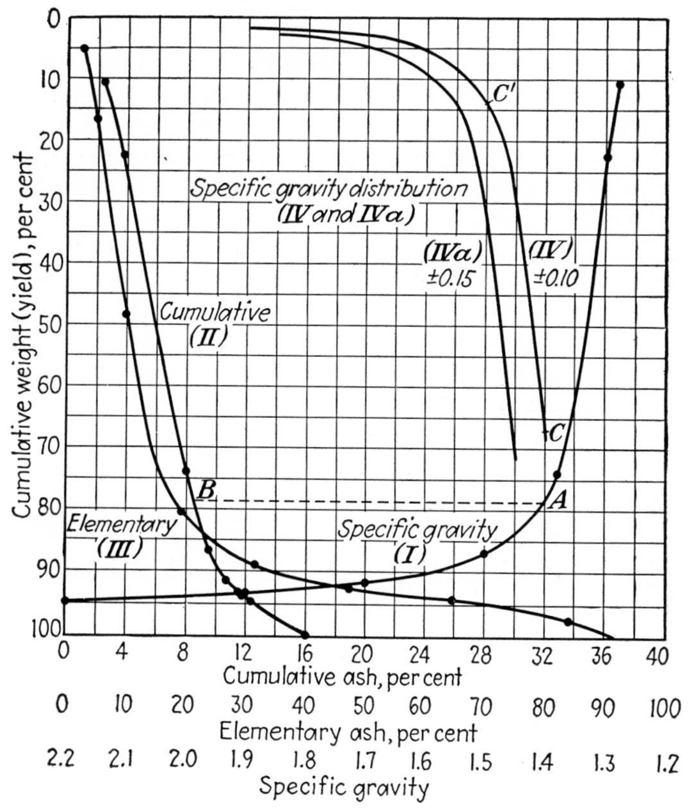

FIG. 127.—Typical washability curves for one coal. (*After Bird, Richardson, and Coe.*)

at any specific gravity.[7,8] Thus if a leeway of 0.10 in specific gravity of individual particle is permissible at 1.40 specific gravity, about 67 per cent of the coal falls between 1.30 and 1.50 specific gravity (point *C* curve IV). This indicates extreme difficulty of practical separation at that specific gravity. If, on the other hand, separation is attempted at a specific gravity of 1.50, and a leeway of 0.10 is permissible, about 14 per cent of the coal falls between 1.40 and 1.60 specific gravity (point C'). This indicates relatively easy separation instead of difficult separation as in the preceding instance. Curve IV*a* is similar to curve IV, but is drawn for a leeway in specific gravity of 0.15 instead of 0.10, *i.e.*, for separation under poorer control.

INDUSTRIAL PROCESSES USING HEAVY LIQUIDS

Three processes have been proposed that use true heavy liquids; the Lessing process, the Bertrand process, and the Du Pont (Nagelvoort) process.

The Lessing and Bertrand processes clean coal in a solution of calcium chloride, and the Du Pont process uses organic liquids of high specific gravity.

Lessing Process.[3] Calcium chloride solution having a specific gravity of approximately 1.4 is used for the separation, which takes place in a cylindrical tank 6 to 10 ft. in diameter with a conical bottom, the total height being nearly 30 ft. Raw coal freed of dust and fines is introduced near the center of the tank after mixing with the separating solution in a mixer. The cleaned coal rises to the top where it is removed by a chain scraper and delivered to draining towers. The slate and bone are lifted from the bottom by a bucket conveyor and dumped in draining towers. After draining, the coal and the slate are washed, the wash liquor returning to the supply of calcium chloride. Further washes are required to free the coal completely of chloride; these go to waste.

Some 320 l. of liquor is withdrawn from the separating tank by each metric ton of raw coal. The specific gravity of the liquor is dropped from 1.4 to about 1.2 by the wash water and the inherent moisture of the coal. This 320 l. of liquor, now increased to about 640 l., is concentrated by evaporation to the original volume. The loss of calcium chloride liquor is of the order of 2 to 3 l. per ton of raw coal.

The Lessing process has been installed in Wales and has produced extremely clean coal; the clean coal is even freer of chloride than the raw coal. It would seem, however, that the cost of thermal concentration of the separating liquor will stand in the way of widespread adoption of the process.

Bertrand Process.[4,5] The Bertrand or Ougrée-Marihaye process also utilizes a calcium chloride solution as separating medium and is applicable only to deslimed feed. In practice, particles varying from 1 to 5 mm. in diameter are treated by this process in Belgium. It differs from the Lessing process in that the raw coal is introduced into the system countercurrent fashion, from water to separating solution, the purified coal and the waste being withdrawn in a similarly countercurrent fashion. There are five circulating liquors, *viz.*, hot water, weak solution, medium solution, strong solution, and separating solution.

The Bertrand process avoids the costly thermal concentration of the diluted leach liquor, but introduces a relatively complex hydrometallurgical flow sheet.

The results obtained by these processes are excellent, coal of extremely high grade being obtained in amounts substantially in agreement with theoretical yields. Coal containing less than 1 per cent ash is said to be obtained by the Bertrand process. Coal of such purity is in demand for the manufacture of special electrode coke, for the preparation of colloidal coal, for hydrogenation, as fuel in Diesel engines of the Rupa type, and as fuel in automotive gas producers.

Du Pont Process.[13a,13b] The Du Pont process, an outgrowth of the Nagelvoort process,[1] is a practical adaptation of the laboratory heavy-liquid separation which has already been described. In basic principle, it does not differ from that laboratory procedure. But several requirements have had to be met in order for the process to become commercial. These are as follows:

1. Low solubility of the "parting liquid" in water and of water in the parting liquid.
2. Low viscosity (high fluidity) of the parting liquid at operating temperatures.
3. Stability, low vapor pressure, and nonflammability of the parting liquid.
4. Prior preparation of the ore to remove fine particles.
5. Prior preparation of the ore with suitable chemicals to make the surface of the particles immune to wetting by the parting liquid.
6. Complete sealing of the separating system to prevent loss of parting liquid by evaporation and to eliminate the health hazard due to the noxious vapors of the parting liquid.

7. Use of a procedure that completely separates the parting liquid from the separated minerals, so as to completely regenerate the parting liquid.

8. Use of a scheme for constantly purifying the parting liquid.

Of these various requirements, the most important was requirement 5, inasmuch as reasonable reagent consumption could not be expected if it were not met.

So-called "active agents" have been devised to keep the minerals wetted by water rather than by parting liquid. In the case of coal, the active agents are starch acetate or tannic acid. The concentration of active agent in water is of the order of 0.01 per cent.

The main expense in the Du Pont process is for the parting liquid, which is a mixture of several halogenated hydrocarbons. The consumption of medium is said to be very low, and often well under 1 lb. per ton of coal treated.

Clearly the separating process although simple in principle requires a number of adjunct operations for the sake of economy in reagents and from a physiological standpoint. The process is not applicable to fine particles. It is therefore limited to the treatment of minerals in a coarse state of subdivision, and such a treatment cannot be expected to be successful unless the ore is coarsely aggregated or if low standards are permissible.

Amber is separated from associated impurities by the use of heavy liquids.[20,25]

INDUSTRIAL PROCESSES USING HEAVY SUSPENSIONS

Industrial processes using heavy suspensions have behind them the record of many years' practice in the washing of coal, but their application to ores in which a separation is to be made at a specific gravity of over 2.6 is still very recent.

These processes include the Chance sand-flotation process, the Vooys barite-and-clay process, and the Wuensch process.

Chance Process.[9,11] For cleaning coal, the Chance process has been in use for about 20 years. The medium consists of a suspension of sand in water. The sand must be of relatively uniform size, $-40 + 80$-mesh being preferred. Coarse sand tends to accumulate in the bottom of the separating vessel, and fine sand is harder to retrieve, as well as likely to accumulate in the upper stratum of the separator.

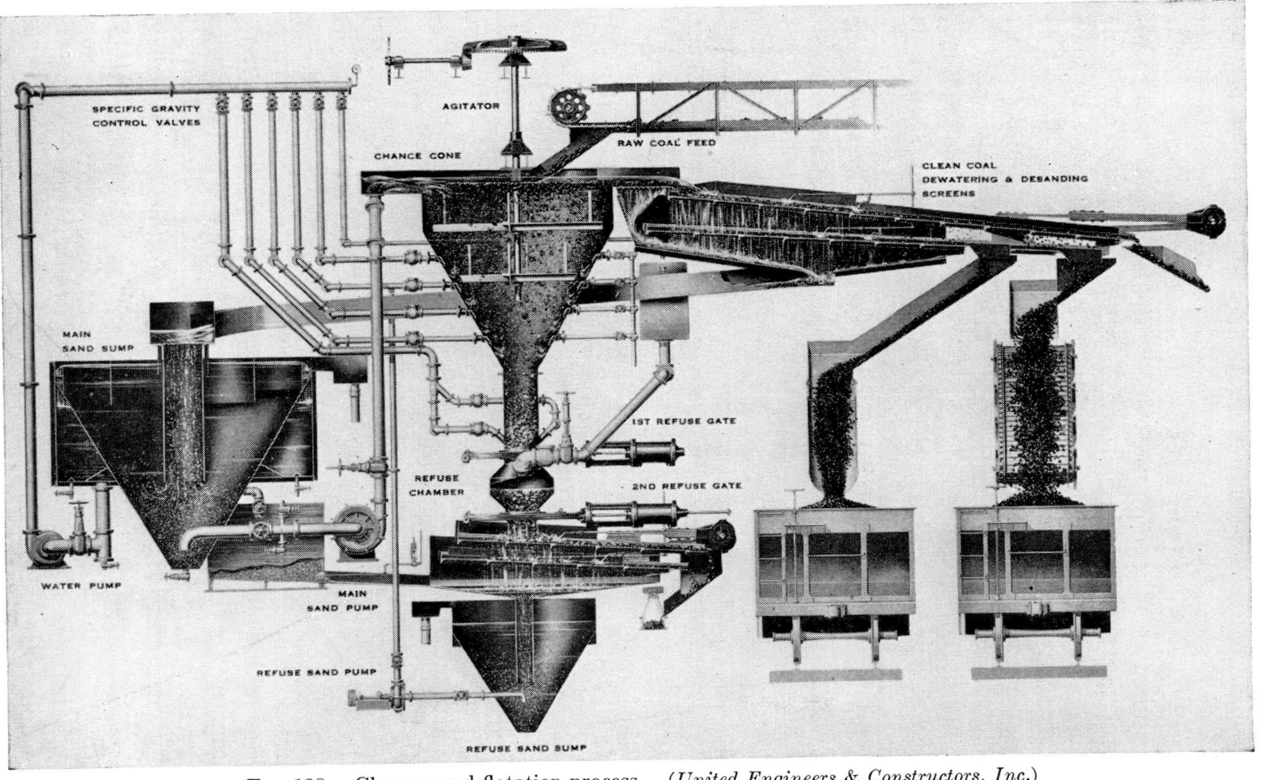

FIG. 128.—Chance sand-flotation process. (*United Engineers & Constructors, Inc.*)

The Chance cleaner (Fig. 128) consists[2,10] of a separating vessel (cone separator) in which the sand suspension moves gently upward. An agitator, by stirring the suspension, prevents packing. The overflow of clean coal and sand passes over clean-coal screens which desand and dewater the coal, spray water being used for desanding. The underflow of the separator passes through refuse valves (two of these valves enclosing a refuse chamber are used to provide a water seal) on refuse screens. These work like the coal screens to desand the refuse. The diluted sand, including sludge coal, is purified in a cone thickener, the sludge coal being wasted. The regenerated medium is returned to the system with new feed.

The specific gravity of the fluid mass is adjusted by varying the proportions of sand and water.

For the cleaning of anthracite, a heavier specific gravity must be maintained in the suspension than is necessary for the cleaning of bituminous coal.

Vooys Process.[3,15] In the Vooys process, the suspension consists of clay and finely ground barite (−150 or −200-mesh) in water. As the process is applied at the Sophia Jacoba mine (Holland), the specific gravity is adjusted at 1.47 and in so far as possible, particles of raw coal finer than 100-mesh are excluded.

Since the solids in the medium are much finer than in the Chance process, the coal that can be treated can also be much finer. This perhaps explains why a coal containing as little as 3.3 to 3.4 per cent ash is steadily produced, with a yield practically equal to the theoretical float-and-sink yield.

Regeneration of the medium requires the use of a thickener; the loss of barite is of the order of 2 lb. per ton of raw coal. The operating cost for the whole plant is given by Berthelot as 1.13 francs per metric ton, or approximately 3 cts. per short ton of raw coal on the basis of treatment of 150 tons per hr. The cost seems lower than for a corresponding jig plant, and the results are better.

Wuensch Process.[26] For the concentration of ores, where the light constituent (waste) has a specific gravity of 2.7 or higher, a mineral having a specific gravity in excess of 5.25 must be used since a suspension containing over 40 per cent solids by volume is too plastic to be utilizable. In fact it would be preferable if the suspension were to contain not over 30 per cent solids by weight, thereby requiring a medium solid of specific gravity 6.7 or higher. Magnetite, hematite, and pyrite are just too light to be used, but galena is suitable.

In the Tri-State (Joplin) district, an installation of 10,000-ton daily capacity to separate chats (locked particles) from waste has recently been completed.

Since galena is relatively valuable, losses of medium must be reduced to the strict minimum. On the other hand, it is imperative to keep the medium clean of degraded ore particles. It is proposed to do this by periodic flotation purification of the medium.

In the Mesabi district of Minnesota there is an experimental plant (Butler Bros.) for the beneficiation of iron ore in which ground ferrosilicon is used as medium. Ferrosilicon was chosen for its magnetic property, its reluctance to oxidize, and its brittleness. Ferrosilicon has a specific gravity ranging from 6.7 to 7.0.[12a] Suspensions having a specific gravity of 3.2 or somewhat higher can be made with this material. The magnetic susceptibility of ferrosilicon makes its choice especially attractive since the installation of magnetic log washers solves the difficult problem of keeping the medium clean.

The prospects are bright for a wider use of heavy-fluid separations not only in the treatment of coal, but also in the treatment of low-grade ores, at least as a preliminary step designed to reject at low cost a large tonnage of coarse, barren tailing while retaining a small tonnage of middling consisting largely of locked particles, suitable as feed for the ultimate dressing process.[23,24,26]

Literature Cited

1. ANON.: Gravity Flotation Process uses Heavy-liquid Separation, *Eng. Mining J.*, **128**, 591 (1929).
2. ANON.: "An Elementary Description of the Chance Sand Flotation Process as Applied to the Washing of Coal," H. M. Chance and Co., Philadelphia (1924).
3. BERTHELOT, CH.: "Épuration, séchage, agglomération et broyage du charbon," Dunod, Paris (1938).
4. BERTRAND, MAURICE: Les cendres du charbon et leur élimination, *Rev. universelle mines*, **10**, 537–544 (1934).
5. BERTRAND, MAURICE F.: Pure Coal and Its Applications, *J. Inst. Fuel*, **8**, 328–336 (1935).

6. Bird, B. M., A. C. Richardson, and G. D. Coe: Washability Studies of the Mary Lee Bed at Hull Mine, Dora, Ala., *U. S. Bur. Mines, Rept. Investigations* 3067 (1931).
7. Bird, B. M.: Interpretation of Float-and-sink Data, *Proc. 2d Intern. Con. Bituminous Coal*, pp. 82–111 (1928).
8. Bird, B. M., B. W. Gandrud, and E. B. Nelson: Washability Studies of the Mary Lee Seam at Lewisburg, Alabama, *U. S. Bur. Mines, Rept. Investigations* 3012 (1930).
9. Chance, T. M.: Application of Sand Flotation Process to the Preparation of Bituminous Coal, *Trans. Am. Inst. Mining Met. Engrs.*, **70,** 740–749 (1924).
10. Chance, T. M.: The Mt. Union Sand-flotation Plant for Preparing Bituminous Coal, *Trans. Am. Inst. Mining Met. Engrs.*, **74,** 573–591 (1927).
11. Chance, Thomas M.: A New Method of Separating Materials of Different Specific Gravities, *Trans. Am. Inst. Mining Engrs.*, **59,** 263–273 (1918).
12. Coghill, Will H.: Degree of Liberation of Minerals in the Alabama Low-grade Red Iron Ores after Grinding, *Trans. Am. Inst. Mining Met. Engrs.*, **75,** 147–165 (1927).
12*a*. DeVaney, F. D., and S. M. Shelton: Properties of suspension media for float-and-sink concentration, *U. S. Bur. Mines, Rept. Investigations* 3469 (1939).
13. Drown, Thomas M.: An Experiment in Coal Washing, *Trans. Am. Inst. Mining Engrs.*, **13,** 341–345 (1884–1885).
13*a*. Foulke, W. B.: Sink-and-float Process for the Beneficiation of Anthracite, read before the *Second Annual Anthracite Conference*, Lehigh University, April (1939).
13*b*. Foulke, W. B.: The Use of Halogenated Hydrocarbon "Parting Liquid" in a Sink-and-float Process, read before the *Am. Inst. Mining Met. Engrs.*, February (1939).
14. Fraser, Thomas, and H. F. Yancey: The Air-sand Process of Cleaning Coal, *Am. Inst. Mining Met. Engrs., Tech. Pub.* 1561-F (1926); abstracted in *Mining and Met.*, **7,** 140 (1926).
15. Gröppel, K.: Mechanical Preparation in a Dense Washing Medium, at Sophia-Jacoba, *Glückauf*, **70,** 429–435 (1934).
16. Holman, B. W.: Water Concentration Tests, *Trans. Inst. Mining Met. (London)*, **39,** 437–445 (1930).
17. McLaughlin, John F.: Control of Chance Cone Operation, *Trans. Am. Inst. Mining Met. Engrs.*, **94,** 324–335 (1931).
18. O'Meara, R. G., and J. Bruce Clemmer: Methods of Preparing and Cleaning Some Common Heavy Liquids Used in Ore Testing, *U. S. Bur. Mines, Rept. Investigations* 2897 (1928).
19. O'Meara, R. G., and B. W. Gandrud: Concentration of Georgia Kyanite Ore., *Trans. Am. Inst. Mining Met. Engrs.*, **129,** 516–519 (1938).
20. Prockat, F.: Amber Mining in Germany, *Eng. Mining J.*, **129,** 305–307 (1930).

21. SULLIVAN, JOHN D.: Heavy Liquids for Mineralogical Analyses, *U. S. Bur. Mines, Tech. Paper* 381 (1927).
22. TOLONEN, F. J.: Sink-and-float Testing, *Eng. Mining J.*, **135,** 161–162 (1934).
23. TOLONEN, F. J.: Heavy-fluid Separation, *Eng. Mining J.*, **137,** 91–92 (1936).
24. TOLONEN, FRANK J.: Experimental Beneficiation of Michigan Iron-bearing Formations, *Mining and Met.*, **18,** 422–424 (1937).
25. TYLER, PAUL M.: Mechanical Preparation of Non-metallic Minerals, *Trans. Am. Inst. Mining Met. Engrs.*, **112,** 789 (1934).
26. WUENSCH, C. ERB: Heavy-fluid Separation, *Eng. Mining J.*, **134,** 320–321 (1933).

CHAPTER XII

JIGGING

Jigging is a process of ore concentration carried out in any fluid and depending for its effectiveness on differences in specific gravity of granular mineral particles. It consists of stratification of the particles into layers of different specific gravities followed by removal of the stratified layers. The stratification is achieved by repeatedly affording a very thick suspension of the mixed particles an opportunity to fall until settled.

Three effects can be distinguished as contributing to the stratification in jigs. They are

1. Hindered-settling classification.
2. Differential acceleration at beginning of fall.
3. Consolidation trickling at end of fall.

Hindered Settling.[26] The essential difference in hindered settling between jigs and classifiers is that in jigging the solid-fluid mixture is so thick as to approximate a loosely packed bed of solids with interstitial fluid instead of a fluid carrying a great number of suspended solid particles; such a thick solid-fluid mixture as is used in jigs cannot be maintained for any length of time, and furthermore it does not allow sufficient play for complete rearrangement of the solids. The jig, by providing an alternately more open and more compact bed, maintains a suspension of very high specific gravity while permitting particle rearrangement during the periods when the bed is open.

The very fact that a jig permits the use of the thickest sort of suspension, a suspension indeed that ceases to be one once during each of the short cycles of pulsation, is one of the reasons for the better mineral separation obtained with it than by the use of sorting classifiers. In Chapter VIII, it was shown that the settling ratio of various solids is a function of the percentage by volume of the solids in the suspension, the settling ratio increasing with increase in solid content, at first slowly as thin suspensions are considered, then rapidly as thick suspensions are considered.

Other things being equal, higher settling ratios are obtainable in jigs than in classifiers.

Differential Acceleration. The theory has been advanced that jigging is more effective than sorting because, in jigging, particles are moving during their accelerating period, and because during that accelerating period, the heavy particles have a greater initial acceleration and speed than the light particles.(29)

It will be recalled that the equation of motion for either viscous or turbulent sedimentation is

$$m \frac{dv}{dt} = (m - m')g - R(v), \qquad \text{[XII.1]}$$

in which v is the velocity of the solid against the fluid, m and m' the masses of the solid and the fluid displaced, and $R(v)$ the resistance exercised by the fluid. Initially, $R(v)$ is nil and

$$\frac{dv}{dt} = \frac{m - m'}{m} g$$

or

$$\frac{dv}{dt} = \left(1 - \frac{\Delta'}{\Delta}\right) g \qquad \text{[XII.2]}$$

if Δ' and Δ are the specific gravities of the solid and fluid.

Clearly, then, galena and quartz in a suspension of specific gravity 2.0 would have initial accelerations in the ratio of 5.5/7.5 to 0.65/2.65, or sensibly 3:1. Their initial speeds, then, would be approximately as 3 is to 1, even though their ultimate speeds were equal, *i.e.*, even though the particles were equal settling.

It follows that if repetition of fall is frequent enough, and the duration of fall short enough, the distance traveled by dissimilar particles should bear more resemblance to their initial accelerations than to their terminal (maximal) velocities. Under those conditions, stratification would take place on the basis of specific gravity alone. The important practical issue is whether such short falls can be realized.

This question can be answered in a qualitative way by referring to Fig. 129 and in a quantitative way by use of Eqs. [VIII.17] and [VIII.18] and Tables 20 and 21, Chapter VIII.

On referring to Fig. 129, whose coordinates are velocity and time, there appear three curves: *OAS* and *OBS* which are the $v - t$ curves of equal-settling particles of different specific gravities, OAS being the curve for the heavy particle; and OCS' which is the $v - t$ curve for some finer particle of the heavy mineral. Clearly, the distances traveled in a given time by equal-settling particles are equal at infinite time only. For a short time, such as t_f, the gain in distance traveled by the heavy mineral may be substantial (Eq. [XII.2]). The finer heavy particle may fall in such a way that in the time t_f it travels as far as the light particle (areas $OCED$ and $OBFD$ equal). From a stratification standpoint, particles C and B are equal jigging, but particles A and B are not.

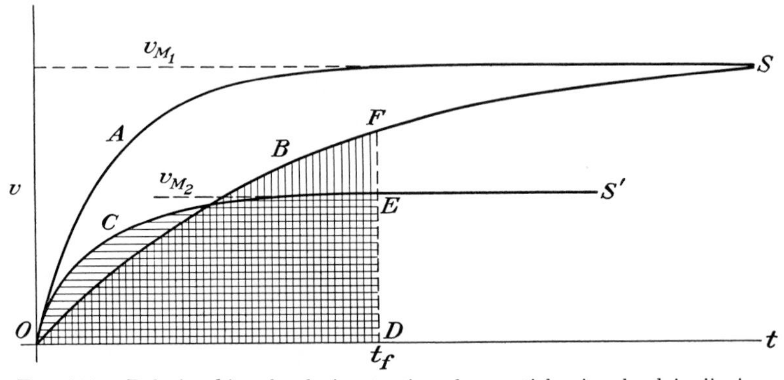

Fig. 129.—Relationship of velocity to time for particles involved in jigging.

Equation [VIII.19] relates the distance traveled by a particle falling under the action of gravity with Newtonian resistance as

$$s = Av_m \ln \frac{2 + e^{t/A} + e^{-t/A}}{4}$$

in which

$$A = \frac{\Delta}{2(\Delta - \Delta')g} v_m.$$

In these relations, v_m is the terminal velocity given by Eq. [VIII.8]:

$$v_m = \sqrt{\frac{8}{3Q} g \frac{\Delta - \Delta'}{\Delta'} r},$$

t is the time, r the particle radius, Δ and Δ' the specific gravities of the particle and fluid, g the acceleration of gravity, all in c.g.s. units, and Q the coefficient of resistance. This coefficient of resistance may well be taken as 0.4 for spheres in the range of sizes in which turbulent resistance prevails; for nonspherical particles, other values may be selected (page 196, Chap. VIII).

From these relationships, it is possible to chart, for a given set of conditions (as to specific gravity and size), the distance traveled per second s_x/t_x if repeated falls of duration t_x are afforded to the particles.

Figure 130, *e.g.*, shows the relationship of s_x/t_x to t_x for quartz spheres 0.8 cm. in radius (curve I), galena spheres equal settling (free-settling) with the quartz spheres and of radius 0.2 cm.

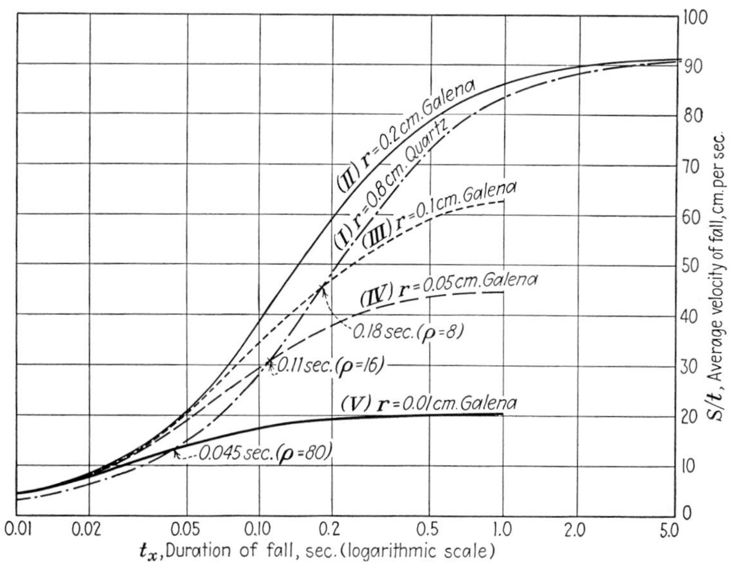

FIG. 130.—Relationship of average distance traveled per second in a jig to duration of each fall, for quartz spheres 1.6 cm. in diameter and for various galena spheres.

(curve II), and galena spheres equal jigging with the quartz spheres for various values of t_x (curves III, IV, V).

It is of interest to note that the *jigging ratio* ρ, which may be defined as the ratio of diameters of equal-jigging particles of different sizes, varies enormously with the duration of fall t_x. This is well shown by Table 36 which presents the results for two cases:

1. Galena and quartz regarded as settling freely in water.
2. Middling particles of specific gravity 2.95, and quartz 2.65 in a heavy fluid of specific gravity 1.80.

TABLE 36.—JIGGING RATIO AS A FUNCTION OF DURATION OF FALL DURING EACH JIGGING CYCLE
(Quartz spheres = 0.8 cm. radius)

Time t_x, seconds	*A* galena and quartz in water	*B* middling particles (Δ = 2.95) and quartz(Δ_2 = 2.65) in fluid (Δ' = 1.80)
∞	4.0	1.35
0.50	4.7	1.45
0.20	7.2	2.0
0.15	10.3	2.8
0.10	18.1	5.0
0.075	32.	8.6
0.05	63.	16.5

These cases may be regarded as practical extremes for easy and difficult jigging.

Clearly, if jigging is practiced on unsized feed or on poorly sized feed, a very short settling time must be used if stratification is to result (if consolidation trickling is temporarily overlooked). If jigging is practiced on feed closely sized by screening, stratification may result even if a long settling time is used.

Rate of Stratification. Figure 130 shows also the distance gained per second by one type of particles over another. Thus, the galena particles 0.2 cm. in radius gain over the equal-settling quartz particles regardless of the duration of each settling period. But the gain in centimeters per second is nil if the time of settling is infinitely small or infinitely large, the maximum occurring at a settling time of about 0.17 sec., and at the rate of 12 cm. per sec. of fall. Galena particles equal jigging with the quartz at 0.18 sec. gain on the quartz if shorter cycles are used—the maximum gain being for settling times of 0.08 sec. and at the rate of 7 cm. per sec.—but lose if longer cycles are used, the loss being, for example, 28 cm. per sec. if a settling time of 0.50 sec. is used.

Figure 130 shows that the spread between curves I and II is large throughout a wide range of values for the duration of fall. Hence if the jig were fed a classified feed, the jigging cycle could be varied within wide limits. If, however, the jig were fed unsized feed, or very poorly sized feed, the duration of the jigging cycle would have to be carefully chosen so as to yield the settling time giving the optimum results.

The rate of gain of the fine heavy mineral is less than the rate of gain of the coarse heavy mineral; so a larger number of settling periods is required if stratification of unsized feeds is to be attained. Thus if a jig treating gangue as coarse as $\frac{5}{8}$ in. were required to save galena as fine as 1 mm., its capacity would be but one-third as great as that of a jig treating classified feed of the same maximum size: the jigging gain of fine galena over quartz *vs.* the jigging gain of equal-settling galena over quartz, both at optimum settling times are, respectively, 4 and 12 cm. per sec. Furthermore, the duration of fall would have to be reduced from 0.18 to 0.05 sec. If, on the other hand, the jig were fed screen-sized material, the capacity would be increased and a much longer settling time (*e.g.*, 0.5 sec.) would be advantageous.

Figure 130 and Table 36 deal with the case in which the quartz has a diameter of 1.6 cm. For other diameters, the same chart and table can be used by making adjustments as follows:

For constant Δ and Δ', the factor Av_m varies directly as v_m^2, or directly as r; if t is chosen to vary as $r^{1/2}$, t/A is unchanged; therefore by changing t_x directly as $r^{1/2}$, s_x/t_x varies also as $r^{1/2}$, and the relationship between s_x/t_x and t_x remains as shown in Fig. 130. So if all particles under consideration are made four times smaller, for example, the gain of one type of particle over another, $(s_1 - s_2)/t_x$ is cut in two, and the time required t_x is also cut in two.

From Table 36, it is seen that if a ratio of size of gangue to valuable particle of 2 is permissible, and if the gangue particles are 1.6 cm. in diameter, it is necessary to reduce the time of settling to 0.20 sec. in order to jig locked particles of specific gravity 2.95 from gangue of specific gravity 2.65 in a pulp of specific gravity 1.8. But if the gangue particles are 0.4 cm. in diameter, results of the same quality should be obtained only if the time of settling were reduced to 0.10 sec. And if the gangue particles are 0.1 cm. in diameter, the time of settling should be reduced further to 0.05 sec. To obtain a jigging ratio of gangue to locked particles of 5:1, and if the gangue particles are 0.1 cm. in diameter, the time of settling would have to be reduced still further to 0.025 sec. The short jigging cycles indicated by this analysis are beyond present-day machines.

Consolidation Trickling. In view of the fact that different particles of either the same or different specific gravities do not travel the same distance during one of the settling periods which is given them, they will come to rest at different instants. A coarse particle may remain in suspension, out of a cycle of 0.30 sec., perhaps only 0.06 sec., although a small particle may remain suspended as long as 0.20 sec. Manifestly, however, a period of time exists during which the fine particles are settled on top of a bed of coarse particles. The coarse particles are bridging against each other and incapable of movement although the fine particles are still free to move. Aside from any velocity that may be imparted to these small particles by the moving fluid, they are bound to settle under the influence of gravity in the passages between the coarse particles. The phenomenon is most pictorially described as *consolidation trickling.*

Consolidation trickling may be regarded as representing a condition under which sedimentation of fine particles continues while coarse particles are "bound and gagged." Of course the fine particles do not settle as rapidly during consolidation trickling as during suspension, but if consolidation trickling can be made to last long enough the effect can be most important.

In order to visualize the occurrence of consolidation trickling, it may be well to consider, first, classification at constant flow: the particles become arranged with the fines at the top and the coarse below. If classification with a pulsating flow is used, such that the flow never becomes nil, the stratification is substantially the same. As the velocity of the flow during the ebbing period decreases, a new phenomenon may appear. If the velocity becomes small enough to permit consolidation of the coarse grains, finer grains may trickle through the meshes of this network. With the return of the flowing period, the particles start with a distribution controlled not only by hindered-settling and initial differential acceleration, but also by consolidation trickling.

Consolidation trickling, then, appears in pulsator jigs (page 266); it is much more marked in jigs using suction along with pulsion. It is most marked in suction-pulsion jigs in which the cycle is such as to give a long time for suction.

Consolidation trickling seems to be a general phenomenon not limited specifically to jigging. Experiments by Dyer,[13] for example, have produced it by vibration of dry solids in a trough. Figure 131 is reproduced from his paper and shows graphically what may be secured by way of stratification when

FIG. 131.—Consolidation trickling. (*After Dyer.*)

(*A*) A mixture of steel and wooden balls of various sizes, before agitation. The wooden balls are bored, to distinguish them from the metal balls.

(*B*) The same mixture after agitation. Layers of homogeneous material have been carefully picked off to disclose what lies underneath.

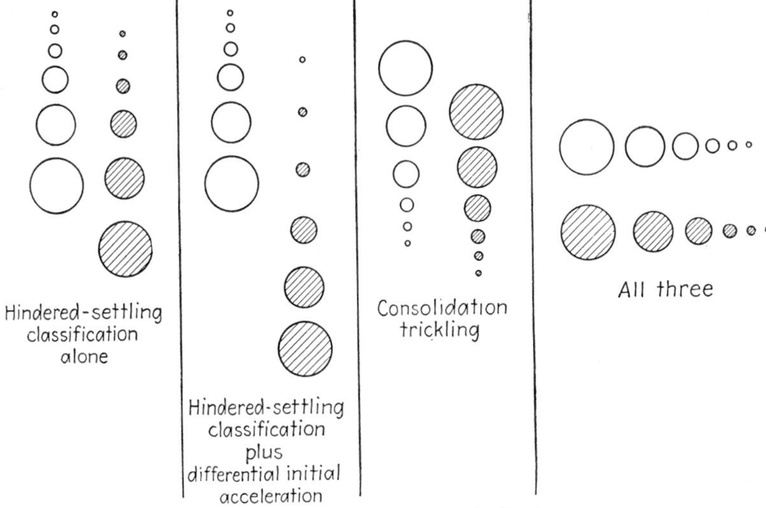

FIG. 132.—Factors effective in jigging.

using only lateral vibration; *i.e.*, if every one of the other stratifying factors used in jigging are excluded. Dyer has termed the phenomenon reverse classification.

To summarize, in jigging, stratification during the stage that the bed is open is essentially controlled by hindered-settling classification as modified by differential acceleration, and during the stage that the bed is tight, it is controlled by consolidation trickling. The first process puts the coarse-heavy grains on the bottom, the fine-light grains at the top, and the coarse-light and fine-heavy grains in the middle. The second process does in a measure the reverse, putting the fine-heavy grains at the bottom, the coarse-light grains at the top, and the coarse-heavy and fine-light grains in the middle. By varying the relative importance of the two, and by varying the importance of differential acceleration, an almost perfect stratification according to specific gravity alone can be obtained. Figure 132 summarizes these ideas.

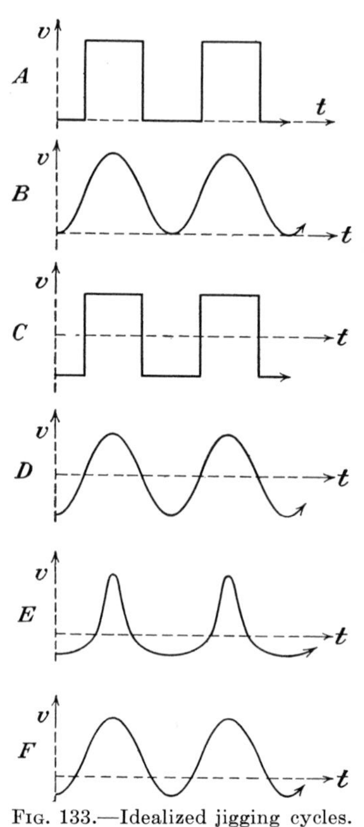

Fig. 133.—Idealized jigging cycles.

Jigging Cycles.[25] Jigging cycles are said to consist of pulsion and suction. In pulsion, the fluid is moving upward and in suction downward with respect to a stationary reference point. All jigs use pulsion, and most jigs suction, but the latter is avoided in some jigs. Figure 133 shows several typical jigging cycles.

Types *A* and *B* use pulsion only.

Types *C* and *D* use pulsion and suction, the motion being wholly symmetrical.

Type *E* uses equal amounts of pulsion and suction in the sense that as much fluid flows up as down, but it is otherwise asymmetrical.

Type *F* uses symmetrical motion, but with more pulsion than suction.

Type A. This jigging cycle (Fig. 134) is theoretical as the velocity changes *ab* and *cd* at times t_1 and t_2 connote infinite acceleration, which cannot be realized.

If the bed of particles is wholly compacted at time t_1, the particles begin to move under the influence of the fluid pulsion. Beginning at time t_1, the particles are moving upward at a velocity indicated by the vector **mp** + **mn.** This sum decreases from time t_1 to time t_2. At time t_2, it suffers a great decrease since **mp** becomes suddenly nil. After time t_2, the particle

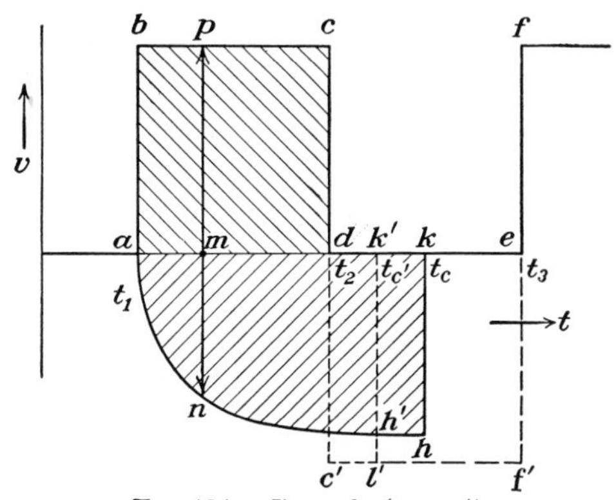

FIG. 134.—Jig cycle (type *A*).

continues to fall until at time t_c the area *ahk*, which represents the downward distance traveled by the particle, equals area *abcd*, which represents the upward travel of the fluid.

Three possibilities present themselves:

1. t_c is greater than t_3. In that case, the jig is working as a pulsator classifier.

2. t_c is smaller than t_3 but greater than t_2. The jig is working properly.

3. t_c is smaller than t_2. This means that the particle has come to rest before completion of the pulsion stroke. The pulsion is too weak or the cycle is too long.

Type B. This theoretical jigging cycle with a sine-wave motion (Fig. 135) can be approximated closely in pulsator jigs.

During the cycle *obcde*, the particle is at first not lifted in spite of pulsion, then as the pulsion becomes sufficiently large, at time t_a, the particle is lifted and it remains suspended until time t_c when it comes to rest at its starting position. The time t_a corresponds to the condition that the acceleration of the particle a_2 equals in absolute value the acceleration a_1 of the fluid (Fig. 135*b*). The particle comes to rest when the distance it has traveled against the fluid (the *v-t* area *fhjgf*) is equal to the distance traveled by the fluid (the *v-t* area *fbcdgf*). Jigging

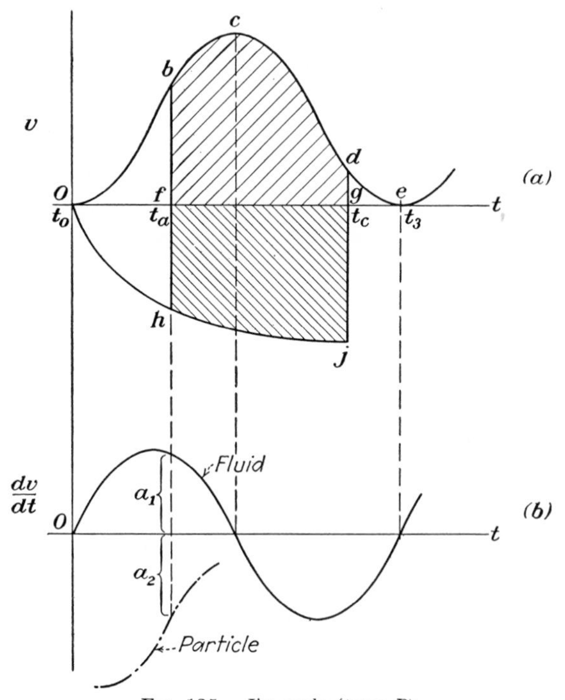

FIG. 135.—Jig cycle (type *B*).

is obtained if t_c is less than t_3, pulsating classification if t_c is greater than t_3.

Type C. This type differs from *A* merely in that the fluid returns to the same position after completion of a full cycle. A full cycle might be represented by *abcdc'f'e* in Fig. 134. The particle comes to rest at some time $t_{c'}$ such that area *anh'k'a* (Fig. 134) equals the algebraic sum of areas *abcda* and *dc'l'k'd* (which are of opposite sign). A cycle of this type cannot operate in the fashion of a pulsator classifier since there is no net upward fluid flow: the first possibility listed under type *A* cannot arise.

Type D. This type differs somewhat from B as may be seen by comparing Fig. 136 with Fig. 135. The essential feature of difference is that acceleration of the particle must be considered to begin at time t_1 when the fluid velocity changes sign to an upward velocity. Since the fluid acceleration at time t_1 is not nil, it is obvious that the time interval $t_a - t_1$ which must elapse is less than the time interval $t_a - t_0$ in the case of jigging cycle B. Indeed, if a_1 is greater than the acceleration of gravity, all

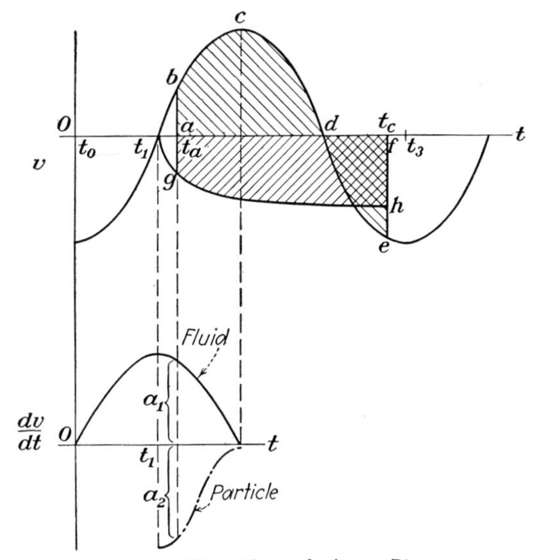

FIG. 136.—Jig cycle (type D).

the particles in the jig simultaneously become suspended at time $t_a = t_1$.

Again with reference to Fig. 136, the particle comes to rest at t_c such that area *aghfa* equals the algebraic sum of areas *abcda* and *defd*.

As in the case of type C, the jig cannot act as a sorting classifier since there is no net upward fluid flow.

In the analysis of these jigging cycles, the effect of fluid acceleration has been neglected, because it must be more or less compensating, this acceleration being positive and then negative. This, however, is merely an assumption as no rigorous yet usefully simple analysis seems possible.

It might be sufficient to indicate that the differential equation of motion instead of being

$$m\frac{dv}{dt} = (m - m')g - R(v)$$

should become modified to a form of the type

$$m\frac{dv}{dt} = (m - m')(g + a(t)) - R(v),$$

in which $a(t)$ indicates that this additional acceleration term is a function of time.

Removal of Stratified Layers. Much has been said about methods followed to achieve stratification of dissimilar minerals. It is now pertinent to inquire into the methods available for removal of the stratified layers.

Removal of the bottom layer is obtained by arranging to pass this layer through the sieve (hutch jigging), by drawing it through a well or a gate on the sieve, or by causing it to travel on the sieve to the discharge end. Progress of the bottom layer may be facilitated by use of a sloping screen. Removal of the top layer is generally obtained through crowding by new feed.

The problem of removal of stratified layers is one that deserves more study than it has received to date. Much of the difference between a successful jig and a poor jig depends upon the application of sound engineering principles to this problem.

Jigging Methods. Jigging on the sieve requires removal of both concentrate and tailing from the jig bed after allowance of enough time and travel to produce the desired stratification. Jigging on the sieve is preferably conducted with a slight deficiency in suction.

Jigging through the sieve requires removal of the heavy product from a hutch below the jig sieve. A mineral bed is required on the sieve of such size as to prevent the passage of the light product into the hutch, while permitting passage of the heavy product into the hutch. The sieve, also, must be coarse enough for ready passage of the heavy product but fine enough to retain the bed material.

Jigging in such a way as to cause travel of the bottom stratum to one end is obtained in movable-sieve jigs. They are characterized by a large capacity.

Design of Jigs. Generally speaking, the fundamental features of jigs were known from antiquity onward, but little progress was made until recent times. The principal features of jig design that require attention are

1. Development of a proper jigging cycle, with ready adjustments as to length of stroke, duration, and character of cycle.
2. Even transmission of jigging motion from point of application to point of utilization of motion.
3. Use of suitable bed material or *ragging*, whenever a hutch product is secured.
4. Rapid evacuation of strata, and conveyance from jig.
5. Design with respect to the relative tonnages of heavy and light strata.

TYPICAL HYDRAULIC JIGS

Hand Jigs. The simplest jig consists of a framed sieve held in hand and actuated by the operator with a reciprocating vertical motion. This is satisfactory for elementary laboratory demonstration.

A simple manually operated movable-sieve jig (Fig. 137) is described by Richards.[26] It consists of a glass tube cut in two at *tt* where a sieve cloth is inserted and held together by wooden bars, bolts, and nuts. The jig is actuated by crank *p* about the fulcrum *k*.

Fixed-sieve Plunger Jigs. The *Harz jig* (Fig. 138) has a fixed sieve. The jigging motion is obtained by a plunger *P* reciprocating in a compartment adjoining the sieve compartment *C*. The bottom layer (usually the concentrate) is removed through the gate *A* after passage into the well or "draw" *B*. The upper layer (usually the tailing) is discharged at the end away from the feed.

Harz jigs are usually built of wood, but construction of concrete has been reported.[24] They are built of several compartments in series, the tailing from one compartment passing as feed into the next compartment. The amplitude of the jigging is greatest in the first cell and least in the last, so as to make concentrate in the first compartment and middling in the others. "Rising water" is added to compensate for excessive suction either above or below the plunger. Variations in the relative use of pulsion and suction are obtained by varying the amount of rising water and by controlling the opening of the hutch draw (*D*, Fig. 138).

The length of stroke usually ranges from 0.5 to 8 cm., depending upon the fineness of the feed, long strokes being used for

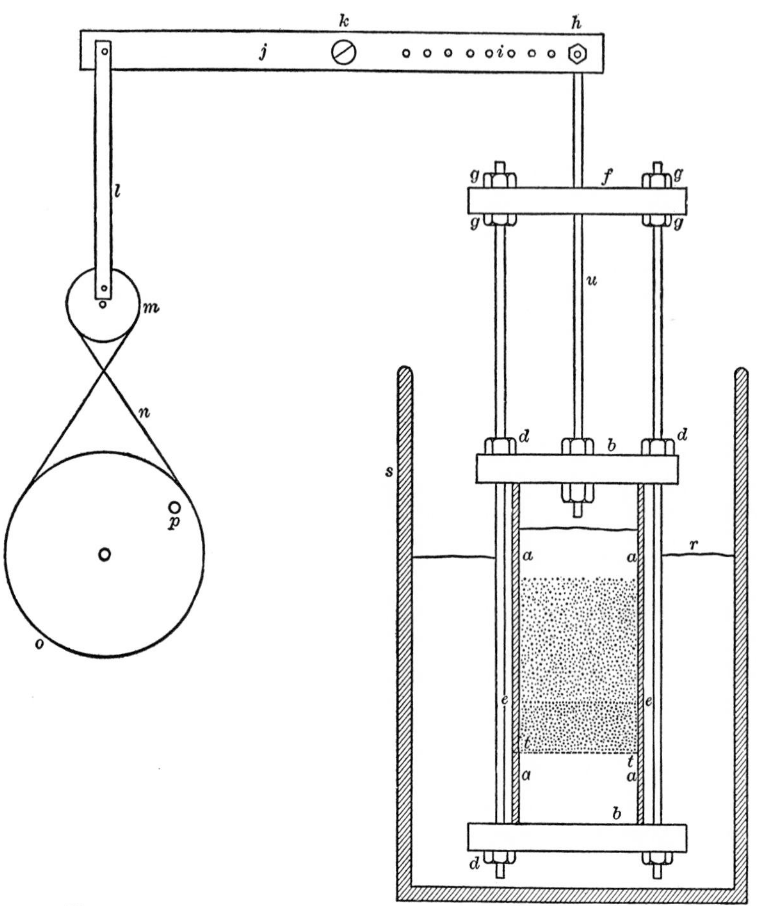

Fig. 137.—Laboratory movable-sieve hand jig. (*After Richards.*)

coarse feed. Jigging cycles range from 0.2 to 0.6 sec. (100 to 300 strokes per minute).

Power requirement ranges from 0.1 to 0.15 hp. per sq. ft. of sieve area. The amount treated ranges from 1 to 4 tons per sq. ft. of sieve area per 24 hr., the larger tonnage being obtained on coarser feed.

Many variants of the Harz jig have been developed. Of these, one of the most widely used has been the *Cooley jig*. In this jig, a relatively coarse screen is used on which a jig bed or "ragging" of suitable size and specific gravity is placed. Most of the concentrate is recovered as a hutch product, but some concentrate or middling usually builds up on the screen. The specific gravity of the ragging substance should be approximately the same as that of the hutch product which it is expected to draw from the jig.

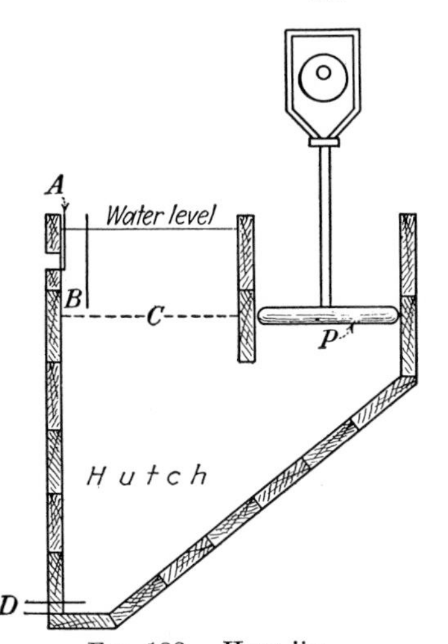

FIG. 138.—Harz jig.

The *Woodbury jig* is a fixed-sieve quick-return plunger jig (asymmetrical cycle).

Fixed-sieve Diaphragm Jigs. The *Bendelari jig* (Fig. 139) differs from the plunger jigs in that the plunger is sealed to the frame by a rubber diaphragm. This prevents leakage of water around the plunger, a frequent difficulty with Harz jigs. Movement of the diaphragm head is vertical, and a long stroke is possible, with the weight of the head carried on the crosshead guide, and not thrown against the rubber, which acts as a seal only. Action of the diaphragm produces positive pressure, and the bed is lifted with less pulsation. Similarly, more suction can be had if desired.

Water from a manifold is admitted below the sieves through check valves on the downstroke of each diaphragm. The amount of water so admitted is regulated by the control valve on each cell, and by manipulation of the valves the action of the jig can be changed from suction to pulsion while it is running. Addition of this water produces a free or loose bed on the downstroke and also results in greater pulsion on the upstroke.

An additional attractive feature of the Bendelari jig is the accessibility of the jigging surface, which is obtained by placing the actuating mechanism at the bottom. This also results in an appreciable saving in floor space and weight.

The *Jeffrey jig* (Fig. 140) is of the fixed-sieve type, with the sieve set at an adjustable angle. It is used for cleaning coal. The motion is imparted to the water by a cam, through adequate levers and rods. The plunger *d*, at the bottom of the hutch, is sealed by live rubber. A float *a* controls the discharge from the jig by actuating at *e* the slate-discharge gate. The length of stroke can be adjusted at *f*.

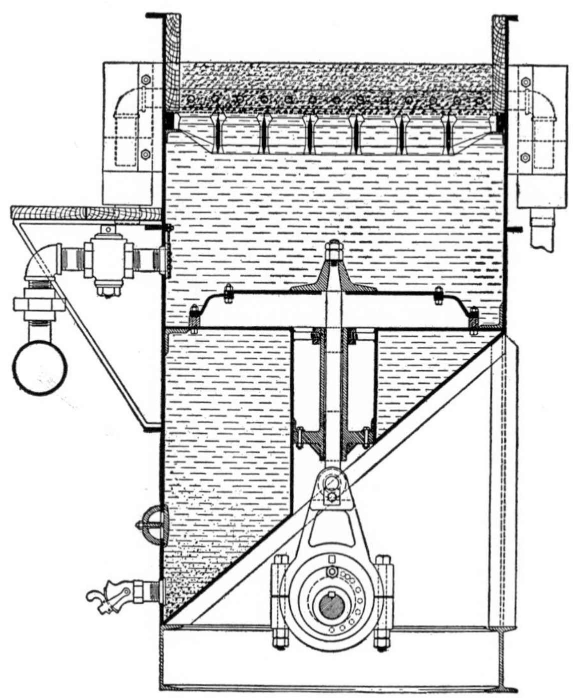

FIG. 139.—Bendelari jig. (*Fred. N. Bendelari.*)

Fixed-sieve Pulsator Jigs. Figure 141 shows a *Richards pulsator jig*. This jig has a rated capacity of 90 tons per day, each of the four compartments having a screen surface only 4 by 4 in. Water is admitted through a rotating valve *V* at 150 to 200 strokes per minute. Water requirement is relatively large, power consumption is low. Clean concentrates are easily obtained, but clean tailings are not the rule.

The *Pan-American pulsating jig* (Crangle jig) is characterized by a new type of valve which permits a higher frequency than other types of jigs, up to 600 per minute (cycles of 0.10 sec.). A general view of the jig is shown in Fig. 142. The valve appears in Fig. 143. Water under a head of 10 to 75 ft. comes in at *A*, pushes the valve *E* off its seat against the pressure of spring *C* and the flexing of the rubber diaphragm *D*, thus passing into the jig at *B*; this reduces the head of the water and the valve closes, only to reopen very soon.

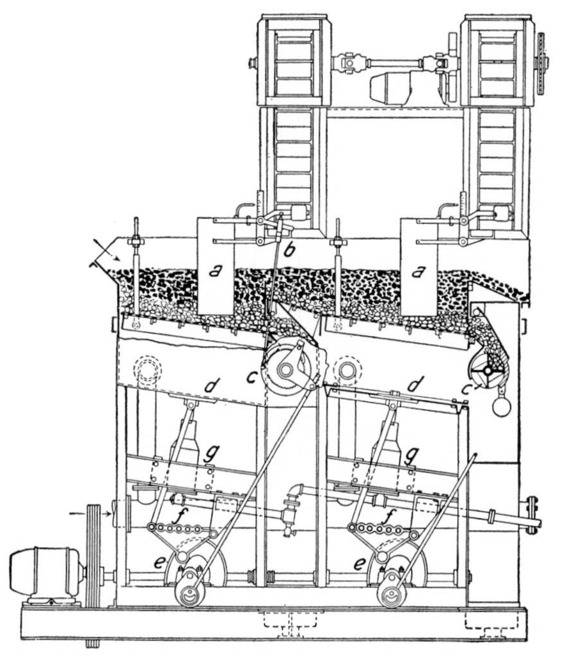

FIG. 140.—Jeffrey diaphragm jig for cleaning coal. (*Jeffrey Mfg. Co.*)

The Pan-American jig is especially designed for the recovery of fine gold from placer operations or as a unit placed within a ball-mill classifier circuit.[16,32]

Fixed-sieved, Air-pulse Jigs. The parent of this type of jig is the *Baum jig*, known in Europe since 1892.

In the *Simon-Carves jig*[12] (Fig. 144), this principle is utilized for cleaning coal. The jig is shaped like a U tube with one side sealed and the other opened. The sealed side alternately receives and exhausts compressed air through a piston valve. This communicates pulsion and suction, respectively, to the other side of the U tube in which perforated plates carry the coal and refuse. The complete machine consists of five cells.

Feed enters at the right end of the machine at *A*, and clean coal leaves at the other end of the machine, at *B*. Tailings dis-

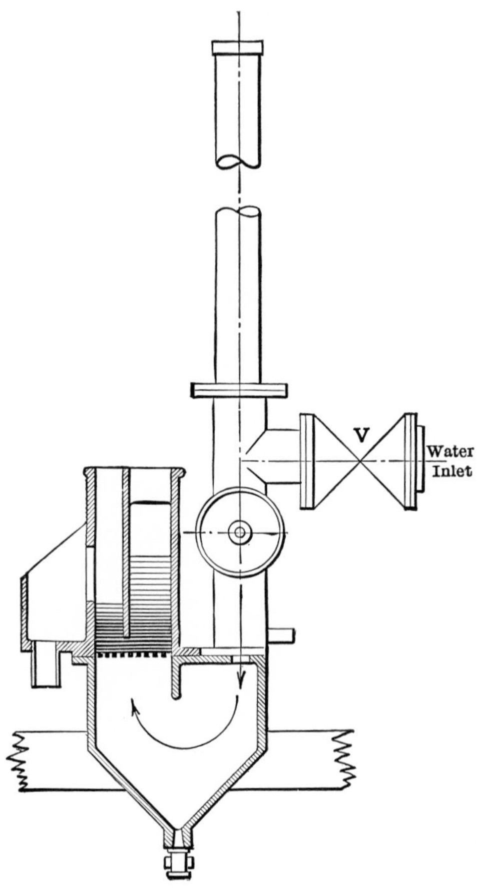

FIG. 141.—Richards pulsator jig. (*After Richards and Locke.*)

charge by a gate under the feed chute to the right of the machine, at *C*; bone is likewise discharged by the gate under the clean-coal discharge, to the left. The hutch product is conveyed to the elevators where it joins the tailing and bone.

A similar jig (Schiechel type) with special concentrate well was devised for the recovery of diamonds not amenable to grease separation.[1]

Movable-Sieve Jigs. The *Hancock jig*[22] (Fig. 145) is typical. It consists of a rectangular tank, a movable sieve, and an actuating mechanism. The motion imparted to the sieve is composed not only of a reciprocating up-and-down motion, but also of a fore-and-aft motion, with greater acceleration at one end than the other. The compound motion is obtained from a cam through a set of levers, links, and rocker

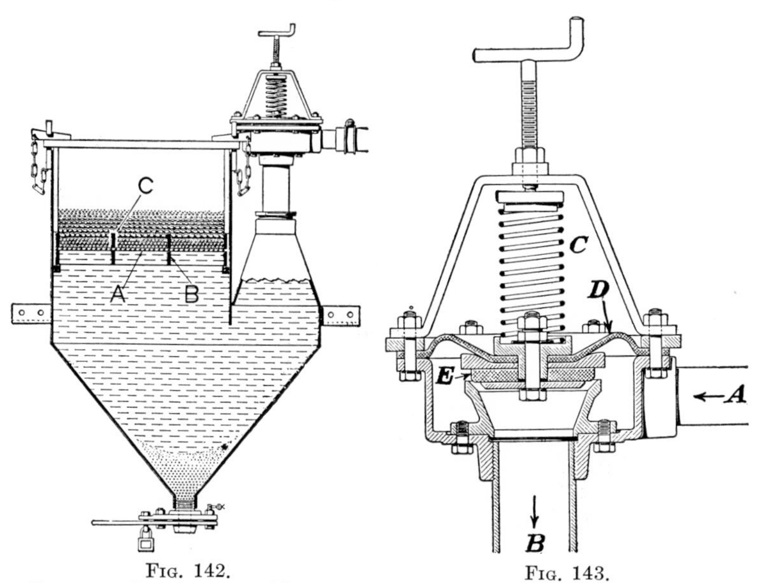

Fig. 142. Fig. 143.

Fig. 142.—Pan-American (Crangle) pulsator jig. Cross section at right angles to direction of pulp flow. (*A*) screen, supporting shot bed; (*B*) grids supporting screen; (*C*) shot-retaining grid, which clamps down the screen. (*Pan-American Engineering Corp.*)

Fig. 143.—Valve of Pan-American (Crangle) jig. (*Pan-American Engineering Corp.*)

arms. The effect of this motion is to cause travel of the solids from one end of the machine to the other.

Jigging is through the sieve, concentrate accumulating in the hutch of the first compartment and a low-grade middling in the hutch of the last compartment. The other compartments make concentrate or tailing according to the characteristics of the ore and requirements.

The Hancock jig is noted for its great capacity which ranges from 300 to 600 tons per day for a machine 25 ft. by 4 ft. 2 in. It is widely used.

Movable-sieve jigs for coal preparation, also known as pan jigs, are typified by the *James coal jig* (Fig. 146). The sieve is moved by a crank mechanism, and water is admitted to the hutches to lessen the suction. Jigging is on the sieve, a bed of very heavy ragging being used to prevent passage of slate in the hutch. Discharge of the coal is from the end of the jig, and of the slate from the side, after passage under a baffle, by a gate-and-dam discharge.

Fig. 144.—Simon-Carves coal-washing jig. (*Link-Belt Company.*)

Automatic Slate Discharge for Coal-cleaning Hydraulic Jigs. Variations in rate of feed and in slate content of feed require constant vigilance on the part of the operator of a coal jig if the products of the jig are to be of uniform quality. This has led to the adoption of automatic control devices for slate discharge. These automatic slate-discharge devices are based on one of three principles:

1. Utilization of the pressure exerted by the slate bed on the gate to open the gate more or less widely (Schuchtermann-Kremer device, extensively employed in Germany).

2. Utilization of the pressure exerted by the water below the jig sieve to control the gate opening. In the Bamag device, the pressure of the water makes or breaks electrical contact in a mercury manometer, thus starting or stopping a small motor which actuates the gate.

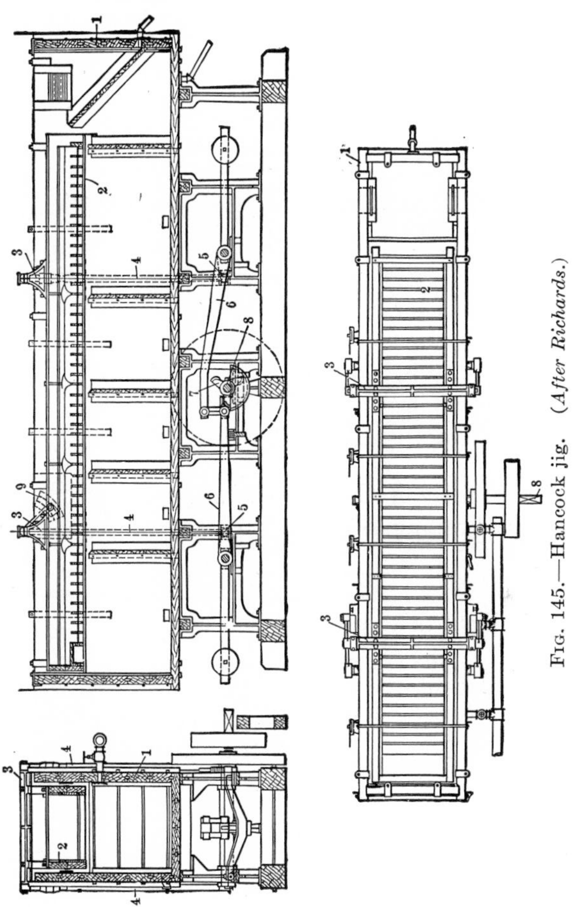

Fig. 145.—Hancock jig. (*After Richards.*)

3. Utilization of the variation in specific gravity of the jig bed with variation in slate level, to raise or lower a float and thereby open or close the gate. This is used in the Simon-Carves, Humboldt-Deutzmotoren, Baum, and other jigs.

Adoption of automatic slate-discharge devices has generally resulted in significant improvements in over-all results: cleaner coal, dirtier waste, and smaller tonnage of middlings.[7]

Use of Heavy Suspensions in Jigging. Attention has been called by Rose (27) to the advantage that would result from the use of heavy suspensions in place of water in jigging. Such a practice is, of course, limited to ore particles coarse in relation to the particles of the suspension. It will result in a definite

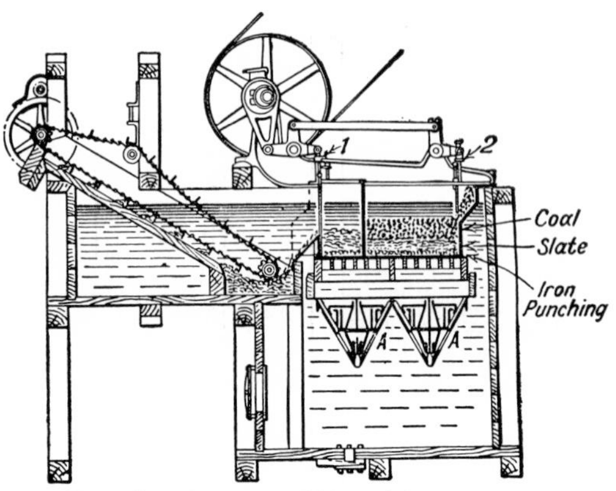

FIG. 146.—James coal jig. (*After Taggart.*)

advantage especially if the spread in specific gravity between concentrate and tailing is small.

PNEUMATIC JIGS

Pneumatic jigs utilize stratification of particles in air instead of in water. They have not been widely used in the treatment of ores and would continue to remain a technological curiosity if it were not for their recent adoption in coal cleaning. The latest estimates, however, place the tonnage of coal treated, the world over, by pneumatic jigs at some 30 to 40 million tons annually.

Because current models of pneumatic jigs use a shallow bed of mineral and are surfaced with riffles in the manner of a shaking table, they are commonly known as *pneumatic tables*. This term is a misnomer, unless the word "table" is used to connote something different from what is known under that name in the art of wet gravity concentration. Actually, the resemblance is considerable between water jigs and pneumatic tables.

All pneumatic jigs or tables produce first a vertical stratification of the particles in order of decreasing specific gravity more or less modified by size. Disposal of the stratified layers can be obtained by removal of strata *in situ* by means of trap gates, concentrate wells, or tailing wells. It can also be obtained by overflow if the strata are fanned out horizontally by the use of suitable auxiliary motions.

Pneumatic Jigs Using No Horizontal Stratification. The *Plumb jig*[31] (Fig. 147) is a pulsator jig controlled by a rotary valve and using 400 to 500 pulsations per min. of 10- to 30-lb. air. Capacity is 0.5 ton per sq. ft. per hr. for the coarsest feed

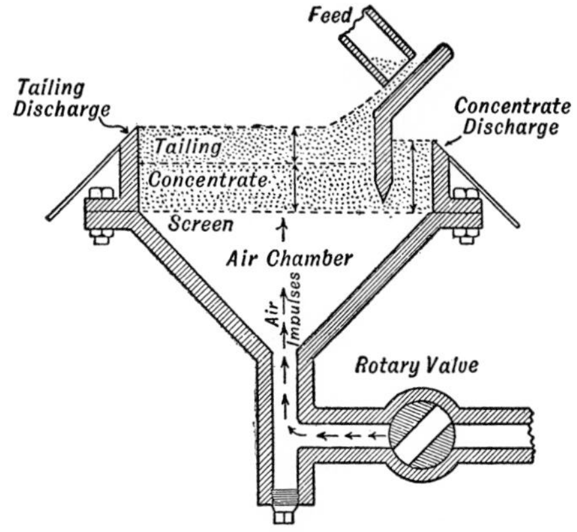

FIG. 147.—Plumb pneumatic jig. (*After Truscott.*)

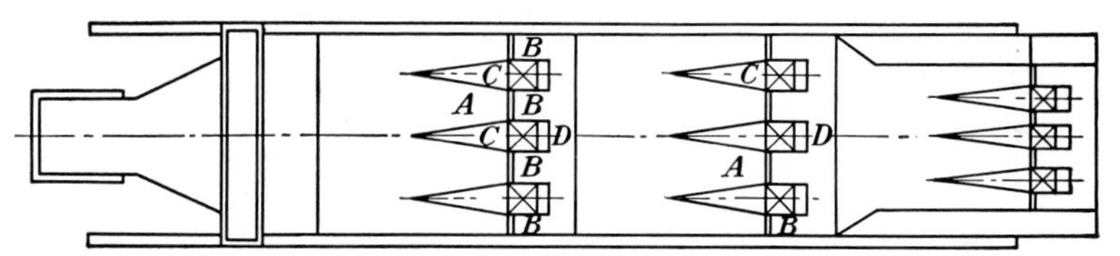

FIG. 148.—Meunier pneumatic table (top view). (*After Berthelot.*)

down to a small fraction of this for fine feed. Close sizing is required. This jig was used for concentration of sulphides.

The *Meunier pneumatic table* is a three-section movable-sieve jig in which removal of concentrate (coal) and of tailing (slate) proceed simultaneously, beginning from the end of the first section of the jig (Fig. 148). This is done by the insertion of V-shaped riffles *A* which crowd the bottom stratum toward the ports *B* while drawing the upper stratum toward the wells *D*. The ports *B* are of course submerged. Regulation of the openings *B*, and of the level at which a cut is made at *D*, controls the amount and grade of the products. Similar withdrawals are made in the second and third sections of the jig.

Pneumatic Jigs Using Horizontal Stratification. Fanning out of the stratified layers in pneumatic jigs prior to their evacuation is accomplished by guiding the strata in different directions. This is accomplished in three different ways.

1. By the use of two superimposed and intersecting sets of riffles which guide the strata to two sides of a fixed-sieve pulsator jig (*Hooper jig*, Fig. 149).

2. By the use of a longitudinally sloping movable-sieve jig, so contrived that the light material (coal) rolls downhill while the heavy material (slate) is guided uphill by the motion of the jig (*Revelart-Berry jig*).

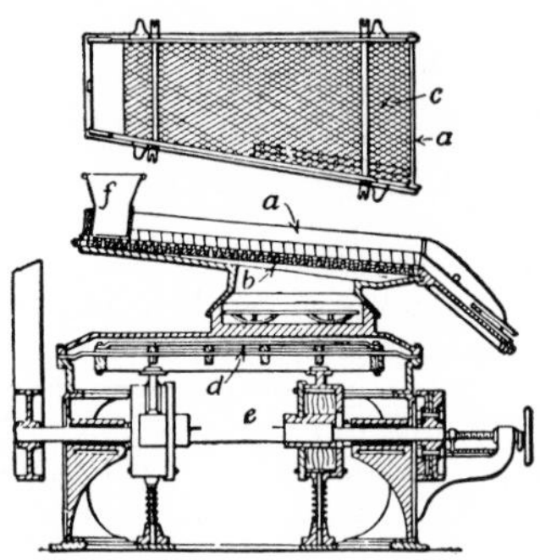

Fig. 149.—Hooper pneumatic pulsator jig. (*After Taggart.*)

3. By the use of a side-sloping movable-sieve jig equipped with tapering riffles so adjusted as to deliver the various strata along its side, from light material at the feed end of the discharge side (coal) to heavy material (slate) at the opposite end. This is the device that is generally termed a pneumatic table. The best known are the *Birtley* (Fig. 150) and the *Sutton-Steele tables* (Fig. 151).

In the Birtley and Sutton-Steele pneumatic tables, the deck is given a longitudinal motion similar to that given to a movable-sieve hydraulic jig with the result that the bed travels lengthwise. In addition, air is blown through openings in the deck, and through the bed. The proper quantity of air yields a fluid, freely flowing bed of solids without any welling-up or geyser action. The combination of the jigging action with the flow of air through the bed results in a pulsating push of the air on the solids which stratify as they would in a hydraulic jig. The side slope permits the upper strata to roll downhill while the riffles lead the lower strata to a different point of delivery. In addition, the close contact between deck and lower strata induces those strata to move as the deck requires, and the "lost motion" between the lower strata and upper strata indicates considerably slower longitudinal progress of the upper strata.

The most important variable [30] in the operation of pneumatic tables is the volume of air supply; next come variables regulating lengthwise travel, such as deck speed and amplitude, and the relative amplitude of the horizontal and vertical components

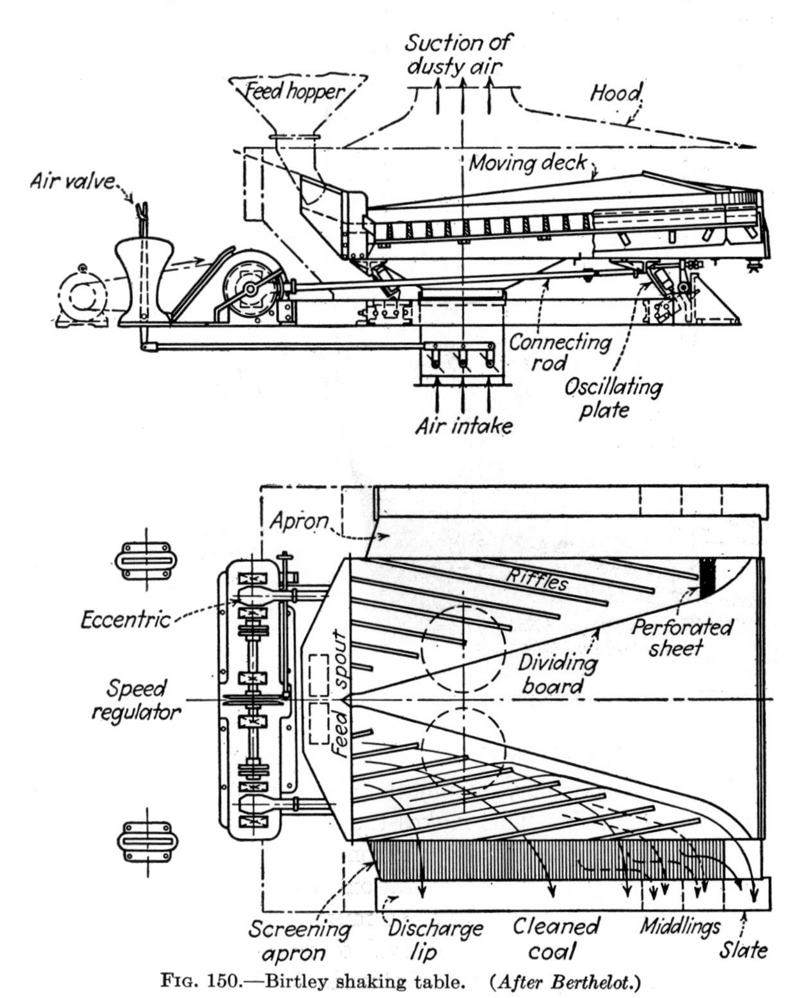

FIG. 150.—Birtley shaking table. (*After Berthelot.*)

of the deck motion; finally the cross slope and longitudinal slope (if any).

Range of Applicability of Pneumatic Jigs. Attention has already been called (Chapter VIII) to the great disadvantage of substituting air, a fluid of specific gravity 0.001, for water, a fluid of specific gravity 1.0, in separating particles of different specific gravity. The same thing is true if the specific gravity of the suspension is considered. Thus with raw coal of specific gravity 1.6, a suspension of 35 per cent solids by volume has a specific gravity of 0.56 if air is the fluid and 1.21 if water is the fluid, so that the settling ratios of clean coal (1.30) and slate (2.6) are respectively $1.39/0.09 = 15.5$ in water and $2.04/0.74 = 2.75$ in air.

Clearly, the use of air instead of water constitutes a great handicap in gravity concentration of coal. To make that process effective, closer sizing is obviously required than in wet jigging. In particular, pneumatic jigging cannot be applied to very fine particles. This makes the dust removal and disposal the two most important steps in pneumatic cleaning. Failure of pneumatic gravity concentration (as in placer treatment, for example) has been due not so much to the handicap of the lighter fluid as to the nonremoval of dust, to the incomplete drying of the feed, and to the attendant adhesion of waste to mineral particles.

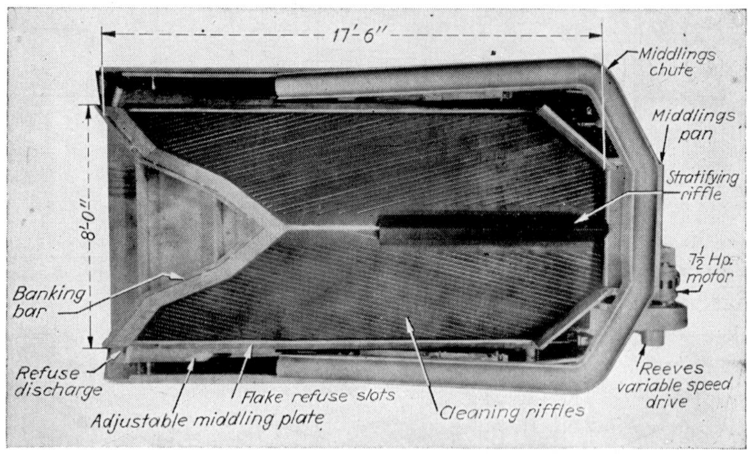

FIG. 151.—Sutton-Steele pneumatic table, plan view. (*American Coal Cleaning Co.*)

Advantages of Pneumatic Jigs.[2] For the treatment of coal, pneumatic jigs have the overwhelming advantage that they yield a dry finished product. In the treatment of coarse coal, this advantage is not so marked since moisture drains readily and in substantially complete manner. But in the treatment of fine coal, the advantage of dryness is considerable.[3,5]

Dry coal, also, has a better appearance; at least an appearance that commands a commercial premium.

Dry-cleaned coal, finally, has advantages over wet-washed coal that arise from its less abrasive and less corrosive action in transit, and the impossibility for it to freeze in railroad cars.

In several instances, operators have found it advantageous to clean their coal partly by a wet and partly by a dry process, this mixed treatment method permitting of synthesizing a blend of suitable moisture content for the best coke making.[4]

Disadvantages of Pneumatic Coal Jigs. Two disadvantages are chargeable to dry-coal cleaning. One of these is the somewhat inferior separation (especially if the coal contains much "bone" or middling particles) and the necessity of removing and disposing of the dust.

USE OF JIGS

Jigs are primarily coarse-mineral concentrating devices. In coal washing, pieces as coarse as 4 to 5 in. can be washed in jigs; in ore concentration, pieces as coarse as 1 in. can be treated. Hydraulic jigs treat coal as fine as ⅛ in. and mineral as fine as 20-mesh. But much finer heavy mineral can be saved in modern hydraulic jigs if a hutch product is made by using "ragging" or "bedding" of suitable gravity and size, and if the tailing particles are not too fine. Gold particles finer than 200-mesh are said to be recovered in Crangle jigs.

Pneumatic jigs can treat mineral as fine as 65-mesh and as coarse as 1 to 1½ in., but one machine cannot treat so wide a range of sizes.

For many centuries, jigs have been the principal concentrating device. They retain this dominant position in coal cleaning in the time-honored form of the hydraulic jig[6,8,9,12,14,34] or in the modern pneumatic table. They also retain a dominant position for the beneficiation of nonmagnetic iron ores and for that of many nonmetallics.

In spite of a few notable exceptions,[11,19,20,22] jigs may be regarded as obsolete in plants for sulphide minerals of the nonferrous metals. This obsolescence arises from the development of new processes, notably flotation and cyanidation, which by their applicability to finely ground products and by their independence of specific gravity have wholly changed the situation. To an extent, also, the obsolescence of jigs may be traced to the development of shaking tables and to the former lack of progressiveness of jig manufacturers—a situation that has changed since jigs lost much of their field of application.

Although jigs are obsolete for the production of sulphide metal concentrates (with some exceptions as already noted), they remain useful for the rejection of waste at a coarse size, the middling product being crushed further or ground, and graded up by flotation.

Jigs are cheap to operate, substantially foolproof (although requiring considerable skill if the specific gravity differential is not great), and easy of access and inspection. On the debit side, it must be noted that jigs require much water, that they must be operated on ore that is frequently not completely freed, and that, since the fines are not treatable in them, they do not provide a complete solution of any dressing problem.

Literature Cited

1. Anon.: A Jig for Concentrating Diamonds, *Eng. Mining J.*, **126**, 19 (1928).
2. Anon.: "Dry versus Wet," American Coal Cleaning Corp., Welch, W. Va.
3. Appleyard, Kenelm C.: Dry Cleaning of Coal in England, *Trans. Am. Inst. Mining Met. Engrs.*, **94**, 235–266 (1931).
4. Arms, Ray W.: Combination Wet and Dry Coal-cleaning Process, *Trans. Am. Inst. Mining Met. Engrs.*, **94**, 267–274 (1931).
5. Arms, Ray W.: Dry Cleaning of Coal, *Trans. Am. Inst. Mining Met. Engrs.*, **70**, 758–774 (1924).
6. Ayres, W. S.: The New Breaker at Cranberry Coal Mine, *Trans. Am. Inst. Mining Engrs.*, **28**, 293–339 (1898).
7. Berthelot, Charles: "Épuration, séchage, agglomération et broyage du charbon," Dunod, Paris (1938).
8. Campbell, J. R.: Mechanical Preparation of Pocahontas Coals—Some Factors in the Problem, *Trans. Am. Inst. Mining Met. Engrs.*, **94**, 275–287 (1931).
9. Campbell, J. R.: Mechanical Separation of Sulfur Minerals from Coal, *Trans. Am. Inst. Mining Met. Engrs.*, **63**, 683–697 (1920).
10. Church, John A.: Recent Improvements in Concentration and Amalgamation, *Trans. Am. Inst. Mining Engrs.*, **8**, 141–155 (1879–1880).
11. Delano, L. A.: The Milling Practice of the St. Joseph Lead Company, *Trans. Am. Inst. Mining Engrs.*, **57**, 420–441 (1917).
12. Dittrick, A. C.: Cleaning Coal by the Simon-Carves Process, *Mining and Met.*, **16**, 217–218 (1935).

13. DYER, FRED C.: The Scope for Reverse Classification by Crowded Settling, *Eng. Mining J.*, **127**, 1030–1033 (1929).
14. HIRST, ARTHUR A.: Coal Cleaning by Gravity Methods, *J. Inst. Fuel*, **8**, 4–10 (1934).
15. JARVIS, ROYAL PRESTON: Investigation on Jigging, *Trans. Am. Inst. Mining Engrs.*, **39**, 451–521 (1908).
16. MALOZEMOFF, PLATO: Jigging Applied to Gold Dredging, *Eng. Mining J.*, **138** [9], 34–37 (1937).
17. MATTSON, V. L.: Disseminated Kyanite Milled Successfully by Celo Mines, *Eng. Mining J.*, **138** [9], 45–46, 94 (1937).
18. MUNROE, H. S.: The English versus the Continental System of Jigging—Is Close Sizing Advantageous?, *Trans. Am. Inst. Mining Engrs.*, **17**, 637–659 (1889).
19. MUNROE, H. S.: The New Dressing-works of the St. Joseph Lead Company at Bonne Terre, Missouri, *Trans. Am. Inst. Mining Engrs.*, **17**, 659–678 (1889).
20. NETZEBAND, W. F., and C. E. HEINZ: Re-treating Tri-state Tailings, *Eng. Mining J.*, **138** [12], 38–43 (1937).
21. OSBORN, WALTER X.: Tabling Tungsten Ore without Water, *Eng. Mining J.*, **123**, 287–289 (1927).
22. RABLING, HAROLD: The Hancock Jig in the Concentration of Lead Ores, *Trans. Am. Inst. Mining Engrs.*, **57**, 309–321 (1917).
23. REEDER, E. C.: Depression Developments in Fluorspar Milling, *Eng. Mining J.*, **135**, 301–304 (1934).
24. REEDER, E. C.: Milling Methods and Costs at the Hillside Fluorspar Mines, Rosiclare, Illinois, *U. S. Bur. Mines, Information Circ.* 6621 (1932).
25. RICHARDS, ROBERT H.: The Cycle of the Plunger Jig, *Trans. Am. Inst. Mining Engrs.*, **26**, 3–32 (1896).
26. RICHARDS, ROBERT H.: Close Sizing before Jigging, *Trans. Am. Inst. Mining Engrs.*, **24**, 409–486 (1894).
27. ROSE, L. A.: Use of Suspensions in Ore Dressing, *Mining and Met.*, **16**, 125–126 (1935).
28. SHEPARD, F. E.: The Richards Pulsator Jig and Classifier, *Colo. Sci. Soc., Proc.*, **9**, 81–98 (1908).
29. SIMONS, THEODORE: Basic Principles of Gravity Concentration—A Mathematical Study, *Trans. Am. Inst. Mining Met. Engrs.*, **68**, 431–462 (1923).
30. TAGGART, ARTHUR F., and R. L. LECHMERE-OERTEL: Elements of Operation of the Pneumatic Table, *Trans. Am. Inst. Mining Met. Engrs.*, **87**, 155–216 (1930).
31. TRUSCOTT, S. J.: "Textbook of Ore Dressing," Macmillan & Company, Ltd., London (1923), pp. 572–583.
32. VEDENSKY, D. N.: Water Pulsator Jig Finds New Application, *Eng. Mining J.*, **138**, 352 (1937).
33. WRIGHT, CLARENCE A.: Ore Dressing Practice in the Joplin District, *Trans. Am. Inst. Mining Engrs.*, **57**, 442–471 (1917).
34. YANCEY, H. F., and THOMAS FRASER: Analysis of Performance of a Coal Jig, *Trans. Am. Inst. Mining Met. Engrs.*, **71**, 1079–1087 (1925).

CHAPTER XIII

FLOWING-FILM CONCENTRATION AND TABLING

Liquid films in laminar flow have a mechanical property that is easily adaptable to the separation of minerals according to specific gravity. This important property is that the velocity of the fluid is not the same at all depths of the film, being nil at the bottom and maximum at or very near the top. This property in turn depends upon the viscosity of the fluid, *i.e.*, upon the existence of internal friction of one layer upon another.

The simplest device for flowing-film concentration consists of an inclined surface or "table" on which assorted particles are subjected to flowing water, the lighter particles being washed off while the heavier particles accumulate. Removal of the heavy particles is intermittent. This is the stationary table known for thousands of years.

In seeking means to make continuous the delivery of the heavy particles, devices have been developed such as vanners and round tables whose basic principle is the same as that of the stationary table. But other devices such as bumping and shaking tables have also been introduced (and they are now far more important) which jointly utilize flowing-film and other principles.

FLOWING-FILM CONCENTRATION

The experimental facts of flowing-film concentration can be summarized as follows: particles at the bottom of a flowing fluid film under the influence of the several forces at play arrange themselves with the lighter, coarser, and rounder particles farther downstream than the heavier, finer, and flatter particles. Broadly speaking the downslope sequence of particles is

1. Fine-heavy particles.
2. Coarse-heavy and fine-light particles.
3. Coarse-light particles.

This arrangement is modified by the shape of the particles, flat and cuboid grains being upstream from rounded grains.

It is interesting to note that flowing-film stratification places the coarse-heavy particles with the fine-light particles. This is the reverse of the stratification that takes place in classification. It suggests the desirability of classifying the feed to a flowing-film concentrator.(25,27)

As the distribution observed in flowing-film concentration appears to be exactly the reverse of what might well be expected (except for the special mechanical property of flowing films to which allusion has already been made), analytical consideration of the forces involved is desirable.

Fluid Velocity in a Flowing Film. A flowing film wetting a substratum moves with a velocity grading from zero at the bottom to a maximum at the top. Actually, the maximum velocity is very slightly below the top because of friction against the air, but as a first approximation this can be neglected.

The velocity does not vary uniformly with depth, but at a constantly decreasing rate. This is on the basic assumption that the flow is wholly laminar. If y is the distance within the flowing film, measured at right angles to it from the substratum-film interface, the fluid velocity v' is given by

$$v' = \frac{\Delta' g \sin \alpha}{2\mu}(2\theta - y)y. \qquad \text{[XIII.1]}$$

In Eq. [XIII.1], Δ' is the specific gravity of the fluid, g the acceleration of gravity, μ the viscosity, θ the film thickness, and α the angle of the film to the horizontal, all in c.g.s. units. This velocity is of course a terminal velocity attained only after the film has ceased to accelerate. This equation is derived as follows:

Hatschek* defines the viscosity coefficient μ as the force required to maintain unit velocity gradient within a fluid.

Therefore (Fig. 152) the force required to maintain a constant difference in velocity dv' between the two sides of a flowing lamina of liquid of thickness dy and cross section A is

$$F_1 = \mu A \frac{dv'}{dy}.$$

This force is provided by gravity and consists of the component in the v' direction of the weight of the liquid resting on the flowing lamina, or

* Hatschek, Emil, "The Viscosity of Liquids," D. Van Nostrand Company, Inc., New York (1928); pp. 5, 59.

$$F_2 = A(\theta - y)\Delta' g \sin \alpha.$$

Hence, when a steady state prevails

$$\mu A \frac{dv'}{dy} = A(\theta - y)\Delta' g \sin \alpha$$

or

$$\frac{dv'}{dy} = \frac{\Delta' g \sin \alpha}{\mu}(\theta - y).$$

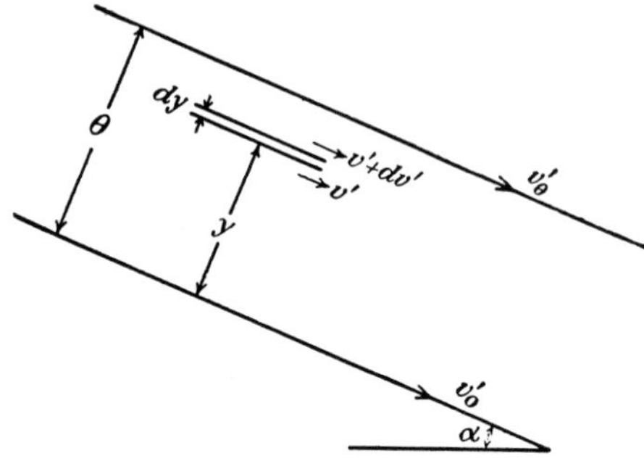

FIG. 152.—Fluid velocity in a film in laminar flow.

Integrating,

$$v' = \frac{\Delta' g \sin \alpha}{\mu}\left(\theta - \frac{y}{2}\right)y + C.$$

The boundary condition $y = 0$, $v' = 0$ requires $C = 0$; hence

$$v' = \frac{\Delta' g \sin \alpha}{2\mu}(2\theta - y)y.$$

The same result can be obtained by starting from the equation of slow motion of an incompressible viscous fluid given by Page[23a], if the boundary condition $\frac{\partial v'}{\partial y} = 0$ for $y = \theta$ is applied.

Rate of Flow in a Flowing Film. The volume W of the fluid flowing per unit time and unit width and the depth of the fluid film θ are related. The relation can be obtained by integration between limits 0 and θ of the element of fluid volume dW:

$$dW = v'dy = \frac{\Delta' g \sin \alpha}{2\mu}(2\theta - y)y\, dy.$$

This gives

$$W = \frac{\Delta' g \sin \alpha}{3\mu} \theta^3 \qquad \text{[XIII.2]}$$

$$\theta = \left[\frac{3\mu \cdot W}{\Delta' g \sin \alpha} \right]^{1/3}. \qquad \text{[XIII.2a]}$$

These relationships show that the depth of the flowing film varies as the cube root of the volume of fluid and inversely as the cube root of the sine of the angle of the film to the horizontal.

Application. Find the volume of water at 20°C. required to form a film 2 mm. deep on a slope of 2°.

$$W = \frac{(1)(980)(0.0349)}{(3)(0.0100)}(0.2)^3 = 9.2 \text{ cc. per sec. per cm.}$$

Effect of Film Thickness on Velocity at a Given Depth. Attention is drawn to the important conclusion that the velocity of the fluid at a given distance from the deck is a function of the film thickness, hence of the rate of liquid flow.

Figure 153 shows the distribution of v' with y and θ for water films of two thicknesses at a definite temperature and slope.

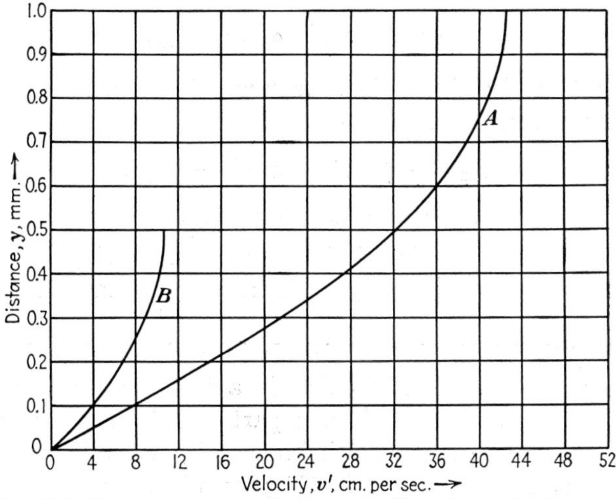

FIG. 153.—Velocity at various depths in water films 1.0 and 0.5 mm. deep (at 20°C. and for a slope of 5°).

Since the velocity at any depth is proportional to $2\theta - y$, an increase in θ at constant y results in a more-than-proportionate increase in velocity. This divergence from strict proportionality between v' and θ is very slight if y/θ is small. In fact, even for large y/θ ratios the divergence is modest. This can be seen from the following numerical illustration:

$$\begin{aligned} \textit{Case } 1: \theta_1 &= 10y;\ v_1' = Cy(20y - y) = 19Cy^2 \\ \theta_2 &= 20y;\ v_2' = Cy(40y - y) = 39Cy^2 \end{aligned}$$

Divergence from proportionality $= 1/39$, or 2.5 per cent.

$$\begin{aligned} \textit{Case } 2: \theta_1 &= y; \qquad v_1' = Cy(2y - y) = Cy^2 \\ \theta_2 &= 2y; \qquad v_2' = Cy(4y - y) = 3Cy^2 \end{aligned}$$

Divergence from proportionality $= 1/3$, or 33 per cent.

This can also be seen by scaling values of v' at given y from Fig. 153.

Forces Acting on a Particle Situated at the Bottom of a Flowing Film. There are three forces acting on such a particle: the pull of gravity, the rub between particle and substratum, and the push (or rub) of the fluid.

Pull of Gravity. For a particle of average radius r, the pull of gravity along the slope is.

$$F_1 = \tfrac{4}{3}\pi r^3 \sin \alpha(\Delta - \Delta')g. \qquad \text{[XIII.3]}$$

Rub between Particle and Substratum. If Φ denotes the specific friction this force amounts to

$$F_2 = -\tfrac{4}{3}\pi r^3 \Phi \cos \alpha(\Delta - \Delta')g. \qquad \text{[XIII.4]}$$

For a stationary particle, Φ may take any value between 0 and the coefficient of static friction Φ_S; for a moving particle, Φ becomes the coefficient of dynamic friction, Φ_D, and is a function of v.

Push of the Fluid. It has already been shown that the resistance to motion of a solid particle in a fluid is a function of the relative velocity of the fluid and particle. The case of motion of a solid particle under the influence of gravity in a still fluid has already been discussed in Chapter VIII. In this case, the particle and the fluid are both moving, and at a velocity that is a function of depth. The net resistance of the particle to the fluid, or fluid push on the particle, can then be obtained only by a summation of the pushes exerted on all the individual elements of the particle.

Since viscous flow has been postulated, it may well be assumed that the resistance between particle and fluid also is a Stokesian force which, it will be remembered, is proportional to the velocity and to the perimeter of the particle, or $R = 3 \times 2\pi r \times \mu \times v$ (Eq. [VIII.2]).

On these assumptions, the net push of the fluid on the particle is

$$F_3 = -\tfrac{9}{2}\pi k\Delta' g \sin\alpha\, r^3 + 6\pi k\Delta' g \sin\alpha \cdot \theta r^2 - 6\pi k\mu r\mathbf{v}, \quad \text{[XIII.5]}$$

in which k is a coefficient (near unity) designed to allow for non-sphericity of the particle, $\mathbf{v}$ is the velocity of translation of the particle with reference to the deck, and the other symbols have the same meaning as heretofore. Equation [XIII.5] can be obtained as follows.

Applying Eq. [VIII.2], and with reference to Fig. 154, the fluid rub $d\mathbf{F}_3$ for an element $r \cdot d\beta$ of the perimeter of the particle is $3k\mu(\mathbf{v}' - \mathbf{v})r \cdot d\beta$ in which $\mathbf{v}'$ represents the fluid velocity at the position β.

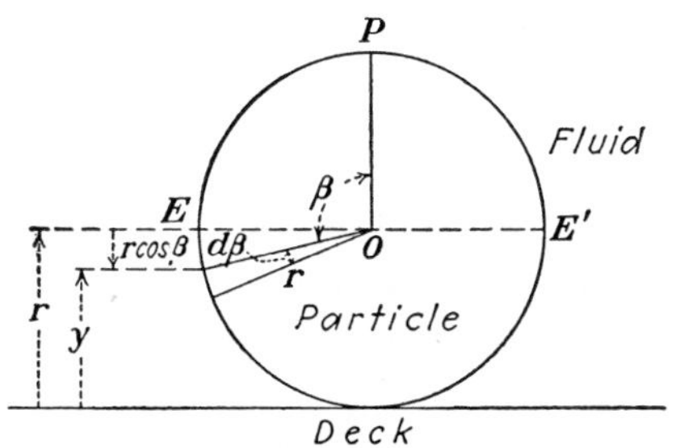

FIG. 154.—Rub of fluid on a particle situated at the bottom of a flowing film (exerted at right angles to the plane of the paper).

According to [XIII.1],

$$\mathbf{v}' = \frac{\Delta' g \sin\alpha}{2\mu}(2\theta - y)y.$$

But Fig. 154 shows that $y = r + r\cos\beta$. Hence

$$\mathbf{v}' = \frac{\Delta' g \sin\alpha}{2\mu}[2\theta r(1 + \cos\beta) - r^2(1 + \cos\beta)^2].$$

Putting $B = 3k\mu$, $A = \dfrac{\Delta' g \sin\alpha}{2\mu}$, the fluid push is

$$d\mathbf{F}_3 = AB[2\theta r(1 + \cos\beta) - r^2(1 + \cos\beta)^2]r \cdot d\beta - Br\mathbf{v} \cdot d\beta$$

$$\mathbf{F}_3 = 2ABr^2\theta \int_0^{2\pi} (1 + \cos\beta)d\beta - ABr^3 \int_0^{2\pi} (1 + \cos\beta)^2 \cdot d\beta - Br\mathbf{v}\int_0^{2\pi} d\beta.$$

Integration of $(1 + \cos \beta)^2 \cdot d\beta$ is obtained by expansion and by use of formula 265 in Pierce[(24)]. The result is

$$\mathbf{F}_3 = 4\pi ABr^2\theta - 3\pi ABr^3 - 2\pi Br\mathbf{v}$$

which after all substitutions gives Eq. [XIII.5].

It might be noted, at this stage, that some question can be raised as to the propriety of considering the Stokesian resistance as applied along the perimeter of the particle; that assumption, however, is the simplest that can be made, and it seems to be reasonable.

Equation of Motion of the Particle. According to the second law of motion, mass times acceleration equals the sum of the forces; or summing up the forces acting along the incline,

$$\frac{4}{3}\pi r^3\Delta\frac{dv}{dt} = \frac{4}{3}\pi r^3(\Delta - \Delta')g \sin \alpha - \frac{4}{3}\pi r^3(\Delta - \Delta')g\Phi \cos \alpha - 6\pi\mu krv + 6\pi k\Delta' g \sin \alpha \cdot \theta r^2 - \frac{9}{2}\pi k\Delta' g \sin \alpha\, r^3,$$

which can be simplified to

$$\frac{dv}{dt} = \frac{\Delta - \Delta'}{\Delta}g(\sin \alpha - \Phi \cos \alpha) - \frac{9}{2}\mu k\frac{v}{\Delta r^2} + \frac{9}{2}k\frac{\Delta'}{\Delta}g \sin \alpha\frac{\theta}{r} - \frac{27}{8}k\frac{\Delta'}{\Delta}g \sin \alpha. \qquad \text{[XIII.6]}$$

Critical Slope Below Which Sliding Motion Does Not Occur. In the preceding derivation, it is implied that the frictional force is fully active. But that may not be the case. Actually, the value of dv/dt remains nil until the second member of [XIII.6] just becomes positive. That condition gives α_{crit} which is a minimum angle for sliding motion.

Putting $dv/dt = 0$ and $v = 0$ in Eq. [XIII.6], we get

$$-\left(\frac{\Delta - \Delta'}{\Delta}\right)g(\sin \alpha - \Phi_s \cos \alpha) = \frac{9}{2}k\frac{\Delta'}{\Delta}g \sin \alpha\frac{\theta}{r} - \frac{27}{8}k\frac{\Delta'}{\Delta}g \sin \alpha$$

from which, by segregating the terms in $\sin \alpha$ and in $\cos \alpha$,

$$\cot \alpha_{crit} = \frac{\frac{9}{2}k\frac{\Delta'}{\Delta - \Delta'}\frac{\theta}{r} - \frac{27}{8}k\frac{\Delta'}{\Delta - \Delta'} + 1}{\Phi_s}. \qquad \text{[XIII.7]}$$

It should be noted that the critical angle below which sliding motion is impossible depends upon the specific gravity of the solid, the size of the solid, the coefficient of friction between solid and deck, and the thickness of the water film. Other things being equal, the critical angle increases with decrease in rate of fluid flow, with decrease in film thickness, with increase in particle size, with increase in specific gravity. As can be seen from Eq. [XIII.7], the relationships are not simple.

It is of interest to note that there is general but not detailed accord between the theoretical results presented here and the experimental results obtained by Richards.[25] The theoretical and experimental results are not wholly comparable, however, as the latter were obtained with irregularly shaped quartz particles and regularly shaped galena particles on a plate of ground glass, *i.e.*, on a roughened surface.

It is also of interest to note that the critical angle is independent of the viscosity of the fluid—a rather unexpected result.

Numerical values of the critical angle for sliding motion are given in Table 37 for the cases described by $\Delta' = 1$, $k = 4/3$, $\Phi_s = 1/3$, $\theta/r = 2, 4, 10, 20$.

TABLE 37.—CALCULATED CRITICAL ANGLES BELOW WHICH MOTION BY SLIDING IS NOT POSSIBLE FOR SOLIDS RESTING ON THE BOTTOM OF A FLOWING WATER FILM

(The coefficient of friction is 0.33, and the out-of-roundness coefficient 1.33)

Specific gravity of mineral	Ratio of film thickness (θ) to particle diameter ($2r$)			
	1	2	5	10
1.3 (bituminous coal)	0°44′	0°18′	0°6′	0°3′
2.65 (quartz)	3°26′	1°30′	0°33′	0°16′
4.0 (sphalerite)	5°26′	2°33′	0°59′	0°29′
7.5 (galena)	8°48′	4°46′	2°0′	1°1′
19 (gold)	13°14′	9°5′	4°40′	2°35′

Motion by Rolling. The values in Table 37 are for sliding friction only. But rolling may take place first provided the bodies are suitably shaped. Consider, *e.g.*, prisms of various regular sections: triangular, square, hexagonal, octagonal, dodecagonal (twelve-sided), icosagonal (twenty-sided).

For rolling to occur prior to sliding, if all the forces are assumed to be applied at the center of gravity, the coefficient of sliding friction would have to exceed 1.732, 1.000, 0.577, 0.414, 0.268, 0.158. With coefficients of friction in the range 0.2 to 0.5, it is plain that the triangular, square, or hexagonal prisms will slide, the icosagonal prism will roll, and the others will roll or slide depending upon the magnitude of the coefficient of friction.

If, however, the force is applied not at the center of gravity, but farther away from it with reference to the surface on which the solid is resting, the pushing forces should be regarded as increased in proportion to the increase in leverarm.

It can be shown that the Stokesian frictional force may be regarded as applied at a point three quarters of the way up from the bottom to the top of the particle, so that the force of the fluid is almost twice as effective in causing rolling as it is in causing sliding. Proof of this is as follows:

The moment dM, with reference to the equatorial plane of the particle EE' (Fig. 154), exerted by an element $d\mathbf{F}_3$ of Stokesian resistance, is the product of $d\mathbf{F}_3$ and the leverarm (referred to the equatorial plane), or

$$dM = r \cos \beta \, d\mathbf{F}_3.$$

For $\mathbf{v} = 0$,

$$dM = AB[2\theta r(1 + \cos \beta) - r^2(1 + \cos \beta)^2]r^2 \cos \beta \, d\beta$$

$$M = (2\theta ABr^3 - ABr^4) \cdot \int_0^{2\pi} \cos \beta \, d\beta + (2\theta ABr^3 - 2ABr^4) \cdot \int_0^{2\pi} \cos^2 \beta \, d\beta - ABr^4 \int_0^{2\pi} \cos^3 \beta \, d\beta.$$

The first and third definite integrals vanish and the second equals π; hence

$$M = 2\pi ABr^3(\theta - r).$$

But the moment is the product of the force $\mathbf{F}_3$ by its leverarm Ω, or

$$M = \Omega \mathbf{F}_3 = \Omega \pi ABr^2(4\theta - 3r).$$

Hence,

$$\Omega = \frac{2\pi ABr^2(\theta - r)}{\pi ABr^2(4\theta - 3r)} = r\frac{2\theta - 2r}{4\theta - 3r}.$$

For very small values of r, Ω approaches $0.5r$; this is a maximum value for Ω. Conversely, for $r = \frac{\theta}{2}$, which is the largest value possible for r, $\Omega = 0.4r$; this is a minimum value for Ω. The fluid push may therefore be regarded as applied about 45 per cent of the way up from the equator or 72 per cent of the way up from the bottom of the particle.

For rolling rather than sliding to occur on a substantially horizontal plane in a fluid film, prisms of regular sections as considered above, *viz.*, triangular, square, hexagonal, octagonal, dodecagonal, and icosagonal, require coefficients of friction of 1.20, 0.69, 0.40, 0.29, 0.185, and 0.110. This shows that all but the triangular, square, and hexagonal prisms would roll readily; the hexagonal prism might also roll.

These deductions are in good agreement with observations made on fluid-film concentrating devices. Galena, for example, is hardly ever seen to roll, and neither is flattish gold, settled mica, or platy ferberite, but irregular gangue particles and roundish particles such as those of phosphate rock are seen to roll.

Effect of Deck Roughness. The foregoing analysis is based on the postulate that the deck is perfectly smooth. If the deck is rough, *i.e.*, if it has at its surface some recesses capable of partly shielding fine particles from the rub of the fluid, the slope required to move the particles by either rolling or sliding will be increased. At the same time such an effect, while present also for large particles, may be so much smaller for them as to be imperceptible. The relationship of critical angle to size obtained above will therefore not hold for rough surfaces. The problem is analytically complex and beyond the scope of this text; it is nevertheless a problem that might well be explored further if a full insight is desired into the mechanism of flowing-film concentration.

Terminal Velocity if Sliding Does Occur. Once the particle has started to move, the coefficient of friction changes to a dynamic coefficient of friction. In fact, because the fluid push on the particles is larger at the top of the particle than at the bottom, the particle rolls, largely according to the shape of the particle and according to the speed. At low speeds, the effective friction is the relatively large coefficient of dynamic sliding friction, and at high speeds it is the lower coefficient of rolling friction. The change probably takes place partly continuously and partly discontinuously. As a first approximation, the dynamic coefficient of friction may, however, be regarded as constant.

Putting $dv/dt = 0$ in Eq. [XIII.6], we get

$$0 = (\Delta - \Delta')g(\sin \alpha - \Phi_D \cos \alpha) - \frac{9}{2}\mu k\frac{v_{max}}{r^2} + \frac{9}{2}k\Delta' g \sin \alpha \cdot \frac{\theta}{r} - \frac{27}{8}k\Delta' g \sin \alpha,$$

Φ_D denoting the coefficient of dynamic friction. This equation can be put in the form

$$v_{max} = r^2 \frac{g \sin \alpha}{\mu}\left[\frac{2(\Delta - \Delta')(1 - \Phi_D \cot \alpha)}{9k} - \frac{3}{4}\Delta'\right] + r\frac{g \sin \alpha}{\mu}\Delta'\theta. \quad \text{[XIII.8]}$$

It is interesting to note that v_{max} is the sum of two terms, one in r^2 the other in r. The term in r is always positive, and the term in r^2 is negative, at least in the range of values of α, Δ, k, and Φ_D that are likely. With increasing size, the downslope velocity of particles increases from zero for particles of infinitesimal size. Depending upon the parameters of Eq. [XIII.8], the velocity may keep on increasing up to the limiting value $2r = \theta$, or it may reach a maximum and decrease thereafter to some intermediate value, or it may reach a maximum and decrease to zero for some value of $2r$ less than θ. The numerical example $\alpha = 5°$, $\Phi_D = 0.2$, $k = 1.33$, $\Delta' = 1.0$ is typical, and is illustrated in Fig. 155A.

If it is attempted to separate particles that behave on a 5° slope as quartz and galena are shown to behave in Fig. 155A, the separation will be successful for particles ranging in size from 0 to θ for galena and from 0.28θ to θ for quartz. Quartz finer than 0.28θ will move with some of the galena particles. If it is attempted to separate coal from slate at the same slope of 5°, the useful range in size of coal would be from 0.60θ to θ, with slate ranging from 0 to θ, *i.e.*, a much less suitable range than in the case of the quartz and galena. This, however, seems due to an unwise choice in slope. In fact the range can be extended to from 0.27 to 0.98θ by changing the slope to 2°, as may be seen by referring to Fig. 155B.

If the feed to a flowing-film concentrator is screen-sized, the useful range in size is *A–B* (Fig. 155A). This range may be small if the specific gravities are close to each other or if the rate of liquid flow and the slope are poorly chosen.

If the feed is a classified product, the range of size is wider, as may be seen by again referring to Fig. 155A. Thus if the quartz and galena were classified with a hindered-settling ratio of 5, galena cannot exceed 0.2θ, and the permissible range in quartz

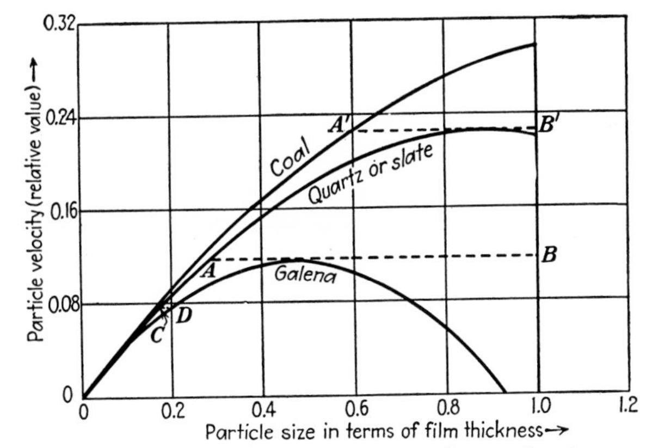

Fig. 155A.—Relative velocities of coal, quartz, slate and galena particles of various sizes on a deck having a slope of 5°.

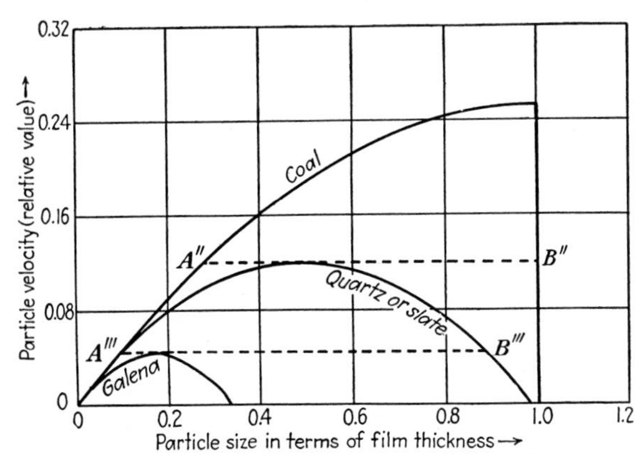

Fig. 155B.—Relative velocities of coal, quartz, slate and galena particles of various sizes on a deck having a slope of 2°.

sizes is from 0.18θ (corresponding to point C) to θ instead of from 0.28θ to θ. This argument is in agreement with those who, like Richards, have advocated classification prior to concentration by flowing-film concentration.

Choice of the most advantageous conditions for flowing-film separation requires adjustment of the slope, and of the film thickness in judicious proportion to the size of the particles. Since the film thickness depends upon both the slope and the rate of flow of fluid, complete control should be obtainable by adjusting the rate of flow and the slope.

Effect of Particle Size on Capacity of Flowing-film Concentrating Devices.—Referring again to Eq. [XIII.8], it is seen that the equation is of the form

$$v_{max} = Mr^2 + Nr\theta,$$

or, expressing θ in terms of r, $r = \rho\theta$

$$v_{max} = Mr^2 + N\frac{r^2}{\rho} = \left(M + \frac{N}{\rho}\right)r^2.$$

For a given pair of minerals of the same particular predetermined size, the slope and film thickness should be chosen so as to make the ratio of the maximum velocities as large as possible. This therefore determines the value not only of M and N, but also of ρ. If particles of the same minerals of finer size are chosen, the maximum velocity of the particles under the same optimum conditions will be reduced in proportion to r^2, *i.e.*, to the square of their size.

Since the velocity of efflux of particles, under similarly satisfying separating conditions, is proportional to the square of the particle size, and since the weight of a sheet of particles one deep is proportional to the particle size, it may be said that the capacity of flowing-film concentrating devices is proportional to the cube of the particle size.

Terminal Velocity If Rolling Motion Prevails. If particles in motion are rolling, the term in Φ_D in Eq. [XIII.8] may be considered to have disappeared, and the terminal velocity becomes

$$v_{max} = r^2\frac{g \sin \alpha}{\mu}\left(\frac{2(\Delta - \Delta')}{9k} - \frac{3}{4}\Delta'\right) + r\frac{g \sin \alpha}{\mu}\Delta'\theta. \quad \text{[XIII.9]}$$

Equation [XIII.9] shows the velocity to increase with increasing specific gravity of the minerals, a result diametrically opposed to that obtained if sliding motion is obtained. Rolling motion of both minerals is inacceptable, but the natural suspicion seems confirmed that optimum conditions for flowing-film concentration involve rolling of the lighter mineral and sliding of the heavier mineral.

Summary of Theory of Flowing-film Concentration.—The behavior of particles at the bottom of a flowing fluid film is affected by the following factors:

1. The slope of the deck.
2. The thickness of the fluid film (or the rate of flow of the fluid).
3. The viscosity of the fluid.
4. The coefficients of friction between the various particles and the deck.
5. The specific gravity of the particles.
6. The shape of the particles.
7. The roughness of the deck.

On a horizontal deck there is no motion of particles. As the deck is tilted, particles begin to move, until all are moving.

On a perfectly smooth deck the critical angle below which motion of a particle does not occur increases with the size of the particle and with the specific gravity. On a rough deck conditions may be different.

The velocity of moving particles depends upon their specific gravity, the viscosity of the fluid, the thickness of the film, the slope of the deck, and the coefficients of friction (static, dynamic-sliding, and dynamic-rolling) of the particle on the deck. Providing sliding motion is obtained, heavier particles move slower than light particles, but not in direct proportion. Other things being equal, the speed of particles is directly proportion to the square of the film thickness and inversely proportional to the viscosity of the fluid. Increasing the slope increases the velocity, but here also not in direct proportion.

The effect of the coefficient of friction is difficult to evaluate as it is a function of velocity, and in fact may cease to have any definite meaning in regard to the jump-skip motion of irregular particles rolling on a sloping deck.

In concluding these remarks it is fitting to caution against indiscriminate application of the equations derived above. Equation [XIII.1] is rigorous, but each succeeding step has implied the use of assumptions; reasonable enough though the formulas seem, that fact must be kept in mind.

AUXILIARY PRINCIPLES UTILIZED IN SHAKING TABLES

The flowing-film concentrators that are known as shaking tables utilize other principles besides those discussed so far. Shaking tables are provided with a reciprocating motion at right angles to the flowing fluid film and directed horizontally. This reciprocating motion of the table deck has an asymmetrical acceleration, the net effect of which is to cause intermittent travel of solids resting on the table. This is one of the auxiliary principles utilized in shaking tables.

Another auxiliary principle derives from the use of riffles which disturb the viscous flow of the fluid across the deck and substitute for it a fluid composed of one top layer flowing more or less by viscous flow and of an eddying bottom layer.

Asymmetrical Acceleration in Vacuum. A particle on a horizontal and horizontally moving deck in vacuum (or in air, for all practical purposes) is subjected to two forces, that due to the acceleration of the deck and the frictional force between the deck and the particle. These forces are oppositely directed. As the acceleration increases, the frictional force increases up to a point corresponding to the coefficient of static friction. The limiting condition is described by

$$m\frac{dv}{dt} = mg\Phi_s,$$

or

$$\frac{\left(\frac{dv}{dt}\right)}{g} = \Phi_s. \qquad \text{[XIII.10]}$$

If the deck is reciprocating, it has an acceleration that changes sign twice per cycle. If the motion is symmetrical (*e.g.*, sine-wave type) the particle may or may not move on the deck according to the relative values of the maximal acceleration and the coefficient of static friction. If the motion is asymmetrical, the particle may move in one direction, yet fail to move in the other—because the maximal acceleration may exceed $g\Phi_s$ in one direction yet not exceed it in the other—or the particle may move in both directions, but at different speeds.

Equation [XIII.10] shows that the drift of the particles occurs at a rate that depends neither on the size nor on the specific gravity of the particles.

Asymmetrical Acceleration in a Nonviscous Fluid. If the vacuum of the preceding case is replaced by a fluid, and provided the resistance of the fluid is neglected, the limiting condition becomes

$$m\frac{dv}{dt} = (m - m')g\Phi_s,$$

or

$$\frac{\left(\frac{dv}{dt}\right)}{g} = \left(\frac{m - m'}{m}\right)\Phi_s, \qquad \text{[XIII.11]}$$

m' being the mass of the fluid displaced by the solid. Clearly, a lower deck acceleration suffices to cause motion.

Equation [XIII.11] can also be expressed in terms of specific gravities as

$$\frac{\left(\frac{dv}{dt}\right)}{g} = \left(\frac{\Delta - \Delta'}{\Delta}\right)\Phi_s. \qquad \text{[XIII.11}a\text{]}$$

Equation [XIII.11*a*] shows that minerals of different specific gravities but having the same coefficients of friction will begin to drift under asymmetrical deck acceleration at different threshold values, light minerals moving at lower threshold accelerations than heavy minerals. Thus it appears that segregation of unlike minerals might be possible on the basis of this phenomenon, considered alone.

Equation [XIII.11] shows that the drift, under the conditions specified, does not depend on the size of the particles.

Asymmetrical Acceleration in a Viscous Fluid. Equation [XIII.11] is but a first approximation as it involves no consideration of the resistance to motion offered by the fluid. The more general equation is

$$m\frac{dv}{dt} = (m - m')g\Phi + R_F. \qquad \text{[XIII.12]}$$

A general solution of this equation is impossible, but specific solutions are perhaps obtainable if the relationships of the resistance to time and to distance from the deck are predetermined. Such specific solutions are not sought in this analysis, but the methods for seeking them are outlined.

In any event, the first requirement is to secure the relationship of the fluid velocity to depth and time. Subsequently, integration of the element of resistance for the depth ranging from y to $y + dy$, within the limits 0 and $2r$ (the particle diameter) should completely determine R_F.

Fluid Resistance. Fluid resistance can be evaluated only if the velocity of the fluid is known at all layers. The general formula is

$$\frac{\partial v'}{\partial t} = \frac{\mu}{\Delta'} \frac{\partial^2 v'}{\partial y^2}. \qquad \text{[XIII.13]}$$

Proof is as follows:

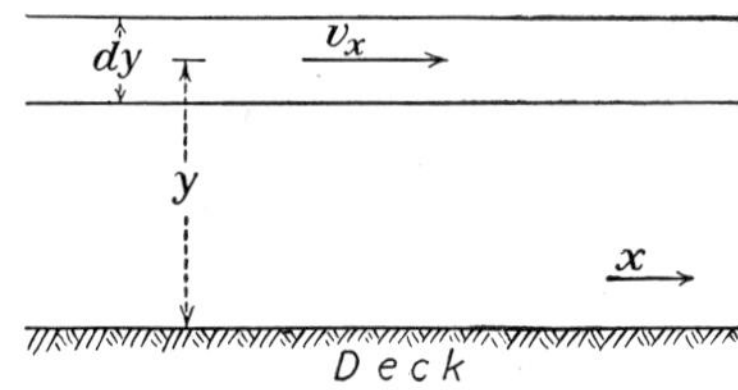

FIG. 156.—Fluid velocity in a fluid layer on a reciprocating deck.

Consider a strip of thickness dy parallel to the deck (Fig. 156). At the bottom of the strip, the fluid exerts on the strip a stress

$$-\left(X_y - \frac{1}{2}\frac{\partial X_y}{\partial y} \cdot dy\right),$$

in which X_y is the stress which is, by the definition of viscosity,

$$X_y = \mu \frac{\partial v}{\partial y}.$$

Likewise at the top of the strip, the fluid exerts on the strip a stress

$$+\left(X_y + \frac{1}{2}\frac{\partial X_y}{\partial y} \cdot dy\right).$$

The force on a strip of area A is then

$$A\mu \cdot dy \cdot \frac{\partial^2 v}{\partial y^2}.$$

The equation of motion of the strip of volume Ady is then

$$A \cdot dy \cdot \Delta' \frac{\partial v}{\partial t} = A\mu \cdot dy \cdot \frac{\partial^2 v}{\partial y^2}$$

or

$$\frac{\partial v}{\partial t} = \frac{\mu}{\Delta'} \frac{\partial^2 v}{\partial y^2}.$$

Unfortunately this equation cannot be integrated unless the equation of motion of the deck is known.

In the case of a simple harmonic motion Bouasse* gives an equation which can be written

$$v_y' = v_o'e^{-y\sqrt{\frac{\pi f \Delta'}{\mu}}} \cos\left(2\pi ft - y\sqrt{\frac{\pi f \Delta'}{\mu}}\right). \qquad \text{[XIII.14]}$$

In this equation v_y' denotes the velocity at distance y from the deck, v_o' the velocity of the deck, t the time, f the frequency, Δ' the specific gravity of the fluid, and μ its viscosity.

In the case of water and of a deck reciprocating 300 times per minute, $f = 5$, $\Delta' = 1$, $\mu = 0.01$, the equation becomes

$$v_y' = v_o'e^{-39.6y} \cos (2\pi ft - 39.6y).$$

The Bouasse formula shows that the maximum velocity at any layer decreases very rapidly with increasing distance from the deck. Furthermore the formula shows that there is a time lag in the transference of motion from one layer of fluid to another.

Table 38 gives the ratio of velocity of water to deck velocity in function of depth.

TABLE 38.—RATIO OF MAXIMUM VELOCITY OF WATER TO MAXIMUM DECK VELOCITY IN FUNCTION OF DEPTH
(Viscous flow is postulated)

Distance from Deck, Centimeters	Relative Velocity
1.0	6.3×10^{-18}
0.2	3.6×10^{-4}
0.1	0.019
0.05 (= 30-mesh)	0.14
0.02 (= 65-mesh)	0.45
0.005 (= 300-mesh)	0.82
0.001 (10 microns)	0.96

This table shows that the fluid velocity becomes negligible at a small distance from the deck. In relation to very small particles (under 50 microns in size), it can be said that the fluid moves with the deck. In relation to coarse particles (over 0.5 mm. in size), it can be said that the fluid is stationary. Toward medium-sized particles the fluid behaves in an intermediate way.

This conclusion is an approximation only, since it is based on the assumption of a deck having a harmonic motion. In the practical case of an asymmetrical deck motion, the equation of

* BOUASSE, H., "Mécanique physique," Ch. Delagrave, Paris (1912); p. 349.

motion of the deck, while still of a trigonometric type, is more complex, and a solution of the basic differential equation [XIII.13] may be difficult to obtain. It seems probable, however, that the decay in maximum amplitude of the motion, from the deck to the surface, is some form of an exponential function of the frequency, specific gravity, and viscosity, with the result that the amplitude decay may well be of the same general form as that numerically suggested by Table 38.

Effect of Size on Rate of Travel on Deck. The equation of motion [XIII.12] bears a strong resemblance to the equation of motion of a particle under the action of gravity, as may be seen by comparing it with Eq. [VIII.3]. In this instance, however, Φ is a function of t, and R_F is a function of both v, the velocity of the particle, and v_D, the velocity of the deck, and v_D is also a function of t.

On the basis of this resemblance it seems likely that the rate of drift of particles of various sizes is as some power of the size, ranging perhaps between 1 and 2. But in view of the incompleteness of the analysis that is outlined here, and of the lack of accurate experiments, the conclusion must be reached that this is a field for further inquiry.

Summary of Observations Concerning Longitudinal Motion on a Shaking Table.

1. In vacuum, or for practical purposes in air, neither size nor specific gravity affects the rate of travel.
2. In an ideal nonviscous liquid, which is practically approximated if very coarse particles and deep water are considered, size does not affect the rate of travel, but specific gravity does.
3. In a viscous liquid, which is practically approximated by shallow water provided fine particles are considered, size and specific gravity both affect the rate of travel.

Use of Riffling. Riffles are of great importance to tabling. They are responsible for the increased capacity of riffled decks over smooth ones. This is realized when it is considered that a smooth deck treats a bed one particle deep and a riffled deck treats a teetering suspension of particles often many particles deep.

Each trough between successive riffles is a place where hindered settling and consolidation trickling occur. These phenomena have already been discussed in connection with jigs and need not be discussed again at this stage. The principles apply to both jigging and tabling.

Because of the loss of motion between successive layers of particles, the stratified material between riffles is acted upon by the lengthwise motion of the deck to a greater extent at depth than at the surface. Since the bottom layer consists of the finest and heaviest particles, these move faster to the discharge end than the coarser and lighter particles. In respect to size, the rate of travel to the discharge end is opposite to that obtained on a bare plane surface.

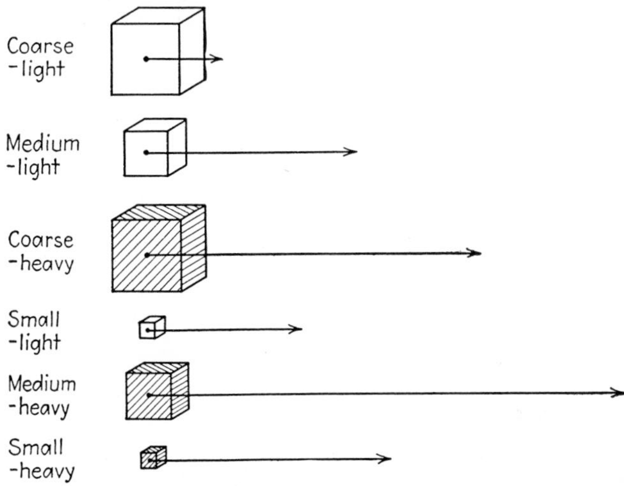

Fig. 157.—Vertical stratification within riffles, and rate of lengthwise travel of particles of three sizes and two specific gravities under the action of asymmetrical deck acceleration.

As a result of the differential response of particle layers to the asymmetrical acceleration of the deck and to differences in response of particles of various sizes and specific gravities, lengthwise travel within the riffles, arranged as from top to bottom, is somewhat as shown in Fig. 157.

Progress of mineral crosswise to the riffles is obtained partly by the crowding action of new feed which forces overflow from one riffle to the next and partly by the taper in riffle depth which is commonly utilized from feed to discharge end. The crowding effect brings the upper stratum within the range of action of the flowing film that tops the riffles. The lighter and coarser particles are thus preferentially washed downstream. The taper of the riffles causes removal of the strata from the top

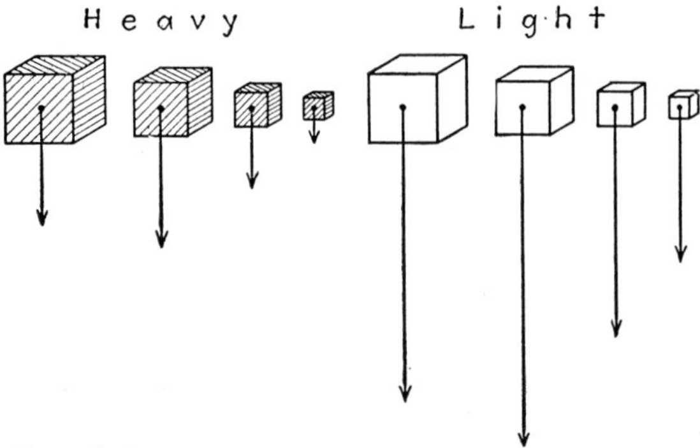

FIG. 158.—Cross-deck travel of particles of four sizes and two specific gravities across the riffles of a riffled deck, under the influence of a flowing fluid.

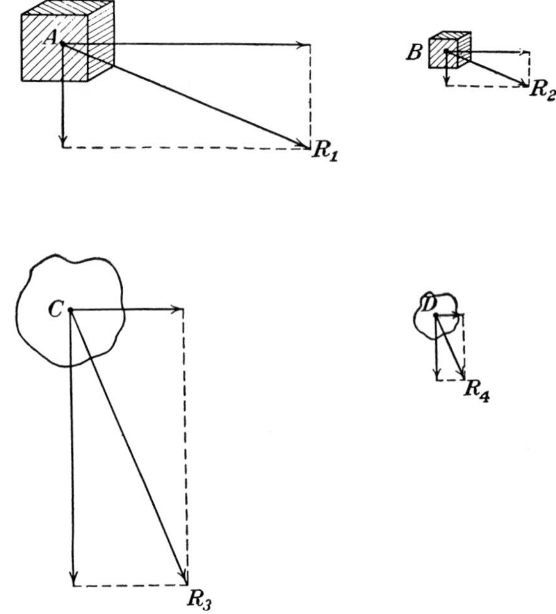

FIG. 159.—Directions of motion, and relative amplitude of motion of particles of two sizes and two specific gravities on a bare-decked table.

downward as the material progresses along the riffles. As a result, again the lighter and coarser particles are washed downstream more than the finer and heavier particles that have settled on the deck. These effects, however, are modified by the response of the particles to eddy currents within each riffle trough.[4] Figure 158 summarizes these effects.

Net Travel of Particles on Smooth and on Riffled Table Decks. By combining all the effects which have already been discussed, it seems that the travel on a bare or riffled deck is along a diagonal line beginning at the feed box. The direction of this diagonal can be obtained from the vectors describing the path elements in the direction of the water flow and crosswise to the water flow.

Particles of the same mineral but of different sizes follow more or less the same diagonal line on a deck (Fig. 159), but at different rates, the rate of discharge from the table being nearly in proportion to the square of the diameter, and the residence on the table varying inversely as the square of the diameter.

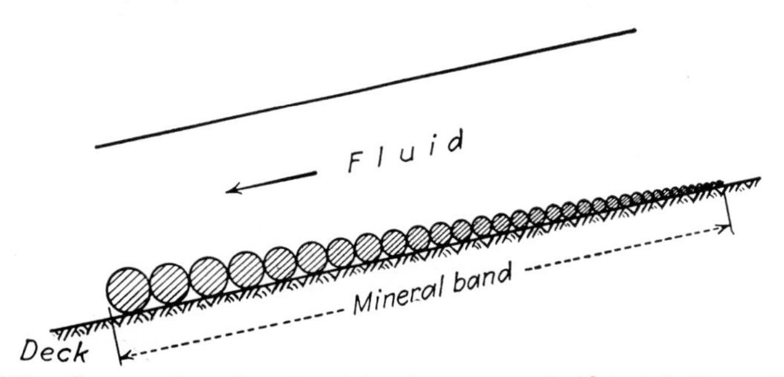

FIG. 160.—Cross section of a mineral band on a smooth-decked shaking table.

It has been suggested that the forward movement proceeds at a rate proportional to a power of the particle size, probably in the range of from 1 to 2. And it has been shown that the cross-deck movement likewise proceeds at a rate proportional to a power of the particle size near 2. The resultant of these movements, for particles of the same specific gravity, is substantially in the same direction, with the fine particles somewhat upstream.

This analysis may not be wholly valid if the deck surface is crowded; in that case, large particles protect small particles against the push of the fluid, at least against the full push of the fluid if the small particles lodge immediately upstream from the coarse particles. A band of particles may then form such as that which is schematically pictured in section in Fig. 160.

On a partly riffled or a fully riffled deck, particles of the same mineral do not emerge at the same place regardless of size because of the vertical stratification and lengthwise travel in the riffles (Fig. 157); the finer particles are farther up on the deck than the coarser particles.

Particles of different minerals of the same size on a smooth deck segregate into bands coming off in various diagonal directions, with the heavier mineral upstream from the lighter minerals. On

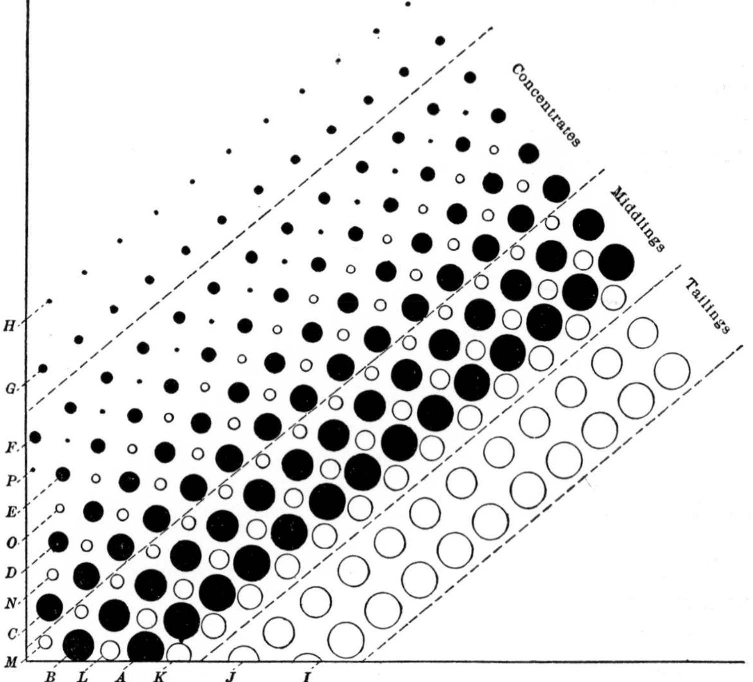

FIG. 161.—Idealized arrangement of particles on a Wilfley table. (*After Richards.*)

a riffled deck, the segregation is partly according to size and partly according to specific gravity with the fine-heavy particles farthest upstream and the coarse-light particles farthest downstream. This is shown well by Fig. 161 which is reproduced from Richards' study of shaking tables.[27]

Variables in the Design and Operation of Shaking Tables. The variables that the designer and the operator may seek to control are

1. The slope (cross tilt).
2. The thickness of the flowing film (amount of water).
3. The relationship between deck acceleration (or throw) and time (speed and throw).
4. The coefficient of friction between deck and particles, in so far as that can be controlled by changing the character of the deck surface.
5. The character of the riffling.

Suitable adjustment of these variables secures the greatest spread between particles of unlike specific gravity and simultaneous treatment of the greatest possible range of particle sizes. Broadly speaking, the heavy particles move least under the influence of the fluid and most under the influence of the deck, and light particles do the reverse. A perfect separation is not generally possible since individual particles may and do present variations in shape and frictional properties, as well as in size, the effect of which is to blur the separation that one might otherwise expect.

Sorting before Tabling. Much has been made of the theoretical desirability of classifying before tabling. Operating evidence does not entirely agree with that theory. The analysis presented here shows that in the case of a bare deck there should be little sizing action, but theoretical argument supporting the advisability of sorting before tabling has already been drawn from Fig. 155A and Fig. 155B, page 291.

It is of course true that bare decks are used in practice only in some instances, and are then limited to a small part of the deck. Where unsized feed is treated on riffled decks, the very fine particles of heavy mineral may be lost because of the turbulence that is required to move coarse particles across the riffles; this loss may be minimized by classifying the feed to the tables. Such a classification may also be instrumental in reducing losses among very coarse particles, as it permits the segregation of coarse middling particles for regrinding.

The whole problem of sorting before tabling *vs.* sizing before tabling and *vs.* tabling raw feed is therefore made up of a number of interdependent subproblems. No general answer is advisable, and the general problem must be examined anew in each instance.

Particle Size Limits in Tabling. Practical and theoretical particle-size limits for a table necessitate consideration of

1. The relative thickness of particle and fluid film.

2. The relative time required for the fluid to be discharged from the table, and for the particles to settle on the table.

Particles should not exceed the film thickness or else they would stick out above the film and fall without the range of its downslope push. That is true of bare-decked rather than of riffle-decked tables. In the case of the latter, the maximum particle size should be riffle depth rather than film depth. Another maximum limit is one-third the riffle width (to avoid bridging).

On the other hand, particles should be heavy and coarse enough to settle on the table well before the water carrying the suspended particles is off the table. As a practical limit, one might well set arbitrarily the condition that particle travel downslope before settling be not more than 6 in. (15 cm.).

The downstream travel, before a particle at the top of the fluid film has settled on the deck, is

$$z = \frac{9}{2} \frac{\mu W}{(\Delta - \Delta')r^2 g \cos \alpha}. \qquad \text{[XIII.15]}$$

In this equation, z is the downstream travel expressed in centimeters, W the rate of fluid flow in cubic centimeters per centimeter of running deck length per second, μ the viscosity, r the particle radius, g the acceleration of gravity, Δ and Δ' the specific gravities of solid and fluid, all in c.g.s. units, and α the deck slope.

Proof is as follows:

Assume (Fig. 162) that at depth y the fluid velocity is v and the particle settling velocity with respect to the fluid is u.

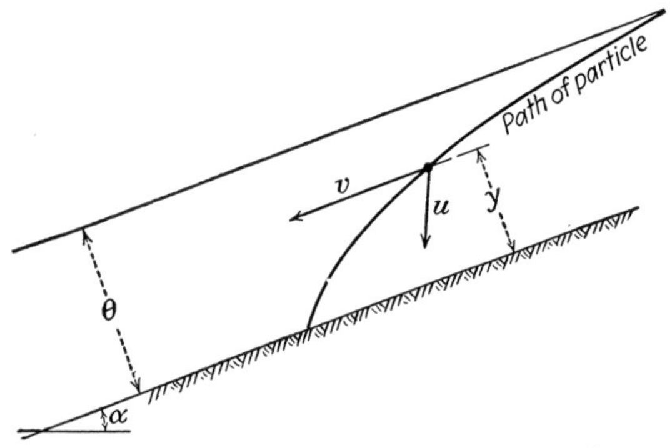

FIG. 162.—Settling of a particle in a flowing film.

A movement of the particle such as to reduce the depth by dy requires a vertical movement of $dy/\cos\alpha$ and requires a time $dt = (1/u)(dy/\cos\alpha)$. The horizontal travel in time dt is

$$dz = v \cdot dt = \frac{v}{u \cos\alpha} \cdot dy.$$

But

$$v = \frac{\Delta' g \sin\alpha}{2\mu}(2\theta - y)y \qquad \text{(Eq. [XIII.1])}$$

$$u = \frac{2}{9}\frac{(\Delta - \Delta')r^2 g}{\mu} \qquad \text{(Eq. [VIII.4])}.$$

Substituting these values in dz and simplifying,

$$dz = \frac{9}{4}\frac{\Delta'}{\Delta - \Delta'}\frac{\tan\alpha}{r^2}(2\theta - y)y \cdot dy,$$

or

$$z = \int_{y=0}^{y=\theta} dz = \frac{3}{2}\frac{\Delta'}{\Delta - \Delta'}\frac{\tan\alpha}{r^2}\theta^3.$$

According to Eq. [XIII.2], the film thickness θ is related to the fluid flow W expressed in cubic centimeters per centimeter per second by

$$\theta^3 = \frac{3\mu}{\Delta' g \sin\alpha} W.$$

Substituting this value of θ^3 in z and simplifying,

$$z = \frac{9}{2}\frac{\mu W}{(\Delta - \Delta')r^2 g \cos\alpha}.$$

It should be noted that in Eq. [XIII.15] the particle radius appears in the denominator raised to the second power. Hence to save very fine particles, a great reduction in flow of fluid W or a great increase in permissible travel z is required. Change in slope is practically immaterial.

Numerical Application. If, as indicated above, the requirement is arbitrarily made that particles to be saved should all settle in the first 6 in. of flow on the deck ($z = 15$ cm.) and if dealing with sphalerite on a deck sloping 5°, and a liquid flow of 10 cc. per cm. per sec., the minimum radius is found to be 30 microns (about 250-mesh).

If the permissible distance is stretched to 1 ft. and the flow of fluid is reduced to 0.5 cc. per cm. per sec., the finest gold particle that can be saved is equivalent to a sphere 6 microns in diameter, *i.e.*, one worth roughly one quarter of one millionth of one cent. This represents the most that can be expected of a flowing-film concentrating operation.

These simple numerical applications show that a flowing-film concentrator can save particles roughly as fine as 10 to 100 microns, depending upon the adjustments and the particular mineral under consideration. Toward the coarse end, an unriffled flowing-film concentrator can treat particles as coarse as about 10-mesh. Thus if $W = 10$ cc. per cm. per sec., $\alpha = 5°$, Eq. [XIII.2] gives $\theta = 0.19$ cm. Riffled flowing-film concentrators can accommodate much coarser feed; in the case of coal cleaning, the riffles can be made so deep that the table resembles a jig more than a flowing-film concentrator. In coarsely riffled decks, on the other hand, the losses become substantial in a size range that might be effectively treated on a finely riffled deck. This is ascribable to the eddy currents generated between riffles.

FLOWING-FILM DEVICES

Devices Utilizing Flowing-film Concentration Only. The simplest kind of flowing-film device is the sloping rectangular table or *buddle* which has been known for several thousand years.

FIG. 163.—Ancient buddle. (*After Agricola.*)

It is now often made of wood or canvas. The slope is adjusted to permit movement of the light minerals while the heavy minerals settle. As the sediment thickens, the slope of the buddle may have to be changed. Eventually the sediment is shoveled or hosed away, and the operation is repeated. To improve the work of the buddle, upslope sweeping during stratification has long been practiced as may be seen from Fig. 163 which is reproduced from Agricola's "De re metallica." Working surface of the buddle is therefore not smooth but somewhat rough, as it consists of an aggregate of settled particles.

FIG. 164.—Gold-saving corduroy, natural size. (*After von Bernewitz.*)

For the recovery of free gold in flotation tailings, the *strake* (*blankets*) very recently has found considerable use. This is a shallow smooth sluice, lined with blanket, canvas, or other roughened surface, in which the pulp is allowed to flow on such a slope that sulphides do not settle. Gold, on the other hand, settles as high-grade concentrate which is removed intermittently after temporary deflection of the pulp flow. "Hosing" of the sediment[3] or replacement of the blanket and smelting of the gold-laden blanket are practiced.[14] Although the pulp is deeper than in other flowing-film devices, the effective part of a strake is the bottom layer to which the principles discussed in this chapter are applicable.

The best covering for a strake or blanket plant is said[3] to be a corduroy with widely spaced ribs (Fig. 164) placed with the ribs crosswise to the flow and with the nap pointing upstream. In this form, a blanket plant is similar to Hungarian riffles.[41] If the gold is coated with oxide slime, gold recovery may be superior to that obtainable by plate amalgamation. The cost, attention, and care for a blanket plant are less, in most cases, than those for a plate-amalgamating plant.

Round Tables.[30] Round tables are circular buddles whose shape permits the operation to be continuous instead of intermittent. Round tables are obtuse-conical smooth surfaces on which the pulp flows from center to periphery, the lighter particles traveling faster. They are of two general types. In stationary circular tables such as the *Linkenbach table*, the buddle is stationary, but a revolving arrangement changes the position of the pulp feed, washwater, and concentrate-removing nozzle.

In the revolving round tables such as the *Evans table* (Fig. 165), the position of the feed, wash water, and concentrate-removing spray are fixed but the table revolves. Percussion round tables

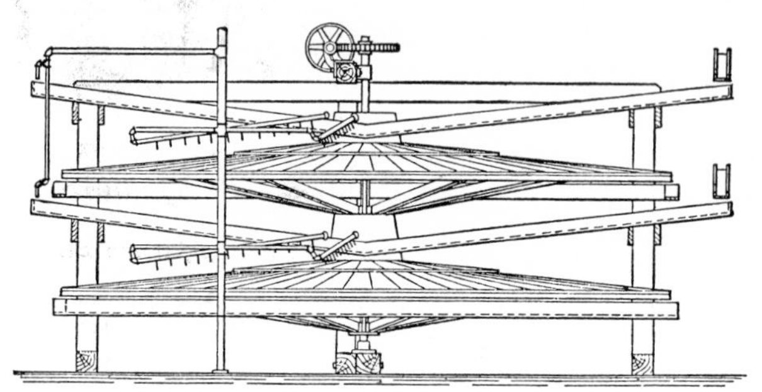

FIG. 165.—Anaconda double-deck Evans table. (*After Simons.*)

(*e.g.*, the *Bartsch table*) were designed to imitate the cross motion of the rectangular shaking tables. *Multiple-deck round tables*, also, were devised and put in operation, especially at Anaconda, but they were made obsolete by the development of flotation.

These round tables did a good job, but their capacity was very low, and they failed, of course, to recover mineral finer than about 20 microns. For certain restricted purposes, round tables are still useful.

Vanners.[6,15] Vanners consist of endless belts of rubber or similar material stretched over rollers and traveling inclined with their upper surface against a flowing film of water. In addition, a sideshake is utilized in the *Johnson*, *Frue*,[21] and *Isbell* vanners, a gyration in the *Senn* vanner, and an end-shake in other vanners.

At first, vanners were made to treat feed as coarse as 1 to 2 mm. Such granular material is better treated on shaking tables. It is now recognized that the field of application of vanners, round tables, and buddles is for the recovery of mineral finer than 0.15 to 0.25 mm. and coarser than 0.01 to 0.02 mm., provided flotation does not offer a more attractive solution of the ore treatment.

Capacity of vanners for this fine-silt recovery ranges from 1 to 3 tons per 24 hr. for a machine 6 ft. wide. Capacity of stationary round tables[30] 32.75 ft. in diameter is about 8 to 9 tons per deck per 24 hr. Capacity of revolving round tables 19 ft. in diameter is about 6 to 7 tons per deck per 24 hr.[8,20] It is likely that these capacities represent operating conditions under which machines were crowded.

Because of the smaller capacity, the cost of treatment on vanners and round tables is higher than the cost of tabling.

Obsolescence of Vanners, Buddles, and Round Tables. Except for the recovery of cassiterite ranging in size from 0.010 to 0.060 mm., vanners, buddles, and round tables are practically obsolete. Reasons for the obsolescence are as follows:

1. These devices fail to treat particles finer than 10 to 20 microns. A large part of the valuable minerals in so-called slime are not recovered by shaking tables, are also not recovered by buddles, vanners, and round tables, but are recovered by flotation.

2. In comparison with flotation, the capacity of flowing-film concentrating devices is extremely small for the treatment of the finer particles.

3. The concentrates obtained are relatively low grade if acceptable recoveries are secured.

4. Blanket plants capable of treating large tonnages are relatively inexpensive to install and operate for the recovery of fine free gold.

Shaking Tables. The *Rittinger* table which is now obsolete has marked a great improvement in design of concentrating machinery. It has introduced the sorting action of asymmetrical deck acceleration. Combination of this effect with flowing-film sorting resulted in a better separation. Since the rate of travel with the fluid and at right angles to it are both functions of the particle diameter of the same general form, use of both forces instead of fluid-film push only will make for a separation that is not markedly a function of particle size. This basic principle was first utilized in the Rittinger table.

The Rittinger table consists of a bare deck about 4 ft. long in the direction of motion and 6 to 8 ft. wide. Motion is obtained by a cam acting against a spring imparting a unidirectional push and is stopped by impact against a bumping block. In later designs, the table was built in pairs.

The *Bilharz-Stein* vanner combines the principle of the Rittinger table with that of an endless rubber belt. It has been widely used in Europe.

Wilfley Table.[27] The Wilfley table (Fig. 166) introduced in 1895–1896 marks the next big advance in the field of flowing-film

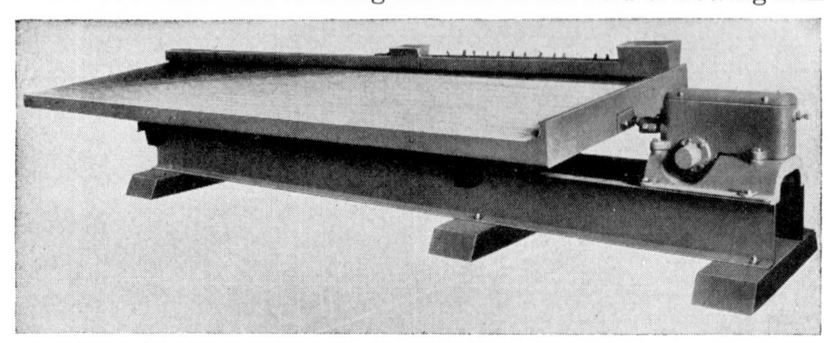

Fig. 166.—Wilfley table, general view. (*Mine and Smelter Supply Co.*)

concentration.[26,38] Its use is still worldwide. The makers claim that at one time over 23,000 Wilfley tables were in use. The advance marked by the Wilfley table consists in the introduction of riffles, which increased the capacity and allowed treatment of coarser feed, and in the introduction of an effective and rugged head motion.

The Wilfley table consists of a four-sided, nearly rectangular deck sloping adjustably toward one of the long sides. This deck is actuated by a pitman and toggle-type head motion (Fig. 167).

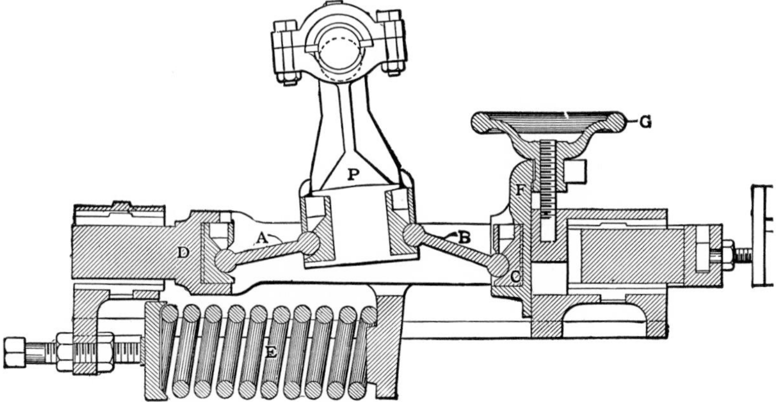

FIG. 167.—Head motion for Wilfley table. (*After Richards and Locke.*)

When the toggles are nearly horizontal, an appreciable vertical movement of the pitman has practically no effect on the position of the outer toggle, but the opposite is true when the toggles are markedly inclined to the horizontal. In this way, a symmetrical

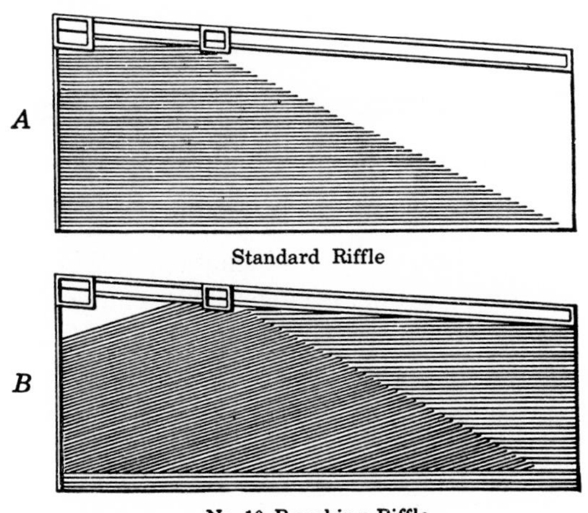

FIG. 168.—Examples of riffling for Wilfley tables. (*Mine and Smelter Supply Co.*)

motion of the pitman becomes an asymmetrical motion for the toggle ends. It is transmitted to the deck which is fastened to one of the toggle ends.

The deck is tilted transversely; the upper side is provided with a feedbox at the corner near the head motion and with a wash-water distributing box. This wash-water distributor consists of a perforated trough fitted with fingers at the perforations for regulation of the water flow. A slight upward tilt, about ½ in. in the length of the table, is provided toward the concentrate-discharge end.

The deck is riffled (Fig. 168*A*) with wooden cleats tacked on the linoleum-covered surface, the cleats tapering in thickness from ½ in. at the feed end to a thin edge near the concentrate end. Their width is usually ¼ in. The end of the riffles forms a diagonal line so that about two-thirds of the deck is riffled and the rest bare. This diagonal line hits the corner of the table diagonally opposite the feedbox. This is standard riffling. Other types of riffling are also used, *e.g.*, the "roughing" riffling shown in Fig. 168*B*.

Other Riffled Shaking Tables. The success of the Wilfley table led to the invention of many other shaking tables. Among these, the following deserve special mention because of their widespread use: the Garfield, Butchart, Card, Deister, James,[18,19] Deister-Overstrom, and Plat-O.

The *Garfield* table differs from the standard Wilfley in that the deck is riffled all the way. There is no unmodified flowing-film concentration since there is no unbroken plane surface on which fluid is flowing. The Garfield table has been widely used for roughing, the tonnage treated per table being large.

The *Butchart* table[7] differs from the Wilfley in the head motion and in that the riffles are bent toward the upper side of the table, for a length of several inches along the table-diagonal starting from the feed box. This riffling compels particles to climb appreciably before discharging into the concentrate. This is designed to yield a cleaner concentrate. One widespread use of Butchart tables has been for roughing, using full-length riffling.

The *Card* table differs from the Wilfley in that the riffles are *cut into* the linoleum to a roughly triangular section, rather than *applied on* linoleum in a rectangular section. The head motion also differs slightly although retaining the basic toggle-and-pitman principle.

The *Deister* and *Deister-Overstrom* tables utilize a different head motion and are roughly rhombohedral in shape. The rhombohedral shape results in some saving of floor space. The action of the head motion is particularly satisfactory. In the Deister-Overstrom coal-washing table,[28,40] every fifth or sixth riffle is higher than the others. This permits deeper pools in the spaces between riffles and therefore allows the treatment of coarser particles.

The *Plat-O* table features two or more plane surfaces or plateaus. The riffle cleats are of constant depth except at the point where the deck surface rises to meet the plane that would be formed by the tops of the riffle cleats. This table is liked particularly for its sturdiness.

Operation of Shaking Tables. Adjustments are provided in all tables for the amount of wash water, the cross tilt, the speed, and the length of the stroke. The speed of the table ranges usually from 180 to 270 strokes per minute, and the strokes are from ½ to 1½ in. long. Generally speaking, the adjustments are as follows:

For a roughing operation—more water, more ore, more tilt, and longer stroke.

For a cleaning operation—less water, less ore, less tilt, a shorter stroke.

For fine feed—less water, less feed, faster reciprocation, shorter stroke.

For coarse feed—more water, more feed, slower reciprocation, longer stroke.

Variations in character of feed require variations in operation. The operator's duty is to take care of them by adjusting the tilt, the wash water, and the position of the splitters that control discharge of table into concentrate, middling, and tailing launders. One man may look after 10 to 100 tables, depending upon the regularity of the feed and the difficulty of the task assigned to the table.

Tonnage treated depends upon

1. Size of feed.
2. Whether the operation is roughing or cleaning.
3. The difference in specific gravity between the minerals that are to be separated.

4. The relative abundance of locked particles.
5. The average specific gravity of the material treated.

A coarse feed can be treated in larger amounts than a fine feed. It would seem that the treatable tonnage increases at least as the square of the average size (theory indicates that it increases as the cube of the particle size).

A roughing operation is preferably conducted on a fully riffled deck. These decks have a greater capacity because the particles are treated throughout the deck in the form of a teetering suspension many particles deep instead of as a restive layer one particle deep. Such decks do not provide flowing-film concentration but some sort of jigging. On the other hand, a cleaning operation is preferably performed on a partly riffled deck.

It is clear that minerals of different specific gravity must be present: the greater the spread in specific gravity between minerals, the greater the capacity since that sort of condition permits crowding without considerable penalty. It is also clear that a bulky substance such as coal, other things being equal, cannot be treated in the same tonnage as a heavy ore. But since coal is usually tabled at a much coarser size, the adverse effect of its bulk is more than counterbalanced by its size advantage.

The effect of locked particles on capacity of tables should also be recognized. These particles behave in a fashion intermediate between that of pure particles of their constituent minerals. It is as if a three-product separation were sought in which one of the products would grade in specific gravity between the two others.

Capacity of Tables. Table capacity may be as high as 200 tons per 24 hr. on a fully riffled deck 4 by 12 ft. treating minus 3-mm. sulphide ore having a specific gravity of about 3.0 (roughing duty), or 500 tons per 24 hr. on a Deister-Overstrom coal table 8 ft. 4 in. by 16 ft. 10 in. treating buckwheat coal. But table capacity may be as low as 5 tons per 24 hr., or even less, for fine ore (minus 0.3 mm.) if there is only a small specific-gravity differential between minerals.

Cost of Tabling. Power requirement per table ranges from 0.5 to 0.8 kw. Most of the energy is expended to move the deck, which must therefore be as light as is consistent with rigidity.

Repairs consist of renewals of riffle cleats and of deck. These must be made at regular intervals. Lost time is not considerable.

Cost of tabling varies considerably in view of the great range through which the elements of the cost vary. Thus labor may range from $0.15 to $3 per table per 24 hr.; power from $0.06 to $0.50 per table per 24 hr., depending upon whether electricity is available at 0.5 or 2.0 cts. per kw-hr. Repairs may range from $0.10 to $0.50 per day. Depending upon the tonnage treated per table, the operating cost may then range from a fraction of 1 ct. to as much as 40 cts. per ton with the average perhaps in the vicinity of 5 to 10 cts. per ton.

Present Applications of Flowing-film Concentration. Flowing-film concentration was more widely used 20 years ago than it is now.(26) Lost fields of application include sulphide ores(16,20) of copper, lead, and zinc and to some extent oxidized ores of gold, copper, and lead. In these fields, flotation and hydrometallurgy have steadily gained ground.

Flowing-film concentration (mostly with shaking tables) is still used for the following purposes:

1. For the concentration of cassiterite.(11)
2. For the concentration of some free-milling gold ores.(29.41)
3. For the beneficiation of many nonmetallics, *e.g.*, glass sand,(17) chromite,(13) tungsten ores.(33,34,35)
4. For the recovery of relatively coarse, pelletlike by-product metal in metallurgical works.
5. For the recovery of a part of the galena and sphalerite in coarsely aggregated lead and zinc ores.(1,9,10,37)
6. For the beneficiation of some iron ores.(12)
7. For the cleaning of fine coal.(4,5,28,39,40)
8. As an adjunct concentrating device for the recovery of free gold in many flotation plants(2,3,14) and in some cyanidation plants.
9. As a pilot and guide to operations in flotation plants.

In addition, tables are being used as a separating device in connection with skin flotation and agglomeration (*cf.* Chapter XV).

Literature Cited

1. Anon.: Lead Mining Today in Southeast Missouri, *Eng. Mining J.*, **136,** 326–330 (1935).
2. Anon.: The Use of Corduroy in Gold Ore Treatment, *Ore Dressing Notes*, Number **3,** American Cyanamid Company, New York (1935).
3. Bernewitz, M. W. von: Corduroy as a Gold Saver, *Eng. Mining J.*, **136,** 63–67 (1935).

4. Bird, B. M., and H. S. Davis: The Role of Stratification in the Separation of Coal and Refuse on a Coal-washing Table, *U. S. Bur. Mines, Rept. Investigations* 2950 (1929).
5. Bird, B. M., and H. F. Yancey: Hindered-settling Classification of Feed to Coal-washing Tables, *Trans. Am. Inst. Mining Met. Engrs.*, **88,** 250–271 (1930).
6. Coggin, F. G.: Copper-slime Treatment, *Trans. Am. Inst. Mining Engrs.*, **12,** 64–68 (1883–1884).
7. Cole, David: Development of the Butchart Riffle System at Morenci, *Trans. Am. Inst. Mining Engrs.*, **51,** 405–423 (1915).
8. Crowfoot, Arthur: Development of the Round Table at Great Falls, *Trans. Am. Inst. Mining Engrs.*, **49,** 417–469 (1914).
9. Delano, L. A.: The Milling Practice of the St. Joseph Lead Company, *Trans. Am. Inst. Mining Engrs.*, **57,** 420–441 (1917).
10. Delano, L. A.: Milling Practice in the Lead Belt, *Eng. Mining J.*, **138,** 286–291 (1937).
11. Deringer, D. C., and John Payne, Jr.: Patiño—Leading Producer of Tin. III. Ore Concentration at Catavi, *Eng. Mining J.*, **138,** 299–306 (1937).
12. DeVaney, F. D., and Will H. Coghill: Concentration Tests on Tailings from the Washing Plants of the Mesabi Range, Minn., *U. S. Bur. Mines, Rept. Investigations* 3052 (1930).
13. Doerner, H. A.: Concentration of Chromite, *U. S. Bur. Mines, Rept. Investigations* 3049 (1930).
14. Engelmann, E. W.: Recovering Gold from Copper-mill Tailing, *Mining and Met.*, **16,** 331–332, 337 (1935).
15. Gahl, Rudolf: The Treatment of Slime on Vanners, *Trans. Am. Inst. Mining Engrs.*, **40,** 517–538 (1909).
16. Hayden, Ralph: Concentration of Slimes at Anaconda, Mont., *Trans. Am. Inst. Mining Engrs.*, **46,** 239–265 (1913).
17. Huttl, John B.: A Glass-sand Enterprise on the Pacific Coast, *Eng. Mining J.*, **138** [12], 29–31 (1937).
18. Krom, S. Arthur: Diagonal-plane Concentrating Table, *Trans. Am. Inst. Mining Engrs.*, **42,** 528–532 (1911).
19. Krom, S. Arthur: The James Diagonal-plane Slimer, *Trans. Am. Inst. Mining Engrs.*, **43,** 427–432 (1912).
20. Laist, Frederick, and Albert E. Wiggin: Flotation Concentrating at Anaconda, Mont., *Trans. Am. Inst. Mining Engrs.*, **55,** 486–526 (1916).
21. McDermott, Walter: The Frue Concentrator, *Trans. Am. Inst. Mining Engrs.*, **3,** 357–360 (1875).
22. Munroe, Henry S.: A Laboratory Slime Table, *Columbia Univ., Sch. Mines Quart.*, **22,** 306–307 (1901).
23. Page, Leigh: "Introduction to Theoretical Physics," D. Van Nostrand Company, Inc., New York (1928); (*a*) p. 229; (*b*) p. 19; (*c*) p. 25.
24. Peirce, B. O.: "A Short Table of Integrals," Ginn and Company, Boston, (1910); formula 265, p. 38.

25. Richards, Robert H.: Sorting before Sizing, *Trans. Am. Inst. Mining Engrs.*, **27**, 76–106 (1897).
26. Richards, Robert H.: The Anaconda Round Table, the Wilfley Table, and the Ten-spigot Classifier, *Mining and Met.*, **15**, 342–343 (1934).
27. Richards, Robert H.: The Wilfley Table, *Trans. Am. Inst. Mining Engrs.*, Part I, **38**, 556–580 (1907); Part II, **39**, 303–315 (1908).
28. Richardson, A. C., and B. W. Gandrud: Retreatment of Sayreton Jig Middlings on Coal-washing Tables, *U. S. Bur. Mines, Rept. Investigations*, 3101 (1931).
29. Scott, W. P.: Milling Methods and Ore-treatment Equipment (at Alaska-Juneau), *Eng. Mining J.*, **133**, 475–482 (1932).
30. Simons, Theodore: The Evolution of the Round Table for the Treatment of Metalliferous Slimes, *Trans. Am. Inst. Mining Engrs.*, **46**, 338–362 (1913).
31. Sperry, Edwin A.: The Sperry Vanning-buddle, *Trans. Am. Inst. Mining Engrs.*, **34**, 572–584 (1903).
32. Taggart, Arthur F.: "Handbook of Ore Dressing," John Wiley & Sons, Inc., New York (1927), pp. 736–737.
33. Vanderburg, W. O.: Methods and Costs of Milling Ferberite Ore at the Wolf Tongue Concentrator, Nederland, Boulder Co., Colo., *U. S. Bur. Mines, Information Circ.* 6685 (1933).
34. Vanderburg, W. O.: Methods and Costs of Concentrating Tungsten Ores at Atolia, San Bernardino Co., Calif., *U. S. Bur. Mines, Information Circ.* 6532 (1931).
35. Vanderburg, W. O.: Methods and Costs of Concentrating Scheelite Ore at the Silver Dike Mill, Mineral Co., Nevada, *U. S. Bur. Mines, Information Circ.*, 6604 (1932).
36. Wiggin, Albert E.: The Great Falls System of Concentration, *Trans. Am. Inst. Mining Engrs.*, **46**, 209–238 (1913).
37. Wright, Clarence A.: Ore Dressing Practice in the Joplin District, *Trans. Am. Inst. Mining Engrs.*, **57**, 442–471 (1917).
38. Wood, Henry E.: The Beginning of the Wilfley Concentrating Table, *Eng. Mining J.*, **125**, 975–977 (1928).
39. Yancey, H. F.: Determination of Shapes of Particles and Their Influence on Treatment of Coal on Tables, *Trans. Am. Inst. Mining Met. Engrs.*, **94**, 355–368 (1931).
40. Yancey, H. F., and C. G. Black: The Effect of Certain Operating Variables on the Efficiency of the Coal-washing Table, *U. S. Bur. Mines, Rept. Investigations* 3111 (1931).
41. Young, G. J.: Gold-ore Mining and Milling, *Eng. Mining J.*, **132**, 195–199 (1931).

CHAPTER XIV

FLOCCULATION AND DISPERSION

So far, a suspension of solids in fluids has been considered as made up of individual particles of solid surrounded by fluid, *i.e.*, as a mixture in the dispersed condition. However, if the particles are fine, a different condition may prevail in which the particles are clustered in flocs: the mixture is said to be flocculated (Fig. 169).

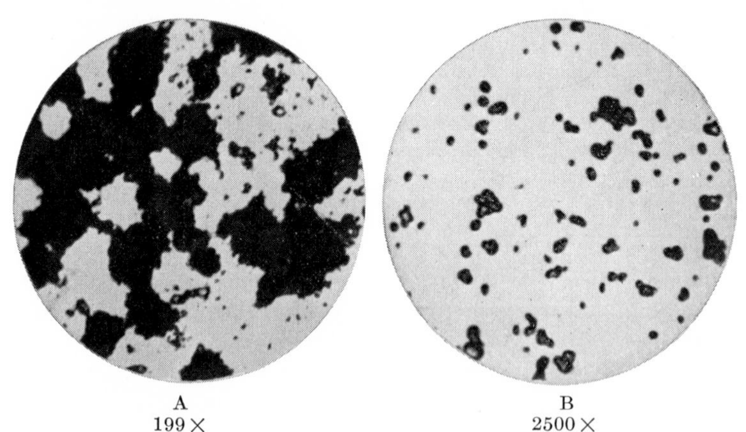

FIG. 169.—Flocculated (A) and deflocculated (B) silver bromide. (*After Sheppard and Lambert*).

The photographs are of the same preparation, at low magnification for the flocculated condition, and at high magnification for the deflocculated condition.

Whether a pulp is dispersed or flocculated is of great importance in many dressing operations, notably in classification, thickening, filtration, and flotation. A single example will illustrate this point. It has already been shown that a comminuted product contains particles much finer than the finest particles that can be settled in water within a reasonable time. Any thoroughly dispersed pulp of a comminuted product will then fail to clear regardless of the settling time used. But if the same pulp is so conditioned that the particles are grouped in flocs of a size large enough to be readily seen with the naked eye, these flocs will settle at a rate sufficient to provide in a reasonable time a clear supernatant liquor above a more or less compacted, flocculent sediment.

The object of the present inquiry is not to evaluate the practical importance of flocculation or dispersion: that is taken up in appropriate relationship to the processes of classification, thickening, filtration, and flotation. It is rather to review the physical and chemical mechanisms that result in flocculation or dispersion.

It may seem at first sight that a dispersed state is a "natural" condition for a suspension and that a flocculated state is accomplished only by a suitable artifice such as the introduction of an electrolyte in addition to the dispersing fluid and the dispersed solids. But as will appear presently, it is at least equally sound to regard the flocculated state as a "natural" state, the dispersed state being attained only if a third entity is artificially introduced. Clearly, the control of fine suspensions requires the use of one nostrum or another; this raises a question as to the ultimate composition of any suspension.

The Solid-fluid Interface. The boundary between a solid particle and the ambient fluid is the seat of physical and chemical forces differing in degree but not in kind from those within the solid and from those within the liquid: boundary forces must partake of the properties of both phases. These forces manifest themselves as excess attraction for some particular constituent of the fluid.

Water itself, even "pure" water, is a mixture of associated molecules, nonassociated molecules, and hydrated ions of various degrees of hydration.[(22)] Laboratory distilled water is always acid because of carbon dioxide dissolved from the air. Good drinking water may contain 100 parts or more of dissolved salts per million parts of water; and mill water is usually an even more complex mixture of many different ions and molecules moving about among water molecules of various sorts. Such a fluid is homogeneous only in a large sense, but is heterogeneous from the standpoint of possible selective behavior at a solid interface.

Every mineral is soluble to some extent in water. With many minerals, this solubility is so low in comparison with the ionic content of the fluid as to be of little significance, but with other minerals the solubility is significant, *e.g.*, with quartz, calcite, fluorite, many silicates. Contributions from the minerals add to the complexity of the fluid phase.

A pulp of finely divided mineral matter in water consists, then, of the mineral matter, water molecules in various stages of association, and a rather large variety of ions. These ions are each more or less hydrated, *i.e.*, associated with one or more molecules of water. Ions and water molecules are constantly attaching themselves to the mineral matter, and conversely, ions from the mineral surfaces are detaching themselves and going off into the liquid phase. At equilibrium, the two actions would exactly balance each other. The important feature of the attachment or detachment of ions and water molecules is that there is no reason to expect each of them to display the same affinity for each of the mineral surfaces and that they, in fact, do not display the same affinity. This phenomenon of *selective ion sorption** is one of the three phenomena basic to all flocculation and dispersion. It bears a resemblance to the phenomena of evaporation and of crystallization. Indeed, the latter may be regarded as that special case of ion sorption in which the sorbed ions are those which constitute the lattice of the crystals.

Brownian Movement. Brownian movement is the second phenomenon basic to flocculation and dispersion. By this name, reference is made to the spontaneous and erratic motion of fine particles suspended in a fluid, a motion that is generally attributed to molecular bombardment. Brownian movement is fundamental in connection with flocculation dispersion phenomena because it provides the means to bring together individual, suspended particles or the means to disintegrate flocs of them without recourse to stirring or to some other form of agitation imposed by outside agencies.

Brownian movement is generally regarded as due to molecular agitation. It has in fact been related quantitatively to many other fundamental physicochemical quantities such as the Avogadro number (number of atoms per gram atom).[32b]

An interesting hypothesis was recently advanced to explain Brownian movement as due[31] to transmission of the thermal agitation of the fluid to the suspended solids by the interionic

* The term "sorption" was introduced by McBain[33c] to avoid the difficulties raised by use of the words *absorption* and *adsorption* to which specific meanings became attached.

forces between the anchored ions and the free ions that comprise the double layer of ions at the surface of the particle.

Surface Energy. Surface energy provides the third basis for flocculation-dispersion phenomena. It is generally accepted that any interface is the seat of energy, surface energy; this energy is of course related to the forces residing at the interface, to which allusion has been made above. It is also well-known that in the words of Hugh S. Taylor[32a]: "*All naturally occurring processes are accompanied by an increase in the entropy of the system.*" This is the second law of thermodynamics or energetics. It simply means that the various forms of energy are of different quality, and that naturally occurring processes display a tendency for energy of some particular quality to form at the expense of energy of a less suitable quality.

In the particular instance of surface energy, there is marked tendency for surface energy to change to kinetic energy (energy of molecular agitation).

Flocculation of discrete particles reduces the interfacial area, hence the surface energy; it proceeds in accord with the second law of energetics and may therefore be expected to occur spontaneously.

The Three Phenomena Basic to Slime Control. The three basic phenomena involved in slime control are then

1. Selective attachment of ions.
2. Particle motion as a result of molecular bombardment.
3. Natural tendency for surface energy to become changed to kinetic energy.

The first phenomenon promotes dispersion as it causes ion-covered particles to repel each other; the third phenomenon promotes flocculation as it corresponds to a naturally occurring energy change; and the second phenomenon is a means of either bringing particles together or of helping them to scatter.

Ion Cover on Particles in Aqueous Solutions. Particles dispersed in water or in aqueous solutions migrate under the influence of an electric field. This migration is more frequently to the positive electrode than to the negative electrode, thus indicating that the particles are electrically charged, negatively more often than positively.

Since the positive charges must equal the negative charges in any electrolyte, the adhesion or sorption of negative charges on particles (as evidenced by their migration in an electric field) must result in superabundance of positive ions in the electrolyte.

Helmholtz[15,33b] postulated that the excess positive ions must arrange themselves as a second layer concentric to the first, and in the immediate vicinity in the fashion of electrical layers in a condenser. The conception of a double layer is inescapable, but it is preferable, in view of the experimental evidence, to consider the second layer as diffuse. This postulate of Gouy[12,19b] visualizes a particle surface with surrounding ion atmosphere as shown in Fig. 170: the solid dots represent cations, the circles anions. For purpose of simplification only excess ions,

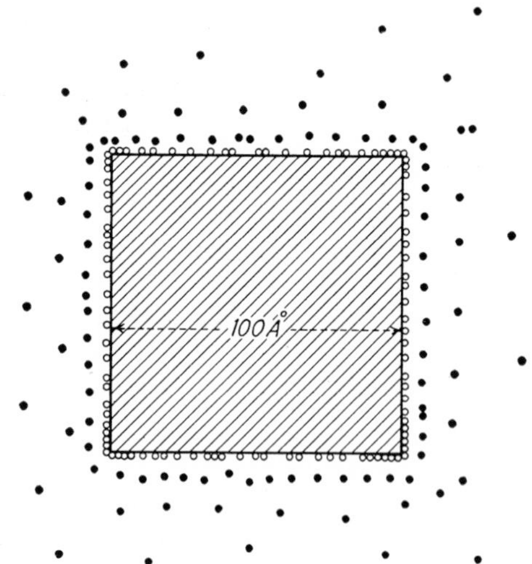

FIG. 170.—Colloidal particle with adhering anions and a diffuse layer of cations.

i.e., those over and above those of the bulk of the electrolyte, are represented.

Electrokinetic Potential. Changes in the composition of the electrolyte result in changes in the proportion of sorbed ions, hence in the net charge of the particles. This can be shown quantitatively by impressing the same e.m.f. to suspensions of the same dispersoid in different electrolytes and measuring the speed of the particles.

More conveniently, this can be expressed in terms of the electrokinetic or zeta potential

$$\zeta = \frac{4\pi\mu u(300)^2}{\kappa X}. \qquad \text{[XIV.1]}$$

In this relationship,* the zeta potential is in volts, the viscosity μ and the dielectric constant κ in c.g.s. units, the particle velocity u in centimeters per second, and the impressed potential gradient X in volts per centimeter.

Changes in electrolyte composition result, then, in changes in zeta potential, sometimes in an increase, sometimes in a decrease, sometimes even in a reversal of sign. Figure 171 after Freundlich and Ettisch[9,10] is typical.

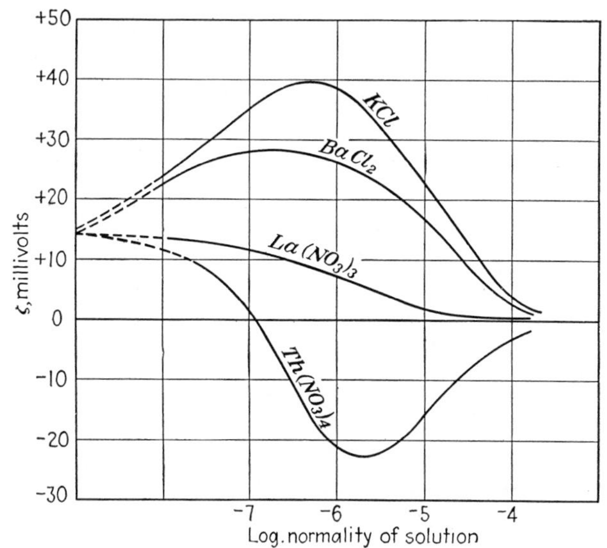

FIG. 171.—Electrokinetic potential of Jena glass against aqueous solutions. (*After Freundlich and Ettisch.*)

The exact quantitative relationship of the zeta potential to flocculation is not wholly clear, nor is the relationship of the zeta potential to the thermodynamic or epsilon potential of Helmholtz ϵ that is commonly employed in electrolysis. But it is clear that there is a qualitative parallelism between flocculation and small zeta potentials, as likewise between dispersion and large zeta potentials.

Flocculation by Addition of Electrolytes. Flocculation can be induced by adding electrolytes in suitable quantity. In dealing with negatively charged dispersions, the cation of the electrolyte is more important, and conversely.

* For derivation and limitations, see Thomas,[33] pp. 225–228.

Ions generally increase in flocculating effectiveness with valency, *i.e.*, trivalent cations are more effective than divalent cations and divalent cations are more effective than monovalent cations. This is explained by Freundlich[8] and others as due to the sorption of these ions at the mineral surface.

If *OABC* represents the sorption isotherm (Fig. 172), *i.e.*, the curve relating the concentration of the solute in the solution to the density of the sorbed solute on the surface of the dispersed particles, it is always found that the density of the sorbed solute increases much more slowly than the concentration of the solution. If salts of the same anion with three different cations are considered such that one of the cations is monovalent,

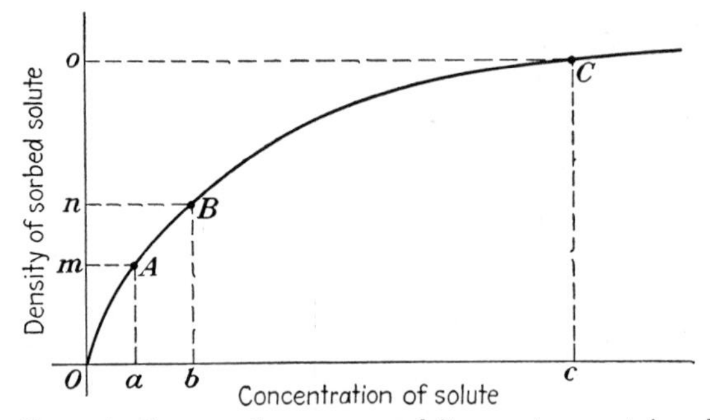

FIG. 172.—Concentration-sorption curve and its use to ascertain relative flocculating concentrations.

the second divalent, and the third trivalent, and if these salts all give the same sorption isotherm (a questionable assumption about which more will be said presently), concentrations *Oa*, *Ob*, *Oc* of the tri-, di-, and mono-valent metal salts will have the same flocculating power: The density of the sorbed layers will be as ⅓, ½, 1, but the charges per cation are as 3, 2, 1, so that the total charge is as 1:1:1.

Many rules have been offered to relate the relative ion concentrations to valency. The oldest is that of Schulze,[28a,b,c] and one of the best known is that of Linder and Picton[21] quoted by Ralston.[25] According to Linder and Picton, the precipitating power of ions of valency 1, 2, 3 is as 1:35:1,023; *i.e.*, trivalent ions are enormously more effective than divalent ions and divalent ions are enormously more effective than monovalent ions.

Many hundreds of investigations have been made on this score to substantiate the Schulze rule or its modifications. Exceptions are so numerous[33a] that the rule is worthless in a quantitative manner. This reconsideration of the rule does not invalidate the Freundlich reasoning to explain the qualitatively more potent flocculating power of polyvalent ions, but it requires a suitable modification. This modification consists in not regarding all ions as giving the same adsorption isotherms. Thus a salt of a monovalent ion such as potassium chloride may give isotherm *OC* (Fig. 173), a salt of another monovalent ion such as toluidine chloride isotherm *OC'*, whereas a salt of a divalent ion such as barium chloride gives isotherm *OB* and that of a

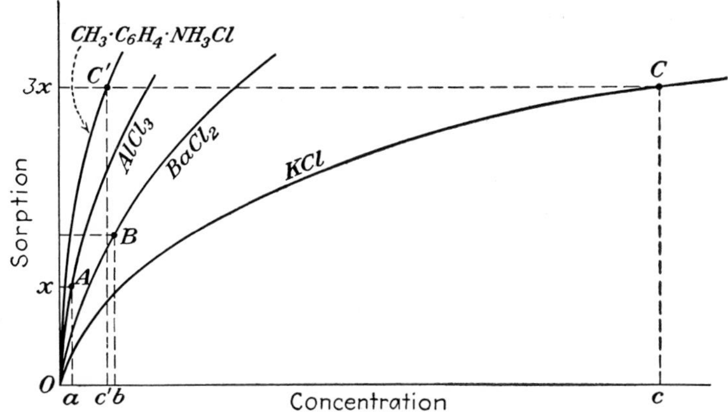

FIG. 173.—Concentration-sorption curves of various electrolytes having the same anion but different cations.

trivalent ion such as aluminum chloride an isotherm *OA*. Equal-flocculating concentrations may well be in the order

$$K^{+} > Ba^{++} > \text{toluidine}^{+} > Al^{+++},$$

and the flocculating power is in the inverse order.

Thomas[33a] believes that the "irregular series" of flocculating electrolytes that form insoluble precipitates with the ions previously engaged in maintaining the system in a dispersed state can probably be explained upon a clearer chemical basis than sorption.

Dispersion by Addition of Electrolytes. Just as electrolytes in properly chosen concentration can flocculate suspensions, so properly chosen concentrations of electrolytes may be used to disperse suspensions. This can be seen, for instance, in Fig. 171 if reference is made to the potassium chloride curve. At a concentration of the order of $10^{-6}N$, the zeta potential is nearly three times as large as at infinite dilution.

Dispersion by the Use of Colloids. Substances in the colloid condition are said to be *lyophobic* colloids if they are dispersed against their natural tendency, and *lyophilic* colloids if they are dispersed because of their natural tendency. Thus a gold sol is a lyophobic colloid because gold certainly has to be coerced into the colloid condition; conversely, gelatin is a lyophilic colloid because it is almost unknown except as a colloid.

Lyophilic colloids can be used to maintain a lyophobic substance in the state of fine dispersion. Hence the surname of protective colloids. It is believed that they operate by coating the lyophobic substance in such a way that the dispersoid appears to the fluid as if wholly made of the protective colloid. To an extent, this is an extreme case of the sorption of ions discussed in a preceding section, the sorbed ions being large or heavily associated (micellar) instead of small and unassociated.

Lyophilic colloids can also be the instrument by which flocculation is emphasized if there is present in the suspension some ion that precipitates the significant (or micellar) ion of the colloid. In that case, addition of the colloid instead of favoring dispersion results in flocculation.

Effect of Acidity or Alkalinity on Flocculation. This is a special case of the effect of electrolytes, and an important one. Fundamentally, it does not differ from the effect of other ions since acidity or alkalinity is merely the expression of presence of excess hydrogen or hydroxyl ions; but it has special significance because these ions are more effective control ions than other monovalent ions, and because they are the ions of water.

Practical Dispersing and Flocculating Agents for Mineral Pulps.[1,11,25,30] For dispersion, the more effective agents are alkali silicates, alkali carbonates, alkali sulphides, alkali hydroxides, alkali cyanides, glue, and gelatin.

For flocculation, the best all-round agent is lime. Its effectiveness is often increased by the joint use of starch.[6] Other flocculating agents include sulphuric acid, alum, gypsum, copper sulphate.

Size Range for Flocculation. It has already been pointed out that flocculation involves two factors, one is the tendency of the particles to stick to each other after they have come together—

this may be regarded as involving a probability of adhesion—the other is the chance of the particles coming in contact as a result of Brownian movement or extraneous agitation—this may be regarded as involving a probability of collision.[19a]

If the probability of adhesion is nil, no amount of waiting will induce flocculation; if on the other hand that probability is unity, flocculation is unavoidable and generally rather rapid. The rate at which flocculation proceeds will depend on the particle population in the fluid, on their size (hence the amplitude of Brownian movement), and on whether or not something else happens first. The size range for flocculation is therefore not expressible without consideration of other factors, principally dilution.

Toward the fine range, colloid phenomena have been noted in gold sols in which the particles are as fine as 20 to 50 Å. This range is much beyond that in which anything is known about mineral-dressing suspensions; it is therefore possible as a first approximation to say that all particles finer than some limiting size are subject to colloid phenomena.

To determine what this limiting coarse size may be, use can be made of an equation due to Einstein[7] and von Smoluchowski[29]

$$\Delta_x = \sqrt{t}\sqrt{\frac{RT}{3N\pi\mu r}}, \qquad \text{[XIV.2]}$$

in which Δ_x is the mean displacement of a particle of radius r in time t. In this equation, N is the Avogadro number, μ the viscosity, R the gas constant, and T the absolute temperature, all in c.g.s. units.

From this equation and by suitable probability considerations,[11] it is found that the statistical time for flocculation θ is (in seconds)

$$\theta = 2.4r^3 \frac{y^2(y+1)}{y+1-\sqrt{y(y+2)}}, \qquad \text{[XIV.3]}$$

y being the ratio of the net average distance between particles to the particle diameter. This ratio y is in turn related to the dilution D (or ratio of volume of water to volume of solids) by

$$y = \sqrt[3]{D+1} - 1. \qquad \text{[XIV.4]}$$

By equating the settling time according to Stokes' law, *e.g.*, [VIII.4] with the time for flocculation θ, and solving for the particle size, the critical size is found above which a particle will settle (a given distance) before flocculating even if the probability of adhesion is unity. The general equation is

$$r = 2.53l^{1/5}\left[\frac{(y+1) - \sqrt{y(y+2)}}{y^2(y+1)}\right]^{1/5}, \qquad [XIV.5]$$

in which l is the settling distance.

Table 39 gives the values of the particle diameter $2r$ for various dilutions and settling distances. This table brings out several points of interest:

1. In spite of enormous variations in pulp density and in settling distances, the range of changes in critical size is small.

TABLE 39.—RELATION OF CRITICAL (MAXIMUM) FLOCCULATING SIZE TO DILUTION AND SETTLING DEPTH (TIME)

Dilution, D	Ratio of interparticular distance to particle diameter, y	Maximum particle size, $2r$ (microns), for flocculation if settling depth is		
		1 cm.	10 cm.	100 cm.
1	0.26	7.2	10.4	18.2
2	0.44	5.5	8.7	13.8
3	0.59	4.7	7.5	11.8
4	0.70	4.2	6.7	10.5
6	0.91	3.6	5.7	9.0
10	1.20	2.9	4.7	7.3
25	1.96	2.2	3.5	5.5
100	3.67	1.42	2.3	3.6
250	5.31	0.94	1.49	2.4
1,000	9.0	0.73	1.16	1.8
10,000	20.6	0.39	0.62	0.98

2. The practical conditions (set out inside the rectangular ruling) indicate that the limiting size toward the coarse end for colloid phenomena is roughly 2 to 7 microns.

Similar conclusions are obtained by Bancroft[2] in the case of pneumatic suspensions.

Equation [XIV.5] shows that the critical velocity is independent of the specific gravity of the solid. This conclusion is probably only an approximation as it is involved in some assumptions made in the derivation of the equation.

If relatively coarse particles are present with fine particles, floccules of fine particles may entrap or enmesh coarse particles which by themselves would settle without flocculating.[27,30] This has two effects: coarse particles behave as if subject to flocculation whereas they are not, of themselves, and coarse particles appreciably increase the weight of floccules without appreciably increasing their volume. This weighting down is of importance in connection with dewatering in thickeners.

Effect of Pulp Density on Flocculation. Table 39 has already shown the importance of pulp density on maximum flocculable size. Increasing the pulp density (or decreasing the pulp dilution) must result in more rapid and complete flocculation if the conditions are ripe for flocculation.

If dispersion is sought, dilution of a recalcitrant pulp is often helpful since this dilution reduces the critical flocculating size and possibly reduces the concentration of the ions that are causing the flocculation.

Dispersion by Molecular Bridges. In a system consisting of finely divided solid and a fluid, insertion of some solutes can lead to the erection of a molecular bridge connecting the dispersoid to the liquid.[13,14,20]

Suitable molecules must be large enough so that the chemical or electrical forces of several special parts of the molecule are sufficiently distant not to affect each other. Such molecules, furthermore, must have a special affinity for the fluid in one part and for the solid in another. This affinity may take the form of executed reaction or merely of potential reaction (sorption). Thus white lead (lead carbonate) can be dispersed in hydrocarbon oil by the use of a fatty acid (*e.g.*, oleic acid): the fatty acid dissolves in the oil, reacts to produce lead oleate, and the lead oleate sticks to the lead carbonate in a special orientation. The oleate part of lead oleate is very soluble in oil, but the lead part is rather insoluble; the lead part is necessarily compatible with lead carbonate, but the oleate part is not. So the lead oleate congregates at the border with the oleate radicals turned toward the oil and the lead atoms toward the lead carbonate. This description is used to explain the increased dispersion obtained in paints[3,4,5,18,23,26] if the vehicle (oil) contains suitable impurities, deliberately or accidentally added. Figure 174 summarizes these ideas.

In dispersion of the molecular-bridge type, there does not seem to be need for electric charges on the dispersed particles. This type of dispersion is perhaps the rule in nonionized liquids, and the ionic type of dispersion is the rule in ionized liquids. It is possible, also, for molecular-bridge dispersion to occur in

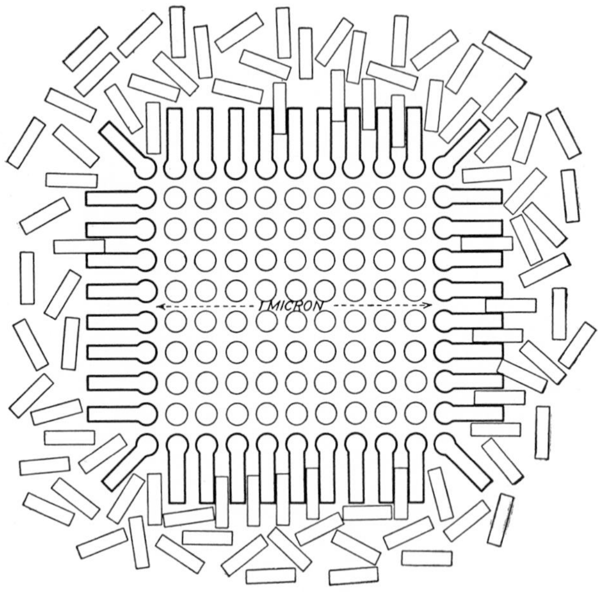

FIG. 174.—Dispersion by molecular bridging. The bridge molecules are shown in bold face connecting an ionic particle (circles) to a hydrocarbon oil (rectangles). Ions and molecules are magnified 400-fold with reference to the scale.

aqueous systems if protective colloids furnish the basis for the dispersion.

One way of viewing molecular-bridge dispersion is to say that the interfacial energy (surface energy) of the solid is reduced to the vanishing point. This removes the incentive for flocculation.[14,24] Yet another way of viewing molecular-bridge dispersion is to say that particles coated with bridging molecules are physically prevented from coming in contact by the fluid molecules adhering to the fluidlike ends of the bridge molecules, or by the fluidlike ends of the bridge molecules themselves.

Although molecular bridges may be monomolecular, as shown in Fig. 174, viscous films many molecules thick may also form.

Molecular bridges are thought to be one of the basic causes for emulsion stability.[16,17,33d] Many, perhaps all, emulsions are stabilized by them (Fig. 175), and it is possible that the dispersion by ionic sorption is just as much a dispersion by

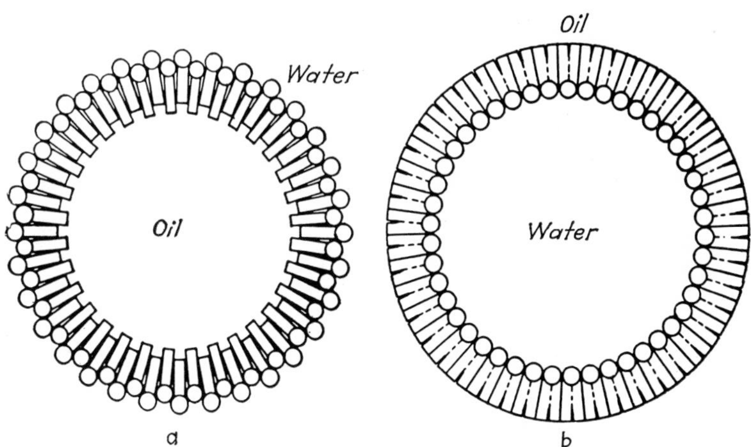

FIG. 175.—Two types of emulsions. Molecular bridge consists of polar-nonpolar molecules. In the case of the oil-in-water emulsion there is some crowding since the bridge molecules are larger toward the polar end. [*After Harkins, courtesy of The Chemical Catalog Co., Inc. (Reinhold Publishing Corporation).*]

molecular bridging, the bridge consisting of the sorbed ions with their train of associated water molecules (hydration molecules).

Literature Cited

1. ASHLEY, HARRISON EVERETT: The Chemical Control of Slimes, *Trans. Am. Inst. Mining Engrs.*, **41**, 380–395 (1910).
2. BANCROFT, WILDER D.: "Applied Colloid Chemistry," McGraw-Hill Book Co., Inc., New York (1932), p. 176, Table XLI.
3. BOWLES, R. F.: Some Observations of Wetting and Dispersion, *Oil Colour Trades J.*, **93**, 211–213 (1938).
4. CLAYTON, WILLIAM: Dispersion of Pigments in Drying-oil Media, *J. Oil Colour Chem. Assoc.*, **21**, 141 (1938).
5. CROUET, JEAN: Unimolecular Layers and Pigment Dispersion without Grinding, *Paint Tech.*, **2**, 55–56, 107–109 (1937).
6. DOWSETT, C. W.: Caustic Starch Improves Settlement of Slime, *Eng. Mining J.*, **137**, 161 (1936).
7. EINSTEIN, ALBERT: Zur Theorie der Brownschen Bewegung, *Ann. Physik* [4], **19**, 371–381 (1906).

8. FREUNDLICH, HERBERT: "Colloid and Capillary Chemistry," translated by H. S. Hatfield, Methuen & Co., Ltd., London (1926).
9. FREUNDLICH, H.: On the Electrokinetic Potential, in "Colloid Symposium Monograph," Vol. III, Chemical Catalog Company, New York (1925); pp. 7–16.
10. FREUNDLICH, H., and G. ETTISCH: The Electrokinetic and the Thermodynamic Potential, *Z. physik. Chem.*, **116**, 401–419 (1925).
11. GAUDIN, A. M.: "Flotation," McGraw-Hill Book Company, Inc., New York (1932), Chap. IV., pp. 44–48.
12. GOUY, M.: Sur la constitution de la charge électrique à la surface d'un électrolyte, *J. Phys.* [4], **9**, 457–468 (1910).
13. HARKINS, W. D., C. H. DAVIES, and G. L. CLARK: The Orientation of Molecules in Surfaces of Liquids, the Energy Relations at Surfaces, Solubility, Adsorption, Emulsification, Molecular Association, and the Effect of Acids and Bases on Interfacial Tension, *J. Am. Chem. Soc.*, **39**, 541–596 (1917).
14. HARKINS, W. D., and H. ZOLLMAN: Interfacial Tension and Emulsification, *J. Am. Chem. Soc.*, **48**, 69–80 (1926).
15. HELMHOLTZ, H.: Studien über elektrische Grenzschichten, [*Wiedemann*] *Ann. Physik*, **7**, 337–382 (1879).
16. HOLMES, H. N., and D. H. CAMERON: Cellulose Nitrate as an Emulsifying Agent, *J. Am. Chem. Soc.*, **44**, 66–70 (1922).
17. HOLMES, H. N., and W. C. CHILD: Gelatin as an Emulsifying Agent, *J. Am. Chem. Soc.*, **42**, 2049–2056 (1920).
18. KHOMIKOVSKY, P., and P. REHBINDER: Dependence of Stabilization and Wetting of Particles Suspended in Oil Medium on the Quantity of Surface-active Substance Adsorbed, *Compt. rend. acad. sci. U.S.S.R.*, **18**, 575–578 (1938).
19. KRUYT, H. R.: "Colloids," translated by H. S. VAN KLOOSTER, John Wiley & Sons, Inc., New York (1927); (*a*) p. 67; (*b*) pp. 92–108.
20. LANGMUIR, IRVING: Constitution and Fundamental Properties of Solids and Liquids, *J. Am. Chem. Soc.*, Part I, **38**, 2221–2295 (1916); Part II, **39**, 1848–1906 (1917).
21. LINDER, S. E., and HAROLD PICTON: Solution and Pseudo-solution. II. Some Physical Properties of Arsenious Sulfide and Other Solutions, *J. Chem. Soc.* (*London*), **67**, 63–74 (1895).
22. PENNYCUICK, S. W.: The Structure of Water, *J. Phys. Chem.*, **32**, 1681–1696 (1928).
23. PRICE, C. W.: Dispersion of Pigments in Drying-oil Media, *J. Oil Colour Chem. Assoc.*, **21**, 63–83 (1938).
24. QUINCKE, G.: Über die physikalische Eigenschaften dünner fester Lamellen, [*Wiedemann*] *Ann. Physik*, **35**, 561–580 (1888).
25. RALSTON, OLIVER C.: The Control of Ore Slimes, *Eng. Mining J.*, **101**, 763–769, 890–894 (1916).
26. REISING, J. A.: Electrostatic Responses of Pigments in Relation to Flooding, Dispersion and Flocculation, *Ind. Eng. Chem.*, **29**, 565–571 (1937).

27. ROBERTS, ELLIOTT, J.: Colloidal Chemistry of Pulp Thickening, *Trans. Am. Inst. Mining Met. Engrs.*, **112**, 178–188 (1934).
28*a*. SCHULZE, HANS: Schwefelarsen in wässriger Lösung, *J. prakt. Chem.*, **25**, 431–452 (1882).
28*b*. SCHULZE, HANS: Antimontrisulfid in wässeriger Lösung, *J. prakt. Chem.*, **27**, 320–332 (1883).
28*c*. SCHULZE, HANS: Über das Verhalten von seleniger zu schwefliger Säure, *J. prakt. Chem.*, **32**, 390–407 (1885).
29. SMOLUCHOWSKI, M. VON: Zur kinetischen Theorie der Brownschen Molekularbewegung und der Suspensionen, *Ann. Physik* [4], **21**, 756–780 (1906).
30. STEWART, R. F., and E. J. ROBERTS: The Sedimentation of Fine Particles in Liquids, *Trans. Inst. Chem. Engrs. (London)*, **11**, 124–137 (1933).
31. TAGGART, A. F., T. C. TAYLOR, and A. F. KNOLL: Chemical Reactions in Flotation, *Trans. Am. Inst. Mining Met. Engrs.*, **87**, 217–260 (1930).
32. TAYLOR, HUGH S.: "A Treatise on Physical Chemistry," D. Van Nostrand Company, Inc., New York (1925); (*a*) Vol. I, p. 63; (*b*) Vol. II, pp. 1279–1284.
33. THOMAS, ARTHUR W.: "Colloid Chemistry," McGraw-Hill Book Company, Inc., New York (1934); (*a*) Chap. VIII, pp. 178–203, with 80 references; (*b*) p. 225; (*c*) p. 272 citing McBain, *Phil. Mag.*, **18**, 916 (1909); (*d*) pp. 412–449.

CHAPTER XV

FLOTATION AND AGGLOMERATION—PHYSICAL ASPECTS

Flotation and agglomeration are processes of concentration based on the adhesion to air of some particles from a pulp and the simultaneous adhesion of other particles to water.

Related processes include greased-deck concentration in which separation is based on selective adhesion of some grains (diamonds) to quasi solid grease with adhesion of other grains to water; bulk-oil flotation and granulation in which separation is based on selective adhesion of some grains to organic liquids with adhesion of other grains to water; and amalgamation in which separation is based on selective adhesion of native gold and silver to mercury with adhesion of other grains to water. Flotation to some extent and agglomeration to a larger extent may include the use of enough organic liquid (oil) to produce a distinct oil phase, in which case selective adhesion of some minerals to oil is involved in addition to selective adhesion to air.

Flotation and agglomeration differ from each other not in the fundamental physicochemical phenomenon of selective adhesion, but in the physical means that are employed to put selective adhesion to practical use. In *froth flotation* or flotation proper, adhesion is effected between gas bubbles and small particles in such a way that the specific gravity of the mineral-air associations is less than that of the pulp so that they rise in that pulp. The floating mineralized froth is then mechanically separated from the pulp. In *skin flotation*, adhesion is effected between a free water surface and particles, usually larger than those involved in froth flotation. In *agglomeration*, loosely bonded associations of particles and bubbles are formed which are heavier than water but lighter and coarser than the particles adhering to water; flowing-film gravity concentration is used to separate the agglomerates from nonagglomerated particles. Agglomerates frequently seem to explode if brought in contact with a free water surface and to become replaced by skin-floating individual particles. The particles that are separated by agglomeration are usually of the same size as those separated by skin flotation, *i.e.*, substantially coarser than those separated by froth flotation, and the two means of separation are utilized jointly.

The crux of the processes of flotation and agglomeration is, then, the existence of a selective tendency for some particles to adhere to air and for other particles to adhere to water. Much ink has been consumed in "explaining" why there is such a tendency. Actually, however, the explanations are merely attempts to correlate physical and chemical phenomena rather than a search for a fundamental cause to the effect of flotation. And yet technological rationalization has been and will continue to be useful in eliminating unpromising avenues of further development in these fields as in others and in opening up promising avenues.

The essential known facts can be summarized as follows:

1. Most minerals if suitably protected from contamination adhere to water, not to air.

2. Paraffin and other hydrocarbon substances adhere to air in preference to water.

3. Some mineral substances adhere to air. It is currently debated whether this is due to surface impurities or to an inherent property of the minerals.

4. Minerals (*e.g.*, most sulphides) can be made to adhere to air by adding a suitable agent to the pulp. The quantity of agent may be well under that required to film the apparent surface of the mineral with a continuous layer one molecule deep, of that magnitude or larger, depending upon the circumstances.

5. Most and perhaps all minerals can be made air adherent or water adherent by the use of the proper agent or combination of agents. But it does not follow that this can be realized while all other minerals are retained in a water-adherent state.

6. Changes in the character of the surfaces of minerals because of oxidation or other processes affect considerably the facility with which they can be made air adherent.

For practical purposes, adhesion to air in preference to water is a property peculiar to hydrocarbon groups of atoms, whether these be part of the mineral itself or part of a coating on the mineral surface. Exceptions to this rule have been shown, more and more convincingly, to be due to contamination by minute quantities of organic substances sometimes traceable to faulty manipulation, sometimes to natural associations.

Physicochemical understanding of the mechanism of flotation requires the use of conceptions of surface energy, surface tension, adsorption, contact angle, polarity, surface reactivity, and of surface condition.

Surface Energy and Surface Tension.(53,15,46) At any interface, there resides a certain amount of energy. This energy is probably intimately related to the unsaturation of the surface atoms and/or the orientation of the surface molecules. Like other forms of potential energy, surface energy is readily convertible into energy of "lower" form, as heat (kinetic energy). This is an application of the second law of thermodynamics (see page 321).

If the phases in apposition are both fluid, surface energy can manifest itself as surface tension (liquid and gas) or as interfacial tension (two liquids).

The largest surface tension accurately known is that of mercury (465 dynes per cm.). Mercury in this respect stands in a class with other liquid metals, differing by a wide margin from other liquids. Water has a rather large surface tension, 72.8 dynes per cm. at 20°C. Organic liquids have surface tensions of the order of 20 to 40 dynes per cm. at room temperature. Interfacial tensions are smaller, sometimes not exceeding a small fraction of 1 dyne per cm.

Surface energy is numerically equal to surface tension; surface energy, of course, is expressed in ergs per square centimeter, not in dynes per centimeter.

As temperature rises, surface tension decreases, becoming nil (or at least very small (54)) at the critical temperature. Likewise, interfacial tension drops with rising temperature, becoming nil at the temperature at which the two liquids become miscible.

The surface tension of aqueous solutions of organic substances is intermediate between that of water and of the solute, but much closer to that of the solute than might be expected. This effect is related to the greater concentration of solute molecules at the surface than within the liquid (adsorption; see page 337) and to the orientation of the organic molecules in this surface film: at any given instant, the proportion of organic molecules that are oriented with their organic group away from the water is enormously greater than average (see Fig. 176). As a result, the surface tension is much nearer that of the low-surface-tension constituent than is indicated by a linear average.

Numerous methods have been developed for the measurement of surface tension. The error of some of the methods is less than 0.10 per cent. Of the various methods, the most attractive seems to be one in which a film of liquid is pulled out of the body of the liquid by a knife-edge or wire upon application of a suitable tension.[13,43] Until recently, the method was theoretically imperfect and gave but relative values. This drawback has

FIG. 176.—Diagram representing a solution of amyl alcohol, $C_5H_{11}OH$, in water. The small objects represent the alcohol molecules; the polar, or OH^- group, being round, and the nonpolar, or $C_5H_{11}^-$ group, being rectangular.

recently been overcome by the adoption of the laborious method of numerical integration to get specific solutions of the fundamental equation of surface tension, a differential equation due to Laplace.[14,22]

Although there is little doubt that solids have surface energy, no direct method of determination is available. Indirect methods have been devised which involve complicated physicochemical quantities. The results indicate surface energies high compared with those liquids; *e.g.*, quartz is said to have a surface energy of 920 ergs per sq. cm.[11] The results are not wholly trustworthy.*

Adsorption.[16,50] This term has had wide use in flotation as in other fields. Unfortunately it has had more than one meaning, and it has been condemned in recent years.

* See also Chapter VI, p. 134.

At first, adsorption meant merely an increase or decrease in concentration of a dissolved substance at the boundary of the phase in which it was dissolved. For example, the increase in concentration of alcohol at the surface of aqueous alcohol. This has also been termed "physical" or reversible adsorption, with the implied thought that the process is not a chemical process. The fallacy of the implication is realized if consideration is given to the specificity of "physical" adsorption.

Later, as evidence accumulated of the chemical specificity of adsorption, the process was viewed from the standpoint of atomic and molecular mechanics. This led to the conception and wide acceptance of the then revolutionary ideas concerning monomolecular films.

Later still, as it became obvious that an exchange often took place between ions in a solid with ions in an ambient fluid, the term "exchange adsorption" was coined and widely used. It is impossible to draw the line between exchange adsorption and metathesis: they grade into each other. Practically speaking, it may be said that if the exchange proceeds to a depth great compared with atomic dimensions the effect is metathesis, and that if the exchange proceeds to a depth of the order of atomic dimensions the effect is exchange adsorption. Additional distinguishing points between exchange adsorption and metathesis are brought out in Chapter XVI.

Irreversible adsorption includes those cases in which an initial reversible adsorption has become converted by some additional reaction that it is as yet unidentified.

As is explained on page 320, the term "sorption" has recently been offered by McBain as a preferable substitute for "adsorption." The terms are used interchangeably in this text.

Contact Angle.[38,41,42] Two fluids come in contact along a surface, three along a line. A section normal to that line of contact shows the line as a point and each two-phase interface as a line. Generally speaking, at equilibrium the contact between three fluids will be as shown in Fig. 177, and

$$\mathbf{T}_{MA} + \mathbf{T}_{MW} + \mathbf{T}_{WA} = 0, \qquad [XV.1]$$

in which $\mathbf{T}_{MA}$, $\mathbf{T}_{MW}$, $\mathbf{T}_{WA}$ are vectors representing the surface tensions acting at O.

If we contrive to make the surface of one phase plane (say M), then Eq. [XV.1] is patently impossible (unless $\mathbf{T}_{AW} = 0$), since there must be of necessity a component of $\mathbf{T}_{AW}$ normal to the surface of M. But the components of the forces parallel to M* can satisfy the condition [XV.1], *i.e.* (Fig. 178),

$$T_{MA} = T_{MW} + T_{AW} \cos \theta. \qquad \text{[XV.2]}$$

In this relation, the various surface tensions are not vectors but scalars. This relationship due to Reynders[34] correlates

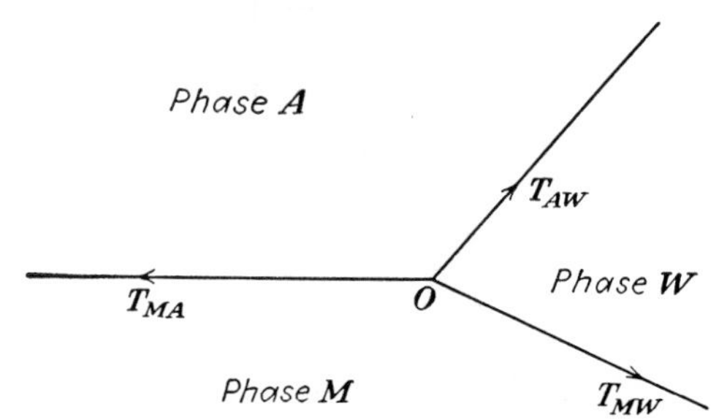

FIG. 177.—Contact of three deformable phases.

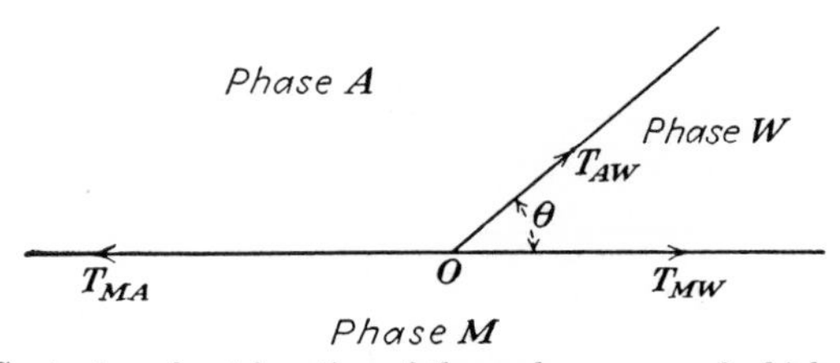

FIG. 178.—Contact angle at junction of three phases, one of which (M) is rigid.

contact angle θ (measured across W, *i.e.*, water) with surface tension and interfacial tensions. It is clear, of course, that for a given trio of phases the contact angle should be independent of any factors other than pressure and temperature.†

Contact angles are not limited to systems in which all the phases are fluid. Instead, M can be a mineral or solid of any sort. It will then be clear that, theoretically at least, contact

* For consideration of the $T_{AW} \cdot \sin \theta$ component, see pp. 351 to 358.

† Sulman[38] claims that contact angles vary through a range which he terms hysteresis; recent work, however, indicates that this effect is merely a frictional effect due to rough surfaces, and that it disappears on polished surfaces.[47]

angle is very important in determining adhesion. Perfect adhesion to water would occur for $\theta = 0°$, *i.e.*, if

$$T_{MA} \geq T_{MW} + T_{AW}, \qquad \text{[XV.3}a\text{]}$$

or

$$E_{MA} \geq E_{MW} + E_{AW}, \qquad \text{[XV.3}b\text{]}$$

and perfect adhesion to air would occur for $\theta = 180°$, *i.e.*, if

$$T_{MA} \leq T_{MW} - T_{AW}, \qquad \text{[XV.4}a\text{]}$$

or

$$E_{MA} \leq E_{MW} - E_{AW} \qquad \text{[XV.4}b\text{]}$$

In these equations, T's denote tensions and E's energies.

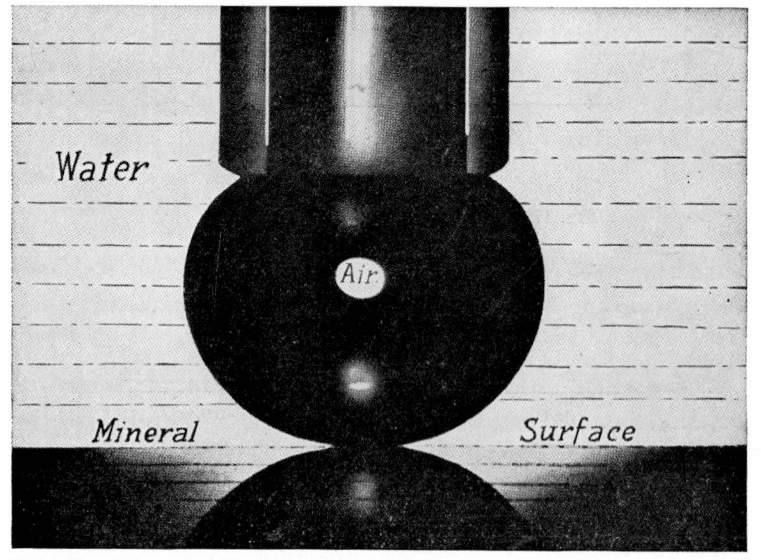

FIG. 179.—An air bubble does not stick to a clean, polished galena surface. (*After Wark.*)

No substance is known for which the air-water contact angle, measured across the water, exceeds 105° (solid paraffin); on the contrary, many are known for which $\theta = 0°$. If θ were known for solids of diverse types and under a wide variety of circumstances, an array of these solids in order of increasing tendency to adhere to air could be prepared.

Figures 179 and 180 show galena-water-air relationships with a clean surface ($\theta = 0°$) and with a surface conditioned by potassium ethyl xanthate ($\theta = 60°$). These are photographs by Wark.[47]

Atomic Bonding and Polarity.[10,17a,26] In the architecture of matter, atoms are arranged in one of four ways which are usually termed the electrovalent bond, the covalent bond, the coordinate bond, and the metallic bond. Of these, the first is also termed polar and the covalent and coordinate bonds are nonpolar.

Fig. 180.—An air bubble adheres to galena immersed in a solution of ethyl xanthate. The contact angle equals 60°. (*After Wark.*)

In the *electrovalent bond*,[23,10] one atom (or group of atoms) gives up one or more electrons to another atom (or group of atoms), each atom (or group) eventually displaying a completed shell of valence electrons, generally eight. The electrovalent bond is typified by common salt, NaCl, whose formula can also be written as shown in Fig. 181*A*, to indicate the transfer of one electron from the valence electron shell of the sodium atom to the valence electron shell of the chlorine atom. The charged atoms are ions.

The electrovalent bond is characterized in the liquid state by repeated association and dissociation of the ions. In the solid state, polar substances are ionized and display the tendency to rearrangement in the form of strong fields (stray fields) of attractive or repulsive force repeated at short intervals in geometric patterns (lattices).

In the *covalent bond*,[25] each atom contributes one or more electrons to a pool. Each atomic nucleus then has at its disposal all the electrons of the pool and acts as if it has a completed shell

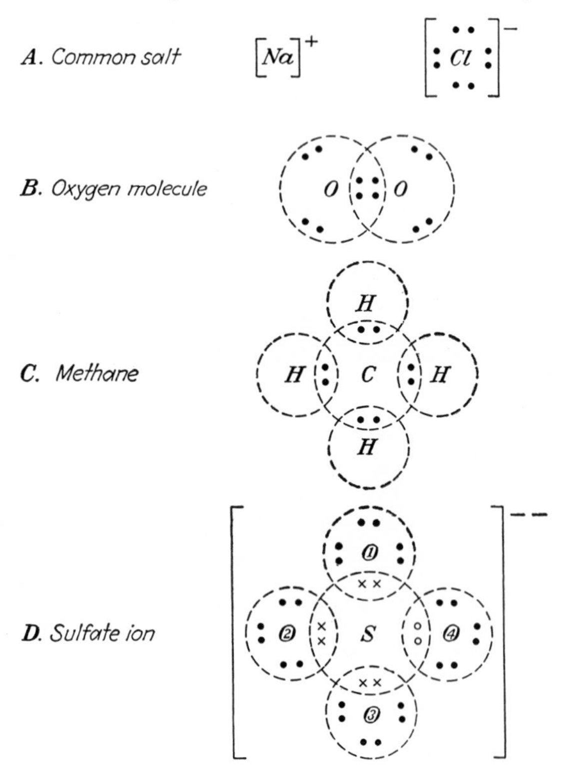

FIG. 181.—Typical linkages between atoms.

of valence electrons, generally eight. The covalent bond is typified by molecular oxygen, O_2, whose formula can be written as in Fig. 181*B*.

From a flotation standpoint, the important covalent bond is that displayed by hydrocarbons and by the hydrocarbon portions of large molecules. The typical example is methane, CH_4 (see Fig. 181*C*).

The covalent bond is characterized by chemical satisfaction, resistance to rupture, and nonreactivity (small "stray" fields).

In the solid state, many atoms rather than two may be involved, the electron pooling often proceeding from one atom to another in atomic chains capable of infinite extension, as in diamond and in other "giant molecules." In such a lattice as that of diamond, it cannot be said that one atom has donated an electron to another, nor can the number of atoms in the molecule be defined. Because a giant-molecule lattice is endless, it is not known whether any peculiar degree of polarity is developed at the surface.

In other cases in covalently bound atoms, the electron pooling proceeds in chains along one or along two directions instead of along three directions. Such a crystal may then be covalently linked in the fashion of oxygen molecules in one direction and covalently linked in the fashion of diamond in other directions. At any rate, the bonding between groups of atoms can be much more marked in some directions than in others, a situation that results in the platy cleavage so common among organic compounds.

The *coordinate bond* resembles the covalent bond in that it is characterized by pooled electrons; it differs from it in that the pooled electrons have all been contributed by one of the atoms, not both.[52,36]

An example of coordinate linkage is afforded by that of sulphur to oxygen in the sulphate ion (Fig. 181*D*). In this diagram, the crosses represent electrons derived from sulphur, the dots those derived from oxygen, and the small circles those acquired from a donor. Oxygen atoms 1, 2, and 3 are linked coordinately; oxygen atom 4 is linked coordinately with the help of two borrowed electrons. These borrowed electrons have been supplied by atoms (now cations) outside the sulphate radical. The polar properties of sulphates arise from the fact that $[SO_4]^=$ is an ion; and the absence of sulphide or sulphite traits in sulphate is due to the strength of the coordinate bonds.

The *metallic linkage* resembles both the giant-molecule covalent linkage and the electrovalent linkage although it can be clearly differentiated from both of these. As in the covalent and coordinate linkages, adjoining atoms are bonded by pooled electrons. But these electrons are not rigidly held in place. In fact, they can and do wander freely across the metal much as if the metal consisted of a nearly rigid lattice of positive ions permeated by an electron gas capable of independent motion under the influence of an impressed difference in electric potential.[8,9]

The investigations of Dean indicate that the surface of a metallic solid, unlike its interior, is characterized by local formation of dipoles. It has a polar character.

Combinations of these various types of linkages are found in various compounds. Examples are presented in the following table:

TABLE 40.—TYPES OF LINKAGE BETWEEN ATOMS IN VARIOUS COMPOUNDS

Compound	Type of linkage			
	Electrovalent	Covalent	Coordinate	Metallic
Sodium chloride (solid)	√			
Sodium sulphate	√	√	√	
Diamond	..	√		
Tetraethyl lead	..	√		
Potassium mercuric iodide, $K_2(HgI_4)$	√	..	√	
Diethyl sulphite, $(C_2H_5O)_2S \rightarrow O$	..	√	√	
Galena	√	√	..	√
Sphalerite	√	√		
Copper	..	..	..	√
Water	√	√		
Methane	..	√		
Acetic acid	√	√		
Stearic acid	√	√		
Potassium ethyl xanthate	√	√		
Nujol	..	√		

Some compounds display jointly several types of linkage because different atoms are connected differently (sodium sulphate), others because of gradational properties (galena), others because tautomeric forms exist (water).

It seems now that only those substances that possess neither metallic nor electrovalent bonds are capable of displaying nonpolar surfaces unless the molecules are so large that parts of the molecules may be considered separately (as in stearic acid). Even among covalent substances, some question might be raised concerning giant molecules. Further experimentation is required.

Water deserves particular consideration because of its use in all adhesion processes of concentration. It occurs in the two forms

$$\underset{\text{covalent}}{H:\ddot{\underset{\cdot\cdot}{O}}:H} \qquad \text{and} \qquad \underset{\text{electrovalent}}{[H, HOH]^{+} \qquad [:\ddot{\underset{\cdot\cdot}{O}}:H, HOH]^{-}}$$

the first of which is covalent, and the second, electrovalent. The covalent form is much more abundant, but the interchangeability of the two forms permits the electrovalent form to assert itself by reaction with dissolved substances and with the surface of suspended substances, as in ionization, in hydration, and in the formation of hydrated surface films. In fact, even the covalent form displays unusual properties in that it is associated (coordinated) to H_4O_2 or H_6O_3.(32)

The essential difference between polar and nonpolar surfaces can best be grasped by viewing polar surfaces as aggregates of geometrically disposed foci or poles of electricity, and viewing nonpolar surfaces as lacking this attribute. These ideas are clearly summarized by Sidgwick(36a) who says:

> In an ionized compound, though the oppositely charged ions are strongly attracted to one another, there is no real bond between them. They are free to take up any relative positions which are convenient. . . .
>
> In a covalent compound the electrical forces are more or less completely satisfied within the molecule, and the external field of force [stray field]* is small, whereas in an ionized molecule the electrical disturbance is greater, and there is a strong field of force outside.

Surface Reactivity and Surface Condition. It is but natural to assume that the surface of a solid particle is like the inside of that particle. This "natural" assumption has been and still is at the root of the difficulty and debate concerning flotation mechanism. A moment's reflection will show the futility of such an assumption.

As soon as a new bit of surface is formed, *e.g.*, by crushing or grinding, it is exposed to water, oxygen, carbon dioxide, and many other substances present in greater or lesser concentration in the ambient fluid. Some kind of action can be expected. It may take the form of adherence to the surface of the solid without obvious chemical action, as in physical adsorption; or of surface action having involved obvious chemical action, as in

* Words in brackets not in original text.

exchange adsorption; it may be evidenced by evolution of heat (heat of wetting), or it may form in short order a visible film. There may even be no reaction. About all of which we can be sure, a priori, is that we cannot be sure the surface is unchanged.

Further than this, it should be obvious that change on the solid requires a corresponding change in the ambient fluid; and a change in the fluid may introduce in it some constituent capable of affecting some other solid.

The problem of surface reactivity and of surface condition is then of the greatest importance, theoretically as well as practically. What a pure surface is, in most if not all cases, is unknown and perhaps immaterial. The characteristics of specially prepared laboratory surfaces depend on the method of preparation. Although they are reproducible, it cannot be affirmed that they are what they are meant to be.

Reaction with water alone takes the form of hydration. Thus quartz which is the anhydride of silicic acid adds water to form silicic acid; metal oxides add water to form hydroxides; the ions of salts hydrate in imitation of dissolved ions. The modified (and polar) surfaces then stick to water just as dissolved ions stick to undissociated water molecules.

Reaction with oxygen alone in the case of sulphide minerals may yield sulphites, sulphates, or other compounds; reaction with water and oxygen is even more likely. In the case of metals and metalloids, oxygen may yield oxide surfaces.

Reaction with carbon dioxide can lead to the formation of superficial carbonate layers.

With organic impurities in air or water, nonpolar surfaces may result.

Natural Floatability. Although the tendency has grown in recent years to deny the existence of an inherent floatability for minerals,[33] and although evidence has been accumulating that this property is much less widespread than was formerly thought to be the case,[37, *contra* 24] absence of inherent floatability for all minerals is not claimed. Thus it is generally agreed that natural hydrocarbons (*e.g.*, ozocerite) and soft coals are naturally floatable and give substantial or even large contact angles.[44] The floatable substance par excellence is solid paraffin.

It is also widely believed that natural sulphur and graphite have large natural floatability. Recent experiments conducted on painstakingly purified natural and synthetic graphite indicate that floatability is due to contamination by organic substances, the removal of which is extremely difficult. Experiments with natural and synthetic crystallized sulphur, however, are discordant, some investigators claiming natural floatability for sulphur,[47,49] others denying it.[44]

Careful laboratory flotation tests show appreciable floatability for some sulphide minerals if a small quantity of frothing agent (page 362) is used; on the other hand, no evidence of native floatability is obtained for these minerals by contact-angle measurements made with a captive bubble on cleaved or polished mineral surfaces.[45,48,51] This is shown by Fig. 180.

For minerals other than those just discussed, it is generally agreed that natural contact angle and floatability are nil.

It is likely that the divergent views are due to several reasons, and that their reconciliation may be anticipated from more careful experiments.

Nonpolarity and Floatability. It will be observed that substances possessing native floatability are nonpolar substances and more specifically, hydrocarbons. The lack of large "stray" electric fields and relative chemical inertness of nonpolar particles when in apposition to the large "stray" electric field and chemical reactivity of water suggest that the boundary must be discontinuous, that there is no transition layer between naturally floatable substances and water.

Formation of a layer between a nonpolar solid and a polar fluid could result only from a reaction enforced by the polar substance, producing a layer that is distinct as a phase from the nonpolar substratum, loosely and discontinuously facing it, but contrariwise transitionally connected to the polar fluid.

Conversely minerals that are polar, *i.e.*, salts like fluorite and sphalerite, lack native floatability. In addition, substances that are structurally nonpolar in bulk acquire a polar coating by reaction with water, *e.g.*, quartz and silicates. Many other substances that are more or less metallic in appearance, *e.g.*, most sulphides, acquire more or less rapidly oxidized coatings which are polar.

Both the minerals that are polar in bulk and those polarly coated are characterized by a lack of natural floatability.

We thus come to the important conclusion that water avidity, absence of contact angle, and nonfloatability are all manifestations of a fundamental accord in polar structure of water (active form) and mineral surface; and that conversely, water repellency, existence of a large contact angle, and floatability are manifestations of discord between the polar structure of water (active form) and the nonpolar structure of the mineral surface.

Acquired Floatability. By suitably coating the surface of one or another of a group of minerals with a film that is nonpolar (at least on the side of the film away from the mineral), particles of the selected groups can be made to act as if nonpolar throughout.

Thus sulphide minerals can be collectively singled out from nonsulphides; galena can be singled out from sphalerite, pyrite, and nonsulphides; chalcopyrite or chalcocite can be singled out from pyrite, pyrrhotite, and nonsulphides; apatite or rock phosphate can be singled out from silicates; quartz and silicates can be singled out from nonsulphide, nonsilicate minerals; it is even possible to single out azurite from malachite, or carbonates of alkaline earths or manganese from base-metal sulphides.

Acquired floatability is the result of the action of a group of reagents termed collecting agents (see pages 367 to 385). Collectors give rise to heteropolar monomolecular films at the surface of the minerals, the films being oriented with the nonpolar ends of the molecules away from the minerals (Fig. 182). However, it seems that in some cases the films are more or less jumbled, and thick; and in other cases, it seems definitely established that the films are incompletely monomolecular. Which subtype of collector film is obtained depends not only upon the collector and the mineral, but also upon the quantity of collector, the condition of the mineral surface, and possibly other factors. It will be seen that in a certain measure the film of lead xanthate on galena in Fig. 182 is exactly the reverse of the molecular bridge discussed on page 329 and diagrammatically represented in Fig. 174. This film of lead xanthate, $Pb(SCSOC_2H_5)_2$, is a molecular barrier between galena and water.

Figure 182 shows also a particle of quartz surrounded by an adsorbed layer of silicate anions and by a second or diffused layer of positive ions required for electrostatic balance, according to Gouy (see page 322, Chapter XIV). The silicate ions adsorbed on the quartz give it its negative charge in an electrical field. These adsorbed ions are also attached to water molecules so that a molecular bridge is established between the quartz and the water. The sulphate, sulphite, hydroxide, and bicarbonate ions have been given off by the galena as is explained in Chapter XVI.

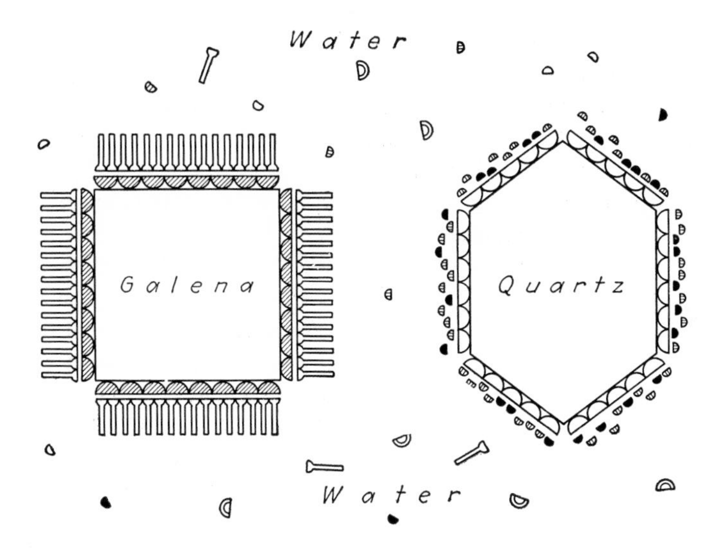

FIG. 182.—Conventionalized representation of a particle of galena and a particle of quartz in a dilute solution of potassium xanthate.

Separation of galena from quartz in a xanthate-treated pulp rests on the production of a molecular barrier between galena and water and on the retention of a molecular bridge between quartz and water.

The floatability obtained through collectors is selective in a broad sense: one group of minerals can be accurately separated from some other group; besides, there are differences in degree of acquired floatability within each group.

Further selectivity in the acquired floatability is obtained by the addition of other agents. These other agents, which may be grouped as "modifiers," perform many different functions. In modern flotation practice, these modifiers are extremely important.

By use of modifiers, it is possible to accomplish the following:

1. Utilize the collectors under the optimum conditions.
2. Prevent or control mutual mineral interaction.
3. Prevent or control action at mineral surfaces by atmospheric or aquatic ingredients.
4. Modify favorably or adversely the ability of some minerals to acquire a floating film.

As a result of the action of these diverse agents, a marvelous flexibility can be made available in the floatability of minerals. This is one of the two major reasons for the success of flotation processes, the other being its applicability to particles much finer than those to which other processes are applicable.

Flotation as a Thermodynamic Operation.(1-5) Attention has already been drawn to the second law of thermodynamics which states in effect that the potential energy of a system always tends of itself to become a minimum; *i.e.*, that potential energy is readily self-transferable into kinetic energy. That law is used to explain why air bubbles are maintained with difficulty in a state of dispersion in water: aggregation to coarse bubbles reduces the interfacial area, hence the potential energy of the system.

Likewise, if solids and gas bubbles are, first, individually suspended in water, subsequent attachment of particles to bubbles *reduces* the solid-liquid and the liquid-gas interfaces each by an area S, but *increases* the solid-gas interface by the same area S.*

The reduction in potential energy of the system Ω, per square centimeter of solid surface, or specific reduction in potential energy, is then

$$\Omega = E_{MW} + E_{AW} - E_{MA}, \qquad [XV.5]$$

in which the various E's denote surface energies correspond-

* This is rigorously true only if there is no distortion in the shape of the phases; but it is always approximately true.

ing to various surface tensions, and in which the subscripts A, M, W, refer, respectively, to water, mineral, and air. But from Eq. [XV.2] and from the equal numerical value of surface tension and surface energy,

$$E_{MA} = E_{MW} + E_{AW} \cos \theta,$$

or

$$E_{MW} - E_{MA} = -E_{AW} \cos \theta$$

whence

$$\Omega = E_{AW}(1 - \cos \theta). \qquad \text{[XV.6]}$$

The quantity Ω is also known as the work of adhesion and is a measure of the tendency to float.[35] Numerically, T_{AW}, the surface tension of the solution, can be substituted for E_{AW} since they are numerically equal, but Ω is of course expressed in ergs per square centimeter.

In the case of an aqueous solution whose surface tension is 70 dynes per cm., and for $\theta = 60°$ (the maximum value obtainable in the case of sulphide minerals if potassium ethyl xanthate is used as a collector),

$$\Omega = 70(1 - \cos 60°) = 35 \text{ ergs per sq. cm.}$$

For cubes of the order of magnitude of 48- to 100-mesh (depending upon their specific gravity), the surface-energy decrease due to particle-bubble adhesion is approximately equivalent to the work of moving the cubes 1 cm. against gravity.

If the contact angle is increased to 90°, Ω is doubled to 70 ergs per sq. cm. and the work of adhesion is doubled.

A flotation operation may be regarded as made up of two steps operating in rapid succession or simultaneously and consisting in

1. Artificially and ephemerally increasing the surface energy of the system by introducing many fine air bubbles in the system.
2. Causing the bubbles to adhere to each other and to suspended minerals whose contact angle is not nil.

Other things being equal, an increase in contact angle results in more rapid flotation.

Lifting Power of Surface Forces.[12] If a thin disk of rigid material is given a surface such that it has a large contact angle, it will float at the surface of water and considerable pressure may be required to sink it. The supporting force equals the weight of the mass of liquid displaced by air (see Fig. 183). This in turn equals the sum of the cylindrical volume $YY'Z'Z$ and the toroid volume $XYZ - X'Y'Z'$. The latter equals the product of the vertical component of surface tension by the perimeter, or $\pi DT_{WA} \sin \theta$. If θ is a right angle (a condition which gives maximum value to the support of surface tension), the total supporting force is $\pi D^2hg/4 + \pi DT_{WA}$.

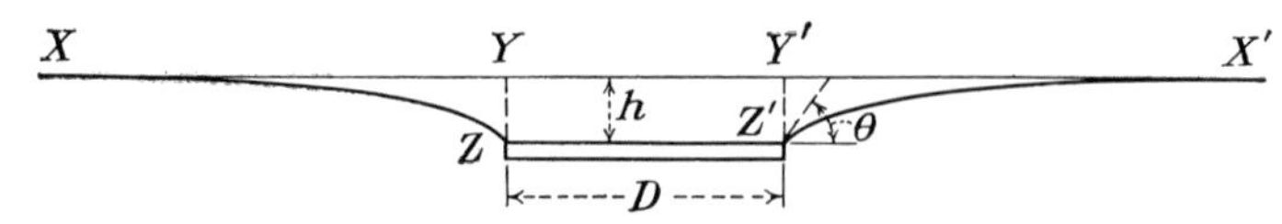

FIG. 183.—Static flotation of a thin disk.

The maximum depth h of the dimple varies with the diameter of the disk from 0 for $D = 0$ to 0.378 cm. for $D = \infty$. Calculations based on numerical integration(14) give for h values that increase with D at first very rapidly, then more and more slowly (see Table 41).

TABLE 41.—MAXIMUM DEPTH OF DIMPLE MADE BY A MINERAL DISK (CONTACT ANGLE OF 90°) IN FUNCTION OF THE DIMPLE DIAMETER

Dimple Diameter, Centimeters	Dimple Depth, Centimeters
0	0
0.001	0.003
0.002	0.006
0.004	0.011
0.010	0.023
0.020 (65-mesh)	0.039
0.040	0.066
0.080	0.105
0.140	0.145
0.200	0.175
0.300	0.212
0.500	0.258
1.0	0.306
2.0	0.335
8.0	0.372
∞	0.378

The supporting force is thus made of two terms, one varying like D^2 and the other like D. For small particles, the term in D^2 becomes negligible and the supporting force may be regarded as πDT_{WA}. Thus for the coarsest particles in sulphide flotation, the term in D^2 is not over one-twentieth of the term in D; and for particles 10 microns in diameter, it is about one-tenth of one per cent of the term in D.

Suspension of Particles at a Free Water Surface. This phenomenon is pictorially shown in Fig. 184 which may be compared with the theoretical diagram, Fig. 183.

It is of interest to compare the supporting force due to surface tension with the apparent weight in water of discoid mineral grains. By varying the size while keeping other factors constant, it is possible to find a size such that the supporting force $\frac{1}{4}\pi D^2hg + \pi DT_{WA} \sin \theta$ equals the static disrupting force

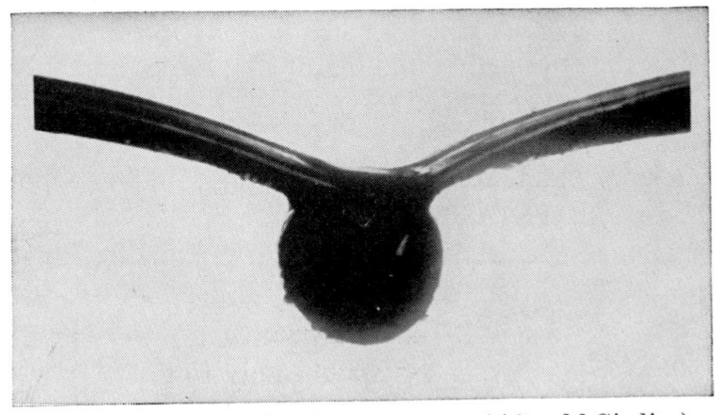

FIG. 184.—A needle floating on water. (*After del Giudice.*)

$(\Delta - 1)g(\pi D^2/4)D'$ in which D' is the thickness of the particle and Δ its specific gravity. Table 42 shows the size of particles for which these forces are equal. The particles the size of which is given in Table 42 are the largest that can be supported at a free water surface. The maximum indicated exceeds by a wide

TABLE 42.—MAXIMUM SIZE OF DISKS (THICKNESS = DIAMETER) SUPPORTABLE AT A FREE WATER SURFACE IF CONTACT ANGLE IS 90°

Mineral	Specific gravity	Diameter (and thickness), centimeters
Gold	19	0.132
Galena	7.5	0.229
Quartz	2.65	0.510
Coal	1.7	0.902
Coal	1.3	1.68

margin the maximum size of particles that can be recovered in practice by froth flotation; it exceeds also, but by a much lesser margin, the maximum size that can be recovered as an air-adhering product by skin flotation or agglomeration.

For a galena disk 30 microns thick and 30 microns in diameter, the supporting force is (the term in D^2 being neglected)

$$\pi(0.0030)(72) \sin \theta \text{ dynes}$$

and the disrupting force of gravity is

$$\frac{\pi}{4}(0.0030)^2(0.0030)(7.5 - 1)(980) \text{ dynes.}$$

Equating the supporting and disrupting forces and solving,

$$\sin \theta = 0.00020, \qquad \text{or} \qquad \theta = 0°0'42''.$$

Clearly, a very small contact angle is sufficient to support even so coarse a particle as a 30-micron disk.

Suspension of Particles to Bubbles in a Swirling Pulp. In a swirling pulp, conditions are materially different from those at a quiet surface. The important difference is that the particle-bubble associations are subjected not only to gravity forces, but also to various frictional and centrifugal forces. Thus, in an agitation-type cell, the maximum angular acceleration with a 24-in. impeller revolving at the rate of 420 r.p.m. (7 r.p.s.) is

$$a = (30)(7 \times 2\pi)^2 = 58{,}200 \text{ cm. per sec.}^2,$$

or roughly sixty times the acceleration of gravity.

If bubble-particle attachment is to persist in a swirling pulp, the contact angle must be much larger than under static conditions.

In the case of the 30-micron galena disk that requires a 0°0′42″ contact angle for static support, a contact angle of 1°19′ is required if the disruption is caused by a centrifugal force 100 times greater than gravity.

For particles of the range of sizes commonly floated (10 to 200 microns), the supporting force may be enormously greater than the force of gravity. Even if the contact angle is but a few degrees, the supporting force may be greater than the disruptive force of gravity for most of the particles.

Table 43 shows the minimum contact angles required for support of galena disks against gravity and against a centrifugal

TABLE 43.—CONTACT ANGLE SUFFICIENT FOR SUPPORT OF GALENA DISKS (DIAMETER = THICKNESS) OF VARIOUS SIZES AGAINST GRAVITY AND AGAINST A CENTRIFUGAL FORCE 100 TIMES GRAVITY

Diameter, microns	Minimum angle required against g	Minimum contact angle required against $100g$
2,290	90°	Impossible
500	3°8′	Impossible
200 (65-mesh)	0°30′	61°
100 (150-mesh)	0°7′30″	12°40′
50	0°1′52″	3°8′
20	0°0′18″	0°30′
10	0°0′4″	0°7′30″

force 100 times as large as gravity. The striking feature about these results is the minuteness of the required contact angles: For all flotation sizes under 65-mesh, a contact angle less than 1° is sufficient against gravity; and for all sizes under 150-mesh, a contact angle less than 13° is sufficient against a centrifugal force 100 times gravity.

Suspension of Particles at Bubble Surfaces. Consider first a spherical particle fast to a spherical bubble of relatively large

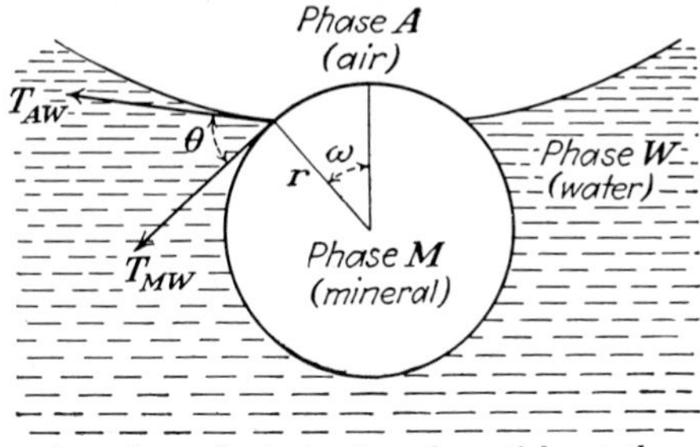

FIG. 185.—Suspension of a spherical mineral particle at the surface of a large bubble.

diameter (Fig. 185). If θ is the contact angle, r is the radius of the particle, and 2ω is the apical angle subtended by the circle of contact, the supporting surface tension makes an angle $(\theta - \omega)$ with the horizontal, and the supporting force is

$$(2\pi r \sin \omega)T_{AW} \sin (\theta - \omega).$$

If this supporting force exceeds the disrupting force of gravity, the particle will move up under its influence, and ω will increase. As ω increases, $f = \sin \omega \cdot \sin (\theta - \omega)$, to which the supporting force is proportional, increases at first, reaches a maximum, and then decreases.

Equilibrium is obtained for

$$(2\pi r \sin \omega) \cdot T_{AW} \cdot \sin (\theta - \omega) = \tfrac{4}{3}\pi r^3(\Delta - 1)g,$$

in which Δ is the specific gravity of the particle. This equation has two roots for ω, the larger of which may be seen geometrically to correspond to stable equilibrium and the smaller to unstable equilibrium.

For example, a mercury sphere 200 microns in radius and with a contact angle of 60° is stably supported against the disruptive force of gravity for $\omega = 56°53'$ and unstably supported for $\omega = 3°7'$.* The maximum supporting force occurs if $\omega = \theta/2 = 30°$, and it is 5.5 times as large as the disruptive force of gravity.

Consider next a disk with rounded edges (Fig. 186) of small radius of curvature, attached to a gas bubble along a circle of radius r. The supporting force is $2\pi r T_{AW} \sin \theta$. If this supporting force exceeds the disruptive force of gravity, the particle will move up, and r will increase. As r increases, the supporting force increases until the curved edge is reached. Then, with no

* The basic equation

$$(2\pi r \sin \omega)(T_{AW}) \sin (\theta - \omega) = 4/3\pi r^3(\Delta - 1)g$$

can be solved by expressing ω and $\theta - \omega$ symmetrically with reference to $\frac{\theta}{2}$:

$$\begin{aligned}\sin \omega \cdot \sin (\theta - \omega) &= \sin (A + x) \cdot \sin (A - x), \text{ where } A = \frac{\theta}{2};\ x = \frac{\theta}{2} - \omega \\ &= \tfrac{1}{2} \cos [(A + x) - (A - x)] - \tfrac{1}{2} \cos [(A + x) + (A - x)] \\ &= \tfrac{1}{2} \cos (\theta - 2\omega) - \tfrac{1}{2} \cos \theta\end{aligned}$$

Then

$$\frac{1}{2} \cos (\theta - 2\omega) - \frac{1}{2} \cos \theta = \frac{1}{2}k, \qquad \text{in which} \qquad k = \frac{4r^2(\Delta - 1)g}{3T_{AW}}$$

and

$$\cos (\theta - 2\omega) = k + \cos \theta$$

appreciable increase in r, a decrease in the supporting force will manifest itself since $\sin \theta$ becomes replaced by $\sin (\theta - \omega)$. Eventually the supporting force equals the disruptive force of gravity, and equilibrium of a stable kind is obtained.

A disk having a sharp edge differs from the rounded disk that has just been considered in a manner of degree only. The

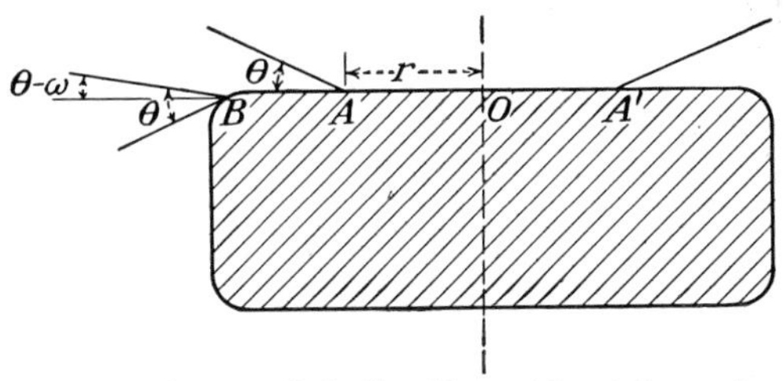

FIG. 186.—Suspension of a rounded, discoid particle at the surface of a bubble.

supporting force then acts from the line that forms the edge *in whatever direction is indicated by the disrupting force.*

The conclusion is thus reached that solid-air attachment is usually bounded by protruding edges or corners on the solid. Regardless of the magnitude of the contact angle, provided it is less than 90°, the appearance of a film of particles in section is then as in Fig. 187*a*. Likewise, Fig. 187*b* shows the appearance of a similar film if the contact angle is over 90°. In drawing Fig. 187, it has been assumed that the particles are so small that the effect of gravity is negligible in comparison to that of surface forces. If it is not negligible, the water-air interface would bulge upward for minerals denser than water. In Fig. 187, it has also been assumed that there are not enough particles to form a complete layer of particles. Mineralized froths consist of double layers of particles held together by a film of water. Clearly froths of types *a* and *b* should differ in appearance. In froths of type *a*, the particles are wet by water, but in froths of type *b* the quantity of water may be reduced to the vanishing point. In practice

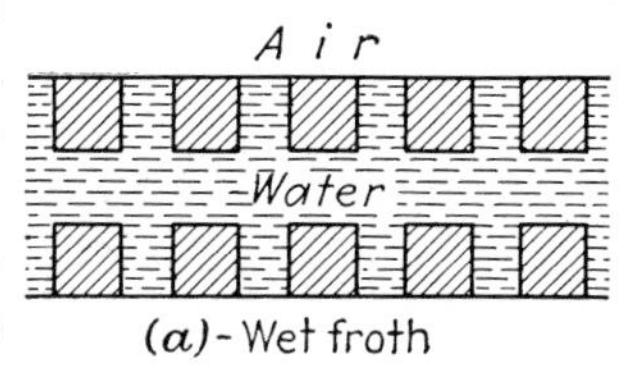

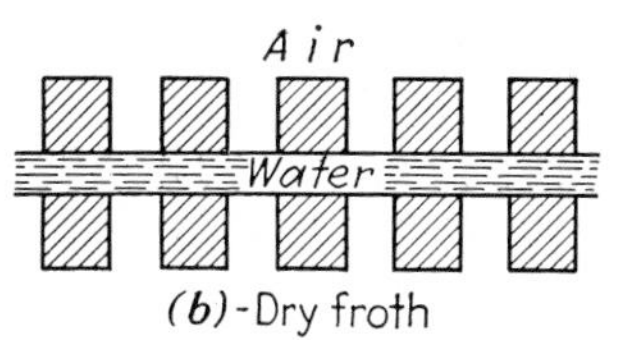

FIG. 187.—Two types of particle disposal at a water-air boundary.

and in laboratory testing, wet (*a*) froths are always used; dry (*b*) froths, however, can be realized if excessive quantities are used of those collectors (such as amyl xanthate and amyl dixanthogen on chalcocite) that give rise to limiting contact angles of more than 90°.[20]

Method of Attachment of Particles to Air. It has been claimed by Taggart[40] that concentration proceeds differently in pneumatic and in agitation machines. According to that theory, there is permanent adhesion of particles to air in agitation-type machines, and that adhesion is the result of selective gas precipitation; according to that theory, also, there is in pneumatic machines merely a preferential, intermittent adhesion of the floated particles at bubble walls, and this result is not related to selective gas precipitation. That theory, however, was obtained many years ago, before the introduction of modern reagents, and it is largely circumstantial.

On the other hand, evidence has been presented to show that adhesion takes place by collision in all types of machines, the differences in appearance of the froths being traced to differences in the amount of air, size of bubbles, speed of bubbles, and magnitude of the mineral-bubble disruptive force.[17c]

In thick suspension, the chances for collision are better than in dilute suspension by more than a proportionate factor because of the increased deviation in the properties of the liquid from those of a perfect liquid. Hence, better recovery of all minerals should be and is obtained in thicker pulps.

Effect of Particle Size on Flotation. Particles of various sizes do not float equally well.[18] Experiments show that as a rule, after flotation has been practically completed, the recovery is maximum for some intermediate size range such as 200- to 2,000-mesh, with a distinct falling off toward the extreme-coarse and extreme-fine ranges. The failure to float the extremely coarse particles arises from (1) incomplete liberation, (2) too small a contact angle, (3) too violent an agitation. The failure to float extremely fine particles is considered below.

Of the floated particles, the coarsest particles float first (Fig. 188). This fact is in line with the direct-encounter hypothesis of gas-solid attachment.[17c]

Experiments also show that pulps of the same mineral mixture ground finer and finer respond less and less well with increasing fineness. The fine particles are not only slow to float, but they become almost unresponsive to reagents, even to exorbitant quantities of reagents; at the same time, the gangue particles are activated until the difference in response between mineral and gangue is wholly lost. To a large extent, the same result is obtained by mere conditioning of mineral and gangue with each other. It is directly attributable to mutual mineral interaction *via* dissolved ions.

The poor response of the finer particles in any flotation pulp seems then to be ascribable [19] not only to the poorer chance for

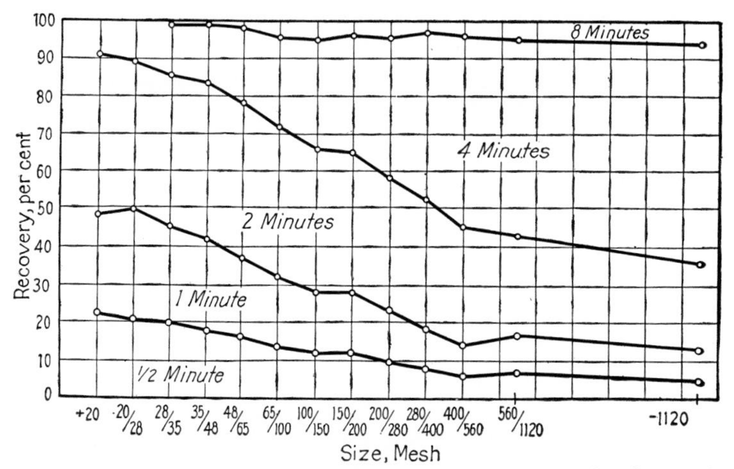

FIG. 188.—Fractional recovery of galena of various sizes in a batch operation and during various periods.

mineral-bubble encounter in the fine sizes,[17c] but also to the fact (which can be proved from considerations of comminution) that finer particles have an older surface than coarse particles, hence one that has been more extensively affected by ions derived from other minerals, by oxygen, or by water.

There seem to be also other factors that act to prevent flotation of very fine particles. It has been observed that to make very fine particles floatable it is effective to add the collector during grinding so as to endow the particles with a nonpolar surface as soon as possible.[20] Selective flocculation (not involving air) or selective granulation with an oil bond is obtained. The favorable effect of these phenomena may be due not only to the chemical protection of the surfaces of the minerals, but also to the physical effect of "making big ones out of little ones."

Bubble Production. Effective production of bubbles requires certain practices whose necessity may not be obvious. These practices include the addition of a suitable frothing agent and the use of agitation.

Pure liquids do not froth; *i.e.*, the introduction of bubbles either by blowing through a tube or by shaking does not lead to the formation of a layer of bubbles, or froth, at the surface. At any rate that is true of water, alcohol, and organic liquids of low or average viscosity.

On the contrary, liquids consisting of two or more constituents which in the pure state differ widely in surface tension and are

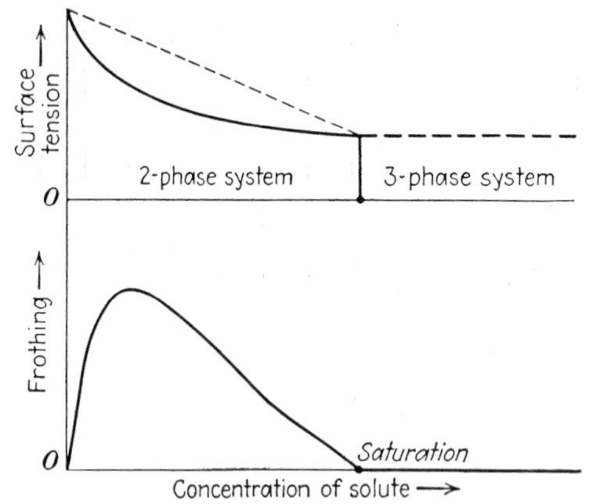

FIG. 189.—Surface tension and frothing of aqueous solutions.

in mutual solution can be made to froth. Large bubbles introduced in a solution can be seen to break up into smaller bubbles if their speed of travel is sufficiently large. If high speed and shearing forces are used, large bubbles can be converted into a fine dispersion.

A study of the frothing of aqueous solutions shows that frothing is nil, both for pure water and for a saturated solution. Maximum frothing is obtained at approximately that concentration at which the surface tension-concentration curve is most distant from the straight line connecting the surface tensions of pure solvent and of saturated solution (Fig. 189).[43]

In that connection, a brief review of the surface tension of aqueous solutions is helpful. Organic substances lower the surface tension of water, and most inorganic substances raise it

The lowering may be considerable and appear out of all proportion to the small quantity causing it, whereas the raising is always very slight and again apparently out of proportion to the large quantity causing it. These facts are interpreted as due to a distribution and an orientation of the molecules in the solution such that the particular molecules giving a low potential energy (surface energy) are at the surface, and in the preferred orientation. The excess (or deficiency) of solute atoms in the surface layer is the adsorption in that layer. Quantitatively, this adsorption is given by the Gibbs theorem[21] which is usually expressed in the form of the differential equation

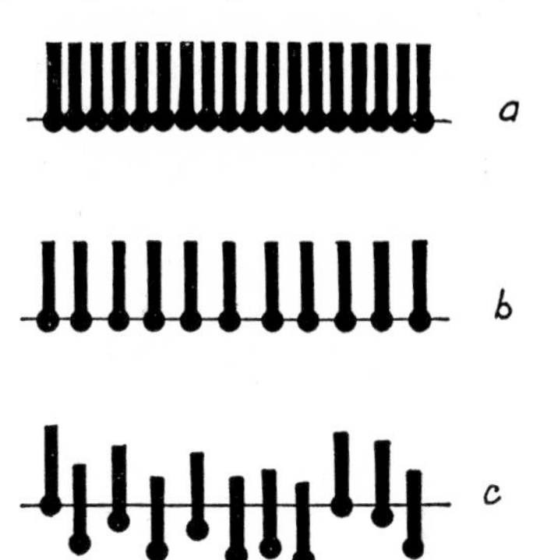

$$u = -\frac{c}{RT}\frac{\partial t}{\partial c},$$

in which u is the adsorption in gram molecules per square centimeter, t is the surface tension in dynes per centimeter, c is the concentration, in any convenient unit, such as percentage or gram molecules per liter, R is the gas constant in ergs per degree centigrade $= 8.32 \times 10^7$, and T is the absolute temperature in degrees Kelvin.

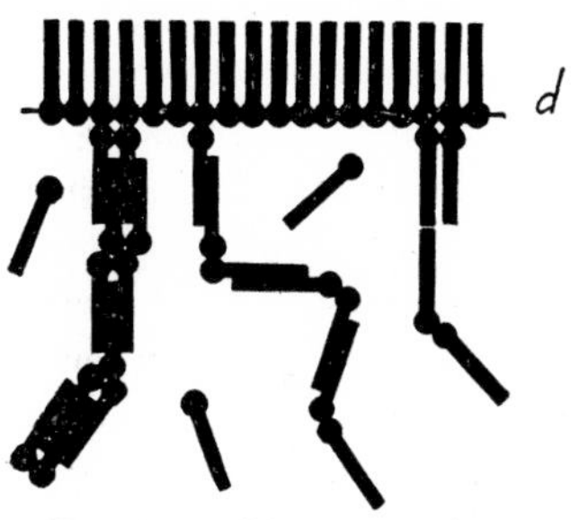

FIG. 190.—Diagrammatic representation of the structure of the surface of an aqueous solution of an organic compound of high molecular weight. (*After McBain and Davies.*)

This Gibbs equation is based on assumptions concerning the shape and behavior of molecules; it must therefore be regarded as a first approximation. Experiments designed to test the Gibbs equation have succeeded in substantiating it in a general way in the case of static equilibrium.[30,31] But in the dynamic equilibrium of relatively concentrated solutions in water of long-chained heteropolar compounds, the Gibbs equation does not seem to hold; the adsorption is greater than indicated by the equation, so that McBain and his associates[27,28,29] regard the surface as possibly covered by monomolecular films, with chains of oriented molecules of solute extending well inside the liquid (Fig. 190). Newer experiments by McBain come closer to checking the Gibbs theorem than the older experiments.

The fact that a very small quantity of an organic substance can appreciably lower the surface tension of water and make it froth where it would not froth in its absence is of great practical value in flotation, as it makes for economy in frothing agents.

Just why frothing occurs with maximum vigor at a concentration for which adsorption is maximum seems related to the optimum elasticity of films in a solution of that composition. As such a film is stretched, "body" fluid reaches the surface and increases the surface tension, thus counteracting the stretching force; conversely, as such a film is compressed, pure water is momentarily ejected from the surface, and the surface tension is decreased.[11,17] These phenomena are not instantaneous, however, and some time is required for the adjustments to take place.

Mineralized Bubbles and Froths. Even in the case of an aqueous solution of maximum frothing ability, the froth is not of considerable duration: its life is a matter of seconds, coalescence of bubbles going on apace. But if the froth is produced in a mineral suspension such that mineral particles adhere to the bubbles, the bubbles coalesce very much slower and the life of the froth may be increased to last hours instead of seconds.[6,7]

Persistent froths are not desired. In the first place, their rigidity prevents the rearrangement of solids, air, and water which goes on in a tender froth; this brings about the unwanted result that mechanically occluded particles are not excluded before evacuation of the froth from the flotation machine. In the second place, froths do not flow, thicken, or filter as readily as pulps; it is therefore desirable that froths be tender enough to break down as soon as they are out of the flotation machines and to display the "life" of constant adjustment while in the flotation machine.

For these reasons, the quantity of frothing agent used is generally a very small fraction of the quantity that yields maximum frothing.

Frothing Agents. Frothers act upon the gas-liquid interface, not at the surface of solids. They are organic compounds whose molecules each contain one polar group and one nonpolar group. The frother molecules are therefore heteropolar; one part of the molecule has an affinity for water, the other has an affinity for gas or a repulsion for water.[39] The molecules concentrate at the gas-liquid interface with the nonpolar part oriented toward the gas and the polar part toward the water (see Fig. 176).

In an homologous series of compounds, solubility decreases and surface activity increases with increase in size of the hydrocarbon group. For each CH_2 group added to the molecule, the solubility decreases about threefold. And for each CH_2 group added to the molecule, the surface activity increases about threefold. This is Traube's rule.[17b,15] This rule is roughly valid so long as the molecules of solute do not exhibit a marked tendency to stick to each other, *i.e.*, so long as colloid properties are not marked.

For these reasons, to be effective, frothers must not be too soluble, or too insoluble. Best range is from 0.2 to 5.0 g. per l.

To lack collecting properties (a desirable feature if flotation operations are to be properly controlled), frothers must not ionize appreciably.

Widely used frothers are pine oils, cresylic acids, "fusel oils" and other "higher" alcohols, eucalyptus oil, camphor. Substances having the alcohol, phenol, ketone, or aldehyde structure are most suitable. But esters, ethers, and to some extent amines, are also effective frothers.[45]

Literature Cited

1. Barsky, George, and S. A. Falconer: Differential Wetting Effects in Flotation, *Trans. Electrochem. Soc.*, **60**, 343–353 (1931).
2. Bartell, F. E., and F. L. Miller: Degree of Wetting of Silica by Crude Petroleum Oils, *Ind. Eng. Chem.*, **20**, 738–742 (1928).
3. Bartell, F. E., and H. J. Osterhof: The Measurement of Adhesion Tension, Solid against Liquid, in "Colloid Symposium Monograph 5," Chemical Catalog Company, New York (1928), pp. 113–134.
4. Bartell, F. E., and H. J. Osterhof: Determination of the Wettability of a Solid by a Liquid, *Ind. Eng. Chem.*, **19**, 1277–1280 (1927).
5. Bartell, F. E., and C. N. Smith: Adhesion Tension Values of Different Types of Carbon Black against Water and against Benzene, *Ind. Eng. Chem.*, **21**, 1102–1106 (1929).
6. Bartsch, Otto: Über Schaumsysteme, *Kolloidchemische Beihefte*, **20**, 1–49 (1925).
7. Bartsch, Otto: Beitrag zur Theorie des Schaumschwimmverfahrens, *Kolloidchemische Beihefte*, **20**, 50–77 (1925).
8. Dean, R. S.: Physicochemical Nature of Metallic Interfaces, *U. S. Bur. Mines, Rept. Investigations* 3400 (1938).

9. Dean, R. S.: Properties of So-called Amorphous Metal, *Chem. Met. Eng.*, **26,** 965–966 (1922).
10. Debye, P.: "Polar Molecules," Chemical Catalog Company, New York (1929).
11. Edser, Edwin: Molecular Attraction and the Physical Properties of Liquids, in "Fourth Report on Colloid Chemistry and Its General and Industrial Application," British Association for the Advancement of Science, London (1922), pp. 40–113.
12. Edser, Edwin: The Concentration of Minerals by Flotation, in "Fourth Report on Colloid Chemistry and Its General and Industrial Applications," British Association for the Advancement of Science, London (1922), pp. 263–326.
13. Fahrenwald, A. W.: Surface-energy and Adsorption in Flotation, *Mining Sci. Press*, **123,** 227–234 (1921).
14. Freud, B. B., and H. Z. Freud: A Theory of the Ring Method for the Determination of Surface Tension, *J. Am. Chem. Soc.*, **52,** 1772–1782 (1930).
15. Freundlich, Herbert: "Colloid and Capillary Chemistry," E. P. Dutton & Co., Inc., New York (1922).
16. Freundlich, Herbert: On the Physical Chemistry of Flotation, *Trans. Electrochem. Soc.*, **60,** 389–394 (1931).
17. Gaudin, A. M.: "Flotation," McGraw-Hill Book Company, Inc., New York (1932); (*a*) pp. 11–15; 32–33; (*b*) p. 57; (*c*) pp. 88–98.
18. Gaudin, A. M., John O. Groh, and H. B. Henderson: Effect of Particle Size on Flotation, *Am. Inst. Mining Met. Engrs., Tech. Pub.* 414, 3–23 (1931).
19. Gaudin, A. M., and Plato Malozemoff: Hypothesis for the Non-flotation of Sulfide Minerals of Near-colloidal Size, *Trans. Am. Inst. Mining Met. Engrs.*, **112,** 303–318 (1934).
20. Gaudin, A. M., and Plato Malozemoff: Recovery by Flotation of Mineral Particles of Colloidal Size, *J. Phys. Chem.*, **37,** 597–607 (1933).
21. Gibbs, J. Willard: On the Equilibrium of Heterogeneous Substances, *Trans. Conn. Acad. Arts Sci.*, New Haven, **3,** 108–248, 343–524 (1874–1878). "The Collected Works of J. Willard Gibbs," Longmans Green & Company, New York (1908).
22. Harkins, William D., and Hubert F. Jordan: A Method for the Determination of Surface and Interfacial Tension from the Maximum Pull on a Ring, *J. Am. Chem. Soc.*, **52,** 1751–1772 (1930).
23. Kossel, W.: Formation of Molecules and Its Dependence on Atomic Structure, *Ann. Physik* [4], **49,** 229–362 (1916).
24. Langmuir, Irving: The Mechanism of the Surface Phenomena of Flotation, *Trans. Faraday Soc.*, **15** [Pt. 3], 62–74 (1920).
25. Lewis, G. N.: The Atom and the Molecule, *J. Am. Chem. Soc.*, **38,** 762–785 (1916).
26. Lewis, G. N.: "Valence and the Structure of Atoms and Molecules," Chemical Catalog Company, New York (1917).
27. McBain, James W.: Structure in Surfaces of Liquids, *Nature*, **120,** 362 (1927).

28. McBain, James W., and G. P. Davies: Experimental Test of the Gibbs Adsorption Theorem: A Study of the Structure of the Surface of Ordinary Solutions, *J. Am. Chem. Soc.*, **49**, 2230–2254 (1927).
29. McBain, James W., and Robert Dubois: Further Experimental Tests of the Gibbs Adsorption Theorem. The Structure of the Surface of Ordinary Solutions, *J. Am. Chem. Soc.*, **51**, 3534–3549 (1929).
30. McBain, James W., and C. W. Humphreys: The Microtome Method of the Determination of the Absolute Amount of Adsorption, *J. Phys. Chem.*, **36**, 300–311 (1932).
31. McBain, James W., and Robert C. Swain: Measurements of Adsorption at the Air-water Interface by the Microtome Method, *Proc. Roy. Soc.* (London) [A], **154**, 608–623 (1936).
32. Pennycuick, S. W.: The Structure of Water, *J. Phys. Chem.*, **32**, 1681–1696 (1928).
33. Pryor, E. J.: The Dressing of Auriferous Ores, *Mining Mag.*, **58**, 153–160; 215–225 (1938).
34. Ralston, Oliver C.: Why Do Minerals Float?, *Mining Sci. Press*, **111**, 623–627 (1915).
35. Shepard, Orson Cutler: Bubble Attachment in Flotation, *Mining and Met.*, **13**, 282–283 (1932).
36. Sidgwick, Nevil Vincent: "The Electronic Theory of Valency," Oxford University Press, London (1929); (*a*) p. 84.
37. Sulman, H. Livingstone: A Contribution to the Study of Flotation, *Bull. Inst. Mining Met.*, **182**, 21–95 (1919).
38. Sulman, H. Livingstone: A Contribution to the Study of Flotation, *Trans. Inst. Mining Engrs. (London)*, **29**, 44–204 (1920).
39. Taggart, Arthur F.: Flotation Reagents, *Trans. Am. Inst. Mining Met. Engrs.*, **79**, 40–49 (1928).
40. Taggart, Arthur F.: "Handbook of Ore Dressing," John Wiley & Sons, Inc., New York (1927), pp. 797–798, 805–807.
41. Taggart, A. F., and Frederick E. Beach: An Explanation of the Flotation Process, *Trans. Am. Inst. Mining Engrs.*, **55**, 547–562 (1916).
42. Taggart, Arthur F., T. C. Taylor, and C. R. Ince: Experiments with Flotation Reagents, *Trans. Am. Inst. Mining Met. Engrs.*, **87**, 285–368 (1930).
43. Taggart, A. F., and A. M. Gaudin: Surface Tension and Adsorption Phenomena in Flotation, *Trans. Am. Inst. Mining Met. Engrs.*, **68**, 479–535 (1923).
44. Taggart, Arthur F., G. R. M. del Giudice, A. M. Sadler, and M. Hassialis: Oil-air Separation of Nonsulphide and Nonmetal Minerals, *Am. Inst. Mining Met. Engrs.*, *Tech. Pub.* 838; also in *Mining Tech.*, **1** [6] (1937).
45. Taggart, A. F., T. C. Taylor, and C. R. Ince: Experiments with Flotation Reagents, *Trans. Am. Inst. Mining Met. Engrs.*, **87**, 285–368 (1930).
46. Taylor, Hugh S.: "Elementary Physical Chemistry," D. Van Nostrand Company, Inc., New York (1927).

47. Wark, Ian W.: "Principles of Flotation," Australasian Institute of Mining and Metallurgy, Melbourne (1938).
48. Wark, Ian W.: Theories of the Action of Flotation Agents, *Proc. Australian and New Zealand Assoc. Adv. Sci.*, **73,** 80 (1935); (*a*) p. 52.
49. Wark, I. W., and A. B. Cox: Physical Chemistry of Flotation. V. Flotation of Graphite and Sulfur by Collectors of the Xanthate Type and Its Bearing on the Theory of Adsorption, *J. Phys. Chem.*, **39,** 551–559 (1935).
50. Wark, Ian William, and Alwyn B. Cox: Principles of Flotation. V. Conception of Adsorption Applied to Flotation Reagents, *Am. Inst. Mining Met. Engrs., Tech. Pub.* 732 (1937).
51. Wark, Ian William, and Alwyn Birchmore Cox: Principles of Flotation. I. An Experimental Study of the Effect of Xanthates on Contact Angles at Mineral Surfaces, *Trans. Am. Inst. Mining Met. Engrs.*, **112,** 189–214 (1934).
52. Werner, Alfred: Beitrag zur Konstitution anorganischer Verbindungen, *Z. anorg. Chem.*, **3,** 267–330 (1893).
53. Willows, R. S., and Emil Hatschek: "Surface Tension and Surface Energy and Their Influence on Chemical Phenomena," J. A. Churchill, London (1915).
54. Winkler, C. A., and O. Maass: An Investigation of the Surface Tension of Liquids near the Critical Temperature, *Can. J. Research*, **9,** 65–79 (1933).

CHAPTER XVI

FLOTATION AND AGGLOMERATION—CHEMICAL ASPECTS

It has already been indicated that flotation and agglomeration are processes that depend on the selective adhesion to air of some minerals and the simultaneous adhesion to water of other minerals. Leaving aside the rather academic question of native mineral floatability, it becomes clear that success is predicated on establishing a water-repellent or air-adhering surface on some minerals, whereas others are prevented from acquiring it.

Agents added to cause air adherence are known as *collectors*. These agents are more or less selective with reference to certain classes of minerals. To make them still more effective, other agents, which taken together may be termed controlling or modifying agents, are used. Accurately speaking, no flotation operation can be carried out without some controlling agent: even in plain water, there are hydrogen and hydroxyl ions, and these ions are among the most effective and important of controllers. An accurate study of collectors is therefore not possible without a simultaneous consideration of modifiers. In order to simplify the study of the chemical aspects of flotation and agglomeration, however, the subject will be artificially subdivided into a study of the action of collectors, and of controllers with collectors.

COLLECTORS

General Requirements.[4a,5,26,27,32] Attention has already been drawn to the following facts:

1. Nonpolar surfaces, *i.e.*, surfaces in which the atoms are covalently linked, are water repellent, and they are the only water-repellent surfaces.

2. The only practical way of securing such a surface is by means of organic compounds, *i.e.*, of compounds containing C_nH_m groups.

It might seem as if hydrocarbons should be ideal collecting agents for polar solids. But they are worthless as experiment abundantly shows: hydrocarbons in the presence of water have no way to attach themselves to a polar solid; they shrink from it much as do air bubbles.

In the case of a mineral already made air adherent by a coat of a suitable collector, or in the case of a nonpolar solid, hydrocarbon oils spread, forming an overcoat. In that fashion, they may be of use, but they are not true collecting agents in the present sense of that term.

Collectors must contain two parts: one that is nonpolar (hydrocarbon), and another that is polar and of such chemical character as to stick to specified minerals.

To a certain extent collectors resemble frothers. The resemblance is in the dual or heteropolar character of their molecules. But they differ in their polar parts: the polar part of collectors should have a specific affinity for specific minerals, whereas the polar part of frothers should have affinity for water only.

An example of resemblance and dissemblance between a frother and a collector is found in alcohols and mercaptans. These compounds are structurally close relatives, so much so in fact that mercaptans are also known as thioalcohols. The molecules of both are heteropolar, and it might seem as if they should have like properties in flotation. But mercaptans are potent precipitants for aqueous base-metal salts such as of lead, copper, and mercury (*mercurium captans* → mercaptans), whereas alcohols do not form water-insoluble compounds with the same metals. The insolubility of base-metal mercaptides is related to the specific affinity of mercaptans for minerals containing mercury, lead, copper, and some other metals; and conversely, the solubility or nonexistence of base-metal alcoholates is related to the nonaffinity of alcohols for base-metal minerals.

Just as a change from amyl alcohol to amyl mercaptan changes a substantially noncollecting agent into one that is a selective collector for minerals of metals forming insoluble amyl mercaptides, so this change affects the frothing properties of the agent. The very polar, water-avid group —OH is replaced by the less water-avid group —SH, and the frothing power is reduced.

It seems that each and every heteropolar compound must have both frothing and collecting properties. In some cases, the collecting action is considerable at a concentration that causes no effective frothing: the agent is a collector. In other cases, the collecting action is faint or nonexistent at a concentration that gives good frothing: the agent is a frother. In other cases, both frothing and collecting are obtained: the agent is a frother-collector.

From a practical standpoint, dissociation of frothing from collecting properties is desirable as it permits the separate control of these variables.

Collector Types. The earliest substances that were found to have collecting properties were some oils substantially insoluble in water, *e.g.*, oleic acid. Since many animal and vegetable oils are more or less rich sources of fatty acids, many oils that were thought to be collectors may have merely contained a small proportion of fatty acid mixed with a large proportion of inactive oil.

Petroleum oils were found to be lacking in collecting properties, unless they contained sulphur. This, perhaps, led to the introduction of sulphur in oils by "reconstructing" them with sulphur, hydrogen sulphide, or phosphorus pentasulphide. It also led to the discovery that organic compounds containing divalent sulphur, and more especially mercaptan sulphur, are excellent selective collectors for sulphide minerals.

A third group of agents arose from the use of oil-refinery waste sludges as a cheap source of "oil." These sludges result from the purification of oils by treatment with sulphuric acid and contain a great many things in unknown quantity and number, but most likely alkyl sulphuric and alkyl sulphonic acids.

The chemical compounds effective in these three groups of compounds are, respectively, alkyl carboxylic acids, alkyl or aryl thio acids, and alkyl or aryl sulphuric and sulphonic acids. This statement should not be taken to mean that these organic acids are useful in the molecular or undissociated form. Indeed, they are frequently used in the form of alkali salts such as soaps and xanthates. It is now practically certain that the effective agents are the ions or micelles that arise when the organic acids or their salts are dissolved in water. Simple as this rule may

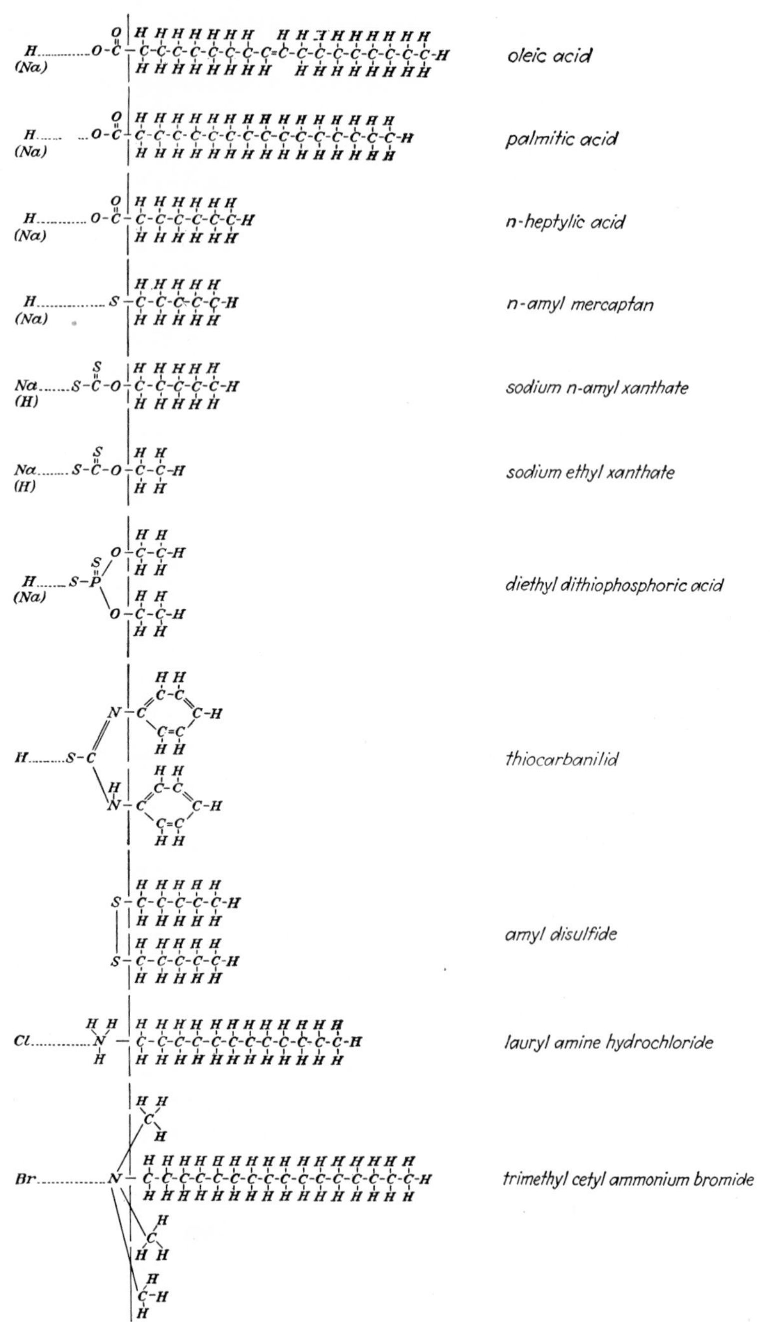

FIG. 191.—Typical collectors, shown with the polar part to the left.

appear now, sight should not be lost of the many difficulties that have lain in the way of its ascertainment.

In the case of each of the collectors discussed so far, the ion containing the hydrocarbon group, the effective ion, is the anion. Collectors in which the effective ion is the cation are amines or amine salts. Their properties promise to be of great interest.

From what has been said, it would seem as if collectors must be able to ionize. That, however, may not be always the case. Examples of nonionized collectors include such effective agents as amyl sulphide, ethyl dixanthogen, and trimethyl cetyl ammonium bromide. Amyl sulphide is an oil very little soluble in water; ethyl dixanthogen (ethyl thioformate disulphide) is a greasy solid melting at about 28°C., substantially insoluble in water. Both are good collectors for sulphide minerals. Neither one is regarded as an ionized compound. Trimethyl cetyl ammonium bromide[44] which is a good collector for quartz and for glass is so little ionized that with silver it fails to give a test for bromide ion.

Figure 191 presents the structural formulas of a number of characteristic collectors.

Collectors Used in Practice. Widely used collectors are

1. Ethyl, propyl, butyl, and amyl xanthates, particularly the ethyl xanthate of sodium or potassium and the amyl xanthates of sodium or potassium.
2. Diethyl and dicresyl dithiophosphates.
3. Oleic and palmitic acids and their sodium soaps.

The first two groups are used for the flotation of sulphide minerals, of oxidized minerals of copper and lead, and for native metals. The third group is used for the flotation of oxide minerals, oxygen-salt minerals, halogen minerals, and silicates. Examples of minerals floated by oleic acid are calcite, rhodochrosite, apatite, pyrolusite, hematite, sylvite, garnet.

The Mechanism of Collection. The property displayed by some flotation agents (which at that time were oils) of causing adhesion of some minerals to air was first thought to be due to the formation of a layer of oil of unknown and undetermined thickness. In view of the relatively large quantity of collector used, and because of erroneous notions as to the extent of the surface of crushed ore—a surface which was very much underestimated—

this collector layer was widely believed to be thousands of molecules thick.

A second idea of the mechanism of collection was introduced in a study by Taggart and Gaudin[33] in which it was shown that soluble agents of the phenol and other types (some now regarded as frothers, others as frother-collectors) are abstracted by minerals. The abstraction was found to be of the order of magnitude that would have been obtained if a monomolecular or partly complete monomolecular film had formed.

These ideas of collection, *viz.*, bulk adhesion of an oil and molecular adsorption of dissolved agents, have had their use in their day. They are, however, of no significance in modern flotation philosophy. Present-day views narrow to a choice between two hypotheses, *viz.*, the chemical-reaction hypothesis and the adsorption hypothesis.

The *chemical-reaction hypothesis* is stated as follows by Taggart, Taylor, and Knoll[36a]: "All dissolved reagents which, in flotation pulps, either by action on the to-be-floated or on the not-to-be-floated particles affect their floatability, function by reason of chemical reactions of well recognized types between the reagent and the particle affected."

"Chemical reactions of well-recognized types" is a statement the meaning of which is bound to differ from one metallurgist to another; it seems, however, that practically speaking the hypothesis amounts to a statement that metathesis takes place between the mineral and the agent with production of a less-soluble reaction product that precipitates at the mineral surface.

The *adsorption hypothesis* that is adopted by Wark[43] may be stated as follows:

All ions dissolved in a flotation pulp liquor adsorb at mineral surfaces. At each mineral surface, the adsorption of each dissolved ion is specific; *i.e.*, it depends on the dissolved ion and on the mineral; this specific ion adsorption is also a function of the concentration of the dissolved ion under consideration and of that of other dissolved ions. If and when a sufficient proportion of the mineral surface is covered by the effective collector ions, the particle becomes floatable.

This view is also that toward which the author leaned as far back as 1928.[5] The evidence for and against these hypotheses will be examined in the next few pages.

Collection of Nonsulphide Minerals. In the case of nonsulphide minerals, it can be regarded as established that a relatively thick crust forms at the mineral surface by metathesis between the substance of the mineral and the collector. Examples are films of lead thiocresylate on cerussite[4b] of calcium oleate on calcite, of calcium palmitate on apatite,[23,24] of copper oleate on malachite, of lead xanthate on cerussite,[8a] of cuprous xanthate plus dixanthogen on azurite[13] (Fig. 192).

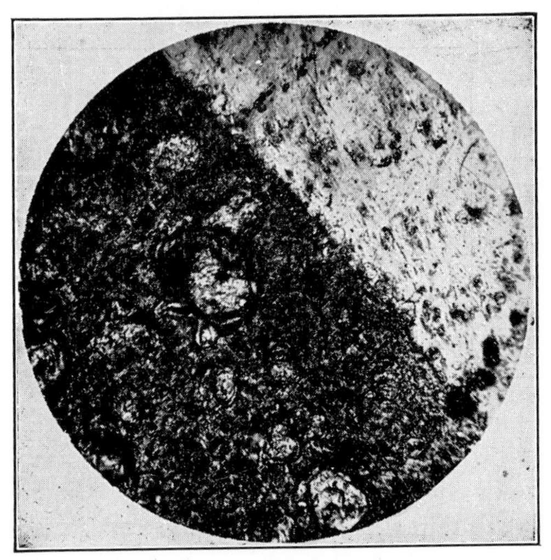

Fig. 192.—Cuprous xanthate and dixanthogen on azurite. Two per cent amyl xanthate solution was allowed to react two minutes on a polished azurite surface. × 40.

Table 44 shows how accurately it is possible to follow the reaction of lead carbonate with potassium xanthate. Wherever data are available, it is clear that the cation of the collector is not involved in the reaction. In its most general form, the reaction is

$$MA + X^- \rightarrow MX + A^- \qquad [XVI.1]$$

M denoting the metal in the mineral, A the anion in the mineral, and X the collector anion.

It is generally believed that the reaction has involved the collector anion in solution and the mineral cation in the solid state. However, there is a distinct possibility, and some evidence for the view, that in some instances the reaction has

TABLE 44.—REMOVAL OF POTASSIUM N-AMYL XANTHATE FROM AQUEOUS SOLUTION (50 CC.) BY CERUSSITE (1 G.) IN 10 MIN. AT 20°C.
(*After Gaudin, Dewey, Duncan, Johnson, and Tangel*)

Test	Potassium n-amyl xanthate added, milligrams	Potassium xanthate left in solution, milligrams	Lead xanthate on cerussite expressed as potassium xanthate, milligrams	Discrepancy, milligrams
1	2.69	0.17	2.59	+0.07
2	7.01	0.17	6.93	+0.09
3	8.08	0.31	7.78	+0.01
4	10.66	2.13	8.50	−0.03
5	11.13	2.40	8.74	+0.01
6	17.06	6.78	10.26	−0.02
7	21.30	11.10	10.20	0.00
8	27.70	17.00	10.55	−0.15
9	53.25	Not determined	10.98	
				Average discrepancy 0.05

proceeded wholly in the liquor between dissolved cation from the mineral and dissolved anion from the reagent, with subsequent mechanical adhesion of the precipitate to the mineral surface.

The coating obtained is more properly a crust: it can often be seen by its color or by the change in luster of the mineral surface; it can often be broken away by rubbing, grinding, or agitation, as in the case of copper xanthate on malachite; it can be extracted with suitable solvents and identified quantitatively. The coating is very much thicker than in the case of sulphide minerals. This thickness varies of course with the mineral and agent involved, but in some cases it seems as if it cannot be less than about 10^{-5} cm. thick and yet retain its effectiveness. This is roughly 100 molecules in thickness.

Clearly, the coating obtained is neither a crude layer of unchanged collector in bulk, nor an elusive monomolecular layer of unchanged collector molecules.

Collection of Sulphide Minerals. Where the mineral is a sulphide, there is again evidence of removal of the effective part of the collector by the mineral; there is also evidence of a reaction but not between the mineral itself and the effective part of the collector; and there is evidence that this reaction does not represent the whole story.

Many researches have been devoted to the elucidation of the mechanism of collection of sulphide minerals by xanthates. The experimental difficulties are very great, and the problem is much more complex than was first thought to be the case; it may therefore be asserted without fear of contradiction that the views which will be presented here are not final but rather in the nature of a progress report. The experiments have dealt principally with galena, pyrite, and chalcocite, and with ethyl and amyl xanthates.

Collection of Galena by Xanthates. The facts that have been ascertained so far are as follows:

1. Xanthate ion is abstracted by galena, but the alkali ion is not abstracted.[8,34,35]
2. Ions are given up by the mineral to the aqueous solutions in quantity equivalent to the quantity of abstracted xanthate ion.[8,36,37] These ions are sulphoxides, sulphate, hydroxide, and carbonate.
3. No sulphide ion is given up to the aqueous solution, and sulphide ion does not seem to be directly involved in the process of collection.[36,37]
4. Lead xanthate can be extracted from a freshly treated galena surface but not in an amount sufficient to account for the xanthate ion abstracted by the galena.[8,50]
5. From an aged, treated galena surface are extracted sulphur and oils with little if any lead xanthate.[8,15]
6. Deoxidation of galena surfaces results in reduced xanthate-ion abstraction and increased floatability.[4c,35]
7. Xanthate ion can be abstracted by galena even under conditions designed to provide complete absence of oxidation at the mineral surface.[43a]

Fact 7 conflicts with the chemical-reaction hypothesis, so does fact 4 if fact 5 is not simultaneously taken into account. Likewise, facts 4 and 6 appear to conflict with the adsorption hypothesis. But if adsorption and chemical reaction proceed jointly, the one on a sulphide surface (already covered by adsorbed inorganic ions, as noted below, and as noted also in Chapter XIV) and the other on an oxidized coating, the facts are in harmony with each other.

It might be noted that if the chemical-reaction hypothesis were operative alone it would not be necessary to postulate, as its proponents have done, that the salts displaced from the galena surface have a solubility lower than that of the same salts in bulk[37a] and that the lead xanthate at the galena surface has a solubility lower than that of lead xanthate in bulk.[36b] These postulates are practically an admission of the correctness, at least in part, of the adsorption hypothesis.

In addition, if the chemical-reaction hypothesis were wholly correct the behavior of cerussite and anglesite should be the same as that of galena. Yet the facts do not support this expectation as may be seen by referring to the work of Wark, and of Wark and Cox on mineral-bubble contact.[43]

Collection of Pyrite by Xanthates. The facts are as follows:

1. Pyrite, pyrrhotite, and marcasite do not show floatability with ethyl xanthate in the same pH range. According to Wark,[43b] under a certain set of conditions of temperature and ethyl-xanthate concentration, pyrite ceases to be floatable at pH 10.5 and pyrrhotite ceases to be floatable at pH 6.0. In other words, these minerals will tolerate hydroxyl-ion concentrations differing by a 30,000-fold factor.

2. Neither siderite (ferrous carbonate), hematite (ferric oxide), magnetite (ferrosoferric oxide), nor limonite (ferric hydroxide) is floated by ethyl xanthate, regardless of the amount of agent added.[4d,34a]

3. Sulphate and other ions are given off by pyrite to the flotation liquor in exchange for the abstracted xanthate.[34a]

4. Xanthate ion is readily abstracted by pyrite, but it is not retrieved by leaching the xanthated pyrite unless extraordinary precautions are taken to prevent oxidation. Even then, the ferric xanthate recovery is very low.[8]

5. Ferric xanthate is very unstable in the presence of oxygen; it hydrolyzes and is oxidized most readily.[8] Ferrous xanthate is relatively very soluble in water.

Fact 1 is consonant with the chemical-reaction hypothesis only if it is postulated that the "anchored" salts are different for pyrite, pyrrhotite, and marcasite.

Fact 2 is consonant with the chemical-reaction hypothesis only if it is postulated that alteration prior to collector addition in the case of pyrite is neither to a carbonate, nor to a hydroxide, even at a pH as high as 10, but to a sulphate. In that connection, it might be noted that the solubility of ferric sulphate (hydrated) is given as 440 g. per 100 cc., *i.e.*, that it is very large, and the ferric ion concentration at pH 10, on the basis of a solubility product of 10^{-37} for ferric hydroxide (the largest value found in the literature), would be only 10^{-25}. If instead of changing to ferric hydroxide, ferric sulphate persists at pH 10, that fact constitutes, indeed, a most extraordinary case of selective adsorption of sulphate ion in preference to hydroxide ion, as the outer ionic layer on pyrite.

On the other hand, facts 1 and 2 are consonant with the adsorption hypothesis. Fact 4 is difficult to explain by the chemical-reaction theory, and fact 5 is embarrassing for either hypothesis.

Collection of Chalcocite by Amyl Xanthate.(8,12) The following facts have been ascertained:

1. Chalcocite abstracts xanthate ion from solution in stoichiometric exchange for various anions particularly hydroxide, carbonate, and sulphate. So does malachite.
2. Cuprous xanthate can be extracted from treated chalcocite; cuprous xanthate and dixanthogen can be extracted from malachite.
3. The cuprous xanthate extractable from the treated chalcocite is always less than that equivalent to the xanthate abstracted from aqueous solution by the mineral. On the contrary, the cuprous xanthate extracted from treated malachite is equivalent to the amount of xanthate abstracted.
4. Chalcocite abstracts cuprous xanthate from benzene solution, but malachite does not.
5. Chalcocite treated with aqueous alkali xanthate and leached with benzene is still air adherent regardless of whether cuprous xanthate is or is not found in the benzene leach. Malachite is not.
6. The benzene-insoluble coating on chalcocite may reach, but cannot exceed, a monomolecular layer.
7. Ample floatability of chalcocite is obtained even if the coating is as little as one-seventh completely monomolecular.

The peculiar forms in which amyl xanthate distributes itself are shown in Fig. 193 after Gaudin and Schuhmann.(12)

The facts presented here do not harmonize with the chemical-reaction hypothesis. Such a hypothesis leaves unexplainable the marked abstraction by chalcocite of molecular cuprous amyl xanthate from benzene solution and the wide difference in properties of malachite and chalcocite. But they harmonize with a dual hypothesis involving both chemical reaction and adsorption in which adsorption takes effect on sulphide surfaces (bare except for a double layer of ions such as is postulated by Helmholtz and Gouy) and chemical reaction on occasional spots where oxidation of the sulphide surface has proceeded far enough to represent a new phase.

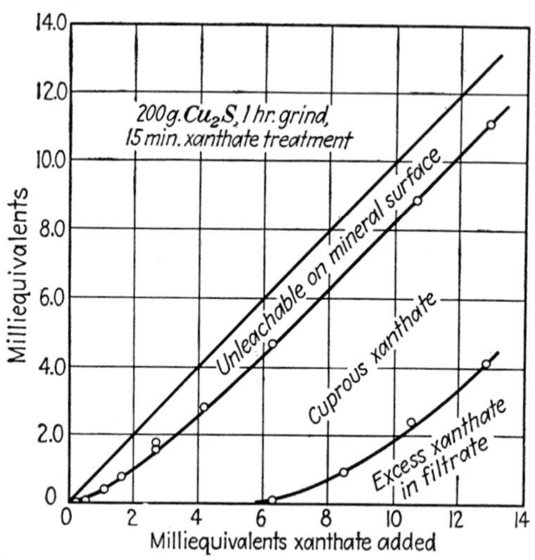

FIG. 193.—Distribution of amyl xanthate added to chalcocite as between alkali xanthate remaining in the liquor, leachable cuprous xanthate on the chalcocite, and unleachable cuprous xanthate on the chalcocite. (*After Gaudin and Schuhmann.*)

Hypothesis Proposed for Collection by Ionized Collectors. In view of the foregoing evidence, it is proposed that the mechanism of collection of sulphides by ionized collectors involves the following:

1. Prior to addition of the collector, all water-wet minerals are the seat of adsorption of ions from solution, a phenomenon that is selective as to the ions involved.

2. Addition of collectors results in adsorption of the effective collector ion as a replacement for some of the ions already there. This substitutional adsorption, or exchange adsorption, is selective also. Its extent is controlled by the concentration of the effective collector ion, the concentration of the displaced

ion, the solubility product for each ion pair, and the relationship between adsorbed-ion density at the mineral surface with subsaturation concentration. (If in this statement the existence of a relationship between adsorbed-ion density and concentration is denied, the present hypothesis becomes the chemical-reaction hypothesis.)

3. The adsorbed-ion density *vs.* concentration relationship is specific to the adsorbent and adsorbate.

Figure 194 shows a diagrammatic sketch of possible relationships between concentration and adsorbed-ion density. *OABX* represents the ideas implied in a chemical-reaction hypothesis

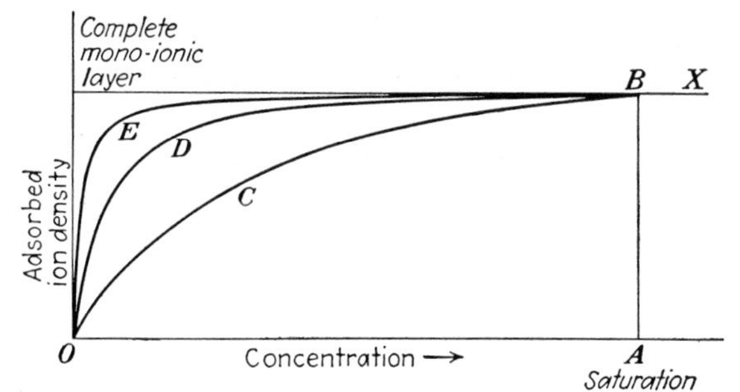

FIG. 194.—Relationship between concentration and adsorption density.

devoid of adsorption. Lines *OCBX*, *ODBX*, and *OEBX* present the ideas implied in the present hypothesis, with increasingly marked adsorption from *C* to *D* and to *E*.

The same general idea that the effective solubility of the collector is a solubility lower than that indicated by the solubility product of the relevant ions is adopted by Wark[43] who refers to it as the adsorption-solubility product and regards it as based on the adsorption theory of Horovitz and Paneth,[17] and in fact the same idea runs through the writings of the arch antiadsorptionists.[34b,36b,37a] Thus, Taggart says:

> We are rather inclined to think that in every case the solubility of a substance as a surface coating is somewhat less than that of the same substance independently put into solution, but I do not know.

Taggart, del Giudice, and Ziehl say:

> But the evidence that surface copper sulfate, soluble though it may be, is not immediately dislodged by water, is the same as with lead

sulfate on galena, but more striking; a water extract of an oxidized copper sulfide surface (air-ground chalcocite) carries a rather small amount of sulfate ion; the same material, leached by xanthate solution for the same time, throws off a tremendously increased quantity of sulfate ion, which must come, clearly, from anchored copper sulfate.

Taylor and Knoll say:

> A set of sulfur-oxygen ions closely related to the galena crystal is postulated as being present on the galena. These are displaced by the sulphydryl radical. These "anchored" lead compounds have a solubility in water different from the gross form of the same salts.

Utilization of Collector Ions. Collector ions are utilized in three ways:

1. To provide the adsorbed ions or the reacted ions, as the case may be.
2. To eliminate precipitating ions present in the liquor.
3. To provide adequate concentration in the liquor.

Reagent consumed in the first way is proportional to mineral surface. Reagent utilized in the other ways is proportional to the volume of liquor. With coarse feeds, effects 2 and 3 are preponderant, with fine feeds, effect 1.

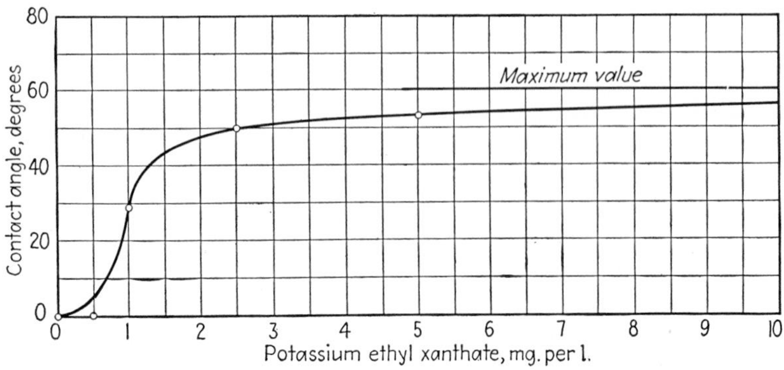

Fig. 195.—Contact angle of galena in potassium ethyl xanthate solutions, after 40-min. to 1-hr. contact. (*From the data of Wark and Cox.*)

Effect of Quantity of Collector. Increasing the quantity of collector results in an increase in the contact angle, which reaches a definite maximal value at a concentration of reagent that is low by usual chemical standards, yet one that is high in relation to the concentration of reagents used in flotation practice. Figure

195, which is drawn from the data of Wark and Cox,[46] is illustrative.

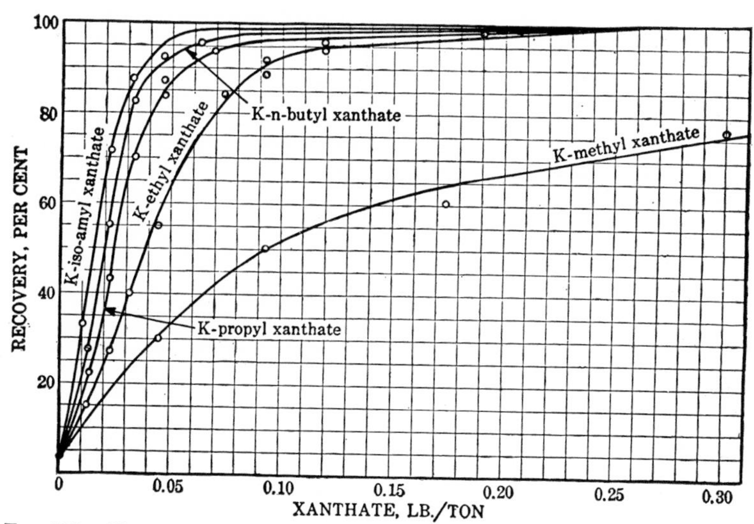

FIG. 196.—Flotation of 100- to 600-mesh particles of galena, with terpineol 0.10 lb. per ton, soda ash 1.0 lb. per ton, and xanthate as shown.

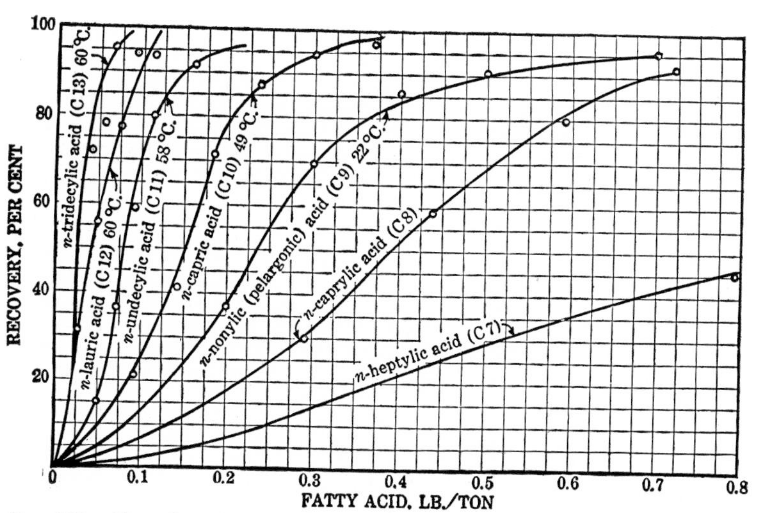

FIG. 197.—Flotation of 100- to 600-mesh particles of galena with saturated fatty acids and terpineol 0.40 lb. per ton.

Increasing the quantity of collector also results in an increase in recovery up to a maximal value.[9] Increasing the quantity of collector is particularly effective in increasing the recovery of very coarse particles. Figures 196 and 197 are illustrative.

The gradual increase in contact angle and in recovery with increase in quantity of collector may be related to one or more effects among which the following can be listed:

1. More rapid reaction at higher concentrations.
2. More rapid approach of the exchange-adsorption equilibrium at higher concentrations.
3. Displacement of the exchange-adsorption equilibrium more and more toward complete collector adsorption as concentration of the collector is increased.

It would seem that the maximum contact angle denotes the attainment of a complete monoionic layer of adsorbed collector ions.

Curves correlating mineral recovery with quantity of reagent, for sized feed, are such as to suggest that for equivalent recovery coarser and coarser feeds require more and more complete formation of the layer of adsorbed collector ions. This is consonant with the expectation that the minimum contact angle required is a function of particle size, and with the expectation that the contact angle varies in function of the concentration of unadsorbed collector.

Increasing the quantity of collector to exorbitant levels may result in decreased flotation. The causes for this effect are not

TABLE 45.—DEPENDENCE OF MAXIMUM CONTACT ANGLE ON MINERAL AND ON STRUCTURE OF COLLECTOR
(*After Wark*)

Mineral	Dithiocarbamate							Piperidinium piperidyl dithioformate	Dimethyl diphenyl thiuram disulphide
	Di-methyl	Di-ethyl	Di-n-butyl	Di-n-amyl	Mono phenyl	Phenyl methyl	Phenyl ethyl		
Galena........	50(*a*)	59	81	91	54	50	61	70(*a*)	50
Sphalerite.....	50(*a*)	59(*a*)	73	80	55(*a*)	50(*a*)	61(*a*)	67	50(*a*)
Pyrite.........	50	58	74	90	54(*p*)	50	61	70	50
Chalcopyrite...	50	59	80	80	....	49(*p*)	62	67	50
Bornite........	Irregular	62	..	..	54	50(*p*)	62	67	50
Average.....	50	59	77	85	54	50	61	68	50

(*a*) Activation necessary.
(*b*) In the presence of a reducing agent (pyrogallol).

yet elucidated. It is possible that this effect is related to the formation of complex ions, or to the adsorption of an additional layer of collector ions or micelles orientated with the polar group toward the water.

Effect of Length of Hydrocarbon Chain. It is interesting and important that the maximum contact angle obtained on any sulphide mineral depends solely on the nonpolar group of the collector. An angle of 60°, for example, is characteristic of the ethyl radical, an angle of 50° of the methyl radical, etc. This is shown by Tables 45 and 46 after Wark.[41,43c]

TABLE 46.—DEPENDENCE OF MAXIMUM ANGLE OF CONTACT ON LENGTH OF HYDROCARBON CHAIN (ALL SULPHIDE MINERALS)
(*After Wark*)

Organic type	Specific nonpolar group					
	Methyl	Ethyl	Butyl	Benzyl	Phenyl	Other cyclic derivatives
Disubstituted dithiocarbamate	50	60	77	..	..	68
Mercaptan	..	60	74	71	70	69 to 71
Xanthate	50	60	74	72	..	71 to 75
Dithiophosphate	..	59	76			
Trithiocarbonate	..	61	74			
Monothiocarbonate	..	61	73			

Increasing the length of the hydrocarbon chain induces the establishment of higher maximum contact angles. This is shown by Tables 45 and 46, and by Fig. 198. It would seem as if this is due to a more and more effective masking of the polar part of the coating on the particles by the increasing thickness of the nonpolar part, until the properties of the coating approach those of bulk paraffin.

Discovery of the dependence of maximum contact angle upon length of hydrocarbon chain, and of independence of maximum contact angle from composition of mineral substratum and from composition of the polar part of the collector is one of the significant contributions to our knowledge of flotation as it makes inescapable the conclusion that the collector film is orientated with the hydrocarbon part of the collector ions toward the water, *i.e.*, precisely in that orientation which will not attach the filmed mineral to water. This discovery brings fitting evidence to crown the brilliant theory of Langmuir.[21,22]

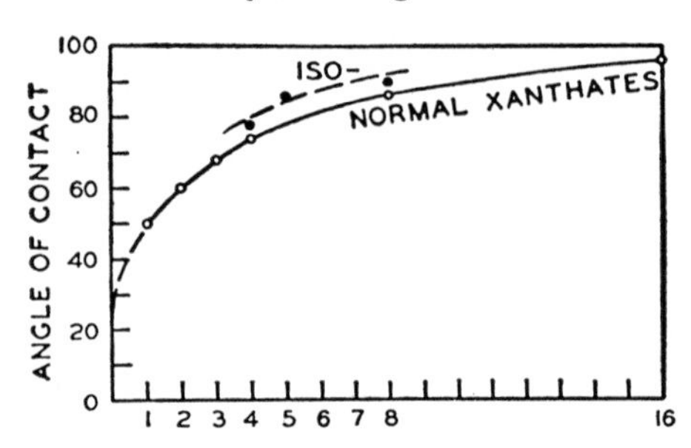

FIG. 198.—Dependence of maximum angle of contact on the nature of the alkyl group of the collector. (*After Wark and Cox.*)

Positive-ion Collectors. Much work remains to be done before these agents will be as well understood as are the xanthates. For their effect on sulphide flotation, the reader is referred to the early experiments by Gaudin and Sorensen[14] and by Gaudin, Haynes, and Haas[10] and to recent captive-bubble studies of Wark and Wark.[42]

Interesting results have recently been obtained with positive-ion collection in the flotation and agglomeration of quartz and other silicates, but little is known of the principles involved.

Nonionized Collectors. Collectors that do not give rise to ions and yet are effective are still mysterious, especially as no systematic study has been made of them.

Of the nonionized collectors, those having a sulphide or disulphide polar group are used in practice for the recovery of sulphide minerals. Laboratory experiments with agents of the dixanthogen type (*e.g.*, ethyl thioformate disulphide) have shown that the mineral becomes covered with reaction products such as would be obtained from ionic agents reduced from the dixanthogens.[8,15]

Hydrocarbon Oils. This term will be used here instead of "collecting oils" so as to exclude compounds of the type of mercaptans, dixanthogens and oleic acid, which although oily in physical appearance do not behave as they do because of their oily physical character.

Hydrocarbon oils have had a definite place in flotation and retain a place in agglomeration by their ability to smear themselves on some surfaces, particularly nonpolar surfaces, this smearing being en masse rather than molecule by molecule or ion by ion.

The smearing of oil droplets on nonpolar surfaces proceeds as does attachment of particles to bubbles and in the direction indicated by the second law of thermodynamics. Subsequently, oiled particles can stick to each other as well as to gas bubbles. The function of the hydrocarbon oil seems to be largely one of protecting the surface films from chemical or physical damage; perhaps also to making possible an agglutination of very fine particles by covering them with a liquid rather than a solid surface film. In the agglomeration of phosphate-rock particles, fuel oil is responsible for a definite economy in reagents and for improved metallurgical results.

Hydrocarbon oils are a large constituent of the old-fashioned collecting oils which preceded the introduction of chemical collectors. Those oils (generally sold under one trade name or another) also contained fractions gifted with frothing properties and other fractions gifted with collecting properties.

CONTROLLERS OR MODIFIERS

Soluble Salts. The view has already been expressed, both in this chapter and in Chapter XIV, that a solid suspended in a liquid and wet by it is surrounded by a more or less extensive double layer of ions consisting of one layer of ions anchored to the solid, and the other diffused and surrounding the first layer a very small distance away.

Thus silver bromide in a solution of silver nitrate is regarded as having an adsorbed layer consisting of silver ions continuing the silver bromide lattice—these are the anchored ions—with a second adsorbed layer consisting of bromide ions (perhaps also hydroxyl ions) that electrically balance the first layer. Contrariwise, silver bromide in potassium bromide solution has an anchored layer of bromide ions surrounded by potassium ions (perhaps also by hydrogen ions). Figure 199 illustrates these concepts.

If silver bromide is placed in potassium iodide or potassium chloride solution, the anchored ions become iodide or chloride. In this case, ions differing from those of the solid phase are adsorbed.

In general, it might well be thought that any polar mineral will be surrounded by a double layer of ions, and that more of the same ions, in dynamic equilibrium with the adsorbed ions, will be present in the liquor.

In the case of the sulphide minerals, oxidation of the sulphides yields sulphoxide and sulphate ions. If several sulphides are present simultaneously, there will be present ions derived from one sulphide and capable of reacting with another sulphide. Thus cupric ions from chalcocite may and do affect sphalerite

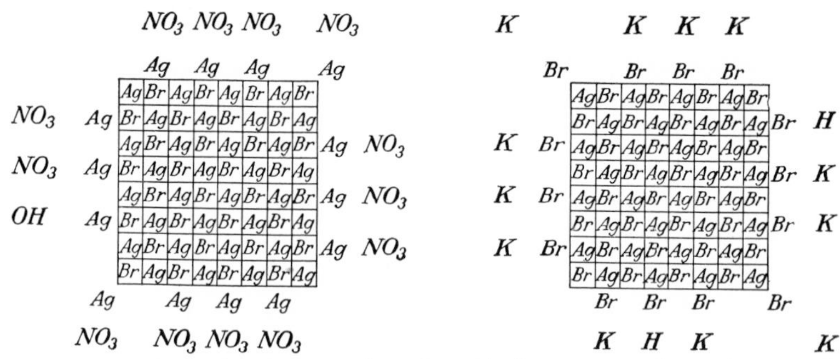

FIG. 199.—Adsorbed layers on silver bromide in aqueous silver nitrate and aqueous potassium bromide solutions.

galena, and pyrrhotite, lead ions from galena may and do affect sphalerite.

In the case of oxide, oxygen-salt, and silicate minerals, which are more soluble than unoxidized sulphides (the corresponding solubility products are much larger), these effects are even more important.

And in the general case of an ore in which many minerals are present, it must be assumed that the mineral surfaces in aqueous pulp are continuously changing. This change may be regarded as in the direction of making unlike minerals more and more alike.[11] Soluble salts are an important class of unintentional flotation controllers.

Hydrogen and Hydroxyl Ions.[3,3a] In an aqueous solution, hydrogen and hydroxyl ions are always present in interdependent amounts defined by

$$[H^+] \cdot [OH^-] = K;$$
$$K = 10^{-14} \text{ at } 20°\text{C}. \quad \text{[XVI.2]}$$

For convenience, the hydrogen-ion concentration is expressed on a logarithmic scale known as the pH scale. The pH of a solution or suspension is the logarithm to the base 10, with the sign changed, of the hydrogen-ion concentration. Thus if $[H^+] = 10^{-8}$, pH = 8.

Our senses are too crude to estimate small variations in pH, and former methods employing paper soaked in litmus or similar vegetable extracts are but slightly better, since they cannot distinguish clearly pH changes smaller than two or three units, *i.e.*, changes in acidity or alkalinity smaller than 100 to 1,000 fold. One pH unit is equivalent to a 10-fold change in $[H^+]$ or $[OH^-]$.

Colorimetric pH methods not only allow the determination of pH changes to about 0.2 pH unit, *i.e.*, to perceive a mere doubling or halving of the hydrogen-ion concentration, but they are available for this service practically across the whole pH range.

Electrometric pH determinations are still more accurate, a precision of 0.02 to 0.03 unit being attainable. For most purposes, such a precision is not necessary.

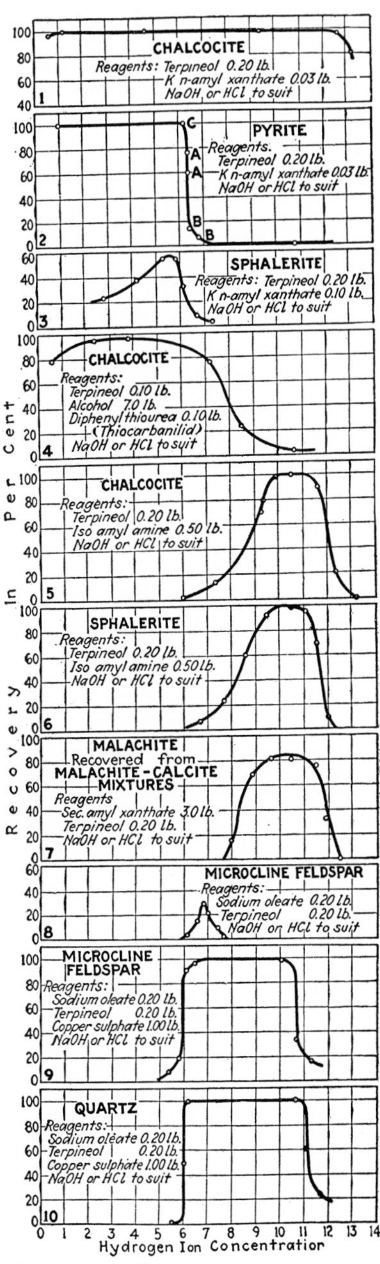

Fig. 200.—Influence of pH on recovery of various pure granular minerals.

The effectiveness of hydrogen-ion concentration as a controlling agent of the very first magnitude is graphically portrayed by Fig. 200.[7]

The effectiveness of high hydroxyl-ion concentrations (high pH values) in preventing the flotation of minerals is interpreted as the result of competition between the effective collector anion and the hydroxyl ion for the mineral surface. Thus in the flotation of galena with potassium ethyl xanthate, for example, it may be thought that hydroxyl ions and xanthate ions (X^-) are in competition for the mineral surface, so that the surface retains both of these ions in proportions defined by the affinities of each ion for lead ion and by the concentration of hydroxyl and xanthate ions in the liquor.

Control of the pH of the circuit is therefore directly important as it may increase or decrease the competition offered to the collector ion for the mineral surface.

Control of the pH is important in two other ways. Firstly, the concentration of soluble salts can be reduced to the point where the tendency for the surfaces to become alike is kept within bounds. This is definitely advantageous, especially if selective flotation is practiced. Secondly, a soluble salt that consumes the collector may be "locked up" as a precipitate, in which form it consumes less collector.

Formation of Complex Ions. A still more effective way of eliminating objectionable ions is to add salts capable of forming complexes that are dissociated to a very small extent. Thus cupric ion can be "locked up" as cuprocyanide ion, ferrous ion as ferrocyanide ion. These complex ions have the further advantage that the individual ions can be regenerated by suitable adjustment of the chemical conditions.

Hydroxyl Ion as a Specific Depressant. Figure 201 illustrates the effect of the concentration of hydroxyl ion and of the collector on the flotation of three sulphide minerals.[48] The data were obtained by bringing a bubble held to the underside of a thistle tube in contact with a polished surface of each mineral. The operations were conducted in specified baths containing the collector sodium diethyl dithiophosphate in various concentrations and at various pH values. The lines (contact curves) in Fig. 201 are the boundaries of the regions in which air-mineral contact was obtained (upper part), and in which air-mineral contact was not obtained.

From Fig. 201, it is clear that if sulphide surfaces unmodified by salts derived from other surfaces are available, a separation should be possible merely by regulating the pH. Thus, at a concentration of sodium diethyl dithiophosphate of 25 mg. per l., chalcopyrite should be floated from galena in the pH range 7 to 9, galena from pyrite in the pH range 4 to 6, and chalcopyrite from pyrite in the pH range 4 to 9.

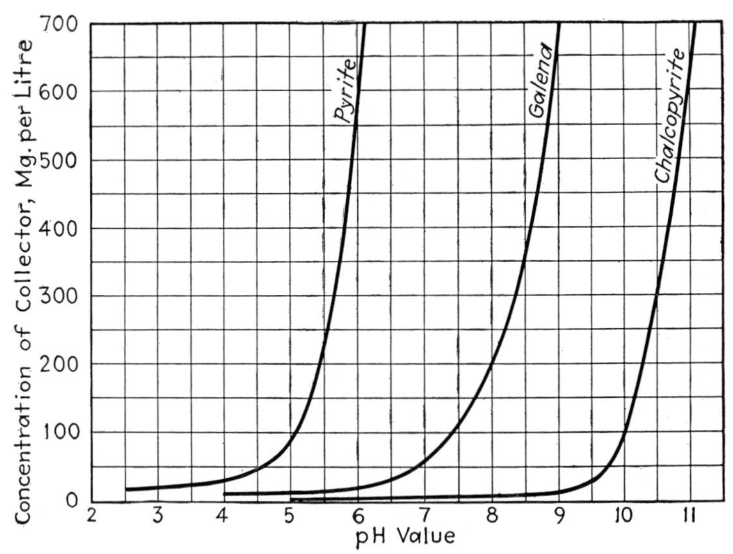

FIG. 201.—Relationship between concentration of sodium diethyl dithiophosphate and critical pH value. (*After Wark and Cox.*)

In accordance with an observation of Barsky[1] on data secured by Wark and Cox, the data of Fig. 201 are such that the concentration of the collector ion $[X^-]$ is approximately proportional to the concentration of the hydroxyl ion. That is, each of the three curves in Fig. 201 is approximately a line such that

$$\frac{[X^-]}{[OH^-]} = \text{constant.} \qquad [XVI.3]$$

The constant is different for each mineral.

This experimental fact is important, especially if it is recalled that similar relationships are found to relate the concentration of the effective collector ion to that of other effective depressing ions. It must be noted, however, that relationship [XVI.3] is but roughly true, and that the discrepancies exceed possible experimental errors, not only by their magnitude, but also by their systematic character.

Other Specific Depressing Ions. Two other depressing ions are of special importance in connection with sulphide minerals, *viz.*, the cyanide ion[2] and the hydrosulphide ion.

Figure 202 shows the contact curves for pyrite, chalcopyrite, and galena in sodium cyanide solutions, using potassium ethyl xanthate as collector in the concentration of 25 mg. per l.[49]

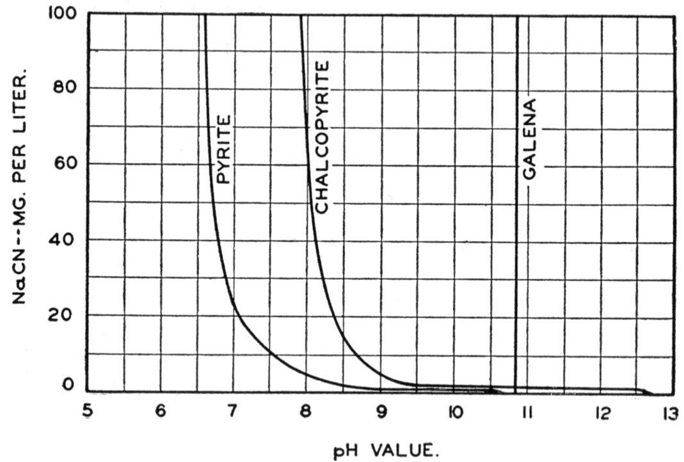

FIG. 202.—Influence of concentration of sodium cyanide on adherence to air of galena, pyrite and chalcopyrite, at 10°C. (with potassium ethyl xanthate, 25 mg. per liter.) (*After Wark and Cox.*)

In the absence of cyanide, galena is not separable from pyrite, but chalcopyrite should be floatable from either galena or pyrite in the vicinity of pH 12. By the addition of sodium cyanide in the concentration of 10 parts per million, galena can be floated from chalcopyrite in the pH range 9 to 10.5 and from pyrite in the range 8 to 10.5. At a concentration of 100 parts NaCN per liter, a wider suitable pH range is available. Figure 202 shows clearly that galena is unaffected by cyanide ion but that both pyrite and chalcopyrite are affected.

Table 47 shows that in spite of wide pH variations the concentration of sodium cyanide that is required to just prevent air-mineral contact is one that will yield a cyanide-ion concen-

TABLE 47.—CONCENTRATIONS OF CYANIDE ION IN SOLUTIONS CONTAINING JUST SUFFICIENT SODIUM CYANIDE TO PREVENT CONTACT AT SURFACES OF CHALCOCITE, COVELLITE, BORNITE, CHALCOPYRITE, AND PYRITE
(After Wark and Cox)

Chalcocite			Covellite			Bornite			Chalcopyrite			Pyrite		
pH	NaCN added, mg. per l.	CN^- mg. per l., calc.	pH	NaCN, mg. per l.	CN^-, mg. per l., calc.	pH	NaCN, mg. per l.	CN^-, mg. per l., calc.	pH	NaCN, mg. per l.	CN^-, mg. per l., calc.	pH	NaCN, mg. per l.	CN^-, mg. per l., calc.
8.7	1000	101	9.0	235	40.0	8.5	100	6.9	7.2	150	0.49	6.8	100	0.157
9.0	600	102	9.5	130	41.2	9.0	37	6.3	7.5	67	0.52	7.0	31	0.077
9.5	480	152	10.0	88	38.6	10.0	17	7.4	8.0	21	0.50	7.5	10	0.078
10.0	450	197	11.0	55	28.6	11.0	13	6.8	8.5	5.0	0.34	8.0	3.5	0.083
11.0	440	229	12.0	40	21.2	12.0	13	6.9	9.0	2.0	0.34	8.5	(1.5)	(0.103)
12.0	435	230	13.0	(7)	(3.7)	13.0	12	6.4	10.0	(1)	(0.44)	9.0	(<1)	(<.17)
13.0	430	228	....	...	....	....	...	...	11.0	(<1)	(<0.52)	10.0	(<1)	(<.4)
Mean	....	177	....	...	33.9	....	...	6.8	....	....	0.44	....	....	0.10

tration substantially the same for any one mineral regardless of the pH, and furthermore that the critical cyanide-ion concentrations are different for different minerals. This table was constructed[47] from contact curves similar to those in Fig. 202 and from accepted values for the dissociation constant of hydrocyanic acid.

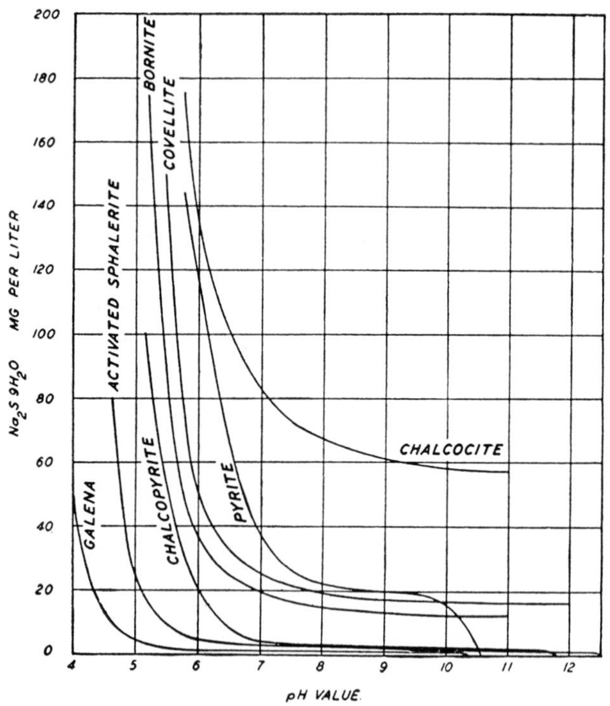

FIG. 203.—Relationship between pH and concentration of sodium sulphide necessary to prevent contact between sulphide minerals and air bubbles in solutions of potassium ethyl xanthate (25 mg. per l.). (*After Wark and Cox.*)

Figure 203 is similar to Fig. 202, except that it deals with sodium sulphide instead of sodium cyanide as depressant.[45] This figure shows that all common sulphide minerals are affected by addition of sodium sulphide, galena being affected most and chalcocite least. Table 48 shows that the effective depressing ion is the hydrosulphide ion rather than the sulphide ion or the undissociated hydrogen sulphide.[45] This conclusion is based on the fact that the critical $[HS^-]$ is practically constant under conditions that give variations for the critical $[S^=]$ and the critical

$[H_2S]$ of a million fold. In conjunction with 25 mg. of potassium ethyl xanthate per liter, the critical hydrosulphide-ion concentration is as follows for the various sulphide minerals: galena, 0.01 mg.; chalcopyrite, 0.30 mg.; bornite, 1.3 mg.; covellite, 1.7 mg.; pyrite, 2.5 mg.; chalcocite, 6.4 mg.

TABLE 48.—CONCENTRATIONS OF SULPHIDE ION, HYDROSULPHIDE ION, AND HYDROGEN SULPHIDE IN SOLUTIONS CONTAINING JUST SUFFICIENT TOTAL SULPHIDE TO PREVENT CONTACT AT A CHALCOPYRITE SURFACE

(*After Wark and Cox*)

pH value	Total sulphide as $Na_2S.9H_2O$, mg. per l.	H_2S, mg. per l.	HS^-, mg. per l.	S^{--}, mg. per l.
5.0	150	21	0.21	4×10^{-11}
5.5	52	7.2	0.22	1.4×10^{-10}
6.0	21	2.7	0.25	6×10^{-10}
6.5	8	0.8	0.27	1.7×10^{-9}
7.0	4	0.3	0.28	5.2×10^{-9}
8.0	3	0.04	0.37	7.8×10^{-8}
9.0	3	0.004	0.41	7.8×10^{-7}
10.0	2.5	0.00035	0.34	6.5×10^{-6}
11.0	2.5	0.000035	0.34	6.5×10^{-5}

Visible Depressing Coatings. The formation of visible depressing coatings has been claimed by Tucker and Head.[38,39] Later work failed to substantiate the claims made by those investigators. Ince[18,35] claims that slime coatings form at the surface of sulphide minerals and are one of the motivating causes for the selective flotation of sulphide minerals.

It seems established, however, that the slime coatings observed by Ince and also by del Giudice[16] are an effect of chemical conditions just as floating or nonfloating of sulphides is an effect of those same chemical conditions.

Practical Depressants. Practical depressants for sulphide flotation include alkali hydroxides, alkali sulphides, alkali cyanides, alkali sulphites, lime, alkali carbonates, and alkali silicates. All these agents supply hydroxyl ions; they act as depressants because of hydroxyl ions but not exclusively on that account. Indeed, in addition to hydroxyl ions, several of these depressants supply other ions capable of an independent depressing action, *e.g.*, hydrosulphide, cyanide, carbonate, and silicate ions. The effects of two of these ions have already been considered. The actions of the others have not yet been fully elucidated, even if a start in that direction has been made.[30]

Several of the more widely used depressants tend to maintain the pulp pH at some preferred range because of variation in degree of hydrolysis with variation in hydrogen-ion concentration.

Depressants in nonsulphide flotation are not yet widely used. The only significant agent is sodium silicate; its action has yet to be investigated with care.

Activation. Activation is a process of alteration of the surface of minerals that permits them to respond to collectors. By this

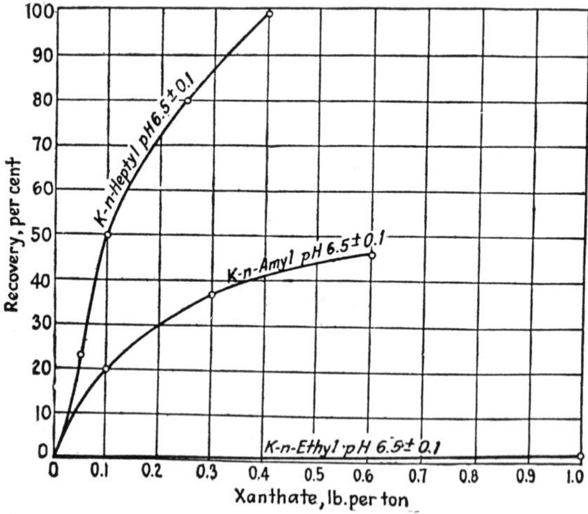

FIG. 204.—Flotation of unactivated sphalerite (100- to 600-mesh) by various xanthates.

process, minerals not amenable to flotation are made responsive; in other instances, the response of the mineral is merely increased.

The classical example is the alteration of the surface of sphalerite by copper sulphate, in connection with the use of collectors of the sulphhydrate type, *e.g.*, with potassium ethyl xanthate.[6,10,29,31] In this connection, it must be recalled that sphalerite cannot be floated with ethyl xanthate as a collector, and is only weakly responsive to a more powerful collector such as amyl xanthate (Fig. 204). But if the sphalerite is properly pretreated with copper sulphate, it is readily collected by ethyl or amyl xanthates. The contrast in the behavior of virgin sphalerite and of activated sphalerite is one of the most striking phenomena in the field of mineral dressing.

When a copper sulphate solution comes in contact with sphalerite, copper ions are abstracted from the solution and zinc ions are given up to it.

$$ZnS + Cu^{++} \rightarrow CuS + Zn^{++} \qquad [XVI.4]$$

The reaction proceeds very rapidly (in relation to the concentration of cupric ion that is available) so long as the cupric sulphide does not form a complete monomolecular layer. After such a layer has formed, however, the reaction is enormously slower, so much so that in order to form a visible coating (perhaps 1,000 atoms thick) treatment in strong, hot copper sulphate solution (*e.g.*, at 150° to 200°C.) is required. This change in rate of reaction may be due to one of several reasons among which may be listed the following:

1. The copper sulphide film on sphalerite has so nearly the same space lattice as zinc sulphide as to effectively halt the reaction. After a monomolecular film has formed, further growth of copper sulphide can result only by diffusion, a process notoriously slow in the solid state.

2. The reaction leading to the formation of the monomolecular film is one of replacement of adsorbed zinc atoms by adsorbed copper atoms, as in Eq. [XVI.5].

$$\text{(ZnS)}^{Zn^{++}} + Cu^{++} \rightarrow \text{(ZnS)}^{Cu^{++}} + Zn^{++} \qquad [XVI.5]$$

When this reaction is completed, reaction [XVI.4] sets in, and it is much slower.

Figure 205 shows a piece of sphalerite partly coated by copper sulphide.

In practice, activation of sphalerite by copper salts is carried out in alkaline circuit, *i.e.*, under conditions that connote a low concentration of cupric ion. Thus at pH 10.0, the solubility product of cupric hydroxide limits the cupric-ion concentration to 2×10^{-5}, which is but 4 per cent of the average addition of copper sulphate for sphalerite flotation. Most of the copper is in a reservoir consisting of cupric hydroxide or of basic copper carbonate on which the sphalerite draws via cupric ions until the reaction is completed.

Activation of sphalerite under those conditions is slow, so that a conditioning period is required. Activation of sphalerite at a relatively high pH is advantageous because some other sulphides are then not consuming copper ion and because they remain in a desirable condition for ulterior depression.

Sphalerite is also activated by other salts, in fact by very many salts.[10,28,40] The activation is easiest and most effective, however, if a copper salt is the activator.

Fig. 205.—Sphalerite coated with copper sulphide by reaction with copper sulphate. One part of the surface was subsequently coated with paraffin, and the specimen was exposed to sodium cyanide; this dissolved the copper sulphide coating where the coating was not protected by paraffin. ×44.

Activation of sphalerite is not an exception but merely a graphic example of a widespread phenomenon. Several sulphides other than sphalerite are activated by copper or lead salts, and many nonsulphides are activated by calcium, barium, copper, lead, and zinc salts.[4,9,20]

Deactivation. Although the net effect of deactivation is the same as that of depression, the mechanism of the process is different. As its name indicates, deactivation is the opposite of activation: it results in the restoration of the mineral surface to its condition prior to activation.

The typical example is the deactivation of copper-activated sphalerite by cyanide ion.(6,29) This is merely a locking up of cupric ion into a cuprocyanide complex that is dissociated to a very small extent. Reaction [XVI.4] is driven back, that is

FIG. 206.—Crystalline cerussite partly coated with lead sulphide as a result of reaction with 0.5 % solution of sodium sulphide. Reaction time, 30 sec. The light-colored area was protected by a paraffin coating which was subsequently removed. ×10.

to the left. The general reaction between sphalerite and copper ion can be written

$$ZnS + Cu^{++} \rightleftarrows CuS + Zn^{++} \qquad [XVI.6]$$

or

$$\text{(ZnS)}^{Zn^{++}} + Cu^{++} \rightleftarrows \text{(ZnS)}^{Cu^{++}} + Zn^{++} \qquad [XVI.7]$$

Attention is drawn to the double arrows which denote reversibility of reaction.

Prevention of Unintentional Activation. Activation can be prevented by the addition of the suitable deactivator; it can also be prevented by the addition of an electrolyte that precipitates the offending activator, thus reducing its concentration to a small value. One example is the use of sodium sulphide to prevent the activation of sphalerite.[19] It does that job just about as well as sodium cyanide. Sodium sulphide in slight excess added prior to grinding permits the selective flotation of some copper-zinc ores. But no amount of sodium sulphide will be effective if it is added after grinding.

A third way of preventing activation is to add a salt of the ion that becomes displaced from the solid surface during activation, thus driving Eq. [XVI.6] or [XVI.7] to the left. It would seem that the use of zinc sulphate in lead-zinc selective flotation is effective at least in part because zinc ions drive to the left the reaction

$$\text{(ZnS)}^{Zn^{++}} + Pb^{++} \rightleftarrows \text{(ZnS)}^{Pb^{++}} + Zn^{++} \qquad \text{[XVI.8]}$$

But this matter has not been investigated.

Sulphidizing. Although oxidized lead, copper, and silver minerals can be floated without first modifying their surfaces to sulphide surfaces, it is more economical to do so.[4,43]

Figure 206 shows a cerussite crystal partly coated by lead sulphide.

Literature Cited

1. Barsky, G.: Discussion of paper by Wark and Cox (ref. 46), *Trans. Am. Inst. Mining Met. Engrs.*, **112**, 236–237 (1934).
2. Brighton, T. B., G. Burgener, and John Gross: Depression by Cyanides in Flotation Circuits, *Eng. Mining J.*, **133**, 276–278 (1932).
3. Britton, H. T. S.: "Hydrogen Ions," D. Van Nostrand Company, Inc., New York, 2d ed. (1932).

3*a*. Clark Mansfield, W.: "The Determination of Hydrogen Ions," Williams & Wilkins Company, Baltimore, 3d ed. (1928).

3*b*. Fahrenwald, A. W.: Surface Reactions in Flotation, *Trans. Am. Inst. Mining Met. Engrs.*, **70**, 647–739 (1924).

4. Gaudin, A. M.: "Flotation," McGraw-Hill Book Company, Inc., New York (1932); (*a*) p. 68; (*b*) p. 295; (*c*) p. 199; (*d*) p. 66.
5. Gaudin, A. M.: Flotation Mechanism—A Discussion of the Functions of Flotation Reagents, *Trans. Am. Inst. Mining Met. Engrs.*, **79**, 50–77 (1928).
6. Gaudin, A. M.: The Effect of Xanthates, Copper Sulfate, and Cyanides on the Flotation of Sphalerite, *Trans. Am. Inst. Mining Met. Engrs.*, **87**, 417–428 (1930).
7. Gaudin, A. M.: The Influence of Hydrogen-ion Concentration on Recovery in Simple Flotation Systems, *Mining and Met.*, **10**, 19–20, (1929).

8. GAUDIN, A. M., FRANKLIN DEWEY, WALTER E. DUNCAN, R. A. JOHNSON, and OSCAR F. TANGEL, JR.: Reactions of Xanthates with Sulfide Minerals, *Trans. Am. Inst. Mining Met. Engrs.*, **112,** 319–347 (1934); (*a*) pp. 321–322.
9. GAUDIN, A. M., HARVEY GLOVER, M. S. HANSEN, and C. W. ORR: Flotation Fundamentals, Part I, *Utah Eng. Exp. Sta., Tech. Paper* 1 (1928).
10. GAUDIN, A. M., C. B. HAYNES, and E. C. HAAS: Flotation Fundamentals, Part IV, *Utah Eng. Exp. Sta., Tech. Paper* 7 (1930).
11. GAUDIN, A. M., and PLATO MALOZEMOFF: Hypothesis for the Non-flotation of Sulfide Minerals of Near-colloidal Size, *Trans. Am. Inst. Mining Met. Engrs.*, **112,** 303–318 (1934).
12. GAUDIN, A. M., and REINHARDT SCHUHMANN, JR.: The Action of Potassium N-amyl Xanthate on Chalcocite, *J. Phys. Chem.*, **40,** 257–275 (1936).
13. GAUDIN, A. M., and J. S. MARTIN: Flotation Fundamentals, Part III, *Utah Eng. Exp. Sta., Tech. Paper* 5 (1928).
14. GAUDIN, A. M., and P. M. SORENSEN: Flotation Fundamentals, Part II, *Utah Eng. Exp. Sta., Tech. Paper* 4 (1928).
15. GAUDIN, A. M., and WALTER D. WILKINSON, JR.: Surface Actions of Some Sulfur-bearing Organic Compounds on some Finely-ground Sulfide Minerals, *J. Phys. Chem.*, **37,** 833–845 (1933).
16. GIUDICE, GUIDO R. M. DEL: A Study of Slime Coatings in Flotation, *Trans. Am. Inst. Mining Met. Engrs.*, **112,** 398–412 (1934).
17. HOROVITZ, K., and F. PANETH: Adsorption Experiments with Radioactive Elements, *Z. physik. Chem.*, **89,** 513–528 (1915).
18. INCE, C. R.: A Study of Differential Flotation, *Trans. Am. Inst. Mining Met. Engrs.*, **87,** 261–284 (1930).
19. JOHNS, J. W., JR.: "Concentration by Flotation of a Copper-zinc Ore from the Mountain Con. Mine, Butte, Montana," Montana School of Mines, Butte, Montana (1936).
20. KRÄBER, L., and A. BOPPEL: Effect of Metallic Salts upon Flotation of Oxidic Minerals, *Metall u. Erz*, **31,** 417–427 (1934).
21. LANGMUIR, IRVING: The Mechanism of the Surface Phenomena of Flotation, *Trans. Faraday Soc.*, **15** [Pt. 3], 62–74 (1920).
22. LANGMUIR, IRVING: The Fundamental Constitution and Properties of Solids and Liquids, *J. Am. Chem. Soc.*, **38,** 2267–2278 (1916).
23. LUYKEN, W., and E. BIERBRAUER; Untersuchungen über die Beziehungen zwischen Adsorption, Benetzbarkeit und Flotation, *Metall u. Erz*, **26,** 197–202 (1929).
24. LUYKEN, W., and E. BIERBRAUER: Untersuchungen zur Theorie der Flotation, *Mitt. Kaiser-Wilhelm Inst. Eisenforschg. Düsseldorf*, **11,** 37–52 (1929).
26. MAYER, E., and H. SCHRANZ: "Flotation," S. Hirzel, Leipzig (1931).
27. PETERSEN, W.: "Schwimmaufbereitung," Theodor Steinkopff, Leipzig (1936).
28. RALSTON, O. C., and WILLIAM C. HUNTER: Activation of Sphalerite for Flotation. *Trans. Am. Inst. Mining Met. Engrs.*, **87,** 401–416, (1930).

29. Ralston, O. C., C. R. King, and F. X. Tartaron: Copper Sulfate as Flotation Activator for Sphalerite, *Trans. Am. Inst. Mining Met. Engrs.*, **87,** 389–400 (1930).
30. Ralston, Oliver C., Leonard Klein, C. R. King, T. F. Mitchell, O. E. Young, F. H. Miller, and L. M. Barker: Reducing and Oxidizing Agents and Lime Consumption in Flotation Pulp, *Trans. Am. Inst. Mining Met. Engrs.*, **87,** 369–388 (1930).
31. Ravitz, S. F., and W. A. Wall: The Adsorption of Copper Sulfate by Sphalerite and Its Relation to Flotation, *J. Phys. Chem.*, **38,** 13–18 (1934).
32. Taggart, Arthur F.: Flotation Reagents, *Trans. Am. Inst. Mining Met. Engrs.*, **79,** 40–49 (1928).
33. Taggart, Arthur F., and A. M. Gaudin: Surface Tension and Adsorption Phenomena in Flotation, *Trans. Am. Inst. Mining Met. Engrs.*, **68,** 479–535 (1923).
34. Taggart, Arthur F., G. R. M. del Giudice, and Othon A. Ziehl: The Case for the Chemical Theory of Flotation, *Trans. Am. Inst. Mining Met. Engrs.*, **112,** 348–381 (1934); (*a*) pp. 357–358; (*b*) p. 359.
35. Taggart, Arthur F., T. C. Taylor, and C. R. Ince: Experiments with Flotation Reagents, *Trans. Am. Inst. Mining Met. Engrs.*, **87,** 285–368 (1930).
36. Taggart, Arthur F., T. C. Taylor, and A. F. Knoll: Chemical Reactions in Flotation, *Trans. Am. Inst. Mining Met. Engrs.*, **87,** 217–260 (1930); (*a*) p. 219; (*b*) p. 253.
37. Taylor, T. Clinton, and A. F. Knoll: Action of Alkali Xanthates on Galena, *Trans. Am. Inst. Mining Met. Engrs.*, **112,** 382–397 (1934); (*a*) pp. 392, 397.
38. Tucker, E. L., and R. E. Head: Effect of Cyanogen Compounds on Floatability of Pure Sulfide Minerals, *Trans. Am. Inst. Mining Met. Engrs.*, **73,** 354–380 (1926).
39. Tucker, E. L., J. F. Gates, and R. E. Head: Effect of Cyanogen Compounds on the Floatability of Pure Sulfide Minerals, *Mining and Met.*, **7,** 126–129 (1926).
40. Wark, E. E., and I. W. Wark: The Physical Chemistry of Flotation. VIII. The Process of Activation, *J. Phys. Chem.*, **40,** 799–810 (1936).
41. Wark, Elsie Evelyn, and Ian William Wark: The Physical Chemistry of Flotation. III. The Relationship between Contact Angle and Constitution of the Collector, *J. Phys. Chem.*, **37**. 805–814 (1933).
42. Wark, Elsie Evelyn, and Ian William Wark: The Physical Chemistry of Flotation. VI. The Adsorption of Amines by Sulfide Minerals, *J. Phys. Chem.*, **39,** 1021–1030 (1935).
43. Wark, Ian William: "Principles of Flotation," The Australasian Institute of Mining and Metallurgy, Melbourne, Australia (1938); (*a*) p. 132; (*b*) p. 187; (*c*) pp. 117–118.
44. Wark, Ian William: The Physical Chemistry of Flotation. VII. Trimethylcetylammonium Bromide as a Flotation Agent, *J. Phys. Chem.*, **40,** 661–668 (1936).

45. WARK, I. W., and A. B. COX: Principles of Flotation. IV. An Experimental Study of the Influence of Sodium Sulfide, Alkalis and Copper Sulfate on Effect of Xanthates at Mineral Surfaces, *Am. Inst. Mining Met. Engrs., Tech. Pub.* 659 (1936).
46. WARK, I. W., and A. B. COX: Principles of Flotation. I. An Experimental Study of the Effect of Xanthates on Contact Angles at Mineral Surfaces, *Trans. Am. Inst. Mining Met. Engrs.*, **112**, 189–244 (1934).
47. WARK, I. W., and ALWYN BIRCHMORE COX: Principles of Flotation. II. An Experimental Study of the Influence of Cyanide, Alkalis and Copper Sulfate on the Effect of Potassium Ethyl Xanthate at Mineral Surfaces, *Trans. Am. Inst. Mining Met. Engrs.*, **112**, 245–266 (1934).
48. WARK, IAN WILLIAM, and ALWYN BIRCHMORE COX: Principles of Flotation. III. An Experimental Study of the Influence of Cyanide, Alkalis and Copper Sulfate on Effect of Sulfur-bearing Collectors at Mineral Surfaces, *Trans. Am. Inst. Mining Met. Engrs.*, **112**, 267–302 (1934).
49. WARK, IAN WILLIAM, and ALWYN BIRCHMORE COX: Principles of Flotation: VI. The Influence of Temperature on Effect of Copper Sulfate, Alkalis and Sodium Cyanide on Adsorption of Xanthates at Mineral Surfaces, *Am. Inst. Mining Met. Engrs., Tech. Pub.* 876; also in *Mining Tech.*, **II**-1 (1938).
50. WILKINSON, WALTER D.: "Untersuchungen über die Wirkung des Xanthogenates auf Bleiglanz," Konrad Triltsch, Wurzburg (1935).

CHAPTER XVII

FLOTATION AND AGGLOMERATION—TECHNOLOGY

The discussion of the physics and chemistry of flotation and agglomeration reviewed in Chapters XV and XVI has shown that successful operation of either process is predicated on having reduced the ore to particles of suitable size and on having established and maintained suitable chemical "weather."

Particle Size. For flotation, the floatable particles must not exceed some maximum size ranging from 10- to 14-mesh in the case of coal, through 48- to 65-mesh for sulphides, to 100- to 150-mesh for free gold. Also, the particles should not be overground as recovery and selection decrease markedly if the particles are finer than 5 to 10 microns. Table 49 and Fig. 207

TABLE 49.—EFFECT OF PARTICLE SIZE ON FLOTATION; SULLIVAN CONCENTRATOR, CONSOLIDATED MINING AND SMELTING COMPANY OF CANADA, LTD. (1930)

	Size, mesh*								
	+100	100/150	150/200	200/280	280/400	400/560	560/1,120	1,120/2,240	−2,240
Chemical composition of lead concentrate:									
Lead, per cent	71.8	72.4	74.0	69.8	69.8	69.4	70.6	71.6	65.8
Zinc, per cent	0.9	2.3	3.2	5.4	6.9	7.0	5.6	3.8	3.7
Iron, per cent	3.9	4.5	5.2	6.0	4.7	4.2	3.5	3.0	4.0
Chemical composition of zinc concentrate:									
Lead, per cent	4.3	4.1	3.6	2.9	2.2	1.8	2.4	4.5	9.2
Zinc, per cent	47.6	46.8	43.7	48.4	52.3	52.9	51.9	49.6	40.3
Iron, per cent	13.6	14.1	17.7	13.8	10.6	9.8	9.7	8.5	8.2
Lead recovery in lead concentrate, per cent	34.3	62.5	83.8	92.4	93.5	93.5	92.4	90.3	74.2
Zinc recovery in zinc concentrate, per cent	23.2	83.6	81.6	95.8	96.9	97.9	96.3	95.9	82.8

* As determined by elutriation and sedimentation. The sizes stated are for quartz particles. Sulphides are naturally finer in accordance with Stokes' law (see Chap. VIII).

illustrate these statements.[4,6] Nonfloating particles may, of course, be much coarser than floating particles, except for possible sanding of cells during an accidental shutdown. An additional size limitation is imposed by the necessity to obtain liberation of dissimilar minerals to a reasonable degree (see Chapter IV).

For agglomeration, the particles to be agglomerated must be coarser than for flotation, the optimum range being 14- to

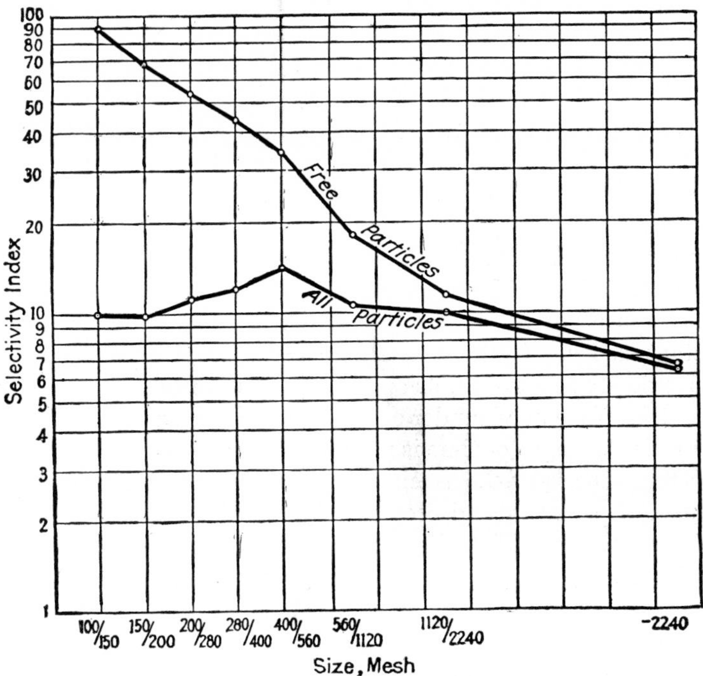

FIG. 207.—Selection of minerals at Sullivan Plant. Upper line, if free particles only are considered; lower line, if all particles are considered. Lead cycle.

100-mesh. The particles that are not to be agglomerated must be coarse enough to settle on the surface to a shaking table, *i.e.*, coarser than about 200-mesh.

Pulp Density. For the separating stage in flotation, the pulp must be dilute enough to permit particle rearrangement to proceed freely. Practically speaking, no pulp thicker than 35 to 40 per cent solids (by weight) can be used, even with heavy sulphide ores. In the case of ores containing an appreciable proportion of aluminous minerals, the pulp may have to be thinned so as to contain not more than about 15 per cent solids. For cleaning operations, a thin pulp is usually desirable,[4] but there appear to be some exceptions to this rule, as was observed by Ralston and King.[29a]

For the preparatory or conditioning stage in connection with either flotation or agglomeration, the thickest kind of pulp is advantageous as it results in an economy of most reagents. The limit is set by the appearance of plastic properties, *i.e.*, of a situation in which proper mixing is impossible.

There is, of course, no metallurgical objection to having the pulp more dilute than 15 to 20 per cent solids during the separating stage; in fact, somewhat better results are obtained in many instances with more dilute pulps. There is, however, an economic objection to overdilute pulps: such pulps require greater quantities of agents, more water, more energy, and more machinery for each ton of raw material or each ton of product.

Conditioning. Many of the effects desired in flotation are obtained so quickly that it frequently is assumed that they are instantaneous. In those cases, no special provision for preparation prior to separation, or *conditioning*, is necessary: the mixing may take place in a launder or a flotation cell. In other cases, an appreciable length of time is required and a mixing, or conditioning tank is required. Such a tank should preferably have an adjustable capacity and provision for aeration.

Temperature. Flotation and agglomeration are usually conducted at the temperature naturally assumed by the pulp, frequently in the range of 12 to 20°C. In some instances, however, the pulp is heated, to accelerate a slow action. An example is the heating of pulp for sphalerite activation.

Reagents. Maintenance of the most advantageous chemical conditions is obtained through the addition of reagents. These belong to the three broad classes of frothers, collectors, and controllers discussed in Chapters XV and XVI. Sometimes one class of agents is left out. For example, collectors may be lacking in the flotation of some soft coals, frothers in agglomeration, and modifiers in collective sulphide flotation. But, broadly speaking, reagents of all three classes are used.

Which reagent to use and in what quantity is still best ascertained by testing and by comparison with current practice. Many published documents give specific details.

Quantity of Reagent. No hard and fast rule can be given as to the quantity of reagent to use. The variations are very great from one ore to another and appreciable from day to day or even hour to hour on any one ore.

Order of magnitude of reagent addition is as follows:

	Pounds per Ton
Frothers	0.025 to 0.25
Collectors	0.05 to 2.5
Hydrocarbon oils	1. to 5.
pH regulators	0.5 to 10.
Activators	0.5 to 2.
Depressants	0.05 to 1.
Deactivators	0.05 to 0.5

In all cases, a good rule to remember is that just because a certain amount of agent is useful, more may *not* be preferable. For example, a certain amount of cyanide is useful in separating chalcopyrite from pyrite, the first remaining floatable with xanthates whereas the second is made unfloatable. But an increase in the quantity of cyanide may make both unfloatable; and an increase in the quantity of xanthate may float both. Conversely, too little cyanide will not give adequate selection, and too little xanthate will not float enough chalcopyrite. In other words, there are optimum quantities of each and every reagent.

These optimum quantities cannot be predicted, but depend upon such a variety of circumstances as to defy classification. However, the following pointers may be useful:

1. Increasing the fineness of grinding requires an increase of all reagents except frothers, which may or may not have to be decreased.

2. Desliming the ore may permit a considerable reduction in the quantity of reagents of that type which are active at the surface of minerals.

3. Use of thick pulps results in some economy of reagents.

4. Changes in type of flotation machine may require some adjustments in the quantity of reagent. The same is likewise true of changes in the place or time or physical method of addition.

Determination of the optimum quantity of reagents is in the last analysis still a matter of cut and try: it is an art, not a science. Experience and interest in one's work count heavily in this field.

Mode and Place of Addition of Reagents. This depends upon the following factors:

1. Object of the reagent.
2. Solubility in water of the reagent.
3. Physical condition of the reagent.

Addition of reagents can be in the solid form, as is the case with lime, thiocarbanilid, and in general with solids not very soluble in water. Usually they are added to the grinding device ahead of flotation. Or the addition can be in the form of a water slurry, as is often the case with lime. Choice between addition in the solid state and as a slurry is largely a matter of local convenience. In the case of compounds readily soluble in water such as alkali xanthates, and cyanides, addition may be in the form of a fairly strong aqueous solution.

Oils are added either as such or as emulsions in water. The tendency is to prefer direct addition, and to depend on action in the grinding mill for emulsification and dissolution. Where several miscible oily liquids are used, as in the Florida phosphate fields, the liquids are often mixed first in order to utilize the solubility or emulsifying property of one (*e.g.*, fatty acids in alkaline aqueous solution) to disperse the other (*e.g.*, hydrocarbon oil).

Reagent Feeders. Solid reagents can be added by a slow-moving belt conveyor of small size, moving at greatly reduced speed and drawing from the open bottom of a hopper-shaped bin resting on the belt. Control of the rate of feed is obtained by raising or lowering the bin gate or by modifying the belt speed.

Aqueous solutions are generally made and stored in definite concentration in storage tanks. Feeding is obtained by drawing the solution with cups mounted at the periphery of a disk and dumped by tripping with an adjustable tripper. By adding or removing cups, by changing their capacity or point of suspension, and by modifying the speed of the disk wide control of feeding is obtained by either continuous or discontinuous increments.

Oily liquids can be fed in the same manner, or else by utilizing the viscosity of the liquids. If the latter alternative is adopted, it is sufficient to provide a rotating disk immersed part way in the oil and a thin gutter to scrape the fluid from the disk. Changing the speed of the disk or the size and position of the gutter permits variations in feeding rate.

Machines for Flotation and Agglomeration.[4b,26a,33] Flotation machines or cells must include

1. Means for receiving and discharging pulp.
2. Means for gas introduction and gas dispersion.
3. Means for settling the pulp away from the froth.
4. Means for discharging the froth and conveying it to the next processing step.
5. Regulation of rates of intake and discharge, of rate of gas introduction and dispersion, and of pulp level.

Accordingly, flotation cells consist of tanks in which pulp is agitated and aerated as it flows through them; the tanks are compartmented to compel circulation, prevent sedimentation, and assure a settling zone.

Flotation machines are often composed of several identical cells arranged in series, *i.e.*, in such a way that one cell receives as feed the defrothed pulp from the preceding cell. Some flotation machines are composed of single cells having great length in comparison with their width.

Flotation cells are of four principal types in accordance with the mode of introduction of the gas.

1. Agitation cells in which air is drawn by a vortex caused by a rotating impeller.
2. Subaeration cells in which air is introduced by suction or by blowing through or to the base of a rotating impeller.
3. Cascade cells in which air is introduced by tumbling of the pulp.
4. Pneumatic cells in which air is introduced directly by blowing.

Combination machines have been designed, featuring more than one mode of gas introduction.

Well-known machines are the Minerals Separation Standard, the Minerals Separation Sub-aeration machine (Fig. 208), the Denver Equipment machine, the Callow pneumatic machine, MacIntosh pneumatic machine, Southwestern matless air-lift machine (Fig. 209). Recent additions include the Booth-Thompson "Agitair" machine, the Fagergren machine (Fig. 210), and the deep Britannia air-lift cell. The first two are of the subaeration type, and the third of the pneumatic type. Deep pneumatic cells have recently been adopted with considerable saving in power.

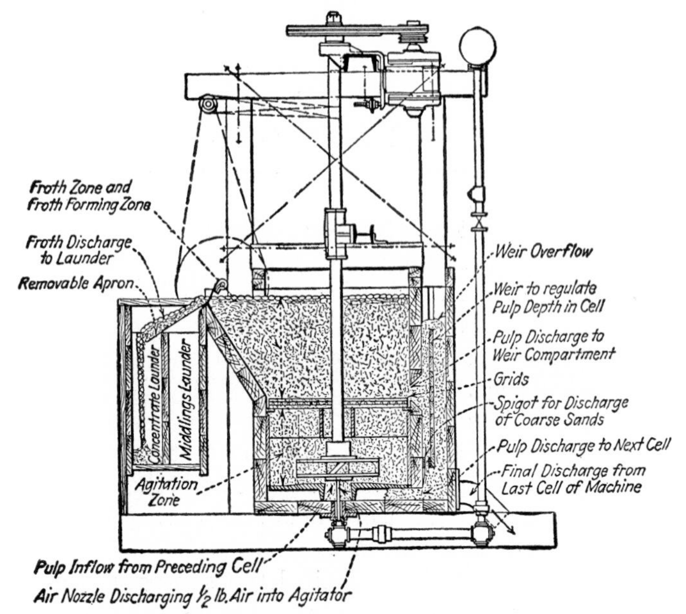

Fig. 208.—Minerals Separation sub-aeration flotation machine. (*Minerals Separation North American Corp.*)

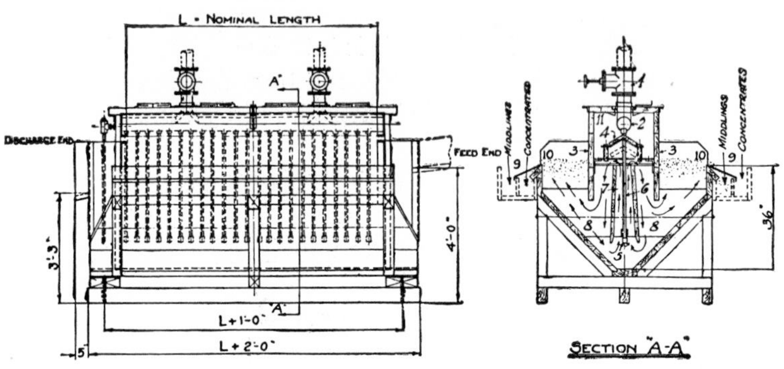

Fig. 209.—Southwestern matless air-lift flotation machine. (*Southwestern Engineering Corp.*)

In the Minerals Separation sub-aeration machine, air is drawn from the bottom immediately below an agitator. The aerated pulp rises in the outer part of the agitation zone, separation of bubbles and pulp taking place above the grids. Froth is scraped from the cells to the concentrate launder by paddles. The method of feeding pulp from one cell to another is clearly shown in the illustration.

The Southwestern matless pneumatic machine consists of a relatively long wooden tank, with semioctagonal bottom, fitted

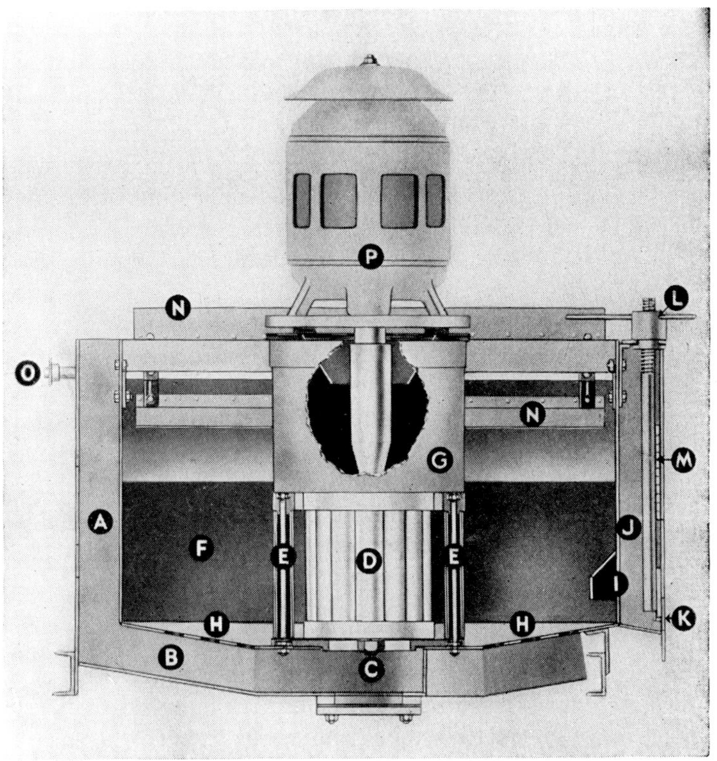

FIG. 210.—Fagergren flotation machine. (*American Cyanamid Co.*)

A, feed compartment
B, feed arm
C, rotor port
D, rotor
E, stator
F, tank
G, standpipe
H, recirculation ports
I, baffled port to discharge compartment
J, discharge compartment
K, sand gate
L, sand gate capstan
M, slat type overflow weir
N, skimmer
O, skimmer sprocket
P, direct connected motor

with longitudinal partitions to regulate circulation, central vertical pipes equipped with regulating valves for gas introduction, a bell cover over the pipe discharge to permit the establishment of quiescent, froth-separating zones on the two sides of the machine, and suitable launders for conveyance of the concentrate. Cross-partitioning of the machine to form short segments prevents by-passing of the pulp, which might otherwise occur, and simulates the conditions of pulp flow obtained in series arrangements such as that in agitation or subaeration machines.

The deep air-lift cell devised at the Britannia mill has recently been described in detail by MacLeod.[15a] At Britannia and also at the new Chelan mill, this cell has produced great economies in power consumption and it has resulted in improved metallurgy. First cost and upkeep are also said to be much lower than with either mechanically or pneumatically agitated machines of older type. The cell is similar to the Southwestern matless cell (Forrester) but is 8 to 10 ft. deep instead of 3 ft. deep.

The glass-celled laboratory Fagergren machine is especially well designed: it is clean, can be kept clean, and the pulp may be observed at all times. The possibility for salting and contamination is reduced to a minimum.

Choice between flotation machines is largely a matter of personal preference. It is true that if the tonnage treated is large, pneumatic cells are generally preferred, whereas if the tonnage is small, mechanical cells are preferred. However, there are notable exceptions to this generalization.

For agglomeration, the device in use is a shaking table, although other gravity-concentrating devices can perhaps be used. The conditioned pulp, prepared in an agitator or conditioning tank in which considerable aeration takes place, flows through the distributing box of a shaking table where it is mixed with water, on to the riffled surface of the table. Loosely bonded agglomerates of particles and air bubbles are lighter, coarser, and rounder than unagglomerated particles, but heavier than water. They therefore stratify vertically above unagglomerated particles and get washed farther downslope (see Chapter XIII). Each passage over a cleat gives a chance to each agglomerate to come in contact with a free air surface. If the agglomerate comes in contact with a free air surface, it "explodes," so to speak, and the particles in discoid aggregates sail downslope, riding at the surface of the water film by skin flotation. Since the surface of the water travels faster than the lower strata (Chapter XIII), this condition makes for a large capacity of the table, as compared with its capacity on unconditioned ore.

Both skin flotation and agglomerate stratification are effective in bringing about the separation.

Time and Cell Arrangement for Flotation. In order for every particle to receive an opportunity to stick to gas bubbles, it is necessary to prolong the floating operation for a certain length of time. This floating time ranges usually between 5 and 15 min. Variations in floating time result from variations in the gas adhesiveness of the floatable minerals, from the proportion of these minerals in the ore, from the pulp density, from the fineness of grinding, and from the degree of agitation secured in the machine. In view of these many variables, it is surprising to find that there is considerable uniformity in the floating time used.

The floating cannot be done continuously in one cell because there would be large by-pass losses, *i.e.*, losses due to pulp passing directly from feed box to tailing launder; it is customary to use 8 to 12 cells in series, froth being drawn from each cell, and finished tailing being discharged from the last cell. This arrangement permits the collection of froth from various cells as different products, which they indeed are from the standpoint of composition. Thus concentrate can be drawn from 4 cells and middling from 8 more cells if a total of 12 cells is used.

Sequence of Mineral Flotation. Table 50 and Fig. 211 illustrate the practice of making a concentrate from the first few cells and a middling from the remaining cells. The table shows that the valuable content drops rapidly from the first to the last cell in the series while one waste constituent after another shows

TABLE 50.—COMPOSITION OF FROTH DRAWN FROM SUCCESSIVE CELLS (SERIES ARRANGEMENT) AT A LARGE COPPER FLOTATION PLANT

Cell	Mineral content, per cent		
	Copper minerals	Pyrite	Silicate gangue
1 to 2	74	25.5	0.5
3 to 5	51.5	47	1.5
6	31	65	4
7 to 9	12	78	10
10 to 15	6	68	26
Tailing	0.5	21.5	78

a disposition to crowd into the froth. The figure shows that all sulphide copper minerals do not float equally readily at Anaconda, enargite floating definitely early and chalcopyrite late under the "chemical weather" available when the samples were taken.

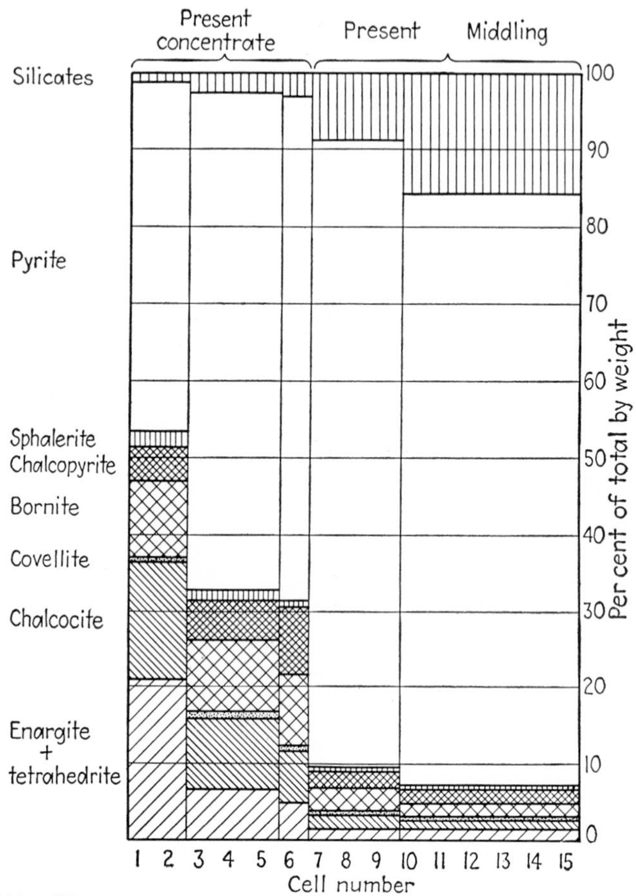

FIG. 211.—Mineral distribution of 100/150-mesh products from Anaconda concentrator (1931). (*After Gaudin and Henderson.*)

The examples presented by Table 50 and Fig. 211 are general. Several studies of this type have been made, but the results are available in only a few instances.(5)

Concentrate Re-treatment and Disposal of Middlings. Retreatment of concentrate under the same or nearly the same chemical conditions as treatment of raw feed is usually practiced. This yields a cleaned or "cleaner" concentrate and a middling or "cleaner tailing." Frequently two or three re-treatment steps are used.

The middling obtained from either the roughing or the cleaning operation generally contains a substantial enrichment in difficult particles so that much of the "nastiness" of an ore can be gathered in the relatively small tonnage of the middling. The difficult particles are locked particles, free particles of odd shape or size, and free particles with damaging surface contamination. Although many of the particles in middling products are of these difficult types, others are free particles which found their way in the middling by mechanical entrapment; they can be eliminated from it by a re-treatment.

Disposal of the middling follows one of three patterns.

1. Return to the flotation circuit. This eliminates most of the ordinary particles accidentally finding their way in the middling.

2. Regrinding and return to the flotation circuit. This takes care not only of the ordinary particles but also of some of the difficult grains.

3. Regrinding and separate re-treatment in a special flotation circuit. This disposes of the same particles as scheme 2 with the further possibility of better control, but with the disadvantage of larger load fluctuations.

Flotation Circuits.[26b] In Fig. 212 are presented schematically some typical flow sheets: 1 represents single-pass flotation, 2 single-pass flotation with splitting of a middling, 3 single-pass flotation with cleaning of the concentrate and production of two middlings, one cleaner concentrate and a tailing. In all the other flow sheets, multiple-pass flotation is used, *i.e.*, there is return with new feed of some pulp that has already passed through the machine. In 4, the two middlings are returned with new feed; in 5 exactly the same procedure as in 4 is used, as careful inspection of the flow sheet will show; in 6 two cleaning steps are used, both multipass, and the first "cleaner tailing" and the "rougher middling" are reground. Flow sheet 7 features separate re-treatment of the primary middling; to that extent, therefore, the circuit is single pass. But multipass flotation is also used in the primary cleaner-recleaner circuit, and in the re-treatment circuit.

Circuits for Selective Flotation. Where several different concentrates are produced, several circuits are required, *e.g.*, a zinc circuit may follow a lead circuit, and be in turn followed by an iron circuit in the selective flotation of pyritic lead-zinc ores. The complexity of flotation flow sheets may then be considerable.

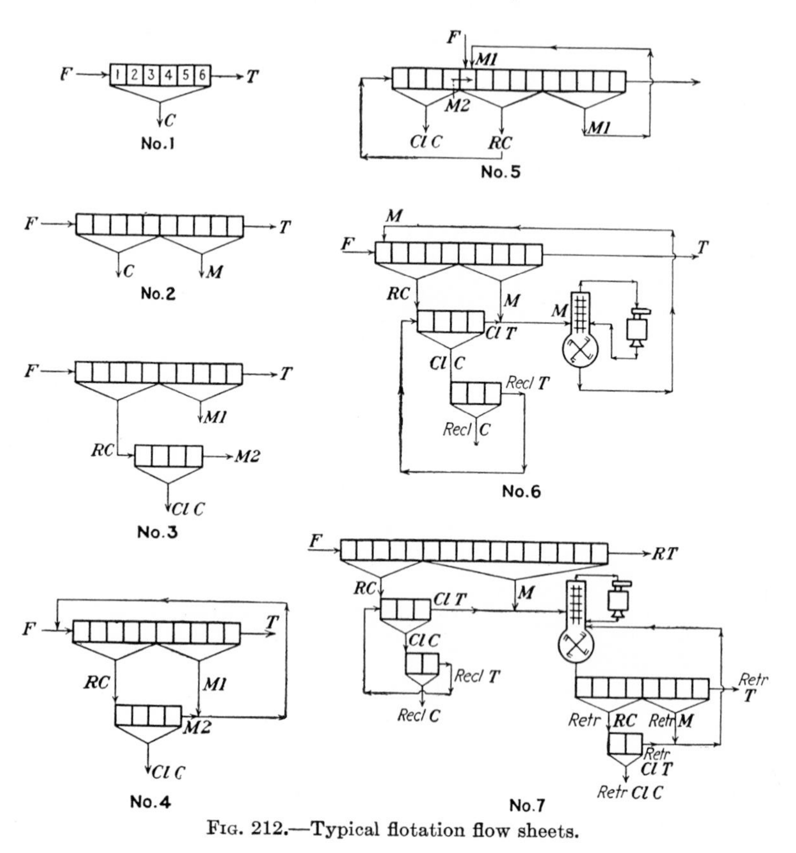

FIG. 212.—Typical flotation flow sheets.

In dealing with the selective flotation of A, B, and C from gangue G, several alternatives are available. Thus A, B, and C together can be floated from G, then re-treated for the successive flotation of A, then B, with C remaining as the re-treatment tailing. Or A, B, and C may be floated in succession, finally leaving G in the tailing. Or a combination scheme may be adopted by which A and B are floated together while leaving C and G in the tailing; subsequently, A is floated and B is dropped, and in another operation C is floated from G. If the quantities of A, B, and C are commensurate, successive flotation is generally preferred; but if one of these constituents is much more abundant than others, part-collective flotation followed by re-treatment is preferred.

EXAMPLES FROM PRACTICE[4,17,25]

Collective Sulphide Flotation. The collective flotation of sulphides has had considerable application. But except for the treatment of sulphide ores in which gold and other precious metals are sought, and except for extraordinary sulphide ores, such as the Roan Antelope ore,[15] there is no place for collective sulphide flotation in modern practice; the premium placed on pure concentrates of each of the base metals has compelled the adoption of selective sulphide flotation.

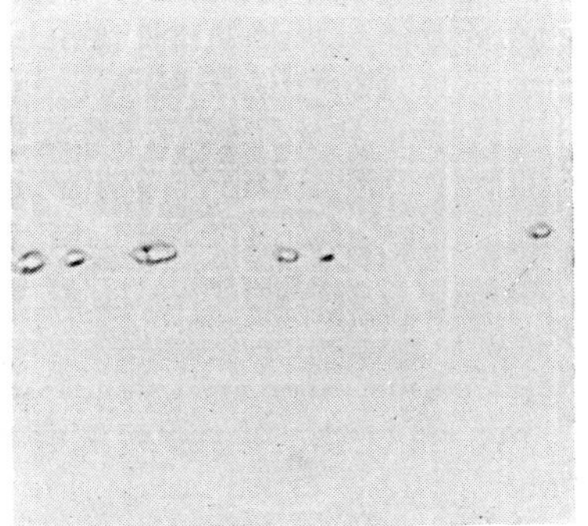

FIG. 213.—Inclusions of gold in pyrite, ×1,000. The largest particle is about 4 microns long. (*After McLachlan.*)

Collective sulphide flotation is still applied to gold and silver ores in which the value of the base metals associated as sulphides with native or telluride gold and native or sulphide silver is small. These ores may be regarded as sulphide gold and silver ores even though the gold does not occur as a sulphide. The principal sulphides are pyrite and arsenopyrite with minor quantities of chalcopyrite and tetrahedrite; the gold occurs as free particles, as particles occluded in sulphides, and to a smaller extent as inclusions in silicates.[9,35]

With ores of this common type, collective flotation of the sulphides with xanthates, particularly butyl or amyl xanthates, often gives a satisfactory recovery of the gold and silver. The pulp is best maintained at a pH near 7 to 8. It seems that under those conditions, free gold[11,12,13,14] as well as gold occluded in sulphides[3,18] (see Fig. 213) is recovered. Gold that is lost is present either as overcoarse particles, as inclusions in silicate gangue, or as particles with an oxide coating. The overcoarse gold and a part of the oxide-coated gold can be recovered on corduroy blankets or by other gravity-concentrating devices treating the flotation tailing. Recovery of gangue-occluded gold requires finer grinding or the use of cyanidation.

If the gold is free, it is often preferred to amalgamate the ore prior to flotation, or to amalgamate the concentrate prior to shipment. In that way, a substantial part of the gold is secured in bullion form and smelter penalties are minimized.

Selective Sulphide-copper Flotation.[20,24] Most copper ores contain not only the usual sulphide copper minerals, but also much worthless pyrite or pyrrhotite (more rarely also arsenopyrite or marcasite). These worthless sulphides must be separated and rejected. If other valuable sulphides such as galena, sphalerite, pentlandite, and molybdenite are also present in sufficient quantity, their separation from the copper minerals is advantageous.

Precious metals in appreciable quantity occur in many copper ores. It is desirable to include them in the copper concentrates, or better still to separate them in bullion form by using amalgamation. If neither procedure is practicable, it may be necessary to include some of the pyrite or pyrrhotite in the concentrate in order to recover particles of those minerals that occlude gold. It should therefore be clear that many considerations must be kept in mind in the selective flotation of copper-bearing sulphides.

For collection of the copper minerals, the xanthates or similar sulphhydrate compounds are suitable, ethyl xanthate being widely used. At the same time, lime in fairly large quantity and an alkali cyanide in small quantity are used to control the rejection of the pyrite (and of other iron sulphides). The pH is generally near 12 for the chalcocite-bornite ores, and in the range 8.5 to 10 for the chalcopyrite ores.

The exact chemistry of the action of lime on pyrite is not completely known, in spite of some significant work by Ralston and his associates.[29] Less yet is known concerning the action of oxygen (which seems to operate jointly with lime). These are some of the problems that still await solution.

Selective Lead-zinc Flotation.[37,38] The treatment of sulphide lead-zinc ores is more complicated than that of sulphide copper ores partly because two or three concentrates are made (one of lead, one of zinc, and perhaps one of gold-bearing pyrite), partly because of the intimacy of association of lead and zinc sulphides, partly because of the substantially complete exclusion of zinc from lead concentrates which is desired by lead smelters, and partly because of the mineralogic and structural distribution of the precious metals.

Collective flotation of lead, zinc, iron, and other sulphides is easy enough, but flotation of the lead with the maximum of copper and silver and minimum of zinc and iron requires delicate reagent control. Among the most effective agents to that end are zinc salts, alkali cyanides, alkali sulphites, and alkali sulphides. The combination of zinc sulphate with sodium cyanide is probably more widely used than all others put together.

The best pH for lead flotation with ethyl xanthate varies from 6.5 to 10, according to the particular subproblem involved. To raise the pH, weak alkalies are preferred, particularly soda ash or sodium phosphate. If lime is used, special care must be exercised.

After floating the galena, say with a combination of ethyl xanthate, sodium cyanide, zinc sulphate, and soda ash, the next step is to float the sphalerite or marmatite. This requires either the use of a collector capable of floating zinc sulphide directly, or of an activator to change the surface of the zinc sulphide plus a collector to float it. The second alternative is generally adopted, the activator being a copper salt and the collector one of the xanthates.

Sphalerite or marmatite activation, although obtained by the addition of copper sulphate, is effected by means of a copper-ion concentration very much smaller than that indicated by the amount of copper sulphate added. This is so because the activator is made to operate at a relatively high pH such as 9 to 11, under which conditions most of the copper added is immediately precipitated as copper hydroxide. It has been found experimentally that at a pH lower than 9 the flotation is not very selective to zinc but results in flotation of considerable iron. It is not known whether that is because of joint activation of the zinc and iron minerals, or for some other reason.

Sphalerite or marmatite activation is also benefited by the use of air (oxygen). It is a slow operation, and its speed is greatly reduced by an increase in pH beyond the practical limit of about 11.

After it is activated, sphalerite is floated with spectacular ease, any one of the common collectors being suitable. Sodium ethyl xanthate is generally used. For alkalinity, lime is employed almost exclusively, even if soda ash has been used in the lead circuit, as lime seems to be specifically able to inhibit pyrite.

Flotation of the gold-bearing pyrite after marmatite flotation is merely a collective flotation made somewhat difficult by the prior steps which were adopted to inhibit the pyrite. Slight acidification of the pulp, or addition of an alkali sulphide, are frequently used together with a generous quantity of a potent collector, such as amyl xanthate.

Collective Flotation of Oxidized Base-metal Ores. In general, oxidized base-metal ores are tremendously complex. This is particularly true of oxidized lead ores that have been derived from the already complex lead-zinc sulphide ores by oxidation of galena to anglesite, cerussite, pyromorphite, mimetite, plumbojarosite, etc.; by oxidation of sphalerite and marmatite to oxidized zinc minerals and to soluble salts that have been partly leached away; by occlusion of precious metals in claylike shells of ferruginous and aluminous character; by oxidation of pyrite; and by precipitation of supergene silica.

In general, lead ores that have undergone oxidation have not been wholly oxidized but retain a sulphide core to many of the grains; and the degree of oxidation varies from place to place to a very great extent.

At present, all that can be hoped of such an ore is a collective float of the lead and whatever copper, silver, and gold can be gathered with it.

Sulphidizing of the oxidized lead minerals with sodium sulphide, hydrogen sulphide, or both, and a powerful collector such as amyl xanthate in relatively large quantity are generally used. Use of a hydrocarbon oil is also often effective in reducing reagent requirement. The tendency is marked for gangue to pass into the froth. The oil with or without a small quantity of waterglass (sodium silicate) is helpful in this respect. The pH should not be high, and if much sulphidization is required sodium hydrosulphide or hydrogen sulphide is advantageous over sodium sulphide.

Fatty acids and soaps are used for the flotation of oxidized copper ores having a siliceous gangue.(30) The results, at best, are mediocre.

Selective Flotation of Constituents of Coals. Besides gross impurities such as slate, which originate in cheap mining, other impurities are present in the high-ash portions of coal usually termed "bone," and as coarse pyritic or marcasitic accretions. Both of these are readily and effectively removed by gravity concentration; they can also be separated by flotation.[36]

In addition, there is a certain group of impurities, *viz.*, fine pyrite, gypsum, organic sulphur, and so-called inherent ash, which are too finely disseminated to permit removal by gravity means. These impurities are not homogeneously distributed, but are concentrated largely as fine but distinct inclusions particularly in and near that constituent of coal known as *attritus*. The constituent known as *anthraxylon* is usually very free of impurities, and the relatively scarce constituent known as *fusain* is extremely variable in impurity content.[32,34]

Anthraxylon is a sort of fossil colloid, high in volatile matter, and is the only constituent that can be coked. It is therefore particularly valuable, and its selective flotation is practiced widely in Europe, particularly in Germany.

Attritus has arisen from the metamorphosis of all sorts of wooden debris, spores, dust, and miscellaneous organic and inorganic substances. This explains its high and variable sulphur and ash content.

Fusain is thought by many students of coals to represent fossil charcoal. It is the least valuable of the carbonaceous constituents of coal.

Selective flotation[8,17] of anthraxylon followed by that of attritus is an operation that does not require perfect separation but it must be cheap. It is perhaps best described, as semi-selective flotation. That result is attained by controlling the quantity of hydrocarbon oil used as smearing collector, by pH control, and by the addition of small quantities of protective colloids such as water glass, starch, glue, etc.

Agglomeration of Florida Land-pebble Phosphate Rock.[1,10,16,21] Florida land-pebble phosphate rock consists of bones and phosphatic nodules of various sizes in a matrix of quartz sand and clay. About half of the phosphate has been and still is recovered by disintegration in water and screening. The other half until recently had been rejected with the sand and clay.

The lost nodules ranged from 8- to 200-mesh and even finer. They are now recovered by agglomeration or by flotation.

The reject from the screening plant is classified for rejection of the clay, and the classifier sand (substantially 14- to 150-mesh) is conditioned and aerated in thick pulp with lye, soap, and a hydrocarbon oil. The quantity of lye is adjusted to give during conditioning a pH of about 9 to 10. Loosely bonded agglomerates or clusters of phosphate grains and air bubbles are formed during conditioning. The hydrocarbon oil coats the soaped particles, makes them stick more tenaciously to air, and makes it less likely for their behavior to change upon subsequent changes in pulp pH.

The conditioned pulp is fed to shaking tables, dilution water is added to the feed, and wash water is also spread as a sheet. The agglomerated particles stratify downstream from unagglomerated particles, and they are also moved downstream by skin flotation. A middling of small bulk is cut between the phosphate concentrate and the quartz tailing and re-treated. If flotation is used instead of agglomeration, the feed is finer.

Other Applications of Flotation and Agglomeration.(2,27,28) The applications of flotation and agglomeration are multiplying steadily. Among the newer fields in which these processes are adopted, there must be mentioned the following:

Beneficiation of limestone for cement manufacture by elimination of coarse quartz and silicates.(19) This results in a more uniform feed to the cement plants, and in a reduction in mining costs.

Purification of potash salts.(31) Sylvite can be floated or agglomerated in natural brine although halite cannot, and vice versa, thus yielding a purified muriate of potash at a lower cost than by the former process of dissolution and fractional crystallization.

Concentration of silicate minerals(4,22,23) such as feldspar, micas, garnets, sillimanite.

Concentration of rhodochrosite, which is separated from sulphides of lead and zinc on the one hand and silicates and quartz on the other.

Table 51 presents the reagents and quantities of reagents for some of the typical flotation operations discussed in the foregoing pages.

TABLE 51.—REAGENTS AND QUANTITIES OF REAGENTS FOR SOME TYPICAL FLOTATION OPERATIONS

	A	*B*	*C*	*D*	*E*	*F*	*G*	*H*
Collectors:								
Potassium ethyl xanthate	....	0.05	....	0.06	0.10			
Potassium amyl xanthate	0.10	....	....	....	....	1.25		
Oleic acid	....	....	....	....	....	....	....	1.5
Sodium diethyl dithio phosphate	....	....	0.04					
Frothers:								
Pine oil	0.15	0.05	....	....	....	0.15		
Cresylic acid	....	....	0.05	0.03	0.07	....	0.05	
Pine-tar oil	....	....	....	....	....	....	....	0.10
Hydrocarbon oils:								
Fuel oil	....	....	....	....	....	....	....	3.0
Heavy coal-tar oil	....	....	....	....	....	....	2.0	
Modifiers:								
Sulphuric acid	....	....	....	....	....	2.0		
Lime	....	5.0	3.0	....	3.0			
Lye	....	....	....	....	....	....	....	0.5
Soda ash	....	....	....	1.5				
Sodium cyanide	....	0.10	0.10	0.20				
Zinc sulphate	....	....	....	1.0				
Copper sulphate	....	....	....	....	1.25			
Sodium sulphide	....	....	....	....	....	5.0		
Starch	....	....	....	....	....	....	0.03	

A. Collective flotation for gold.
B. Selective flotation for chalcocite from pyritic ore.
C. Selective flotation for chalcopyrite from heavy pyrite ore.
D. Selective lead-zinc ore, lead circuit.
E. Selective lead-zinc ore, zinc circuit.
F. Oxidized lead ore.
G. Coal.
H. Deslimed phosphate rock.

Literature Cited

1. ANON.: Selective Film Flotation on Tables Applied to Potash Ores, *Eng. Mining J.*, **136,** 182 (1935).
2. COGHILL, WILL H., and J. BRUCE CLEMMER: Soap Flotation of the Nonsulfides, *Trans. Am. Inst. Mining Met. Engrs.*, **112,** 449–465 (1934).
3. DENNY, J. J.: McIntyre Metallurgy—Flotation and Cyanidation, *Eng. Mining J.*, **134,** 472–484 (1933).
4. GAUDIN, A. M.: "Flotation," McGraw-Hill Book Company, Inc., New York (1932); (*a*) p. 147; (*b*) pp. 387–401.

5. Gaudin, A. M.: Unusual Minerals in Flotation Products at Cananea Mill Studied Quantitatively by the Microscope, *Eng. Mining J.*, **134,** 523–527 (1933).
6. Gaudin, A. M., John O. Groh, and H. B. Henderson: Effect of Particle Size on Flotation, *Am. Inst. Mining Met. Engrs., Tech. Pub.* 414 (1931).
7. Gaudin, A. M., and H. B. Henderson: "Quantitative Microscope Study of the Flotation Products of the Anaconda Copper Concentrator," Montana School of Mines (1933).
8. Götte, A.: Grundlagen der Steinkohlenflotation, *Glückauf*, **70,** 293–297 (1934).
9. Haycock, Maurice H.: The Role of the Microscope in the Study of Gold Ores, *Trans. Can. Inst. Mining Met.*, **40,** 405–414 (1937).
10. Heinrichs, Charles E.: Economic Results of the New Technique in Phosphate Recovery, *Mining and Met.*, **14,** 329–332, 350 (1933).
11. Johns, J. W., Jr.: Further Tests in Flotation of Free Gold, *Eng. Mining J.*, **136,** 498–499 (1935).
12. Kraut, Max.: Floating Gold on the Mother Lode, *Mining and Met.*, **13,** 175–176 (1932).
13. Lange, L. H.: More Facts on the Flotation of Free Gold, *Eng. Mining J.*, **136,** 116–118 (1935).
14. Leaver, E. S., and J. A. Woolf: Flotation of Minor Gold in Large-scale Copper Concentrators, *Am. Inst. Mining Met. Engrs., Tech. Pub.* 410 (1931).
15. Littleford, J. W.: Concentrating Operations at the Roan Antelope Copper Mines, Ltd., *Trans. Am. Inst. Mining Met. Engrs.*, **112,** 935–951 (1934).
15*a*. Macleod, N. A.: Deep Air-flotation at Britannia, *Trans. Can. Inst. Mining Met.*, **41,** 473–480 (1938).
16. Martin, H. S., and James Wilding: "Industrial Minerals and Rocks," Section on "Phosphate Rock," American Institute of Mining and Metallurgical Engineers, New York (1937).
17. Mayer, E., and H. Schranz: "Flotation," S. Hirzel, Leipzig (1931).
18. McLachlan, C. G.: Increasing Gold Recovery from Noranda's Milling Ore, *Trans. Am. Inst. Mining Met. Engrs.*, **112,** 570–596 (1934).
19. Miller, Benjamin L., and Charles H. Breerwood: Flotation Processing of Limestone, *Am. Inst. Mining Met. Engrs., Tech. Pub.* 606 (1935).
20. Morrow, B. S.: Both Copper and Zinc Ores Treated by Selective Flotation in Concentrators at Anaconda, Montana, *Eng. Mining J.*, **128,** 295–300 (1929).
21. Pamplin, J. W.: Ore Dressing Practice with Florida Pebble Phosphates, Southern Phosphate Corp., *Am. Inst. Met. Engrs., Tech. Pub.* 881 (1938).
22. Patek, John M.: Colloidal Depressors in Soap Flotation, *Eng. Mining J.*, **135,** 558 (1934).
23. Patek, John M.: Relative Floatability of Silicate Minerals, *Trans. Am. Inst. Mining Met. Engrs.*, **112,** 486–508 (1934).

24. PEARSE, H. A.: Three-product Flotation at the Britannia, B. C., Mill, *Mining and Met.*, **15**, 379–383 (1934).
25. PETERSEN, W.: "Schwimmaufbereitung," Theodor Steinkopff, Leipzig (1936).
26. RABONE, PHILIP: "Flotation Plant Practice," Mining Publications, Ltd., London (1932); (*a*) pp. 79–99; (*b*) pp. 100–106.
27. RALSTON, O. C.: Froth Flotation and Agglomerate Tabling of Non-metallic Minerals, *Trans. Can. Inst. Mining Met.*, **40**, 691–726 (1937).
28. RALSTON, O. C.: Flotation and Agglomerate Concentration of Non-metallic Minerals, *U. S. Bur. Mines, Rept. Investigations* 3397 (1938) (with 222 refs.).
29. RALSTON, OLIVER C., LEONARD KLEIN, C. R. KING, T. F. MITCHELL, O. E. YOUNG, F. H. MILLER, and L. M. BARKER: Reducing and Oxidizing Agents and Lime Consumption in Flotation Pulp, *Trans. Am. Inst. Mining Met. Engrs.*, **89**, 369–388 (1930).
29*a*. RALSTON, O. C., and C. R. KING: Some Effects of Diluting a Flotation Pulp, *Mining Met.*, **16**, 332–333 (1935).
30. REY, MAURICE: Fatty Acid and Soap Flotation Applied to Oxidized Copper Ore, *Eng. Mining J.*, **136**, 221–222 (1935).
31. SMITH, HOWARD I.: Developments Affecting the American Potash Industry, *Am. Inst. Mining Met. Engrs., Tech. Pub.* 722 (1936).
32. STACH, ERICH: "Lehrbuch der Kohlenpetrographie," Gebrüder Borntraeger, Berlin (1935).
33. TAGGART, ARTHUR F.: "Handbook of Ore Dressing," John Wiley & Sons, Inc., New York (1927).
34. THIESSEN, REINHARDT: "What is Coal?," United States Bureau of Mines (1937).
35. WARREN, HARRY V., and JOHN M. CUMMINGS: Textural Relations in Gold Ores of British Columbia, *Am. Inst. Mining Met. Engrs., Tech. Pub.* 777; also in *Mining Tech.* **I**-2 (1937).
36. YANCEY, H. F., and J. A. TAYLOR: Froth Flotation of Coal—Sulfur and Ash Reduction, *U. S. Bur. Mines, Rept. Investigations* 3263 (1935).
37. YOUNG, A. B., and W. J. MCKENNA: Selective Flotation of Lead-zinc Ores at Tooele, Utah, *Eng. Mining J.*, **128**, 291–294 (1929).
38. YOUNG, GEORGE J.: Selective Lead-zinc Flotation at Kimberley, B. C., *Eng. Mining J.*, **131**, 313–315 (1931).

CHAPTER XVIII

MAGNETIC SEPARATION

Concentration of certain mineral products can be achieved by utilizing magnetic force. Different minerals have widely different magnetic properties. By combining the different forces of magnets with gravitational or frictional forces a separation of mineral particles is possible. Two or more products are obtained. Depending upon their commercial value, these products are concentrates, middlings, or tailings.

Magnetism. A magnet is surrounded by a field of force, known as a magnetic field, which in many respects resembles a gravitational or electric field of force.

Any substance placed within a magnetic field is affected in some way. With most substances, this effect is so slight as to become evident only after careful measurements, but with others it is very great. Substances may be classified into two broad classes, according to their attraction or repulsion by an ordinary magnet. To be more precise, a *diamagnetic* substance may be defined as one that is repelled along the lines of force of a magnetic field to points where the magnetic field intensity is smaller; on the other hand, a *paramagnetic* substance is attracted along the line of force of a magnetic field to points of greater magnetic field intensity. Diamagnetism, except possibly in the case of metallic bismuth, involves magnetic forces too small to be of use for the purposes of magnetic separation. Paramagnetism frequently generates magnetic forces great enough to make possible the practical application of special, high-intensity magnetic separators. For a very few substances, notably some of those containing iron, the forces manifested in a magnetic field are so very large that new magnetic phenomena may be said to exist. These phenomena are collectively known as *ferromagnetism.*

It is believed that the effect of a magnetic field on a foreign body consists fundamentally in compelling specific molecular or intramolecular orientation within the body. The subject, however fascinating, must remain outside the scope of this text; the inquisitive reader is referred to modern texts on the theory of magnetism.[e.g., 6,48] It should be noted that modern magnetic theory has shown that diamagnetism and paramagnetism are fundamentally different both in origin and in behavior. It has also been shown that ferromagnetism is a special case of paramagnetism. The characteristic usually considered as distinguishing ferromagnetism from paramagnetism, other than mere intensity of effect, is the retention of magnetism after the substance is taken out of the field. This property of residual magnetism is often called *remanence*, and gives rise to the permanent magnets.

It is customary to speak of the attraction of one magnet for another on the one hand, and of a magnet for iron, magnetite, or some of the magnetic minerals on the other. There is no essential difference between these two types of magnetic attraction if the material that is influenced by the magnet is regarded as becoming a temporary magnet.

THEORY

A completely satisfying theory of magnetic separation has yet to be devised.[16,23,24,41] This is because of the great complexity of the problem and of the interdependence of the effective variables. The theory developed in the following pages is, of course, premised on simplifying assumptions. Its conclusions should not be regarded as more than qualitatively indicative.

Magnetic Force. A magnet behaves as if the magnetic field which it generates emanates from two poles. These poles are frequently regarded as mathematical points. This is but a convenient approximation; rigorous analysis would require consideration of polar regions rather than of polar points. The poles are equal in absolute strength and are of opposite sign. Unlike poles attract and like poles repel each other. The attraction or repulsion is quantitatively expressed as

$$F = \frac{1}{\mu} \frac{m_1 m_2}{d^2}, \qquad \text{[XVIII.1]}$$

where the force F is expressed in dynes, m_1 and m_2 are the strengths of the poles expressed in terms of unit polar strengths,

d is the distance between the poles, expressed in centimeters, and μ is a constant depending on the medium.

The constant μ is the *magnetic permeability*. It is unity for a vacuum, slightly larger or smaller than unity for most substances, and extremely large under some circumstances for ferromagnetic substances.*

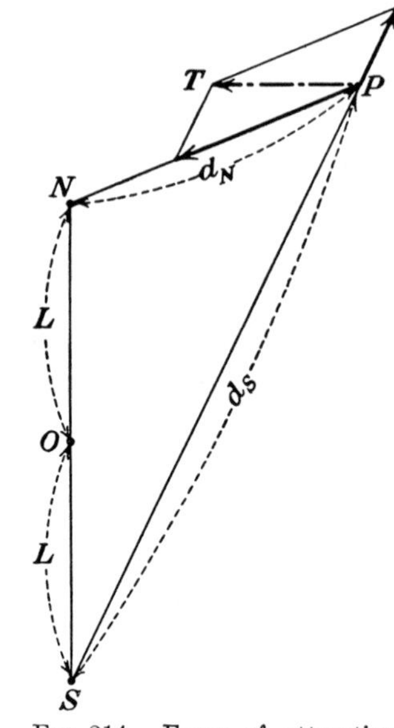

FIG. 214.—Force of attraction on a magnetic pole.

From Eq. [XVIII.1], it is possible to derive most of what we know in a quantitative manner concerning magnetism.

Magnetic Moment.[43] The magnetic moment M of a magnet is the product of its polar length $2L$ by the strength of pole m.

$$M = 2Lm. \qquad [\text{XVIII.2}]$$

It is equal to the maximum torque acting on the magnet when the magnet is placed in a uniform field of unit strength—a condition that arises if the axis of the small magnet is at right angles to the direction of the magnetic field in which it is placed.

Force of Attraction on a Magnet Pole. Consider a magnetic south pole at P of strength m_2 in the field of the magnet NS of polar strength m and length $2L$ (Fig. 214).

The attraction of N for P is

$$F_A = \frac{1}{\mu} \frac{mm_2}{(d_N)^2}.$$

The repulsion of S for P is

$$F_R = \frac{1}{\mu} \frac{mm_2}{(d_S)^2}.$$

* Permeabilities of 200 to 2,000 are common with ferromagnetic metals and alloys. Much larger permeabilities have recently been obtained with special alloys. For example, $\mu_{\max} = 162{,}000$ for an alloy of iron, silicon, and aluminum called sendust.[30] To date, the highest permeability is said to be approximately 600,000.[6] Magnetites range in permeability up to 12. Magnetically separable paramagnetic substances range down to 1.0001.

Both of these forces are vectorial; they are directed along **PN** and **SP,** respectively.

Vectorial summation shows that the force on P is directed in the direction **PT,** and that its magnitude is given by the suitable diagonal in the parallelogram of forces drawn from P.

If m_2 is made equal to unity, *i.e.*, if the test pole P is of unit strength, the values obtained for F are numerical measures of the intensity of the field. The field intensity, or *field*, is usually denoted by the letter H. In c.g.s. units, the field is expressed in oersteds. The general expression for the field is given in Eq. [XVIII.3.]

$$H = \frac{m}{\mu}\left[\frac{1}{(\mathbf{d_N})^2} + \frac{1}{(\mathbf{d_S})^2}\right]. \qquad \text{[XVIII.3]}$$

In this formula, $1/(\mathbf{d_N})^2$ and $1/(\mathbf{d_S})^2$ denote vectors.

In special cases, the solution is geometrically simple. Thus, in the case of a point P between the north and south poles of a large magnet (Fig. 215), so shaped that the body of the magnet is well removed from the field between the poles, the field is

$$H_P = \frac{m}{\mu(L - D)^2} + \frac{m}{\mu(L + D)^2} = \frac{2m(L^2 + D^2)}{\mu(L^2 - D^2)^2}.$$

FIG. 215.—Force of attraction on a magnet pole.

FIG. 216.—Force of attraction on a dipole.

Expressed in terms of the moment of the large magnet

$$M = 2Lm,$$

the field becomes

$$H_P = \frac{M}{\mu}\frac{L^2 + D^2}{L(L^2 - D^2)^2}. \qquad \text{[XVIII.4]}$$

Force of Attraction on a Magnetized Particle. In the case of a material, magnetized particle of polar length $2r$, in air ($\mu = 1.0$), situated axially in the field between the north and south poles of a large magnet, the fields at the two poles are (Fig. 216)

$$H_{P_S} = \frac{M[L^2 + (D + r)^2]}{L[L^2 - (D + r)^2]^2}$$

$$H_{P_N} = \frac{M[L^2 + (D - r)^2]}{L[L^2 - (D - r)^2]^2}.$$

The N-wise attraction of a south pole of strength m_2 situated at P_S is then

$$F_N = Mm_2\frac{L^2 + (D + r)^2}{[L^2 - (D + r)^2]^2}.$$

In like manner, the N-wise attraction of a north pole of strength m_2 situated at P_N is

$$F_S = -Mm_2\frac{L^2 + (D - r)^2}{[L^2 - (D - r)^2]^2}.$$

The total or net attraction on the particle becomes

$$F = F_N + F_S = Mm_2\left\{\frac{L^2 + (D + r)^2}{[L^2 - (D + r)^2]^2} - \frac{L^2 + (D - r)^2}{[L^2 - (D - r)^2]^2}\right\}$$

or expressing D and r in terms of L,

$$D = \delta L; \qquad r = \rho L;$$

and putting $M' = 2rm_2$,

$$F = \frac{MM'}{2\rho L^4}\left\{\frac{1 + (\delta + \rho)^2}{[1 - (\delta + \rho)^2]^2} - \frac{1 + (\delta - \rho)^2}{[1 - (\delta - \rho)^2]^2}\right\}. \qquad \text{[XVIII.5]}$$

This formula shows that in the idealized magnetic field under consideration, the force of attraction on a particle varies as the product of the magnetic moments of the magnet and particle, inversely as the fourth power of the polar distance of the magnet, and as the complicated function

$$\Phi = \frac{1}{\rho}\left\{\frac{1 + (\delta + \rho)^2}{[1 - (\delta + \rho)^2]^2} - \frac{1 + (\delta - \rho)^2}{[1 - (\delta - \rho)^2]^2}\right\}, \qquad \text{[XVIII.6]}$$

in which ρ and δ represent the size and position of the particle in relation to the polar distance of the magnet. The quantity Φ may be termed the relative force of attraction.

Numerical results for Φ are tabulated in Tables 52 and 53. Table 52 shows that there is a very great variation in the relative force of attraction with variation in position in the magnetic

TABLE 52.—RELATIONSHIP BETWEEN POSITION OF A SMALL PARTICLE IN THE AXIS OF THE FIELD OF A SYMMETRICAL IDEAL MAGNET, AND THE FORCE OF ATTRACTION EXERTED BY THE MAGNET

Distance from Center of Particle to Center of Symmetry of Magnet δ, Expressed as Fractions of the Half-polar Length of the Magnet	Relative Force of Attraction, Φ
0.000	0.000
0.001	0.012
0.01	0.12
0.1	1.25
0.3	5.0
0.5	15.2
0.7	74.0
0.8	249.
1.0	∞

TABLE 53.—RELATIONSHIP BETWEEN SIZE OF A PARTICLE SITUATED IN THE AXIS OF THE FIELD OF A SYMMETRICAL IDEAL MAGNET AND SPECIFIC FORCE OF ATTRACTION EXERTED BY THE MAGNET

Distance from center of particle to center of symmetry of magnet, expressed as fractions of the half-polar length of the magnet	Size of particle ρ, expressed as fractions of the half-polar length of the magnet	Relative force of attraction, Φ
0.3	0.01	5.0
	0.05	5.0
	0.1	5.2
	0.2	5.9
	0.3	7.7
0.5	0.01	15.3
	0.02	15.5
	0.05	15.7
	0.1	16.8
	0.2	22.1
0.7	0.01	74.0
	0.02	74.2
	0.05	77.8
	0.1	93.3
	0.2	239.7

field, the force being negligible near the center of symmetry of the magnet and enormous near the poles. It is clear, of course, that the force is equally large but of opposite sign on the other side of the center of symmetry.

The field of an ideal magnet consisting of two unconnected point poles converges rapidly near the poles and is uniform at the plane of symmetry and at infinity. On paramagnetic material particles, there is attraction everywhere, within such a field, except at the points in space where the field is uniform. An ideal magnet is approximately realized by horseshoe-type magnets. They, however, do not have point poles and must in fact depart considerably from the most ideal shape if they are to handle an appreciable quantity of ore. The tabular data presented here are therefore merely indicative and are not quantitatively applicable to realizable magnets.

Table 53 shows that there are but slight variations in specific force of attraction, *i.e.*, in force of attraction per unit volume, as between particles of various sizes so long as the size of the particles is small compared with the distance of the particles to the poles of the magnet. But if a particle is large enough and so situated that one of its poles is appreciably closer to the near pole of the large magnet than to the far pole, the particle is attracted more forcefully than a smaller particle whose center is in the same situation.

Induced Magnetic Moment. The quantity M' introduced in Eq. [XVIII.5] is the temporary magnetic moment induced in the particle by the field in which it is placed.

It is the product of the volume of the particle V, the field H' across the particle, and the magnetic susceptibility κ of the material of which the particle is made.

$$M' = VH'\kappa. \qquad [XVIII.7]$$

It may seem at first sight that the effective field in determining the intensity of magnetization of a magnetizable particle is the external magnetizing field H. This, however, is not so since the particle acts as a magnet itself; its own field H'' is directed against the magnetizing field H:

$$H' = H - H''.$$

It has been shown[43b] that for a spherical particle in a uniform field

$$H' = \frac{3\mu_1}{2\mu_1 + \mu} H, \qquad \text{[XVIII.8]}$$

in which μ_1 is the permeability of the fluid and μ the permeability of the particle. The field which is to be used in evaluating M' is H'.

The magnetic *susceptibility*[43a] κ is defined as the ratio of the intensity of magnetization to the magnetic field in the object. It is related to the magnetic permeability μ by the following equation:

$$\mu = 1 + 4\pi\kappa, \qquad \text{or} \qquad \kappa = \frac{\mu - 1}{4\pi}. \qquad \text{[XVIII.9]}$$

Either μ or κ can be determined experimentally, and the other can be calculated from it.

The magnetic susceptibility and permeability are at the basis of magnetic separation, for if μ were always 1, or κ nil, M' would be nil and there would be no attraction. The actual force on a particle, per unit volume of particle, varies directly not as the permeability of its substance but as M'/V, or as

$$\frac{3\mu_1}{2\mu_1 + \mu} \cdot H \cdot \frac{\mu - 1}{4\pi}.$$

In the case of air, μ_1 is sensibly unity so that the force of attraction varies as

$$\frac{\mu - 1}{\mu + 2} \cdot H.$$

If the permeability is very small, $\mu + 2$ is practically constant at 3, and the force of attraction varies as $(\mu - 1)H$, or as $4\pi\kappa H$, *i.e.*, in direct proportion to the susceptibility and to the intensity of the field. If the permeability is large, the attraction is proportional to H and independent of the particular value taken on by μ.

$$\frac{\mu - 1}{\mu + 2} \rightarrow 1.0.$$

By making use of Eq. [XVIII.4] with μ_1 (air) $= 1$, and by substituting δL for D, the induced magnetic moment per unit volume becomes

$$\frac{M'}{V} = \frac{\mu - 1}{\mu + 2}H = \frac{\mu - 1}{\mu + 2}M\frac{L^2 + D^2}{L(L^2 - D^2)^2} = \frac{\mu - 1}{\mu + 2} \cdot \frac{1}{L^3} \cdot \frac{1 + \delta^2}{(1 - \delta^2)^2}$$

and the force of attraction F per unit volume (Eq. [XVIII.5]) becomes

$$\frac{F}{V} = \frac{M^2}{2L^7} \cdot \frac{\mu - 1}{\mu + 2} \cdot \Phi \cdot \Phi', \qquad \text{[XVIII.10]}$$

in which Φ is given by Eq. [XVIII.6] and Φ' by Eq. [XVIII.11].

$$\Phi' = \frac{1 + \delta^2}{(1 - \delta^2)^2}. \qquad \text{[XVIII.11]}$$

Since Φ' increases from unity without limit as δ increases from 0 to 1, it is seen that the effect of δ on F is even more marked than appeared from Table 52.

Table 54 presents actual values of the relative force of attraction $\Phi \cdot \Phi'$ if the magnetic moment induced in the particle is

TABLE 54.—RELATIONSHIP BETWEEN POSITION OF A SMALL PARTICLE IN THE AXIS OF THE FIELD OF A SYMMETRICAL IDEAL MAGNET AND THE FORCE OF ATTRACTION EXERTED BY THE MAGNET, AFTER ALLOWING FOR THE EFFECT OF POSITION ON THE MAGNETIC MOMENT INDUCED IN THE PARTICLE

Distance from center of particle to center of symmetry of magnet δ, expressed as fractions of the half-polar length of the magnet	Φ'	Relative force of attraction, $\Phi \cdot \Phi'$
0.000	1.0000	0.000
0.001	1.0000	0.012
0.01	1.0003	0.12
0.1	1.03	1.29
0.3	1.32	6.6
0.5	2.22	33.8
0.7	5.73	424.
0.8	12.7	3152.
1.0	∞	∞

taken into account. The enormous variation in relative attraction with position of particle is well brought out.

Equation [XVIII.10] brings out a somewhat unexpected result, *viz.*, that for a given magnetic moment M of the separator magnet, and all other things being equal, the force of attraction varies as the *seventh* power of the polar distance.

Practical Deviations from the Theoretical Magnetic Field. To avoid the acquisition of erroneous impressions, it is desirable to emphasize at this stage that the magnetic field on which the foregoing analysis is based can be but crudely approximated. In practice, point poles do not exist; instead, magnets have polar regions and these polar regions are not at the surface of the substances of the magnet, but somewhat inside of it: $\delta = D/L$ can rarely exceed 0.8. Since point poles are in fact replaced by polar regions, the convergence of the field, although not materially different from what it is in theory if δ is small, becomes very much smaller than the theoretical values as δ approaches its practical limit of 0.8.

It is impractical to make magnets as point polar as possible, since such an attempt would narrow the field very greatly and would induce great lateral variations in intensity and convergence of field. This practical requirement spells further deviation between practice and theory.

MAGNETIC PROPERTIES OF SUBSTANCES

Magnetic Susceptibility and Permeability. The impression may have been gained from the foregoing that magnetic susceptibility and permeability are constants characteristic of various substances. This is true provided measurements are made at a definite field intensity. But it is far from being true if measurements are made at different field intensities.

Indeed, the permeability varies enormously with the field intensity, at least in the case of ferromagnetic substances. Generally speaking the permeability starts from a fairly large value at zero field, increases to a maximum, then decreases to unity. In the case of soft iron, for example, the permeability in very weak fields may be of the order of 100; from that level it increases to 2,000 to 5,000 for fields of intermediate strength, then decreases rapidly to approach unity for very intense fields.[18,25]

Figure 217 shows a typical *B-H* curve and the corresponding μ-*H* curve for a ferromagnetic material. The *B-H* curve relates the field intensity to the induced magnetization. This curve

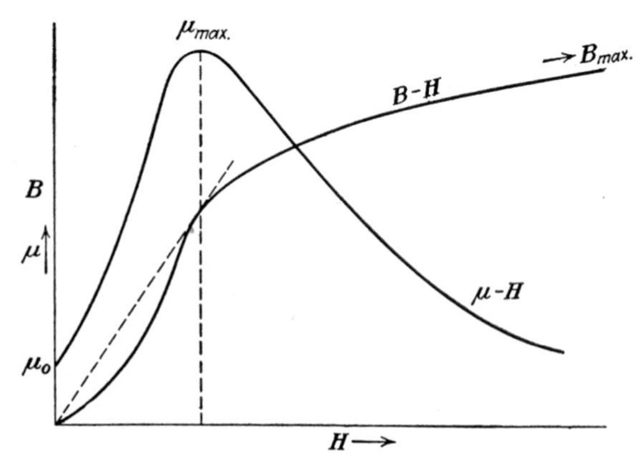

Fig. 217.—Relationship of the permeability (μ) and of the induction (B) to the field intensity (H) for ferromagnetic substances.

shows that increasing the field increases the induction B (*i.e.*, the product $B = \mu H$ of the field by the permeability) until the latter approaches a maximum value. Ferromagnetic substances behave as if they can become magnetically saturated.

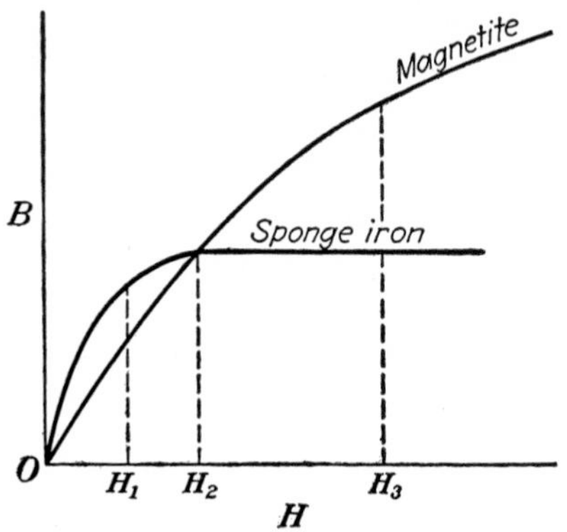

Fig. 218.—Relationship of the induction to the field strength for two ferromagnetic substances. (*After Dean and Davis.*)

The saturation value of the induction is approximately the same for all irons and steels, but the permeabilities may reach vastly different maximum values, or have vastly different initial values, depending on the heat-treatment or composition of the metal.

The intensity of magnetization (or induction) does not bear the same relationship to the field strength for all substances. That is shown by Fig. 218 from which it appears that at low field strengths (H_1) sponge iron is more magnetic than magnetite and at high field strengths (H_3) the reverse is the case.[14]

Weakly paramagnetic substances do not display such marked changes in permeability with changes in field strength as do the ferromagnetic substances. It is not known whether this is because paramagnetic substances intrinsically do not display such changes or whether the range of experimentation has not been carried to fields of adequate strength. Generally speaking, the permeability and susceptibility of weakly paramagnetic substances are regarded as independent of the field.

Magnetic Susceptibility and Permeability of Minerals. Good data on the magnetic susceptibility and permeability of minerals are well-nigh unavailable because of

1. The great influence exerted by impurities mechanically held in the mineral grains.
2. The often great influence of dissolved impurities.
3. The influence of grain size and state of crystal aggregation.
4. The difficulty of making measurements on fine particles.
5. The effect of intensity of the magnetic field in which the determinations are being made.

However, an idea of the magnetic permeability of various minerals may be had from Table 55 which is reproduced from Crane's data.[8] Table 55 also gives the relative force of attraction exerted on the minerals, 100 being arbitrarily taken as the force of attraction on iron filings. These relative forces of attraction were evaluated by Davis[11a] from Crane's data. The data of Table 55 have relative value only since the determinations were not made at constant field or at constant polar distance.

The great variation in magnetic susceptibility and in force of attraction from ferromagnetic minerals to nonferromagnetic minerals makes it plain that the occlusion or adhesion of even a minute particle of iron or magnetite to a relative large particle of substantially nonmagnetic material may affect considerably its response to a magnetic field. The effect of dissolved impurities is also large, but the way in which they are effective is not

TABLE 55.—MAGNETIC PERMEABILITY AND RELATIVE ATTRACTABILITY OF VARIOUS MINERALS

Material	Permeability (after Crane)	Force of attraction (after Davis), iron = 100
Iron	2.16	100.00
Magnetite	1.47	40.18
Franklinite	1.41	35.38
Ilmenite	1.28	24.70
Pyrrhotite	1.078	6.69
Siderite	1.022	1.82
Hematite	1.008 to 1.024	1.32
Zircon	1.002 to 1.029	1.01
Limonite	1.0088 to 1.0099	0.84
Corundum	1.0018 to 1.025	0.83
Pyrolusite	1.0078 to 1.0088	0.71
Manganite	1.0061	0.52
Garnet	1.0047	0.40
Quartz	1.0022 to 1.0055	0.37
Rutile	1.0030 to 1.0053	0.37
Pyrite	1.0007 to 1.0064	0.23
Sphalerite	1.0007 to 1.0057	0.23
Dolomite	1.0015 to 1.0056	0.22
Apatite	1.0026	0.21
Willemite	1.0024	0.21
Talc	1.0008	0.15
Arsenopyrite	1.0017	0.15
Chalcopyrite	1.0016	0.14
Gypsum	1.0005 to 1.0033	0.12
Fluorite	1.0010 to 1.0017	0.11
Zincite	1.0012	0.10
Orthoclase	1.0001 to 1.0011	0.05
Calcite	1.0004	0.03

altogether clear. An example is the influence of dissolved iron on the magnetic properties of ferruginous sphalerite (marmatite).

Variations in magnetic susceptibility have been found to result from variations in particle size. It is, however, not agreed whether these variations are due to size per se or to some chemical change, *e.g.*, to surface oxidation, *i.e.*, a function of surface exposed, hence of size. The variations are small. As a first approximation, they can be neglected.

Hysteresis, Remanence, and Coercive Force. If a ferromagnetic substance is acted upon by a magnetic field H_x, and if this field is then withdrawn, the substance retains a certain magnetism The residual magnetic intensity is known as the *remanence* B_r. To overcome it, it is necessary to impress a certain field $-H_c$ having a direction opposite to that of the original field. This field $-H_c$ is known as the *coercive force* of the substance tested.

Upon application of a field of intensity H_x of sign opposite to that of the initial field, the substance acquires an induction $-B_y$ equal in magnitude and of opposite sign to B_y. If the field is again withdrawn, the substance has remanence $-B_r$; and to overcome this remanence, a coercive force $+H_c$ is required (see Fig. 219).

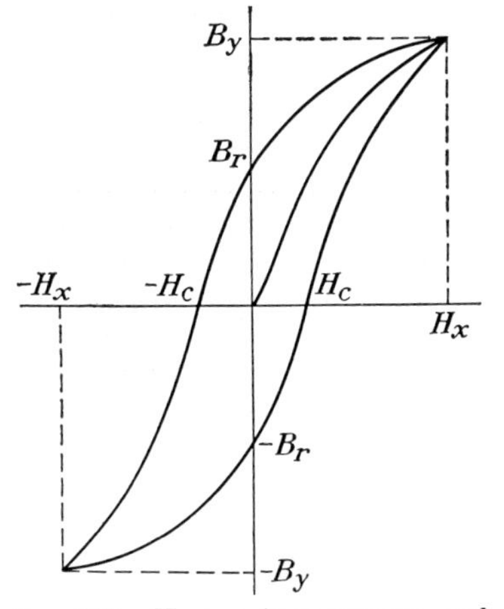

Fig. 219.—Hysteresis, remanence, and coercive force.

The closed curve obtained by connecting B and H values for a full cycle of variation of H from $+H_x$ to $-H_x$ and back to $+H_x$ is called a hysteresis loop. The area within the loop is proportional to the energy dissipated as heat when a ferromagnetic substance is put through a magnetic cycle; this is also known as the hysteresis loss.

Great variations in remanence and in coercive force occur for various substances. Some experimental results, according to Dean and Davis,(14) are presented in Table 56. One of the outstanding relationships is the direct proportion of mineral surface (external surface plus intercrystalline surface) and coercive force.(19) This relationship has afforded an additional proof of the correctness of the Rittinger law of crushing and grinding(13); it may be also of some commercial use.(15)

Hysteresis effects do not play a prominent part in the usual magnetic separations. Recently, however, a process based on utilization of coercive force has been suggested. It is discussed briefly on page 452.

TABLE 56.—MAGNETIC CONSTANTS FOR MINERAL POWDERS* AT A FIELD STRENGTH H OF 2,000 OERSTEDS
(*After Dean and Davis*)

Substances	Intensity of magnetization, $I_{2,000}$	Susceptibility, $\kappa_{2,000}$	Coercive force, H_c	Remanence, B_r
Tool steel, solid	1,200	0.600	12	874
Tool steel, 60-mesh cuttings	214	0.011	12	20
Magnetite, crystal	560	0.280	2	4
Magnetite, 100 per cent, powder	222	0.111	6	56
Magnetite, pure, −325-mesh	199	0.100	49	24
Mineville magnetite	200	0.100	19	9
Mineville magnetite, grains oriented	220	0.110	19	13
Reduced hematite	72	0.036	400	34
Lodestone, California	92	0.046	269	39.1
Lodestone, titaniferous	137	0.069	142	39.0
Lepidocrocite, dehydrated	33	0.017	43	6.8
Lepidocrocite, reduced	100	0.050	88	10.3
Ilmenite, treated	4.8	0.0024	335	1.8
Pyrrhotite	4.6	0.0023	220	1.9

* Powders at a packing density of 2.5–2.8 g. per cc. and −60 + 250-mesh (except one designated as −325-mesh).

Tool steel, solid and as hack-saw cuttings, and pure crystallized magnetite have been included for comparison.

Materials Usually Separable by Magnetic Separation. General use is made of magnetic means for the removal of "tramp" iron, and for the concentration of magnetite. Magnetic separation is also employed for the concentration of iron minerals rendered ferromagnetic by conversion to magnetite, *e.g.*, of roasted hematite, goethite, limonite, and siderite, and for the concentration of pyrrhotite, siderite, marmatite, franklinite, spinel, ilmenite, chromite, garnet, wolframite, monazite, rutile, and manganese minerals. It is also widely used for the removal of minor quantities of iron or ferruginous minerals from ceramic raw materials.

With a few exceptions,(17) the magnetic property is traceable to the presence of iron either as such, as magnetite, or in some other ferromagnetic form. The reason for this limitation is readily apparent from the wide spread in magnetic permeability and attractability of ferromagnetic and nonferromagnetic minerals. In spite of the preeminence of iron, it is well to keep in mind that a few other elements are notably paramagnetic.[41a] These elements are nickel, cobalt, manganese, chromium, cerium, titanium, oxygen, and the platinum metals. Many of the compounds of these elements are also markedly paramagnetic.

Effect of Magnetized Particles on Each Other. Introduction in a magnetic field of a particle of appreciably greater permeability than the fluid in the field facilitates the passage of lines of force through the particle. Toward a second susceptible particle, the first particle acts as a magnet, since lines of force converge toward it, as toward the poles of the primary magnet, and vice versa. This is well shown by some photographs by Davis.[11b] The net effect is the development of forces of attraction from one particle to another, as of the primary magnet for each particle. In this connection, it may be useful to think of the magnetic field as being puckered by magnetized particles just as the surface of water is puckered by floating particles, with the result that the magnetized particles move about until they come together just as the floating particles slide downhill toward each other until they have come together.

The aggregation or flocculation of magnetically susceptible minerals is observed particularly well if the particles are small, if the permeability is high, and if the field is intense. The effect is also observed on minerals having a substantial remanence or a large coercive force even in the absence of any other magnetic field than that of the Earth. It is seen, for example, in the elutriation of fine mill products containing magnetite or pyrrhotite.

The separation of "magnetic" from "nonmagnetic" minerals may then be either a separation of individual particles from each other or of aggregates of magnetized particles from nonaggregated nonmagnetized particles, depending upon the size and specific characteristics of the minerals. This phenomenon is of great practical importance.

Magnetic Separation of a Moving Stream of Particles. If the operation of a magnetic separator is to be continuous, the process must be carried out on a moving stream of particles passing into and through a magnetic field. The duration of application of the field is therefore limited to a short time, usually a fraction of one second. Furthermore, the intensity of the field varies during this short period in which it is applied. Movement of a particle within the field is then controlled not only by its magnetic response, but also by its momentum as it is crossing the field. This momentum depends on the apparent specific gravity of the particle, upon its speed, and upon its volume. Since the force of attraction is also proportional to the volume, the effective factors in controlling particle response are magnetic susceptibility, speed, and specific gravity. By reducing the rate of passage of the particles, it is possible, within limits, to affect less susceptible particles. Thus[20a,28] by reducing the speed of passage of a mixture of magnetite, rhodonite, and marmatite from 300 to 150 and 100 ft. per min. in a Mechernich separator, removal of the three minerals was obtained, in the order stated.

MAGNETIC SEPARATORS

Elements of Design of Magnetic Separators. The following elements seem essential to the design of a successful magnetic separator; if they are kept in mind, the apparently bewildering variety[14,20,28,29,37,41,44,45] of devices that have been built will become classified into a few simple forms. The important elements of design of a magnetic separator are the following:

1. Production of a suitably converging magnetic field.
2. Easy regulation of the intensity of the field.
3. Even feeding of ore particles as a stream or sheet.
4. Control of speed of passage of ore through the magnetic field.
5. Avoidance and/or correction of occlusion of nonmagnetic material between or within magnetic flocs.
6. Provision of suitable means for disposing of separated products.
7. Provision for production of a middling.
8. Elimination, or reduction to a minimum, of moving (wearing) parts.

It has already been indicated that the field of a magnet consisting of two point poles (Fig. 220*a*) is strongly convergent near the poles and nearly uniform at the equator. If a magnet is used, that resembles this theoretical type, the field near each pole, only, is suitable A field consisting of two flat poles near each other (Fig. 220*b*) gives a nearly uniform field and is not suitable. On the other hand, a field of one flat and one pointed or beveled pole (Fig. 220*c*) gives a suitable converging field.

If one pole is made in laminations alternately of iron and of a nonmagnetic material, such as wood, the field in the immediate vicinity of the laminated pole is strongly converging (Fig. 221). This type of magnet is frequently utilized, especially in the form of a laminated drum.

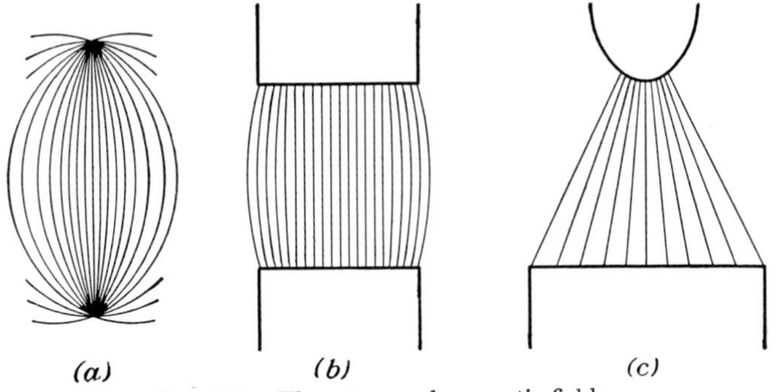

FIG. 220.—Three types of magnetic fields.

Regulation of the intensity of the field requires the use of electromagnets. In addition, control of the field is obtained by provision for ready adjustment of the polar distance.

Uniform feeding in the case of dry separators means that the feed must indeed be dry and not damp, and that a suitable feeding gate and bin are associated with, or are an integral part of, the separator.

Control of speed of passage of the particles in the field excludes free fall of particles as a suitable means, since that is not subject to control. Belt or conveyor-passing devices have the edge, but drum-passing devices are also widely used.

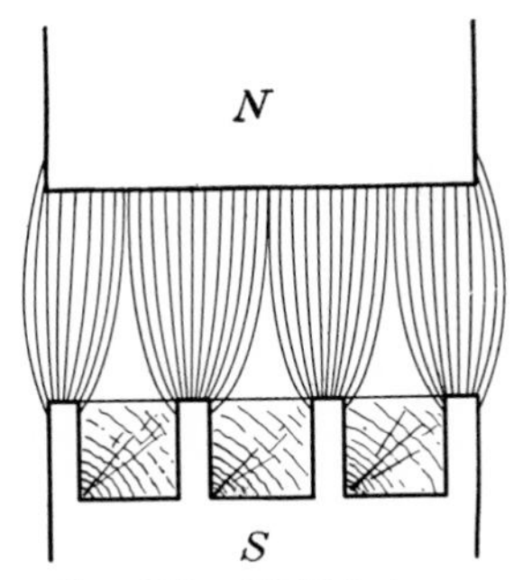

FIG. 221.—Field between a block pole-piece and a laminated pole-piece.

Occlusion of nonmagnetic material within magnetic flocs is especially serious in dry separators working on fine, highly magnetic material. It seems to be caused in part by a sudden rush of the highly magnetic material toward the zone in the field that has the highest density, and in part by the instant formation of chains and flocs of magnetized particles which occlude and entrap nonmagnetized particles. If the ore could be fed as a layer one particle deep, this occlusion would be much less noticeable, but such practice is clearly impossible on fine material, if the capacity of the machine is to be reasonably large.

If the magnetic material has low coercive force (*e.g.*, natural magnetite), the occlusion may be reduced by compelling the magnetized material to pass alternately from a region where it is attracted to a north magnetic pole to a region where it is attracted to a south magnetic pole, etc. The ensuing flip-flops of the magnetized particles result in a rearrangement of the flocs and in a reduced detention of the nonmagnetic occlusions. That, however, is not readily achieved if the magnetized particles have a high coercive force; in that case it is necessary, as C. W. Davis has pointed out, to break the d-c field in such a way that a damped oscillatory discharge is set up through the magnet windings.

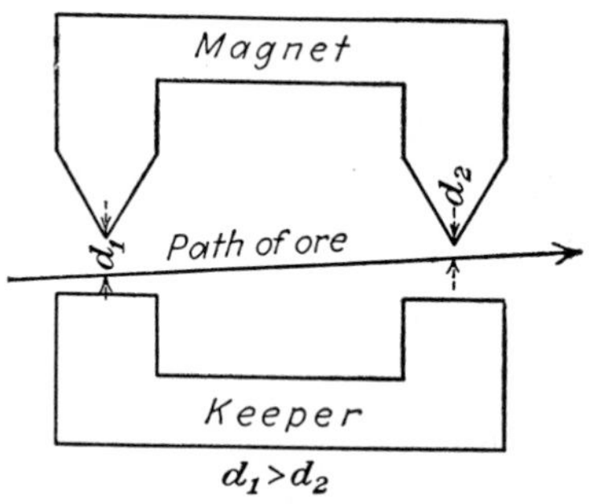

FIG. 222.—Diagram of a magnetic separator designed to make a concentrate, a middling, and a tailing.

Occlusion is not important in dealing with well-sized coarse material. This is because (1) the sheets or beds of material are but one or two particles deep, (2) coarse particles cannot plug openings in flocs in the effective way in which fine particles plug them, and (3) the would-be occlusions are heavy.

In wet separators, there is no great rush of magnetically susceptible particles toward the region of denser magnetic field because of the relatively great resistance to motion of the particles that is exerted by the water. Occlusion of nonmagnetic particles in magnetic flocs is therefore less serious than in dry separators. If occlusion within flocs does occur, it is overcome by reversing the polarity of the field from place to place.

Disposal of the nonmagnetic product is simple: either free fall onto a belt or conveyance on the feed belt is suitable. The magnetic product, on the other hand, must be detached from the magnet. Rather, it is preferable to so design the separator that the magnetic product does not stick to the magnet but to some pulley or belt, which by its motion readily places that product beyond the effective field. and thus drops it.

A middling is readily obtained by using a more intense field after removal of the concentrate. This is facilitated by the use of a "keeper" to close the magnetic circuit. In Fig. 222, for example, the field at d_2 is denser than at d_1 as well as more strongly convergent. As a result, a concentrate is deflected from the ore path at d_1 and a middling at d_2.

Electromagnets. Because of their convenience, temporary electromagnets rather than permanent magnets are used exclusively for magnetic separation.* Electromagnets consist of a coil of wire wound around a core of permeable material, *e.g.*, soft iron.

Reduced to its simplest element, an electromagnet consists of a single loop of coil in air. At the center of the loop, the field is $2\pi i/10r$. In this relationship, r is the radius of the coil and i the intensity of the current, expressed in amperes. (The factor 10 in the denominator is required to convert absolute electromagnetic units to practical electromagnetic units.)

By suitable integration,(9a) it is found that for a flat coil of relatively large radius consisting of N turns the field is

$$H = \frac{2\pi iN}{10r}$$

and for a long coil of small radius containing $n = N/l$ turns per centimeter,

$$H = \frac{4\pi in}{10} = \frac{4\pi iN}{10l}.$$

The latter formula is true only at the center of a solenoidal coil of great length. At the ends, H is only half as large.

If a permeable core is added to the solenoidal coil, the field within the core is much increased over what it would be in air. In the case of a uniformly wound ring solenoid, or torus, the flux density (*i.e.*, the induction) at the center of the core is

$$H = \frac{4\pi in\mu}{10} = \frac{4\pi iN}{10} \cdot \frac{\mu}{l}, \qquad \text{[XVIII.12]}$$

μ being the permeability of the core substance.

* But some such alloys as "alnico" may come into use in the near future for the making of permanent magnets (V. H. Gottschalk, private communication).

If the core is wound over but a fraction of its length, H can still be determined by the preceding formula, using the second form, $4\pi iN/10 \cdot \mu/l$. And if the core consists of several sections of lengths l_1, l_2, l_3, . . . , permeabilities μ_1, μ_2, μ_3, . . . , and cross-sectional areas S_1, S_2, S_3, . . .

$$H_1 = \frac{4\pi iN}{10S_1} \cdot \frac{1}{\dfrac{l_1}{\mu_1 S_1} + \dfrac{l_2}{\mu_2 S_2} + \dfrac{l_3}{\mu_3 S_3} + \cdots} \qquad \text{[XVIII.13]}$$

S_1 being the cross-sectional area of the magnetic circuit in the part which is wound.

This relationship shows that the strength of the magnetic field is proportional to iN, the ampere-turns, or product of the intensity of the exciting current by the number of turns in the winding. The magnetic moment M of the electromagnet is therefore also proportional to iN, and the force of attraction, Eq. [XVIII.10], is proportional to $(iN)^2$.

The induction for a current of given intensity decreases rapidly with increase in air gap (term in $l_2/\mu_2 S_2$ in Eq. [XVIII.13]); hence the magnetic moment decreases with increase in the air gap, and the force of attraction decreases even more rapidly with increase in the air gap.

To get a suitable field at the lowest operating cost, the following recommendations are generally in order:

1. Reduce air gaps to a minimum.

2. Reduce the reluctance $\sum \frac{l}{\mu S}$ of the metallic part of magnetic circuit to a minimum by using highly permeable iron or alloys and by utilizing a large cross section and a short length.

3. Make the permeability of the core (or in the case of extremely powerful electromagnets the saturation magnetization) as high as possible.

4. Secure the necessary ampere-turns by using many turns and a small current. A limit is set on this by the resistance of the wire, and by heating within the wire.

Classification of Magnetic Separators. Magnetic separators can be classified according to the medium in which the separation is made, the mode of presentation of the feed, the mode of disposal of the products, and whether the magnets are stationary or moving.

Table 57.—Classification of Some Well-known Magnetic Separators (For descriptions see Taggart, "Handbook of Ore Dressing"; Gunther, "Electromagnetic Ore Separation"; Truscott, "Textbook of Ore Dressing")

Medium	Presentation of feed	Disposal of magnetic product by	Magnets	Type
I. Air	1. Gravity fall	Gravity fall	Stationary	Edison
	2. Flow on chute	Gravity fall	Moving	Mechernich
			Stationary	Motor
				Ubaldi
				Humboldt ring
	3. Flow around drum or roll	Gravity fall	Moving	Pulley
			Stationary	Dings; Stearns
				Ball-Norton (drum)
				International
				Wetherill F
				Wenstrom
				Heberle
				Primosigh
				Payne
	4. Horizontal belt	Gravity fall	Stationary	Humboldt-Wetherill VI
				Rapid
		Mechanical scraping	Moving	Cleveland-Knowles
		Longitudinal belt	Stationary	Ball-Norton (belt)
		Cross belt	Stationary	Ferraris
				Wetherill-Rowand
	5. Shaking tray (or belt)	Gravity fall	Stationary	Dings tray
				Ullrich
		Longitudinal belt	Stationary	Knowles magnetic belt
II. Water	1. Agitated pulp	Scraping conveyors	Stationary	Davis magnetic log washer
		Wash sprays	Moving	Stern
	2. Launder	Gravity	Stationary	Ullrich
		Wash sprays	Moving	Ericksson
	3. Vertically flowing pulp	Gravity	Stationary	Heberle
				Forsgren
		Wash sprays	Moving	Dellvik-Groendal
	4. Horizontally flowing pulp	Gravity	Stationary	Groendal
		Differential settling of magnetic flocs		Groendal slime
		Longitudinal belt		Crockett
	5. Round table	Wash sprays	Stationary	Froeding
				Leuschner
	6. Shaking table	Cross-water wash	Stationary	Weatherby
		Gravity	Stationary	Wiser-Chino
	7. Vanner belt	Belt speed	Stationary	Dings-Roche
				Stearns

Table 57 presents such a classification.

Of magnetic separators, the following will be described briefly, as typical: Edison gravity-fall separator, Ball-Norton drum separator, Dings induced-roll separator, Wetherill-Rowand cross-belt separator, Davis magnetic log washer, Groendal wet separator, and Dings-Roche magnetic vanner.

Edison Separator.[7,20b] Reduced to one of its elements, the Edison separator consists of a bar magnet. The ore as thin streams falls in front of the poles, susceptible particles being

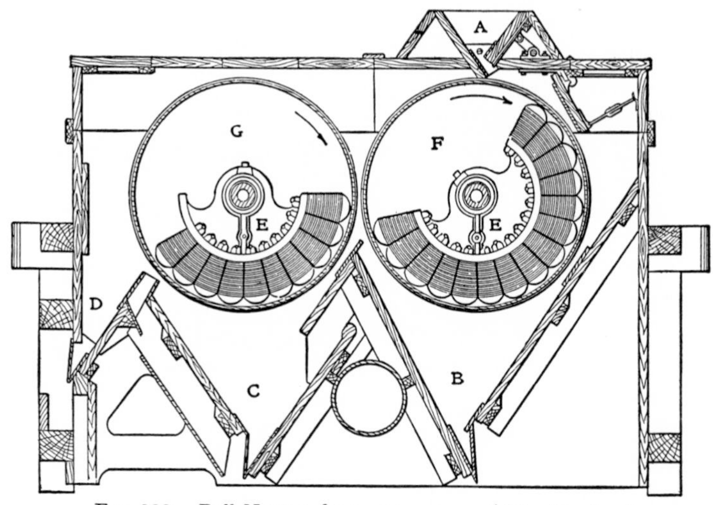

FIG. 223.—Ball-Norton drum separator. (*After Gunther.*)

deflected inwards and nonsusceptible particles continuing to fall, undeflected.

Actually, the separator consists of a series of such magnets one above the other, arranged in order of downwardly increasing strength for roughing work and of downwardly decreasing strength for cleaning.

This separator was not wholly successful because of inability to control flow of solids in a thin sheet, and because of lack of control over the speed of the falling particles.

Ball-Norton Drum Separator.[4,5,20] This separator consists of one or two rotating drums of nonmagnetic metal, surrounding a number of fixed magnets arranged so that consecutive poles are of opposite sign (Fig. 223). Much of the magnetic field passes directly from one pole to the other, inside of the drums, and is thereby wasted, but enough comes out through the drums to attract and hold strongly magnetic particles until the rotation of the drums brings the magnetic particles opposite the place where there are no magnets. The particles are then released and drop in another compartment. In the two-drum machine, the second drum revolves at higher speed and has weaker magnets. This permits separation of an ore into three products, concentrate at *D*, middling at *C*, and tailing at *B*.

Dings Induced-roll Separator.* This, one of the better-known induced-roll separators, consists (Fig. 224) of a horseshoe magnet *A* faced by an iron keeper *B* and of two rolls *C*, one opposite each pole. The magnetic circuit is thus completely in iron except for the very small clearance between the rolls and keeper and for the gaps between the rolls and poles.

The rolls are laminated, *i.e.*, they consist of alternate laminae of permeable and impermeable material. As a result, the magnetic field is strongly convergent toward the roll at the place where the roll is in closest proximity to the primary pole. The roll becomes an assemblage of a large number of secondary poles. The strength of these secondary poles varies with time as the roll revolves, being nil twice per revolution.

As the ore passes over a roll, the susceptible particles are drawn to the laminated roll from which they fall only when they are in a position such that the strength of the adjoining secondary pole is nil, *i.e.*, they fall later than nonsusceptible particles.

If the separator is used to separate feebly magnetic from nonmagnetic material, a scalping roll is used, as shown in the cut at *D*. This is obtained by merely extending the first primary pole so as to have a branch or auxiliary primary pole. A scalping roll permits primary removal at *D* of the most strongly magnetic particles, such as of abraded iron, magnetite, etc.

The forces involved in making the separation are not only magnetic force but also that of gravity, centrifugal force, and the friction of the air against the particles. The last is especially important in that it makes particles of different sizes behave differently. Although not absolutely essential, close sizing is desirable especially if the spread in magnetic susceptibilities is

* The Stearns Magnetic Manufacturing Company also makes an induced-roll separator.

not large (*e.g.*, as between locked particles high in magnetite content and locked particles low in magnetite content).

This machine is especially suitable for the separation of granular to coarse material of medium to low susceptibility. It is being used successfully on material as low in susceptibility as mica and manganese dioxide.

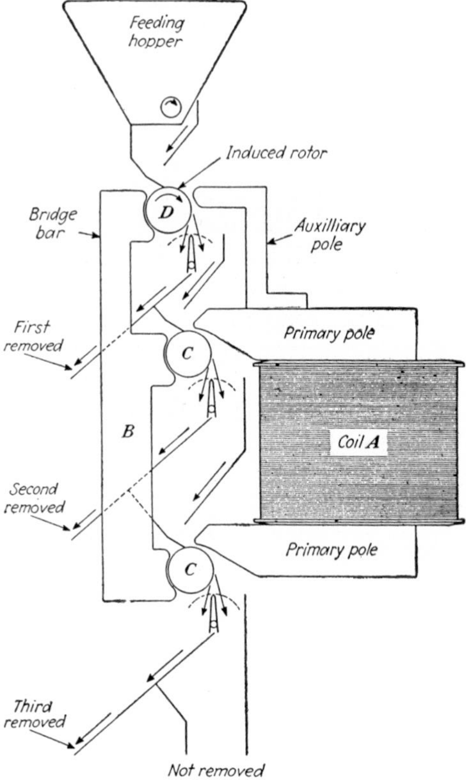

FIG. 224.—Dings induced-roll separator. (*Dings Magnetic Separator Co.*)

Wetherill-Rowand Separator.(47) This machine consists of one or more pairs of powerful horseshoe electromagnets arranged with poles of opposite sign in apposition, one electromagnet being below and the other above the feed belt, and the air gap being small (Fig. 225). This produces, then, a strong magnetic circuit. Convergence of the field is obtained by pointing or rounding the poles of the upper magnet while making the poles of the lower magnet flat. The ore passes on a longitudinal belt between the poles; crossbelts, also termed take-off belts, prevent the susceptible particles from sticking to the upper poles as they jump from the lower belt to the region of denser magnetic field.

In this type of separator, magnetic force is opposed by gravity. Separation depends not only on the magnetic variables, such as field intensity, shape of pole pieces, distance apart of pole pieces, but also on the specific gravity of the particles and on the speed of the feed belt.

Fig. 225.—Diagrammatic illustration of operation of Wetherill-Rowand crossbelt separator. (*Dings Magnetic Separator Co.*)

This separator has been employed with success in the concentration of the ores from Franklin Furnace, N. J., containing franklinite (the zinc ferrate), zinkite (the zinc oxide), and willemite (the zinc silicate), together with garnets, calcite, magnetite, and other minerals.

Magnetic Log Washer.(46) The Davis magnetic washer consists of a log washer or of a spiral-ribbon classifier resting over magnets. The pole pieces are so shaped and disposed that the pulp is in the magnetic field. Magnetite, to which this device has been applied, becomes magnetically flocculated, sinks and is carried up by the blades of the log or by the spiral ribbon. Davis states that the finer the ore is ground the better the results.

Taggart(44a) reports excellent results with this machine, as well as low cost and little wear, provided the feed is −48-mesh

Groendal Wet Separator.[20] This machine (Fig. 226) consists of a drum of nonmagnetic material enclosing a number of magnets (6) arranged radially and with alternate polarity across a sector of the drum, much as in the Ball-Norton cobber. This drum is placed over a tank made in two compartments; pulp flows from compartment (2) to compartment (5). The drum revolves so that, as it dips in the pulp, it proceeds from (2) to (5). New pulp is fed by launder at (1), and feed water is forced up at

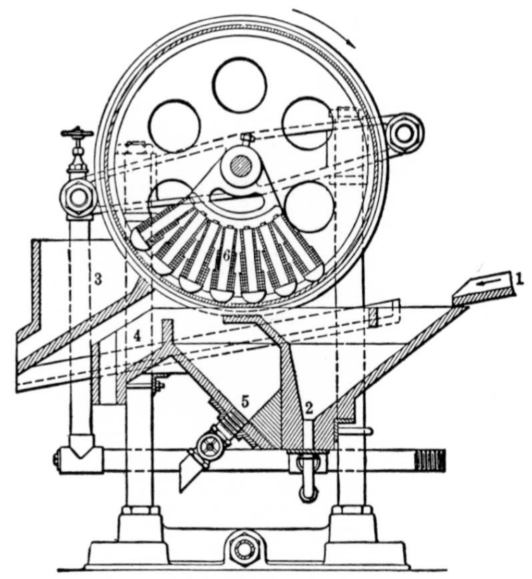

Fig. 226.—Groendal wet magnetic separator. (*After Gunther.*)

(2). Tailing is drawn at (5), middling at (4), and concentrate at (3).

Groendal separators have been used extensively in Europe.

Dings-Roche Magnetic Vanner.[39,40] This is essentially a steeply sloping vanner in which the belt travels against the flow of pulp, carrying up with it magnetic particles. The space under the belt is occupied by a bank of waterproofed and ventilated magnets arranged with poles of alternate polarity.

This machine has given good results in the treatment of the New Jersey magnetites.

Crockett (Dings Type K) Wet Magnetic Separator.[3] This device consists of a compartmented tank arranged to receive pulp at one end (right, Fig. 227), to receive water under pressure in the first compartment, and to discharge tailing, middling, water, and concentrate from the second, third, fourth, and fifth compartments, respectively. Dipping in the tank there is an endless belt that travels up against a bank of magnets. A 20-pole bank of magnets is used for that purpose. The magnetic particles are drawn upward against the belt which carries them to the concentrate discharge (left, Fig. 227). The nonmagnetic material is discharged in the second, or tailing, compartment. The material of intermediate magnetic property is dropped in the middling compartment (by the "winnowing"

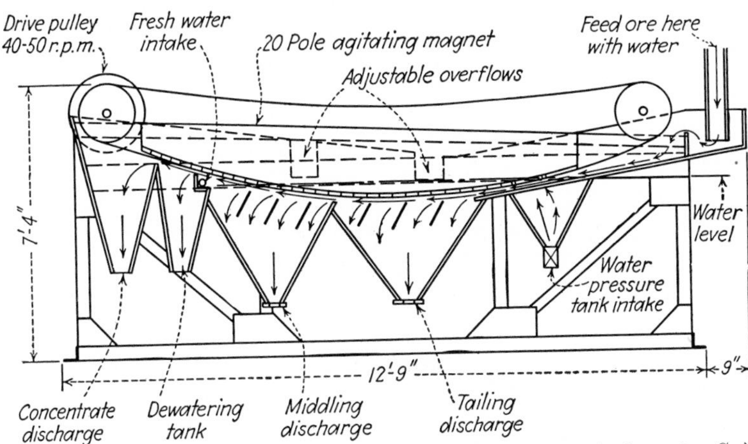

FIG. 227.—Crockett wet magnetic separator. (*Dings Magnetic Separator Co.*)

action of specially designed alternating-pole magnets and directed water currents). This machine is one of the latest to be introduced. It has given excellent results.

Demagnetizer. After subjection to a magnetic field and withdrawal from the field, susceptible minerals retain a certain residual magnetism. Residual magnetism in rougher concentrates or in middlings is objectionable since it prevents rearrangement of the particles and exclusion of mechanically entrapped nonmagnetic particles. A demagnetizer is required to eliminate permanent magnetism.

If the residually magnetized body is sufficiently long, the strength of that body, as a permanent magnet, is determined by the remanence of its substance. This is not true of small particles. In that case, the retention of magnetism is almost entirely a matter of coercive force. Substances that have high remanence and low coercive force fall into another group.

One type of demagnetizer has been proposed by E. W. Davis.[11c] It is similar to that used for demagnetizing watches. It consists of an a-c coil through which the dry ore or the pulp is made to flow. This coil has a gradually reducing diameter so that the alternating magnetic field to which the ore is exposed has a gradually reducing intensity (Fig. 228). The alternation in the direction of the field together with the reduction in the

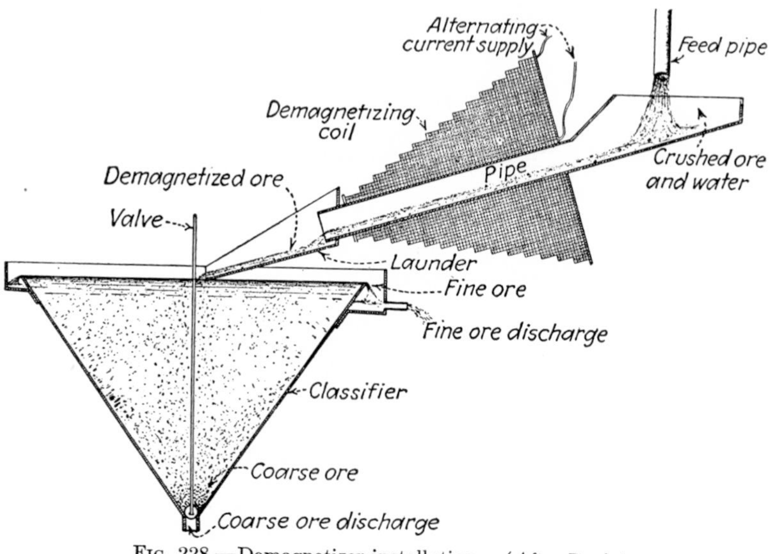

FIG. 228.—Demagnetizer installation. (*After Davis.*)

intensity of the field is effective in wearing the permanent magnetism out of the particles.

This type of demagnetizer works well on magnetites that have low coercive force, *i.e.*, on relatively coarse magnetites, but it does not work well on artificial magnetites (reduced hematites), which are very finely crystallized and have a very high coercive force. According to C. W. Davis, the only really effective means of demagnetizing material having a high coercive force is to use a damped oscillatory field of fairly high frequency.

Dean-Davis Coercive-force Magnetic Separator.[14] This separator operates on a principle different from that of other separators. It utilizes the remanence and coercive force of magnetic minerals to alternately attract and repel particles of high coercive force, thereby causing rapid vibration of particles of high coercive force, but no vibration of particles of low coercive force, even if highly permeable. On an inclined surface,

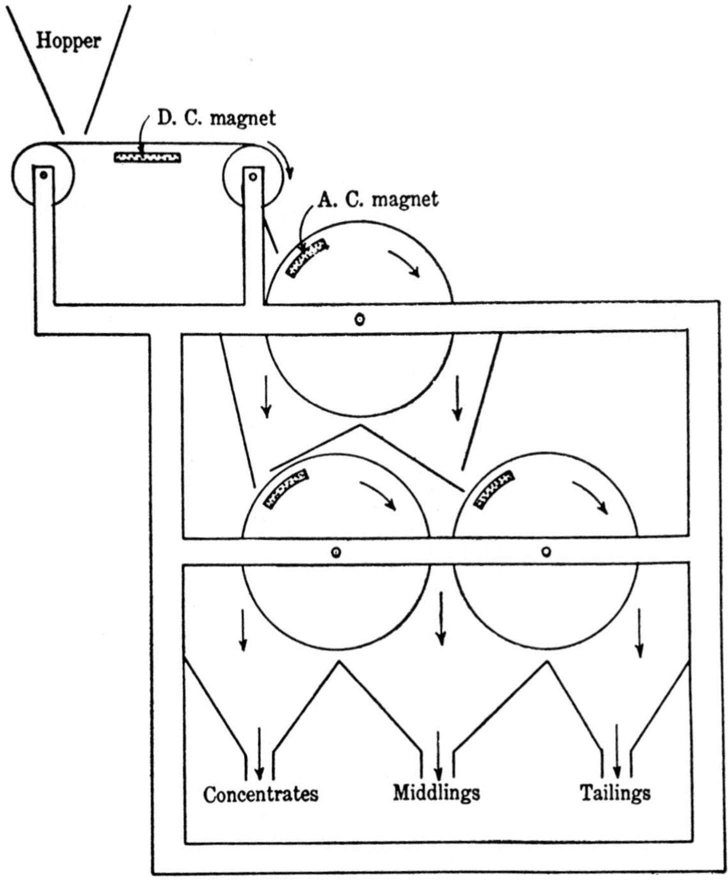

FIG. 229.—Schematic view of coercive-force magnetic separator. (*After Dean and Davis.*)

particles having high coercive force are caused to slide, but particles having low coercive force do not slide. The separation requires a field smaller than the coercive force of the material of high coercive force.

The ore is first activated by being magnetized in a d-c field. If B_r is the remanence and H_c is the coercive force (see Fig. 219), the maximum repelling force obtainable by adjustment of the a-c magnetic field to the optimum point is approximately $(B_r \cdot H_c)/4$. The Dean-Davis separator (Fig. 229) consists of one d-c magnet for excitation and one or more a-c magnets for separation. If roughing only is used, one a-c magnet is used for separation. If both the rougher concentrate and the rougher tailing are cleaned, the separator requires three a-c magnets as shown in the illustration.

Hysteretic Repulsion. Shilling and Johnson[42] have made some experiments on the separation of hematite by hysteretic repulsion. Dean and Davis have pointed out the differences between separation by hysteretic repulsion and by coercive-force alternation.[42, *discussion*]

Fig. 229*A*.—Magnetic pulley for the protection of crushing machinery. (*Dings Magnetic Separator Co.*)

Magnetic Pulley. Figure 229*A* is an illustration of a common type of magnet that is used to protect dressing machinery against breakage by tramp iron.

APPLICATIONS OF MAGNETIC SEPARATION

Magnetism is used for five distinct types of separation:

1. For the removal of tramp iron in coarse and intermediate-crushing circuits, as a protection to the crushing machinery.

2. For the concentration of magnetite ores.

3. For the concentration of iron ores other than of magnetite, after preliminary conversion of the iron minerals to artificial magnetite by suitable roasting.

4. For the removal of small quantities of iron or iron minerals from ceramic raw materials.

5. For the concentration of slightly paramagnetic minerals such as those of manganese, tungsten, chromium, tin, zinc, titanium, cerium, etc.

For the first type of separation, magnets of the inverted-mushroom type are suspended above conveyor belts. Iron nails, hammers, blocks, etc., adhere to the magnets and are intermittently removed by hand. Use of this protection against breakage of coarse-crushing machinery is almost universal in large-scale operations. Magnetic head pulleys for conveyor belts are also widely used. They make removal of iron materials a continuous process.

For the next two types of separation and in connection with some ceramic operations, low-intensity magnetic separators are used.

For the concentration of slightly paramagnetic minerals and in connection with other ceramic operations, high-intensity separators are used.

Concentration of magnetite ores is the principal application of magnetism to mineral dressing. Interest in this subject has been widespread for well over 50 years, and excellent results are now obtained.

Magnetite Ores.[26,33,34,40] The principal aims in the concentration of magnetite ores are to increase the iron content of the concentrate and to reduce the sulphur and the phosphorus.[1,27,31,38] Sulphur is usually present as pyrite and chalcopyrite, phosphorus as apatite. Some of the iron is usually present in nonmagnetic form, *e.g.*, as a silicate or ferric oxide, and must be lost. Proper technologic evaluation of the operation is then on the basis of the content of magnetic iron in the ore, concentrate, and tailing. Almost any degree of perfection of separation is possible, provided the liberation is adequate, because of the enormous spread in magnetic properties of the magnetite on the one hand, and all the other minerals on the other.*

Finely divided iron concentrates must be sintered, and sintering is a relatively expensive operation. (The cost of sintering

* That, however, may not be true if titanium is present, usually in the form of ilmenite.

consists largely of the cost of fuel. The most reliable estimates place this cost at from $0.50 per ton for 6-mesh concentrate to $1.25 per ton for −65-mesh concentrate.) Accordingly, it is often preferable to make a poor separation by concentrating with dry cobbers at a coarse size than to make a much better separation by concentrating with wet separators at a fine size. If much sulphur or phosphorus is present, wet magnetic separation is almost compulsory in order to reject a sufficient amount of these objectionable impurities. From ore containing 30 to 45 per cent magnetic iron, a concentrate with 60 to 68 per cent magnetic iron and tailing with 1 to 4 per cent magnetic iron can usually be obtained. If copper (as chalcopyrite) and phosphorus are present, the tailing from the magnetic separator may have some value. In that case, concentration of the magnetic tailing by flotation or by gravity and flotation may have interesting possibilities.

Other Iron Ores. Siderite, hematite, goethite, and limonite are usually considered nonferromagnetic. By controlled oxidizing roasting in the case of siderite or controlled reducing roasting in the case of hematite, goethite, or limonite, the iron mineral is converted into magnetite pseudomorphic after the parent mineral.[2,10,12,36] According to the best figures, the cost of roasting hematite ores is of the order of magnitude of 65 cts. per ton of feed, of which some 85 per cent is for fuel.

Artificial magnetite has as high a permeability as natural magnetite. In magnetic separators, it responds therefore just as readily as natural magnetite. But artificial magnetite has a high coercive force. This makes it difficult to clean it in a d-c separator; on the other hand, artificial magnetite works well in an a-c separator, whereas natural magnetite does not.

The roasted ore is much more friable than the natural ore, indicating a much-reduced cost of crushing and grinding and a greater tendency for fracturing to take place along grain boundaries. Artificial magnetite has also one advantage over natural magnetite: it is porous and reduction in the blast furnace is more rapid.[1]

Because of the relative friability of artificial magnetite, use of magnetic separation after a magnetizing roast is almost certain to require sintering or briquetting of at least a substantial part of the concentrate. This is an expensive procedure (about $1 per ton) which it is desirable to avoid if at all possible.

Ceramic Raw Materials.[32,35] For the glass or porcelain industry, purity of color of the product has definite commercial value. Iron discolors ceramic products and must be reduced to a small fraction of one per cent if quality is to be attained. Metallic iron is usually removed by low-intensity wet separators after the grinding or crushing of the raw materials. Iron minerals such as biotite and hematite are usually removed by high-intensity separators operating on dry, sized feed.

Concentration of Weakly Paramagnetic Minerals. These minerals usually have a specific gravity appreciably higher than the common gangue minerals, *viz.*, quartz, calcite, dolomite, and silicates low in iron content. Considerable concentration is then possible by gravity methods at a cost that cannot be approached by magnetic methods. As a result, magnetic concentration of weakly paramagnetic minerals is usually applied not to the ore but to gravity concentrates.[21] This has the further advantage of bringing to the magnetic separators a product already deslimed. There are of course exceptions, as in the concentration of manganese oxides in which the unconcentrated but sized and dedusted ore is treated directly in magnetic separators (Philipsburg, Mont.) and in the concentration of the zinc ores from Franklin Furnace, N. J.

Typical applications include the following: separation of cassiterite from tungsten minerals (wolframite), of monazite from rutile and garnets, of garnet from hornblende, of chromite from silicates, and the elimination of magnetic minerals from corundum, barite, or other concentrates.

Formerly, magnetic separation was important for the separation of sphalerite from pyrite, of roasted sphalerite from marcasite,[22] of pyrite from cassiterite, and of sphalerite from rhodonite and garnet. For these purposes, however, magnetic separation has been supplanted by flotation.

Literature Cited

1. AGNEW, C. E.: Benefits from the Use of High-Iron Concentrates in a Blast Furnace, *Am. Inst. Mining Met. Engrs., Tech. Pub.* 956; also in *Metals Tech.*, **V-5** (1938).
2. ANON.: Experimental Roaster for the Mesabi, *Eng. Mining J.*, **135,** 445 (1934).

3. ANON.: Tests with New Wet Belt Magnetic Separator Show Interesting Results, *Eng. Mining J.*, **134,** 123–124 (1933).
4. BALL, C. M.: The Ball-Norton Electromagnetic Separator, *Trans. Am. Inst. Mining Engrs.*, **19,** 187–194 (1891).
5. BALL, CLINTON M.: The Magnetic Separation of Iron-Ore, *Trans. Am. Inst. Mining Engrs.*, **25,** 533–551 (1895).
6. BITTER, FRANCIS: "Introduction to Ferromagnetism," McGraw-Hill Book Company, Inc., New York (1937).
7. BIRKENBINE, JOHN, and THOMAS A. EDISON: The Concentration of Iron-Ore, *Trans. Am. Inst. Mining Engrs.*, **17,** 728–744 (1889).
8. CRANE, WALTER R.: Investigations of Magnetic Fields with Reference to Ore-Concentration, *Trans. Am. Inst. Mining Engrs.*, **31,** 405–446 (1901).
9. CULVER, C. A.: "Electricity and Magnetism," The Macmillan Company, New York (1930); (*a*) pp. 183–187.
10. DOAN, DONALD J.: Effect of Lattice Discontinuities on the Magnetic Properties of Magnetite, *U. S. Bur. Mines, Rept. Investigations* 3400, 65–86 (1938).
11. DAVIS, EDWARD W.: Magnetic Concentration of Iron Ore, *Univ. Minn., Mines Exp. Sta., Bull.* 9 (1921); (*a*) p. 34; (*b*) p. 23, Fig. 17; (*c*) pp. 77–80.
12. DAVIS, EDWARD W.: Magnetic Roasting of Iron-Ore, *Univ. Minn., Mines Exp. Sta., Bull.* 13 (1937).
13. DEAN, R. S.: Magnetite as a Standard Material for Measuring Grinding Efficiency, *Am. Inst. Mining Met. Engrs., Tech. Pub.* 660 (1936).
14. DEAN, R. S., and C. W. DAVIS: Magnetic Concentration of Ores, *Trans. Am. Inst. Mining Met. Engrs.*, **112,** 509–537 (1934).
15. DEAN, R. S., and C. W. DAVIS: Physical Properties of Magnetite and Its Possible Uses as an Industrial Mineral, *Am. Inst. Mining Met. Engrs., Tech. Pub.* 795; also in *Mining Tech.*, 1–2 (1937).
16. EUCKEN, A., and M. JAKOB: "Der Chemie Ingenieur. Ein Handbuch der physikalischen Arbeitsmethoden in chemischen und verwandten Industriebetrieben," Bd. I, Teil 4, "Magnetische Trennungsverfahren," by G. Stein, Akademische Verlagsgesellschaft, Leipzig (1933).
17. FRANTZ, SAMUEL GIBSON, and G. W. JARMAN, JR.: Magnetic Beneficiation of Nonmetallics, *Trans. Am. Inst. Mining Met. Engrs.*, **102,** 122–130 (1932).
18. GILBERT, NORMAN E.: "Electricity and Magnetism," The Macmillan Company, New York (1932), p. 186.
19. GOTTSCHALK, V. H.: The Coercive Force of Magnetite Powders, *U. S. Bur. Mines, Rept. Investigations* 3268, 83–90 (1935).
20. GUNTHER, C. GODFREY: "Electro-Magnetic Ore Separation," McGraw-Hill Book Company, Inc., New York (1909); (*a*) p. 10; (*b*) p. 38; (*c*) p. 24; (*d*) p. 30.
21. HEIZER, OTT F.: Concentration of Tungsten Ore by the Nevada-Massachusetts Company, *Trans. Am. Inst. Mining Met. Engrs.*, **112,** 833–840 (1934).

22. HOFMAN, H. O., and H. L. NORTON: Roasting and Magnetic Separation of a Blende-Marcasite Concentrate, *Trans. Am. Inst. Mining Engrs.*, **35**, 928–947 (1905).
23. HOLMAN, BERNARD W.: Magnetic Separation Tests, *Mining Mag.*, **37**, 73–86 (1927).
24. HOLMAN, BERNARD W.: The Theory of Magnetic Separation, *Trans. Inst. Chem. Engrs. (London)*, **6**, 26–34 (1928).
25. JACKSON, D. C., and J. P. JACKSON: "An Elementary Book on Electricity and Magnetism and Their Applications," The Macmillan Company, New York (1916), pp. 131–132.
26. KEISER, H. D.: The Forest of Dean Mine, *Eng. Mining J.*, **127**, 872–875 (1929).
27. LANGDON, N. M.: The Use of Magnetic Concentrates in the Port Henry Blast-Furnaces, *Trans. Am. Inst. Mining Engrs.*, **20**, 599–602 (1891).
28. LANGGUTH, F.: "Elektromagnetische Aufbereitung," in "Handbuch der Elektrochemie," W. Knapp, Halle (1903), p. 16.
29. LOUIS, HENRY: "The Dressing of Minerals," Longmans, Green & Company, New York (1909).
30. MASUMOTO, HAKAR: On a New Alloy "Sendust" and Its Magnetic and Electric Properties, *Kótaró Honda Anniv. Vol., Science Reports, Tôhoku Imp. Univ., Sendai, Japan*, 388–402 (1936).
31. MCCLURKIN, ROBERT: Sinter in Blast-furnace Burdens, *Trans. Am. Inst. Mining Met. Engrs.*, **100**, 47–56 (1932).
32. MCDERMID, A. J.: Feldspar and Silica Processing, *Eng. Mining J.*, **134**, 105 (1933).
33. MYNERS, T. F.: Magnetic Concentration Methods and Costs of Witherbee, Sherman & Company, Mineville, N. Y., *U. S. Bur. Mines, Information Circ.* 6624 (1932).
34. NORTON, S., and S. LEFEVRE: The Magnetic Concentration of Low-Grade Iron Ores, *Trans. Am. Inst. Mining Engrs.*, **56**, 892–916 (1917).
35. PEDDRICK, C. H., JR.: Magnetic Separation Purifies Feldspar, *Eng. Mining J.*, **130**, 613–614 (1930).
36. PHILLIPS, WILLIAM B.: Notes on Magnetization and Concentration of Iron-Ore, *Trans. Am. Inst. Mining Engrs.*, **25**, 399–423 (1895).
37. RATEL, C.: "Préparation mécanique des minerais," Dunod, Paris (1908).
38. READ, T. T., T. L. JOSEPH, and P. H. ROYSTER: The Effect of Silica in Iron-Ore on Cost of Pig Iron Production, *U. S. Bur. Mines, Rept. Investigations* 2560 (1924).
39. ROCHE, H. M., and R. E. CROCKETT: Evolution of Magnetic Milling at Scrub Oak, *Eng. Mining J.*, **134**, 241–244 (1933).
40. ROCHE, H. M., and R. E. CROCKETT: Magnetic Separation, An Up-to-Date Mill, *Eng. Mining J.*, **134**, 273–277 (1933).
41. ROUX-BRAHIC, J.: "Préparation mécanique des minerais," Dunod, Paris, (1922); (*a*) p. 609, quoting experiments by Faraday.
42. SCHILLING, E. W., and HARWICK JOHNSON: Separation of Hematite by Means of Hysteretic Repulsion, *Am. Inst. Mining Met. Engrs., Tech. Pub.*, 654 (1936).

43. Starling, S. G.: "Electricity and Magnetism," Longmans, Green & Company, New York (1921); (a) p. 234; (b) p. 267.
44. Taggart, A. F.: "Handbook of Ore Dressing," John Wiley & Sons, Inc., New York (1927); (a) pp. 930–931.
45. Truscott, S. J.: "Textbook of Ore Dressing," Macmillan & Company, Ltd., London (1923).
46. Wade, H. H.: Magnetic Log Washer in Iron-Ore Concentration, *Eng. Mining J.*, **113,** 769–771 (1922).
47. Wilkens, H. A. J., and H. B. C. Nitze: The Magnetic Separation of Non-magnetic Material, *Trans. Am. Inst. Mining Engrs.*, **26,** 351–370 (1896).
48. Williams, S. R.: "Magnetic Phenomena," McGraw-Hill Book Company, Inc., New York (1931).

CHAPTER XIX

MISCELLANEOUS PROCESSES

The processes of ore concentration that have been discussed so far depend upon the following physical properties of the particles: size, shape, specific gravity, magnetic susceptibility, and relative adherence to air and water. In addition, there are a number of other less important processes that depend on (1) color and general appearance, (2) heat properties, (3) electric properties, (4) hardness, (5) adherence to oils or mercury in preference to adherence to water.

PROCESSES DEPENDING ON COLOR AND GENERAL APPEARANCE

Hand Sorting. Hand picking, or hand sorting, is a time-honored method of ore concentration.[18,23,33] It consists in sorting the ore into pieces of various grades, choice being based on color, heft, and other features of the appearance or "feel" of each lump. Clearly, it is a laborious method suitable only where labor is cheap, reasonably industrious, and effective on simple tasks. Since a definite length of time is required to examine one piece of ore and since the number of pieces of ore per ton increases inversely as the cube of the size, the cost of the process must increase tremendously with fineness. On the other side, a limit is placed by the weight of a single piece that can be handled, and by the increased locking with coarseness. Accordingly, picking is practicable only for ore pieces and lumps exceeding 2 to 3 in. in diameter, and up to about 10 in.

The material picked out, whether ore or waste, should be that of smaller weight. Lighting should be uniform and designed to bring out the differences in appearance of the lumps to be separated. The ore is preferably washed before sorting, and must be sized sufficiently to remove the fines. More generally, the attempt should be made to make the pieces to be separated as unlike as possible. Washing is effective as it removes adhering

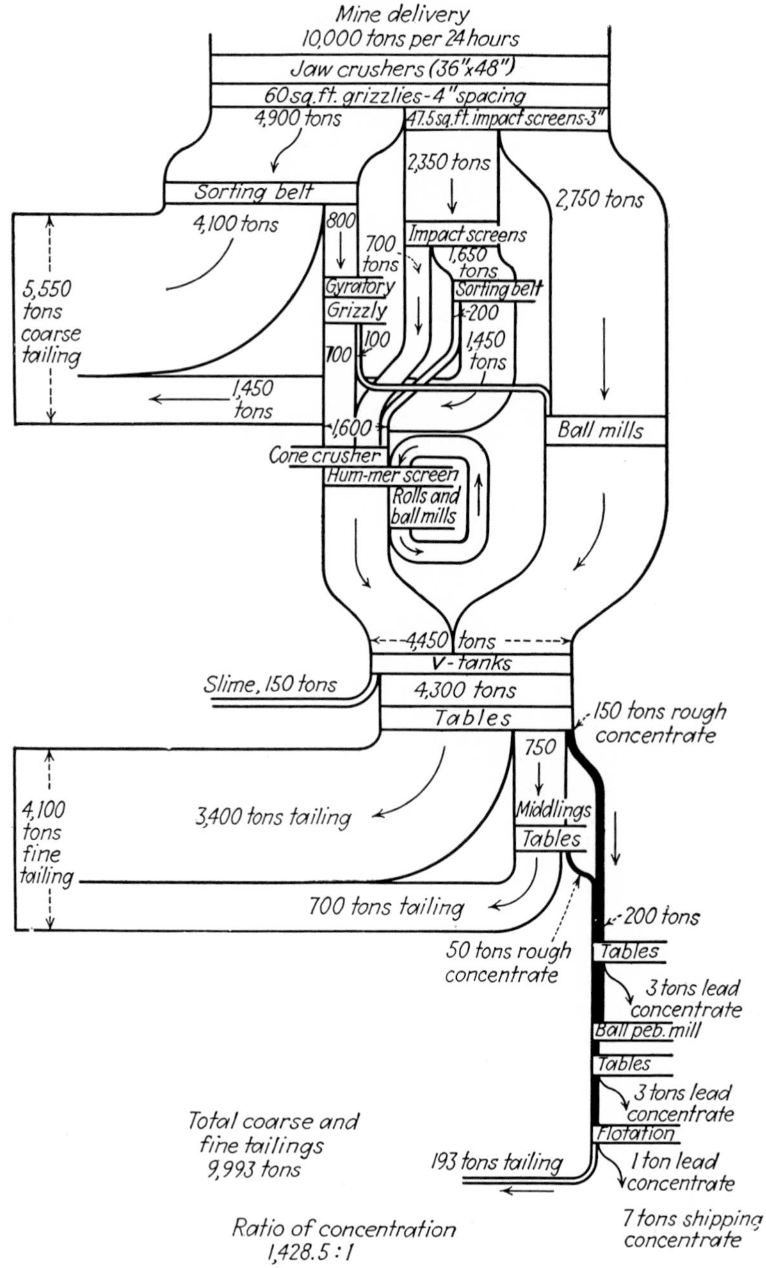

FIG. 230.—Tonnage flow sheet at Alaska-Juneau, showing the use of hand sorting.

fines and leaves a film of water on the lumps. This changes the reflectivity of the lumps so as to increase the relative brightness of the bright lumps. Prior chemical treatment of the lumps may change their color; or ultraviolet light can be used to make phosphorescent lumps glow with ephemeral splendor: an example of this practice is the use of ultraviolet lighting to make possible the picking of fluorescent willemite from nonfluorescent associated minerals.[5]

The sorting surface may be stationary, as in the case of a sorting floor or table, but it is preferably moving, to provide mechanical disposal of the material left behind. This is conveniently obtained by belt, pan, or shaking conveyors, or by revolving picking tables. For smaller pieces (less than 3 to 3½ in.), the picked pieces are usually flicked away from the operator; for larger pieces, even if only one hand is required, the picked pieces are drawn to the operator.

Tendency is to use hand picking less and less. This is traceable to

1. Increased cost of labor.
2. Decreased grade of feed.
3. Increased efficiency of mechanical-dressing methods. A good review of the advantages and disadvantages of hand picking is brought out in an article by Handy, and in the discussion that follows.[10]

An outstanding exception to the trend of eliminating hand sorting from ore-treatment flow sheets is the modern practice at the Alaska-Juneau mill where a white gold-bearing quartz is enriched by sorting from the barren, dark country rock with which the quartz is associated.[1,27] Figure 230 is a flow sheet with tonnages, from which it appears that 55 per cent of the tonnage mined is discarded by hand sorting.

Use of Electrical Eye. Mechanized picking is a distinct possibility for future operation. In the development of such processes, photoelectric cells would take the place of eyes; vacuum tubes and magnetic relays, of the brain; and compressed air, or some other mechanical contrivance, of the hand. Laboratory investigations have already been made.[2] Mechanized picking would undoubtedly be applicable to much finer particles than hand picking, and would probably be cheaper.

PROCESSES DEPENDING ON HEAT PROPERTIES

Two different effects of heat have been made the basis of processes of ore concentration. One effect is based on differences in degree of decrepitation and the other on differences in specific heat and in heat absorption.

Decrepitation. Decrepitation is the disintegration of crystals or of crystal aggregates on sudden heating or cooling. Many crystals contain occluded water; upon heating, a tremendous internal pressure is created which explodes the crystals. This is one cause for decrepitation. Another cause is present if the crystals are poor conductors of heat; the sudden expansion or contraction, due to large and sudden changes in temperature, are localized and result in very great strains and in explosion. Besides these causes for decrepitation, there is a predisposition toward it in substances that have a strong cleavage, in substances that form large crystals, and in substances that exist in several allotropic forms. Notoriously decrepitating minerals include barite, fluorite, calcite, quartz (beta → alpha) and some micas.

Concentration by decrepitation involves the following steps:

1. Sizing of the feed.
2. Heat-treatment of each size.
3. Sizing of each heat-treated size to yield the decrepitated minerals as undersize and the unaffected minerals as oversize.

The principal device is an oven in which the heat-treatment is carried out. This oven, as *e.g.*, in the Heusschen process, is designed to permit continuous operation.[26]

Decrepitation of a sized feed followed by screening of the decrepitated product can be used according to Tyler[37] and to Riddle[24] to concentrate kyanite and sillimanite. The process depends upon the sudden transformation of quartz from the beta to the alpha form.

Use of Other Heat Properties. Concentration of minerals by utilizing their heat properties is reported by E. Kirchberg.[16] Sulphides such as gray-copper minerals will proceed through a layer of snow more rapidly than gangue minerals such as barite. This passage results from melting of the snow by the sulphides, and is in turn related to their greater absorption of solar radiation and greater heat conductivity. Other heat properties such as specific heat are also believed to be involved. Viewed in the light of the effectiveness of modern methods, this process of concentration seems to have merely scientific quaintness.

PROCESSES DEPENDING ON ELECTRICAL PROPERTIES

Three different types of processes falling within this classification have been proposed, *viz.*, electrostatic separation, dielectric separation, and electrosmosis.

Electrostatic Separation. Electrostatic separation has been used for the separation of sulphide zinc from sulphide lead.[38,39] It was practically abandoned after the introduction of flotation, but it seems to have recently received a new lease of life for the separation of nonmetallic minerals.[13,14,15] Electrostatic separation is a means for separating from each other minerals of different electrical conductivity. It is based on the attraction for each other of unlike electrical charges, and the repulsion of like charges.

In one type of machine (Blake-Morscher type), particles of various conductivities are brought on a conducting, charged surface. Conducting particles become charged by induction, the total charge on each particle being nil, but the distribution being such as to concentrate close to the conducting surface electricity of a sign opposite to that on the surface. The net effect of this induction is attraction of the conducting particles to the surface. As soon as contact takes place, the opposite charges at the place of contact neutralize each other, with the result that the particles and surface now have the same charge and repel each other. Nonconducting particles are unaffected.[39]

In the Blake-Morscher machine, the conducting surface is a slowly revolving roll which draws feed from a bin; the conducting particles are thrown clear of the path of free fall which is adopted by the nonconducting particles (Fig. 231).

In the Huff electrostatic separator, the same fundamental property is used in a somewhat different way. The same type of feeding on a slowly revolving roll is used. In this case, however, the roll is grounded and the charged element is a fine wire situated at one side of the roll. In this case, the "electrical spray" which emanates from the wire sprinkles charges on all particles; the conducting particles lose these charges quickly by passing them to the grounded roll, and they fall down undeflected; the nonconducting particles acquire a charge, and this induces in the grounded roll a charge of opposite sign, thereby fostering their adhesion to the roll. The nonconducting particles gradually drop off because of leakage of the charge, or are scraped off by a brush (Fig. 232).

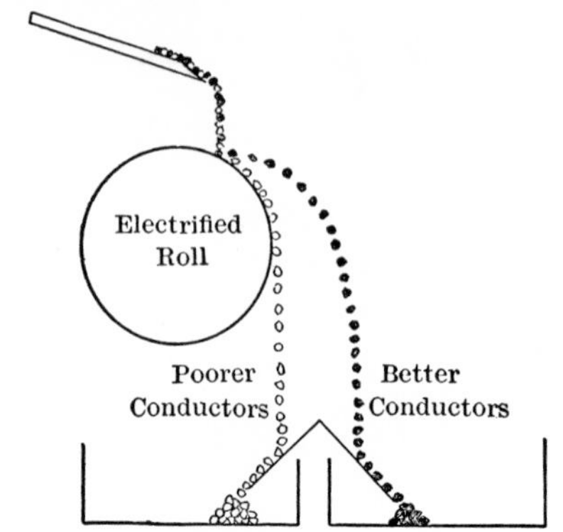

FIG. 231.—Simplest form of electrostatic separator. (*After Wentworth.*)

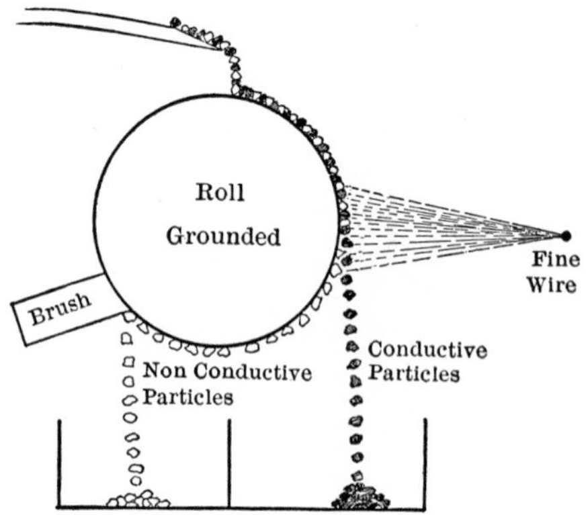

FIG. 232.—Principle of Huff electrostatic separator. (*After Wentworth.*)

The variations in conductivity of minerals are enormous from one species to another; great variations also occur within a single species because of variations in solid solution or because of variable occlusion of moisture. One would then expect greater variations in behavior in electrostatic separators than are generally reported. This indicates one of two things: that present-day machines are poorly designed or that the fundamentals of the process are not as pictured above.[4] According to Truscott,[36] the relative conductivities of minerals and metals (expressed in $ohm^{-1} \cdot cm.^{-1}$) are as per Table 58. To the data of Truscott have been added some figures from current handbooks of chemical and physical quantities.

TABLE 58.—RELATIVE CONDUCTIVITY OF MINERALS

Mineral	Conductivity
Copper	0.634×10^6
Gold	0.455×10^6
Covellite	8×10^3
Galena	3.35×10^3
Graphite	0.7×10^3
Pyrrhotite	119
Chalcocite	91
Pyrite	41.7
Magnetite	1.2
Chalcopyrite	0.98
Cuprite	25×10^{-3}
Siderite	0.14×10^{-3}
Marble	10^{-9} to 10^{-11}
Mica	10^{-13} to 10^{-17}
Quartz	10^{-14} to 10^{-19}
Sulphur	10^{-17}

In view of the extremely small conductivity of many minerals, it is likely that the presence of minute quantities of moisture or of dissolved or adsorbed impurities might affect enormously their conductivity and their behavior in electrostatic machines. The recent work of Johnson[14, 15] discloses the existence of preferential polarity in minerals treated in electrostatic fields, but little is known concerning the utilization of this property. In a current article Fraas and Ralston[41] summarize the principles and present status of electrostatic separation.

Present practice of electrostatic separation requires that the ore be dried prior to separation.

Dielectric Separation. Dielectric separation is a relatively untried electrical-separation process which bears great theoretical resemblance to magnetic separation. It was invented by H. S. Hatfield[11,12] and has recently been used by geologists and mineralogists[3,8,25,40] as a research tool in the study of minerals. No industrial applications have been announced.

The process consists in suspending a mixture of minerals of various dielectric constants[29] in a nonconducting fluid whose dielectric constant is intermediate between that of two groups of minerals, and in setting up a converging electrical field within the suspension. As a result, the particles of dielectric constant higher than the fluid travel in the direction of most rapid increase in electric field and the particles of dielectric constant lower than the fluid travel in the opposite direction.

A laboratory-type separator can be built according to Fig. 233 (Hatfield). It consists of a vessel *NK* of rectangular section provided with a two-way outflow at *P* or *M*. *C* and *D* are brass

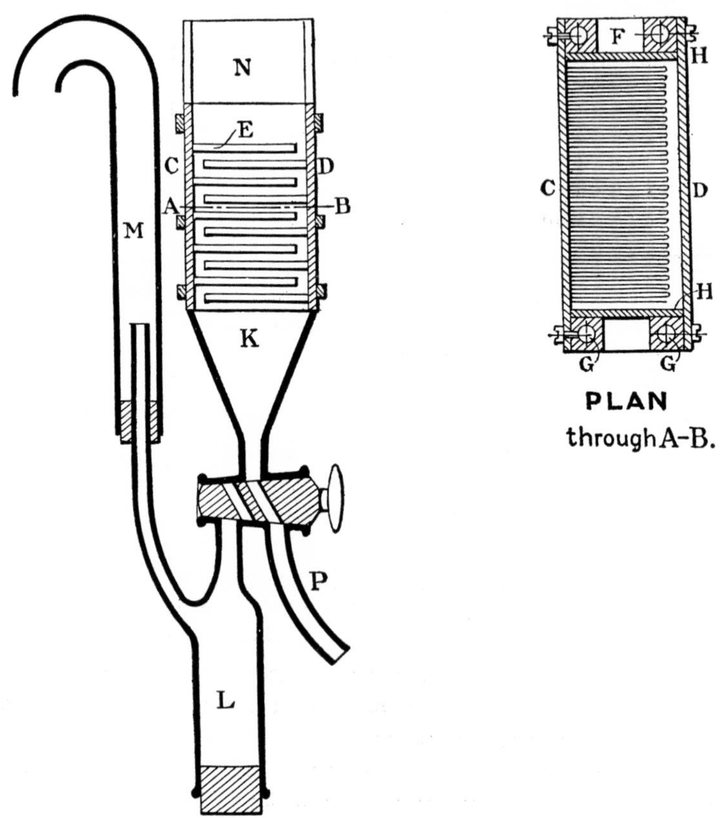

FIG. 233.—Laboratory-type dielectric separator. (*After Hatfield.*)

plates screwed to wood ends *FF* provided with glass windows *HH*. *GG* are take-up rods for rigidity. The brass plates are insulated from each other. Zigzagged strips of metal *EE* alternate between the metal plates. The sharp edges provide the necessary points for the establishment of convergent electrical fields. The strips *EE* and the plates are connected to opposite poles of an alternator.

The vessel is filled with the suspension of the finely ground ore in the dielectric (*e.g.*, a mixture of kerosene and nitrobenzene), and the electric current is turned on. The fluid is allowed to flow through until the strips are loaded with particles of high dielectric constant, when the fluid is turned into the other discharge channel (P), the current is turned off, and wash fluid flushed through.

Table 59 presents the dielectric constants of several fluids. A fluid of intermediate dielectric constant can be obtained by proportionate mixing of pure liquids.

TABLE 59.—DIELECTRIC CONSTANTS OF SOME LIQUIDS
(Air = 1.00)

Liquid	Dielectric constant
Water	81
Dimethyl sulphate	55
Nitrobenzene	36.5
Ethyl alcohol	26.8
Pyridine	12.4
Phenol	9.7
Chloroform	5.2
Olive oil	3.0
Benzene	2.3
Octane	1.9

Dielectric constants of minerals are not well known; besides, considerable variation with crystal orientation and with porosity and moisture content should be expected. However, some data are available.[25] Generally speaking, sulphides and minerals of high conductivity behave as if the dielectric constant were higher than that of water (81); other minerals have intermediate dielectric constants (*e.g.*, cassiterite 27.7, cuprite 16.2), but most minerals have constants under 10, *e.g.*, quartz (3 to 6.5), corundum (3 to 5.3), biotite (5 to 9.3), calcite (3 to 8.5).

The process is said to be applicable to particles finer than 60-mesh, including particles of slime sizes. Required steps include:

1. Elimination of water.
2. Use of an artificial dispersant in the dielectric fluid, *e.g.*, a mixture of oleic acid and aniline.
3. Filtration of concentrate and tailing.
4. Distillation and reclamation of the fluid remaining in the filter cake.

Compared with other mineral-dressing processes, it would appear as if this process should be very expensive. The estimated treatment cost of $2.50 per ton given by Hatfield has been regarded by Sulman[30] as highly optimistic.

Electrosmosis. Another electrical process that has been suggested for the treatment of slimes suspended in water is the process of electrosmosis. In principle, this is analogous to electrolysis, with suspended solid particles taking the place of ions. Because of the very great size without commensurately larger charge of suspended particles as compared with ions, the transfer of solid particles must be extremely slow. The process is said to depend upon the anchoring of ions at solid surfaces. It is wholly untried in a practical way, and is in fact stated to be a failure by Hatfield.[11]

PROCESSES UTILIZING DIFFERENTIAL HARDNESS

Minerals display widely differing resistances to wear. This is in fact the basis of hardness tests such as are used by mineralogists. They display also differences in ability to flow under pressure. This is the basis of hardness tests such as are used by metallurgists. They also display differences in brittleness or shattering ability. Very little is known concerning these properties of matter and their possible utilization.

Differential Resistance to Wear. Differential resistance to wear (not to impact) has been utilized in the laboratory[9] and in practice. It was shown that diamonds, although as brittle as quartz, will outwear quartz many thousandfold in a grinding mill equipped with pebbles or balls small enough to be unable to nip the particles, the process being carried out either wet or dry. The method is equally applicable to the concentration of mineral pairs other than diamond-quartz provided there is a sufficient spread in resistance to wear between the minerals. It is said that this method is used in Brazil for the recovery of black diamonds whose density is the same as that of the associated gravel.

Differential Malleability. Impact ruptures brittle materials but merely deforms malleable materials. A mixture of brittle and malleable particles upon grinding yields coarse malleable particles with fine, slimed, brittle particles. Classification or screening completes the concentration. To a considerable extent, this procedure is utilized in the recovery of native gold, silver, and copper.

Differential Weathering. Differences in weathering quality are somewhat related to differences in hardness. Some ores are aggregates of large grains of one mineral embedded in a matrix that is a fine aggregate of other crystals. Disintegration, after weathering, results in the production of a few large particles of one mineral and of a vast multitude of fine particles of other minerals. Concentration by differential weathering and sizing has been used for many years in the South African diamond fields as the first step in the recovery of the diamonds. The concentrate from the weathering and washing operations is concentrated further by gravity and adhesion processes.

The same process of concentration by weathering and washing is also applicable in other instances where the valuable constituents are more valuable in the coarse state than in the fine state, *e.g.*, in the recovery of any of the other precious stones.

PROCESSES UTILIZING ADHESION TO OIL

The important adhesion processes, *viz.*, flotation and agglomeration, have already been dealt with. In addition to these, there are two adhesion processes that have been and are still employed to some extent in ore dressing: one utilizes selective adhesion to oil for the formation of oil-mineral granules, which are then separated by gravity from ungranulated constituents; the other utilizes selective adhesion of some minerals to quasi solid grease and the subsequent separation of the grease and adhering valuable minerals by intermittent scraping.

Adhesion to Oil and Granulation. Adhesion of minerals to oil has historic importance rather than practical importance now, as it led to the discovery of flotation. The process was studied and improved by Cattermole.(6) In one of Cattermole's procedures, the acidified ore pulp was mixed with a quantity of fatty acid just sufficient to fill the pores between adjacent grains of valuable mineral. In one instance, the quantity of oil was of the order of magnitude of 2 to 4 per cent or more on the ore, by weight, *i.e.*, of the order of magnitude of 10 to 15 per cent of the granulable mineral, by weight. Although the oil-mineral granules contained occasionally some air, the aim was to make a relatively hard, compact association of mineral particles and oil which could then be gathered as a heavy sediment in classifiers.

The Cattermole process was a scientific success, but it did not stand up under the constant incentive of reducing costs by reducing the quantity of oil used. As a result, air was used in place of oil in the granules, and the lightened granules ceased to sink and began to float. Thus the reduction in the amount of oil led to an abandonment of the Cattermole idea of granulation as a basic means of ore concentration.

Several years later, substantially the same idea was applied by Trent[35] in connection with coal. The Trent process depends on the production of relatively large granules of oil (largely a hydrocarbon oil) with coal and on the separation of these granules (or "caviar") from ungranulated waste by screening or classification. The Trent process gives a very good recovery of coal provided the grinding is fine. In comparative tests, it has also yielded a better grade of concentrate than flotation.[21]

In the Christensen[7] process, the same idea of granulation was extended to very finely ground oxidized minerals, the separation of the granules from ungranulated particles being effected by froth flotation.

The drawback of the Trent process, like that of the related Cattermole process, and also of the Christensen process, is of course the great quantity of oil that must be employed and regenerated.

Adhesion to Grease. Adhesion to grease in preference to water has been the object of several patents. However, except for the final recovery of diamonds in South Africa, it does not seem to have been employed industrially. In the diamond fields, the concentrate obtained from the disintegration of the so-called "blue ground," to which reference has been made above, is concentrated by gravity means, mostly jigging, until a product is obtained that contains the diamonds together with other heavy minerals. Further concentration by gravity means is not possible, and the picking of the diamonds from this relatively low-grade, yet high-priced concentrate, would be laborious. The mixture is passed on a shaking table, the surface of which is smeared with a heavy grease. The diamonds adhere to the grease, and the other minerals do not adhere to the grease, so that after a period of time, it is possible to scrape off the grease with adhering diamonds as a much purer concentrate. Adhesion of the diamonds to grease is no doubt related to their nonpolar structure and to the likewise nonpolar structure of the hydrocarbon molecules of which the grease is made.

Concentration by adhesion to grease can be applied to particles coarser than those recovered by either flotation or agglomeration, perhaps up to ½ in. in diameter, or even more.

AMALGAMATION

Theory. The time-honored process of recovering gold by treatment with liquid mercury is usually regarded as a metallurgical process involving dissolution of gold in mercury with formation of a gold-mercury alloy or amalgam. But amalgamation can be viewed with at least equal propriety as an adhesion process; and as such, it resembles oil-mineral and air-mineral adhesion processes.

Gold and mercury are said to form two compounds Au_2Hg and $AuHg_2$ both melting with a peritectic reaction at 400 to 420°C. and 310°C., respectively. Gold is insoluble in solid mercury but somewhat soluble in liquid mercury. The solubility in liquid mercury increases with temperature and is of the order of 0.1 per cent at ordinary working temperatures.[31,32]

The amalgams formed in the recovery of gold by amalgamation are complex mixtures, not at equilibrium, consisting of gold, one or several gold-mercury compounds, and a solution of gold in mercury.

That which controls recovery of gold by mercury is not the completion of reaction of the metals with each other, but the tendency toward that reaction, a tendency which finds expression in wetting of gold by mercury.[19,20,32] This process is analogous to the wetting of minerals by oil or by air, and can be viewed as a reduction in surface energy of the system resulting from substitution of a gold-mercury interface for two interfaces, one between gold and water and the other between mercury and water.

Experiments have shown that the contact angle is of the order of 160°, measured across the water. The configuration of a drop of mercury on a gold foil under water is then somewhat as shown in Fig. 234.

Other things being equal, this shows that the tendency for gold to be wetted by mercury is much greater than the tendency for paraffin or any mineral to be wetted by air, when water is the competing phase.

From these considerations, it is obvious that the amalgamability of native gold is very much subject to the condition of its surface and to the condition of the surface of the mercury, just as in flotation and in agglomeration the character of mineral surface and air-water interface controls the operation of the process. Here also, reagents may be effective in favoring or hindering the wetting of gold by mercury. Electric charges are also effective in modifying the contact angle of mercury at a gold surface.[20a]

Fig. 234.—Contact angle at a gold-water-mercury junction.

What is said of gold is also true of other metals that can be amalgamated, particularly of silver and copper, although the conditions under which they may be amalgamated differ in degree.

Amalgamation Inhibitors. Amalgamation inhibitors include substances dissolved in water, *e.g.*, some salts; substances suspended in water which tend to spread at metal surfaces, *e.g.*, organic substances; and substances mechanically driven into the gold surfaces by the pounding of the grinding devices, or by agencies of weathering.

Among dissolved substances, alkali sulphides and some flotation agents seem to be particularly offensive. Their effect is probably related to the formation of surface coatings, adsorbed or chemically bonded on either or both gold and mercury.

Among undissolved substances are some sulphide minerals, especially sulphides containing antimony and arsenic, oils, and other organic contaminants. Both dissolved and undissolved inhibitors tend to flour the mercury to fine droplets and to some extent to tarnish the gold.

The ill behavior of "rusty" gold, *i.e.*, gold apparently covered with adherent iron oxides, is an example of the third type of inhibitors.

Amalgamation Nostrums. Many remedies have been proposed to overcome the action of inhibitors. In most cases, however, application of the remedies is not preceded by the proper ascertainment of the facts, so that improved results may or may not be obtained.

Among the more reputable nostrums are the use of alkalies or alkali cyanides in the pulp, the use of sodium or zinc amalgam in place of mercury, the auxiliary use of electric current passing from pulp to amalgam.[28]

The object of each is to provide better wetting of the metal by the mercury; that object is realized more or less, according to circumstances.

The viewing of amalgamation as an adhesion process makes it easy to understand that many substances, particularly substances in aqueous solution, must have an effect on the contact angle at the contact of the amalgam, gold, and water. Just as depressing agents exist in flotation, so there exist amalgamation inhibitors; and amalgamation nostrums are the counterpart of flotation activators and collectors.

Practice of Amalgamation.[17,22,34] The mercury for amalgamation may be either free as liquid mercury moving around in a mill or drum, or it may be present as a film at the surface of a copper plate or of a silvered copper plate. It is often stated that amalgam is a better amalgamator than mercury. This is perhaps because a relatively thick layer of the mushy mix of gold, gold-mercury alloys, and mercury can be applied easily, whereas clear mercury is too fluid; but there may be some other reason for the well-substantiated observation that amalgam is a better amalgamator than mercury.

Until recently, the trend in amalgamation has been to wider and wider use of *plate amalgamation.* Plate amalgamation, or outside amalgamation, consists in allowing the pulp to flow on an inclined copper plate or an inclined silvered copper plate covered with a layer of soft amalgam. Passage of pulp should be by surges or with ripples so as to turn the particles over as they flow over the mercury. Velocity of flow, thickness of pulp layer, length of plates, pulp density, and other variables should be regulated so as to give the particles of gold a chance to come in contact while avoiding permanent settlement of the gangue on the mercury. Fine particles of gold have a relatively poorer chance to become amalgamated than coarse particles. Locked particles are largely lost.

It is often advantageous to break the flow of pulp over the plate by a cascade from one plate to another. This turns over the pulp better than any other simple means. The surges in pulp flow produced by a stamp mill have been widely regarded as beneficial.

Inside amalgamation has also been used, the mercury being placed inside of the grinding device which is subjected to periodic cleanups. This practice is almost obsolete.

Recently, however, inside amalgamation has been revived for the treatment of flotation and table concentrates. It is done in barrels, the time of treatment being relatively long.

In addition to barrel and plate amalgamators, there have been devised an almost innumerable variety of contraptions designed to ensure contact of the fine gold particles with the mercury. Most of these devices are in fashion for a short time only as their performance is rarely better than that of plates or barrels.

Amalgam Cleanup. Plate amalgamators are scraped with a rubber tool every few hours to gather the amalgam. Care is taken not to scrape off all of the amalgam; then mercury or sodium amalgam is added, the surface of the plate being worked to a soft, mushy condition. Hard amalgam is to be avoided at all times as it is sometimes gouged from the surface by the larger pieces of gangue, and lost.

Barrel amalgamators are operated on batches, the amalgam being recovered by dumping the barrel, settling the pulp, and washing with clear water.

Separation of the gold from the mercury is obtained by distillation after squeezing through soft canvas or in a press.

A low red heat is sufficient to distill the mercury; a higher temperature is used at the finish to eliminate other impurities and to melt the spongy gold. Mercury and gold losses can be maintained very low.

Applicability of Amalgamation. Amalgamation is widely used for the recovery of gold; a substantial proportion of the gold produced the world over is recovered by that process. The process is applicable only to so-called free-milling ores in which the metal is coarse and native. Gold ores in which the metal is finely divided in silicates, oxides, or sulphides, and gold ores in which the gold is present as telluride are not treated with success by amalgamation.

Former practice was to treat an ore by amalgamation only, but the recent trend is to combine that process with gravity concentration, cyanidation, and flotation, the attempt being made to recover as much gold as possible by amalgamation. The advantage of amalgamation over flotation or gravity concentration is that it yields finished bullion, and the advantage over cyanidation is that clean-up is easier and presents fewer chances for a loss. Amalgamation of auriferous concentrates has recently assumed increasing importance.[8a] Because of the effect of flotation agents on amalgamability of gold, amalgamation of concentrates differs somewhat from the amalgamation of ores.

Amalgamation was also used for the treatment of silver ores. Native silver, however, is relatively uncommon; the other silver minerals that are amenable to some extent to amalgamation, silver chloride, for example, consume much mercury, perhaps by a reaction that precedes amalgamation. Refractory silver ores were formerly treated by a chloridizing roast followed by amalgamation.

Literature Cited

1. ANON.: Alaska-Juneau Milling System, *Eng. Mining J.*, **131,** 369 (1931).
2. ANON.: Color and Luster as Aids in Mechanical Separation of Minerals, *Eng. Mining J.*, **124,** 520 (1927).
3. BERG, G. A.: Dielectric Separation of Mineral Grains, *J. Sed. Petrol.*, **6,** 23–27 (1936).
4. BIBOLINI, ALDO: "Note sur la séparation électrostatique des minerais à propos d'une nouvelle électro-trieuse," Congrès Scientifique de l'Association des Ingénieurs Sortis de l'École de Liège (1923).
5. BRUNTON, DAVID W.: Modern Progress in Mining and Metallurgy in the Western United States, *Trans. Am. Inst. Mining Engrs.*, **40,** 543–561 (1909).
6. CATTERMOLE, ARTHUR EDWARD: U. S. Patents 763,259; 763,260; 777,-273; 777,274 (1904).
7. CHRISTENSEN, N. C.: U. S. Patent 1,467,354 (1923).
8. DERKACH, V. G.: Use of Dielectric Properties of Minerals in Ore Dressing, *Inst. Mekhanicheskoi Obrabotki Poleznuikh Iskopaemuikh "Mekhanobr," 15 yrs. Socialistic Ind. Service*, **1,** 543–556 (1935). Abstracted in *Chem. Abs.*, **30,** 3372 (1936).
8a. FLYNN, A. E.: Amalgamation of Auriferous Concentrates, *Trans. Can. Inst. Min. Met.*, **42,** 150–163 (1939).

9. Gaudin, A. M.: Resistance to Wear a Basis for Concentrating Ores: *Eng. Mining J.*, **123,** 245–246 (1927).
10. Handy, R. S.: Hand-sorting of Mill Feed, *Trans. Am. Inst. Mining Met. Engrs.,* **61,** 224–236 (1919).
11. Hatfield, H. S.: Dielectric Separation: A New Method for the Treatment of Ores, *Trans. Inst. Mining Met.* (*London*), **33,** 335–342, 350–370 (1924).
12. Holman, Bernard W., and St. J. R. C. Sheperd: Dielectric Minerals Separation: Notes on Laboratory Work, *Trans. Inst. Mining Met.*, **33,** 343–349 (1924).
13. Jarman, G. W., Jr.: Special Methods for Concentrating and Purifying Industrial Minerals, *Am. Inst. Mining Met. Engrs., Tech. Pub.* 959; also in *Mining Tech.*, **II**-4 (1938).
14. Johnson, Herbert Banks: Electrostatic Separation, *Eng. Mining J.*, **139** [No. 9], 37–41, 51 (1938); **139** [No. 10], 42–43, 52 (1938); **139** [No. 12], 41–45 (1938).
15. Johnson, Herbert Banks: Selective Electrostatic Separation, *Am. Inst. Mining Met. Engrs., Tech. Pub.*, 877; also in *Mining Tech.*, **II**-1 (1938).
16. Kirchberg, H.: Aufbereitung von Mineralien auf Grund ihrer Wärmeeigenschaften, *Metall u. Erz*, **34,** 301–318 (1937).
17. Louis, Henry: "A Handbook of Gold Milling," The Macmillan Company, New York (1902).
18. Louis, Henry: "The Dressing of Minerals," Longmans, Green & Company, New York (1909).
19. Plaskin, I. N.: Theory of the Process of Amalgamation, *Sovietskaya Zolotoprom* [Nos. 9–10], 20–25 (1933).
20. Plaskin, I. N., and M. A. Kozhukhova: The Physico-chemical basis for the Amalgamation Process, *Ann. Inst. Chim. gén.* (*U.S.S.R.*), *Secteur Platine* [No. 13], 95–111 (1936).
20*a*. Plaskin, I. N., and M. A. Kozhukhova: The Change in Cosine of Contact Angle and Time of Wetting in the Electroamalgamation of Gold, *Ann. Inst. Chim. gén.* (U.S.S.R.), *Secteur Platine* [No. 15], 101–111 (1938).
21. Ralston, Oliver C.: Comparison of Froth with Trent Process, *Coal Age*, **22,** 911–914 (1922).
22. Read, T. T.: The Amalgamation of Gold Ores, *Trans. Am. Inst. Mining Engrs.*, **37,** 56–84 (1907).
23. Richards, Robert H.: "Textbook of Ore Dressing," McGraw-Hill Book Company, Inc., New York (1909), pp. 192–200.
24. Riddle, Frank Harwood: Mining and Treatment of the Sillimanite Group of Minerals and their Use in Ceramic Products, *Trans. Am. Inst. Mining Met. Engrs.*, **102,** 131–154 (1932).
25. Rosenholtz, Joseph L., and Dudley T. Smith: The Dielectric Constant of Mineral Powders, *Am. Mineral.*, **21,** 115–120 (1936).
26. Roux-Brahic, J.: "Préparation mécanique des minerais," Dunod, Paris (1922), pp. 35–40.

27. Scott, W. P.: Milling Methods and Ore-treatment Equipment (at Alaska-Juneau), *Eng. Mining J.*, **133**, 475–482 (1932).
28. Shepard, Orson Cutler: Electrocapillary Amalgamation, *Am. Inst. Mining Met. Engrs., Tech. Pub.* 676 (1936).
29. Smyth, Chas. P.: "Dielectric Constant and Molecular Structure," Reinhold Publishing Corporation, New York (1931).
30. Sulman, H. L.: Discussion of article by Hatfield, ref. 11, *Trans. Inst. Mining Met. (London)*, **33**, 350–370 (1924).
31. Sunier, Arthur A., and C. B. Hess: The Solubility of Silver in Mercury, *J. Am. Chem. Soc.*, **50**, 662–668 (1928).
32. Sunier, Arthur A., and Chester M. White: Solubility of Gold in Mercury, *J. Am. Chem. Soc.*, **52**, 1842–1850 (1930).
33. Taggart, Arthur F.: "Handbook of Ore Dressing," John Wiley & Sons, Inc., New York (1927).
34. Thomson, Francis A.: "Stamp Milling and Cyaniding," McGraw-Hill Book Company, Inc., New York (1915).
35. Trent, W. E.: U. S. Patents 1,420,164; 1,421,862 (1922).
36. Truscott, S. J.: "Textbook of Ore Dressing," Macmillan & Company, Ltd., London (1923).
37. Tyler, Paul M.: Mechanical Preparation of Non-metallic Minerals, *Trans. Am. Inst. Mining Met. Engrs.*, **112**, 809 (1934).
38. Wentworth, H. A.: Electrostatic Separation at Midvale, *Trans. Am. Inst. Mining Engrs.*, **49**, 809–813 (1914).
39. Wentworth, Henry A.: Electrostatic Concentration or Separation of Ores, *Trans. Am. Inst. Mining Engrs.*, **43**, 411–426 (1912).
40. Williams, J. W., and J. L. Oncley: Dielectric Constant and Particle Size, *Physics*, **3**, 314–323 (1932).
41. Fraas, Foster, and Oliver C. Ralston: Electrostatic separations of solids, *Ind. Eng. Chem.*, **32**, 600–604 (1940).

CHAPTER XX

SEPARATION OF SOLIDS FROM FLUIDS

The separation of solids from fluids is one of the important yet often neglected subjects in the field of mineral dressing. In treating coarse solids, the separation is so readily accomplished by settling that the very fact that a separation is being made escapes attention. In treating fine solids, the importance of dewatering muds or of settling dusts becomes considerable.[12] These procedures are important not only in mineral dressing, but also in water purification, in mining and smelting operations, and in almost all industrial processes.

Mechanical separation of solids from liquids is accomplished by one of two general procedures, thickening and filtration. Ultimate moisture removal is obtained by drying. Drying partakes of industrial chemistry rather than of mineral dressing since a change of phase is involved for one of the products. It is therefore merely indicated in this book.

Separation of solids from gases is accomplished likewise by one of two general procedures, settling and filtration, which are more or less exact counterparts of the corresponding procedures involving liquids.

THICKENING

In thickening, a suspension of solids in a liquid is allowed to settle until a clear liquid layer tops a mud layer; the clear liquid is withdrawn from the top of the thickener as likewise the mud from the bottom; the operation is continuous. For its success, thickening depends upon execution of the following steps:

1. Flocculation of the minute solid particles so as to form aggregates or flocs of many particles.
2. Sedimentation of the liquid-laden flocs, leaving clear supernatant liquid.
3. Compaction of the sedimented flocs.
4. Elimination of the clear fluid and of the thickened mud.

The first step has already been discussed in Chapter XIV.

Sedimentation of Floccules. Floccules or flocs are rounded aggregates of solid particles. It may therefore be thought that their sedimentation is substantially the same as that of individual solid spheres, such as were considered in Chapter VIII. This would be a gross error because the flocs consist of solid particles with interstitial fluid. Actual determinations of the fluid content of floccules do not seem to have been made; yet in many cases the individual loose floccule will consist of 10 or more volumes of fluid for each volume of solid; and the tightly compressed and packed floccules discharged as a thickener underflow rarely consist of less than two to two and one-half volumes of fluid for each volume of solid.(19)

The specific gravity of a floccule, fluid and solid being considered together, is therefore very near that of the fluid, and the settling velocity is small, even at a fairly large size. Thus the settling velocity in water of spherical floccules 0.4 mm. in diameter (35-mesh) consisting, by volume, of 10 per cent solid of specific gravity 2.7, and 90 per cent water is, according to Eq. [VIII.4],

$$v_{max} = \frac{2}{9} \frac{[(0.10)(2.7) + (0.90)(1) - 1] \times (0.02)^2 \times 980}{0.01} = 1.48 \text{ cm. per sec.}$$

This calculation is tantamount to an assumption that the water within the floccule moves with the solid of the floccule. Actually, it is almost certain that there is movement of fluid with respect to solid within the floccule; this must result in an increased friction and decreased floccule velocity. To support this statement, it might be pointed out that flocs 1 mm. in diameter or more are often observed to settle as slowly as 0.1 to 0.2 cm. per sec. If the slow descent of these flocs is not due in part to their porosity, they must indeed be extremely dilute.

These considerations lead one to accept, but only with reservations, the first approximation that individual flocs settle as individual spheres. But flocs, per se, do not occur as individuals. In fact, in a pulp having the same water content in the dispersed and in the flocculated condition, hindered settling may prevail if the pulp is flocculated, although substantially free settling would occur if it were dispersed. For example, if a pulp containing 5 per cent solids by volume is flocculated to relatively dense flocs that contain four volumes of water to one of solids, the flocculated pulp is made of 25 parts flocs with 75 parts of water, by volume, and the dispersed pulp is made of 5 parts solids to 95 parts water. The flocculated pulp is settling by hindered settling, but the dispersed pulp is settling almost freely. Even so thin a pulp as one containing 1 per cent solids by volume may, after flocculation, settle by hindered settling, provided the flocs are dilute enough.

An interesting case occurs with a pulp containing 20 per cent solids by volume. If the flocs are again of the 4:1 type, such a pulp in the flocculated state should contain no free liquid. Yet it is known that such a pulp appears to settle. Actually, it is more accurate to consider that the pulp sheds water by a process different from hindered settling. It will presently be demonstrated that this phenomenon is closely related to filtration.

Dewatering of Flocs by Exudation. After the *interfloccular* liquid has gathered above the settled flocs, the *intrafloccular liquid*

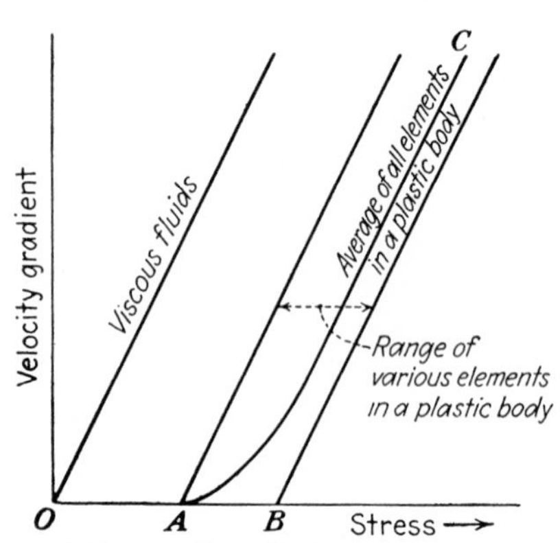

FIG. 235.—Stress-velocity gradient diagram for viscous fluids and for plastic bodies. (*Partly after Bingham and after Houwink.*)

is exuded through the pores of the flocs. Deformation of the structural unit constituted by the floc is a plastic phenomenon[3,4,10a] requiring the application of a force exceeding the yield point of the structure (see Fig. 235). This yield point (OA, OB) is not the same for all floccules, since the floccules are not exactly alike. Statistically, then, the exudation will begin at a certain low yield point and will increase as the force is increased.

The available exuding force per unit area is the difference in pressure exerted at a certain depth by the pulp and by the water in the pulp. So long as the flocs are settling freely, there is no such force; as the flocs are hindered in their settling there is some exudation, the exuding force being represented by the blows received by the particle (probably a small but steadily increasing effect). After hindered settling is completed, exudation alone is effective.

As the process of exudation continues, floccules of low yield point are replaced by floccules of higher and higher yield point until the floccules are all so rigid as to be beyond the reach of the exuding force, and that process is terminated. Further dewatering requires mechanical rupture of structures having a high yield point, or the application of greater exuding forces. Both of these effects are used, one in gluten-type thickeners, the other in filters.

Processes Involved in Dewatering Flocculated Pulps. The dewatering of a flocculated suspension involves then the following processes:

1. Free settling of the flocs.
2. Hindered settling of the flocs.
3. Exudation of water from the settled flocs, under the influence of pulp pressure.
4. Exudation of water from the settled and naturally exuded flocs by application of outside pressure or suction.

In the thickening of very dilute suspensions, the first three processes are operating; in the thickening of medium-thick suspensions, processes 2 and 3 are operating; in the thickening of heavy suspensions, process 3 only is operating. Process 4 is characteristic of filtration.

Free Settling of Flocs. The flocs of a flocculated pulp are more nearly alike than the individual particles of a dispersed pulp, at least in respect to settling velocity. But they are not absolutely alike. Consequently, free-settling flocs quickly arrange themselves with the fast flocs at the bottom and the slow flocs at the top. The dewatering is then controlled by the slowest flocs. Since free settling is found only in very dilute suspensions and since it is found that the flocs are consequently very watery, the settling rate or rate of clarification may be disappointingly slow. This is where the most effective flocculating agents must be used. Except for water clarification, free settling flocs are not found in mineral pulps.

For any individual floc, the free-settling rate is constant.

Hindered Settling of Flocs. Under conditions of hindered settling, relatively fast-settling flocs must open their way between slower settling flocs; and if unable to do so, they may lodge between them. In any event, fast-settling flocs are retarded more than slow-settling flocs, and if the hindrance is sufficient, they must travel with the crowd. Conversely, slow-settling flocs are accelerated by the blows of their faster moving companions, and also move with the crowd. So, under hindered-settling conditions, there is practical uniformity in settling rate. In spite of the generally slower settling of the mass as a whole, the settling rate may be faster than that of the slowest flocs in a free-settling pulp.

The settling rate decreases with time as the hindered character of the settling grows more marked.

Exudation of Water by Pulp Pressure. Exudation of water results in water oozing out at the top of the settled flocs and moving upward through the hindered-settling column. This continues while the hindered-settling column is settling, until that column is gone. During this period, the settling of the pulp, as measured by the movement of the top of the mud, is at a constant or at a slowly decreasing rate.

After the hindered-settling column has completely disappeared, exudation cannot proceed in ever-fresh layers as heretofore, but by further consolidation of the same layers. As the structures of the flocculent material have higher and higher yield points, with less and less space for rearrangement, this second half of the exudation process, or filtration stage, proceeds at a steadily diminishing rate.

These phenomena are summarized in Fig. 236.

Observation of natural pulps(5,8,8a,13,14,15,18) shows that they follow closely the behavior that has been outlined. Figure 237 shows in time sequence the clear liquid *A*, the hindered-settling zone *B*, settled sand (black dots), zone of exudation *D* with water channels (wiggly vertical lines), and shallow transition zone *C*. The critical point (cylinder I) corresponds to the disappearance of the hindered-settling zone, and to point *C* in Fig. 236.

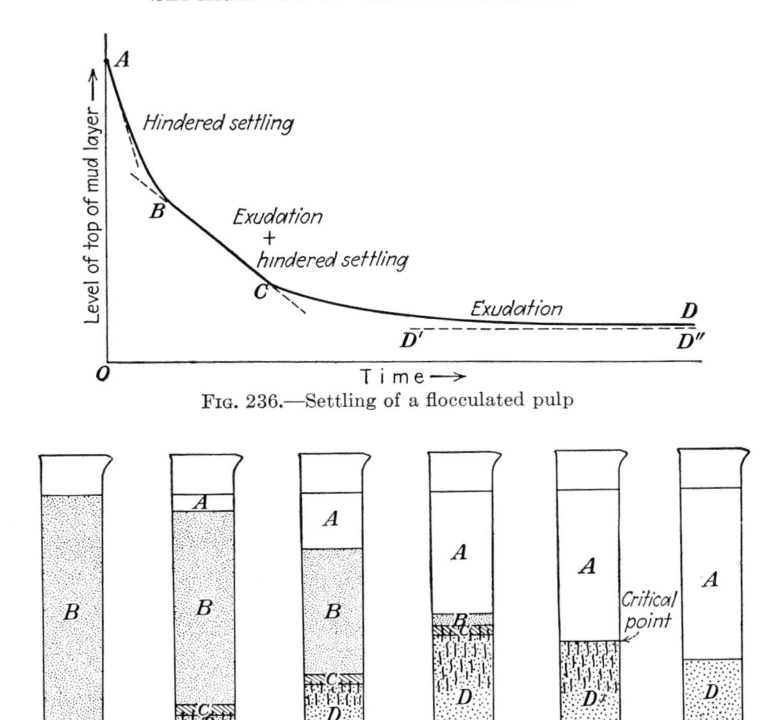

FIG. 236.—Settling of a flocculated pulp

FIG. 237.—Experiment showing various stages of slime-settling. (*After Coe and Clevenger.*)

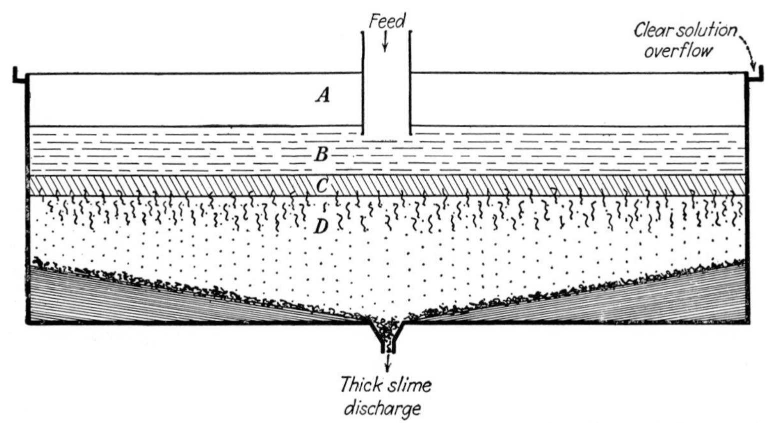

FIG. 238.—Slime-settling zones in Dorr thickener. (*After Coe and Clevenger.*)

A continuously operated thickener contains in vertical strata all the layers recognized by Coe and Clevenger.[5] This is diagrammatically shown in Fig. 238.

Mechanical Working of Thickener Sludge. It has been observed that a substantial reduction in the water content of a thickened sludge can be obtained if gentle stirring is used to break the gel-like structures that form. Stewart and Roberts[19] state that a suspension of milk of magnesia cannot thicken without mechanical work to more than 4 to 7 per cent solids, by weight, but that gentle stirring will raise the solid content to 25 per cent.

One of the functions of thickener rakes is to work the sludge mechanically. This effect is emphasized in the gluten-type thickener which is equipped with a stirring mechanism analogous in construction to a picket fence.[21]

Elimination of Clear Liquid and of Compacted Flocs. Clear liquid is naturally eliminated by overflow. The rate of overflow, in turn, is controlled by the rate of intake in the settling tank.

Compacted flocs can be eliminated by one of two means. The simpler in principle is by natural flow through a pipe equipped with a gooseneck discharge arrangement, the rate of flow being controlled by a valve and by the difference in head at the bottom of the tank due to the height of pulp in the tank and of the compacted flocs within the gooseneck outside the tank. This scheme although simple in principle is difficult to operate satisfactorily because of the variations that occur in height of the pulp level and in the density of the thickened pulp.

A more suitable arrangement includes discharge by a diaphragm pump. This arrangement is steadier in operation and should be preferred. It has the further advantage of delivering the mud high enough to flow readily into another thickener set up at the same level.

Thickeners. The typical thickener is the Dorr thickener. It consists of a cylindrical tank some 10 ft. deep and of relatively large diameter with a slowly revolving central shaft on which sweeping paddles are fastened (Fig. 239). The shaft is driven from a motor through speed reducers, the motor being either at the center and mounted on a permanent superstructure, or at the edge, moving on a circular track and driving the shaft through a revolving truss. The centrally driven type is preferred for relatively small tanks, and the traction type is preferred for large tanks.

In the Dorr tray thickener, several shallow tanks are superimposed and driven by one mechanism. These thickeners can be used for either series or parallel flow of pulp. Although less expensive per square foot of settling area than standard thickeners, they are generally less effective and more difficult to control.

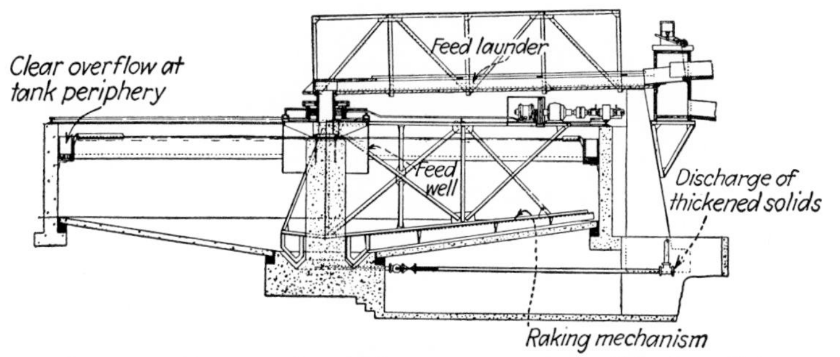

Fig. 239.—Dorr traction thickener cross section. (*The Dorr Co.*)

Removal of the clear liquid is by overflow around the periphery. Removal of the thickened mud is by means of a diaphragm suction pump (Fig. 240) capable of lifting the underflow 6 to 8 ft. The pump consists of a flexible diaphragm which divides the pump body into a lower suction chamber and an upper discharge bowl. Two one-way valves, one above the other, are placed in the floor of the suction chamber and in the diaphragm lift yoke. The stroke of the diaphragm may be varied by means of an adjustable eccentric on the drive shaft.

Wherever possible, thickeners are placed outdoors, as that eliminates the cost of a building. If the volume of pulp is very large, this is practicable even in fairly cold climates.

Thickener tanks are made of wood, metal, or concrete. For large installations, concrete is preferred, and for small installations, metal. Wood is preferable only if the chemical character of the pulp precludes the use of metal or concrete.

Cost of thickening is largely a capital cost, as power, attendance (lubrication only), and repairs are practically nonexistent.

Even for small installations, the operating cost rarely exceeds 1 ct. per ton of solids.

Recently, centrifugal force has been substituted for gravity in thickening (Bird solid-bowl centrifugal). Great saving in floor space is accomplished, but the first cost is somewhat higher, and the operating cost is much higher than with gravity thickening.

FIG. 240.—Diaphragm pump (duplex) for discharge of the underflow from a thickener. (*The Dorr Co.*)

FILTRATION

Filtration[20] is that separation of finely divided solid particles from a fluid which is accomplished by driving the pulp to a membrane or *septum*, porous to the fluid but impervious to the solid, through which the fluid passes, and by the removal of the solid cake from the septum.

Mechanism of Filtration. The simplest type of filter consists of a tube of small bore through which the fluid is sucked while the solid particles accumulate at the entrance.[11] As the device is operated, solids at first pass through the tube, but they quickly arch or bridge across the opening, allowing only clear liquid to pass. The septum consists of a layer of particles derived from the suspension which is being filtered (Fig. 241). The role

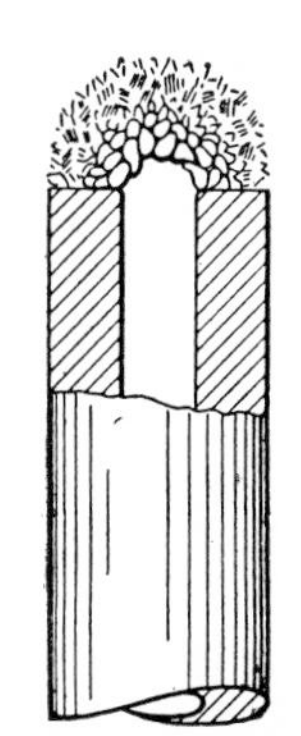

Fig. 241.—Structure of a filter cake. (*After Hixson, Work, and Odell.*)

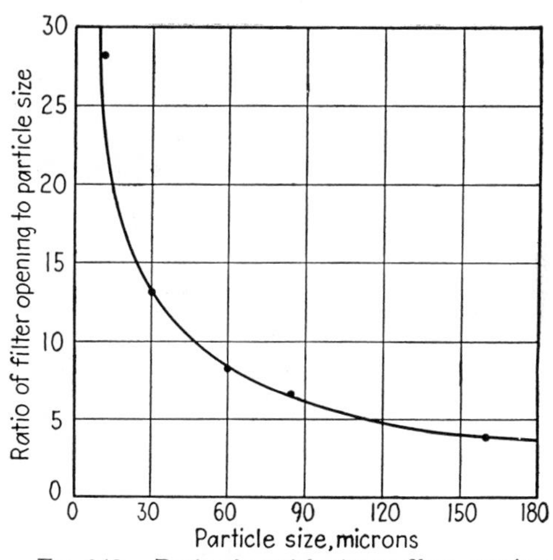

Fig. 242.—Ratio of particle size to filter opening in relation to the size of the particles (20 to 60 per cent solids). (*After Hixson, Work, and Odell.*)

of the tube is merely to act as a framework on which the bed of solids is supported.

Hixson, Work, and Odell[11] have shown that for bridging of a pore by solids, the diameter of the coarsest solid particles must exceed a certain minimum size; this minimum size is one-third of the pore opening if dealing with coarse material, but it may range to a small fraction of this ratio if the particles are fine (Fig. 242).

It seems that the data of Hixson, Work, and Odell dealt with pulps in the dispersed state, or at any rate with particles so coarse as to make flocculation phenomena wholly secondary. Yet filtration is usually, and more effectively, carried out with flocculated pulps. It is possible, of course, that the increase in bridging with fineness noted by Hixson and his associates was due to initiation of flocculation at the entrance of the experimental thistle tube.

Rate of Filtration. The rate of filtration depends on

1. The filtering area.
2. The difference in pressure between the two sides of the filter.
3. The average cross section of the pores within the filter cake.
4. The number of pores per unit area of the septum.
5. The thickness of the filter cake.

The effect of each of these factors is obvious, *i.e.*, the rate of filtration increases directly with (1), (2), and (4), inversely as (5), and directly as some power of (3).

Factors (1) and (2) are controlled by the device in use, (5) by the way in which the device is operated, and (3) and (4) by the character of the pulp.

If flow of fluid through a filter cake is regarded as made up of flow through many capillaries, the flow through each capillary is proportional to the fourth power of the radius. This is the famous hydrodynamical relationship due to Poiseuille, and it can be derived from the same fundamentals as Stokes' law. For a given percentage of pore space in a filter cake, the number of capillaries per unit area is inversely proportional to the square of the radius of the capillaries. The net result is that the rate of flow through a filter bed, per unit area of the bed, is proportional to the square of the diameter (or radius) of the pores (factors 3 and 4 combined).

If the pulp is dispersed, the particles in the cake pack very firmly so as to make the pores exceedingly small, each large pore becoming rapidly clogged by finer and finer particles; if on the other hand the pulp is flocculated, the cake is porous, the pores remain relatively large, and filtration is rapid. As an example, it might be observed that filtration of a dispersed slime may be fifty to two hundred times slower than filtration of the same slime after it has been flocculated. Thus the state of aggregation of a pulp is the largest single factor affecting rate of filtration. It is fortunate that flocculation is required in both filtration and thickening since these operations are usually carried out sequentially. An overflocculated pulp, although filtering readily, may yield a wet cake; an underflocculated pulp filters slowly to a drier cake.

As in the case of thickening, mechanical work on the filter cake results in a drier cake. In this instance, however, instead of gentle stirring, mechanical work takes the form of vigorous tamping and slapping of the cake.

Types of Filters. Filters can be classified according to whether pressure is applied on the pulp to push the liquid through the filter bed or whether suction is applied to pull it through the bed. The familiar analytical filters used in chemical laboratories are suction filters. A sand bed on canvas or other porous bottom is a pressure filter in which the pressure equals the head of the pulp on the bed.

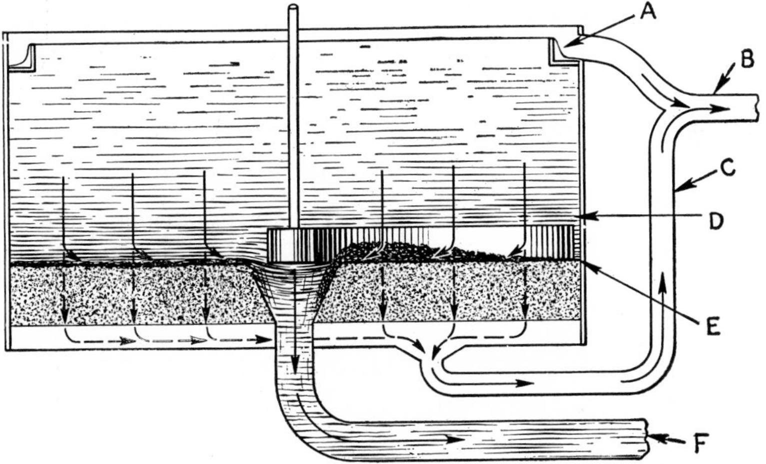

FIG. 243.—Principle of operation of the Hardinge filter thickener. (*The Hardinge Co.*)

For hydrometallurgical work in which interest centers on the filtrate, as in cyanidation, filters of the pressure type are generally used for the final clarification of the filtrate; for mineral dressing, in which interest centers on the cake, filters of the suction type are used exclusively.

Filter Thickeners. Filter thickeners are devices that combine filtration with thickening. An example is afforded by the Hardinge superthickener (Fig. 243) in which pressure filtration on a sand bed is used. This device gives two clarified liquors, one overflowing the rim of the tank and the other as a filtrate from the porous bed. Its capacity is necessarily somewhat larger than that of a thickener of the same diameter.

Another filter thickener is the Genter thickener (Fig. 244). In this device, filtration is of the suction type. Suction is exerted through numerous cylindrical elements dipping in the tank. The cake that forms on the outside of these filtering elements is detached at intervals of several minutes by the sudden and short-lived establishment of counterflow of the fil-

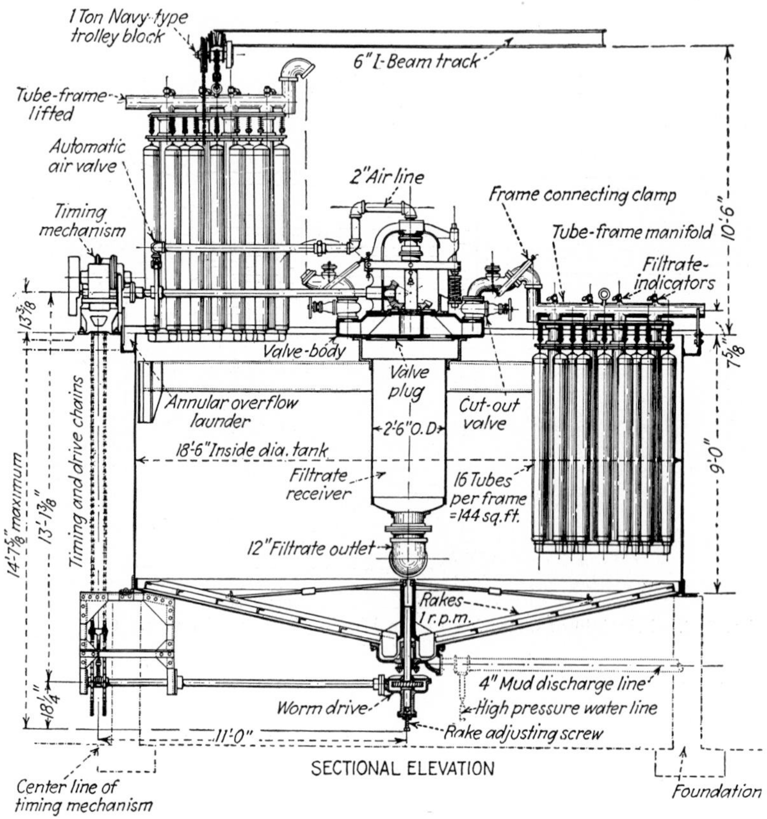

Fig. 244.—General arrangement of Genter thickener. (*General Engineering Co.*)

trate. Pressure is exerted on the average for 1 or 2 per cent of the time and suction for 98 to 99 per cent. This alternation of suction and pressure is obtained by the use of a special central valve. The Genter filter thickener is more of a filter than a thickener as the major part of the fluid, if not all, is removed by the filtering elements, and a small part only as overflow. In common with thickeners, however, it delivers a mud instead of a cake. The makers claim great reduction in floor space, an increase in density of underflow, and a lowered total cost, as compared with sedimentation-type thickeners.

Suction Filters. Suction filters consist of rotating drums or disks covered with cloth partly immersed in the material to be filtered. Suction is applied so as to cause flow of filtrate into piping within the rotating drums or disks. Pressure is also

FIG. 245.—Oliver low-submergence continuous filter. (*Oliver-United Filters, Inc.*)

applied continuously to whatever part of the revolving filtering surface comes in front of a special stationary sector, thus permitting the cake of solid to be blown away from the filtering surface and on to a belt conveyor.

The Portland and the Oliver filters are of the drum type with the outer face of the drum acting as the filtering surface; the Dorrco filter is of the drum type with the inner face of the drum acting as the filtering surface; and the United filter is of the disk type.

The mechanisms in all these filters are relatively complicated and expensive. In all cases, the "heart" of the device is the valve system which at a predetermined position causes a change from suction to pressure (or "blowback").

Figure 245 is a general view of the Oliver low-submergence continuous filter. This is a new filter which it is claimed has higher capacity and produces a drier cake by free discharge (no blowback; merely suspension of suction) than the standard Oliver filters. This filter is characterized by the low pulp level and therefore short cake-forming period, with long cake-drying period.

FIG. 246.—General arrangement for disk filter: *A*, electric motor; *C*, filter; *D*, vacuum receiver; *E*, vacuum release valve; *F*, dry vacuum pump; *G*, motor-driven filtrate pump; *H*, air receiver; *I*, air-pressure control valve. (*Oliver-United Filters, Inc.*)

Figure 246 shows the general arrangement for a small Oliver filter. The layout includes the filter proper *C*, motor *A*, speed reducer, dry-vacuum pump *F*, air receiver *H*, air-pressure control valve *I*, vacuum receiver *D*, vacuum release valve *E*, and motor-driven filtrate pump *G*.

General appearance of disk filters is shown in Fig. 247.

Other Types of Filters. *Plate-and-frame* pressure or suction *filters* are widely used for industrial purposes where the bulk of solids is small and the volume of fluid large. Their principal advantages are higher fluid capacity per square foot of filtering surface, better control of cake formation, possibility of better washing of cake. This is offset by the great disadvantage that they are intermittent instead of continuous. The best known are the Shriver filter press, the Sperry filter press, the Butters and Moore vacuum filter.

Centrifugal filters have been used for dewatering granular coal. They consist of a revolving basket of cylindrical or cylindro-

FIG. 247.—General view, six-disk American filter. (*Oliver-United Filters, Inc.*)

conical shape lined with a screen or other filtering surface. Centrifugal force causes the suspension to press on the filtering basket and the liquid to pass through the septum. They are used principally for dewatering granular coal to a low-moisture content.

Electrophoretic Dewatering. Clays may be dewatered by electrophoresis instead of by thickening and filtration.[6a] The suspension of clay in water (or, more properly speaking, in the suitable electrolyte) is subjected to d-c current. The kaolin and other clay particles migrate to one of the electrodes where they are deposited as a layer. The filter cake is then removed, intermittently as in filter presses, or continuously as in drum-type suction filters. If fine quartz is also present in the suspension, some preferential settling out of the quartz may take place, and thus yield a somewhat purer dewatered clay.

DUST ELIMINATION

Dust elimination[1] is a counterpart of clarification of fluids. The means used to eliminate dust are similar to those used to clarify solutions, with some notable differences. Dust elimination, although practiced in a number of leading mills, is by no means widespread, but it promises to develop considerably in the future, not only as a means of effecting additional savings, but also from the standpoints of bettering working conditions and of bettering relations between workers and management.

Since the viscosity of air is roughly one-sixtieth that of water, and since its specific gravity is negligible, particles that remain suspended in air for any length of time are roughly one-eighth to one-tenth the diameter of particles that remain suspended in water for any length of time (Eq. [VIII.4]). Practically speaking, dusts are made up of particles all finer than 10 microns. Of these, the particles in the size range from ½ to 2 microns seem to be the worst from a physiological standpoint, and the bulk of the weight is in particles coarser than 2 microns.

Dust elimination can be obtained by settling—a process analogous to classification—or by flocculating and settling—a process analogous to thickening—or by passage through bags—a process of filtration.

Dust Elimination by Settling. This can be practiced by mere passage of the dusty gases through an enlarged settling chamber. It permits removal of the coarsest particles and may therefore be economically sound, but as it does not remove most of the fine particles, it may fail from a physiologic standpoint: removal of the coarser dust particles takes place in nasal passages anyway.

An improved means of dust removal by settling is one in which centrifugal force is used along with gravity. If centrifugal force of the order of magnitude of one hundred times gravity is used, settling of dispersed dusts is readily obtained down to 1 micron and finer, a result that signifies elimination of many of the particles that are physiologically harmful.

Devices used to accomplish this end are known as *cyclones.* They consist of a cylindrical tank that receives the dust-laden air in a tangential stream at the top of the tank and delivers the dedusted gases as a rising stream along the axis of the tank. The centrifugal force is

$$f = m\frac{v^2}{r}, \qquad [XX.1]$$

in which f is force, m, mass, v, velocity of air stream, and r the radius of the cyclone. That is, the force is directly proportional to the tangential gas velocity squared, and inversely proportional to the radius of the cyclone. For effective dedusting, large velocities and small diameters are required. Large velocities in turn mean relatively large losses in gas pressure by friction (loss in "head") in the pipes.

A suitable modern cyclone is the small-diameter multiple-unit cyclone, or "multiclone" of which one unit is shown in Fig. 248.

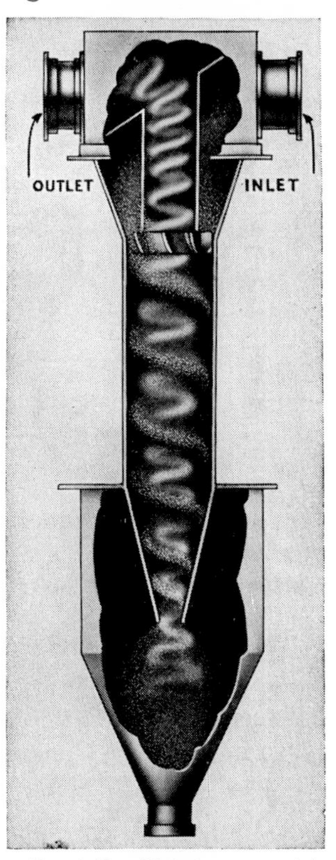

FIG. 248.—Unit from a multiple-unit cyclone of the *multiclone* type. (*Western Precipitation Corporation.*)

Dust Elimination by Flocculation and Settling. In view of the nonconducting character of air and gases in general, the electrical counterpart of flocculating electrolytes has to be somewhat different. The Cottrell treater[6] for furnace gases is the outstanding example of dust elimination by flocculation and settling.

In the Cottrell treater, the dust- or fume-laden gas is passed in a large chamber where an electrical field of steep gradient is set up. The flying ions have the effect of flocculating the dust which then settles rapidly to the bottom of the chamber, or on the electrodes toward which they are carried by ionic charges. Flocculation is obtained by a-c, continuous d-c, or intermittent d-c current. With d-c current, the results are better, especially since the flocculating action is supplemented by dust drift to the collecting electrode where deposition can take place in spite of substantial air currents in the chamber as a whole. Intermittent shaking of the electrodes results in settlement of all or substantially all the dust. Figure 249 is a typical installation.

The efficiency of electrical precipitation depends on the time the dust-laden gas remains in the electrical field, on the distance

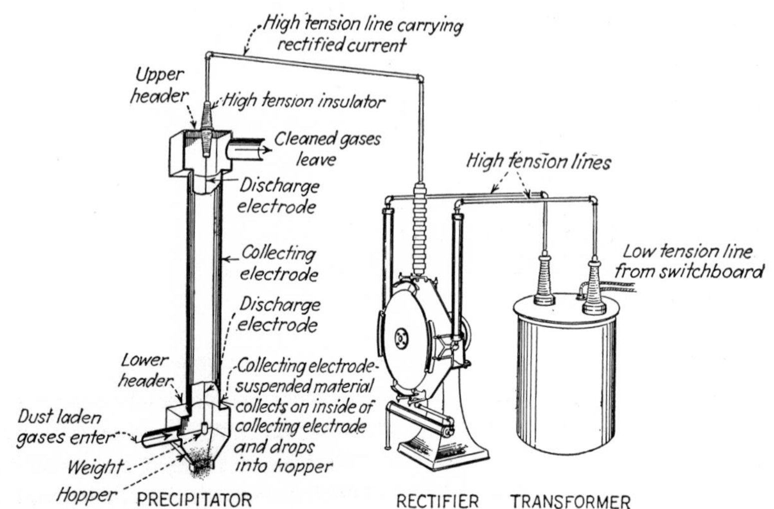

Fig. 249.—Schematic diagram of Cottrell dust precipitator. (*Western Precipitation Corporation.*)

between electrodes, and on the electrical-field gradient. For a given installation, the efficiency E can be expressed[1a] as

$$E = 1 - K^t, \qquad [XX.2]$$

in which K is the precipitation factor, depending on the nature of the gas, of the suspended matter, and on the type of precipitator; t is, of course, the time the gas remains in the electrical field.

A new method of eliminating dust by flocculation has been proposed in which ultrasonic waves are utilized. Ultrasonic waves are material waves too short to be heard and of very high frequency. The ultrasonic waves concentrate the dust at their nodes; the dust particles are brought in close proximity, and flocculation follows.[9]

Dust Elimination by Synthetic Fogs and Rains. This, an imitation of nature's way of eliminating dust, is very attractive as well as practical. The water acts as a bond between particles, much as oil acts as a bond in agglomeration (*cf.* Chapter XIX). The practical problems are to create a curtain of rain across a flowing gas stream, to collect the muddy water, free it of solid content and return it to the rainmaker.

If the temperature of the gases is not far from that of the water, the scheme is easy to carry out. Air conditioning is simultane-

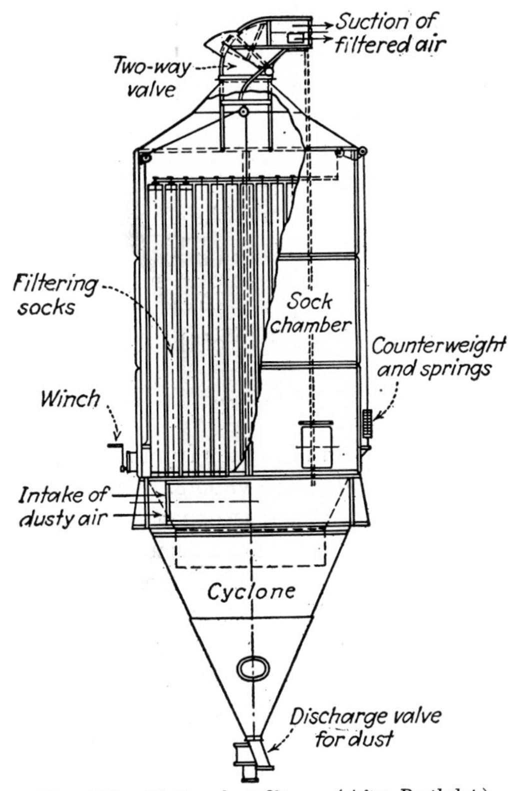

Fig. 250.—Birtley dust filter. (*After Berthelot.*)

ously obtained, a result that has tangible commercial value at the same time since it induces more efficient work.

A simple device consists of a revolving drum that picks up water and sprays it across the gas stream; the water then falls on a flowing stream of water.

Another simple device is the Laurent hydrocaptor consisting of a hollow drum fitted with screen surfaces and revolving so as to dip in and pick up water which drops as a curtain across the dusty air.[2b]

Dust Elimination by Filtration.[7] Baghouses are a familiar installation in many smelters. They are filters for the separation of dusts and fumes from the gases in which they are suspended. Baghouses are regarded as a necessary evil, as they are both bulky and expensive, and as their product is disagreeable material to handle. Fortunately, they are not yet necessary in mills except perhaps for the pneumatic dressing of coal,[10,17] *e.g.*, in plants using the Birtley process[2a] (Fig. 250). The bottom part of the Birtley filter is a cyclone that operates to remove the coarser particles, as has already been explained. Above are placed a number of filter socks that retain the fines. Every 2 hr. or so, suction is cut off and the upper part of the filter is connected with the atmosphere. This reversal of pressure, equivalent to the blowback of pulp in suction filters, clears the clogged pores. Discharge of the settled dust is semiautomatic, a flap valve opening when the weight of the dust becomes excessive.

DRYING

Drying is the separation of liquids from solids by evaporation of the liquid and removal of the vapors. Compared with thickening or with filtration, drying is expensive, *i.e.*, if the cost is figured per unit of fluid removed. This is because a great quantity of energy has to be supplied to heat and evaporate the liquid, whereas in filtration a relatively small quantity of energy is required to overcome friction of the filter bed, and in thickening a quantity of energy even smaller is required to overcome mechanical friction in the thickener.

Where the liquid is water, as is always the case in mineral-dressing operations, the quantity of energy per unit mass of liquid is particularly large because the specific heat and latent heat of evaporation of water are very large.

In spite of the objection to drying which has just been stated, there is a definite place for that process because filtration cannot reduce the moisture content below a minimum which may be too high for some ulterior processing, or for marketing purposes. In some cases also, the saving in freight creditable to drying may exceed the cost of drying.

Methods of drying include flash drying, which is accomplished by dropping the material in a tower against a rising current of hot gases; rotary drying, which is accomplished by turning the material over and over in a cylinder against or with a current of hot gases; and rubble-hearth drying, which is accomplished by mechanically turning over the material on a horizontal hearth in a current of hot gases.

Heating is indirect in most cases, but directly heated rotary driers are also used.

There are a number of installations for the heat-drying of washed coal. Parmley[16] gives costs.

Literature Cited

1. ANON.: "Methods for Determination of Velocity, Volume, Dust and Mist Content of Gases," *Bull. W.P.* 50, Western Precipitation Corporation, Los Angeles (1936).

1*a*. ANON.: Catalog W.P. 1, Western Precipitation Corporation, Los Angeles.

2. BERTHELOT, CH.: "Épuration, séchage, agglomération et brovage du charbon," Dunod, Paris (1938); (*a*) p. 106; (*b*) p. 137.

3. BINGHAM, E. C., and BAXTER LOWE: The Nature of Flow, in "Colloid Symposium Annual," Vol. VII, John Wiley & Sons, Inc., New York (1930), pp. 205–212.

4. BINGHAM, EUGENE C.: Plasticity in Colloid Control, in "Colloid Symposium Monograph," Vol. II, Reinhold Publishing Corporation, New York (1925), pp. 106–113.

5. COE, H. S., and G. H. CLEVENGER: Methods for Determining the Capacities of Slime-settling Tanks, *Trans. Am. Inst. Mining Engrs.*, **55,** 356–384 (1916).

6. COTTRELL, F. G.: Electrical Fume Precipitation, *Trans. Am. Inst. Mining Engrs.*, **43,** 512–520 (1912).

6*a*. CURTIS, CARL E.: The Electrical Dewatering of Clay Suspensions, *J. Am. Ceram. Soc.*, **14,** 219–263 (1931).

7. EILERS, A.: Notes on Bag Filtration Plants, *Trans. Am. Inst. Mining Engrs.*, **44,** 708–735 (1912).

8. FREE, E. E.: Colloids and Colloidal Slimes, *Eng. Mining J.*, **101,** 681–686 (1916).

8*a*. FREE, E. E.: Rate of Slime Settling, *Eng. Mining J.*, **101,** 429–432, 509–513 (1916).

9. GOTTSCHALK, V. H., and H. W. ST. CLAIR: Use of Sound and Supersonic Waves in Metallurgy, *Mining and Met.*, **18,** 244–247 (1937).

10. HEBLEY, HENRY F.: The Dedusting of Coal, *Trans. Am. Inst. Mining Met. Engrs.*, **108,** 88–127 (1934).

10*a*. HOUWINK, R.: "Elasticity, Plasticity, and Structure of Matter," Cambridge University Press, Cambridge (1937).

11. Hixson, Arthur W., Lincoln T. Work, and Isaac H. Odell, Jr.: Mechanism of Filtration, *Trans. Am. Inst. Mining Met. Engrs.*, **73,** 225–238 (1926).
12. Laist, Frederick, and Albert E. Wiggin: The Slime Concentrating Plant at Anaconda, *Trans. Am. Inst. Mining Engrs.*, **49,** 470–484 (1914).
13. Mishler, R. T.: Settling Slimes at the Tigre Mill, *Eng. Mining J.*, **94,** 643–646 (1912).
14. Mishler, R. T.: Methods for Determining the Capacities of Slime-thickening Tanks, *Trans. Am. Inst. Mining Engrs.*, **58,** 102–125 (1918).
15. Nichols, H. S.: Theory of the Settlement of Slime, *Mining Sci. Press*, **97,** 54–56 (1908).
16. Parmley, S. M.: Heat Drying of Washed Coal, *Trans. Am. Inst. Mining Met. Engrs.*, **94,** 336–350 (1931).
17. Patterson, Charles H. J.: Dust Collection in Pneumatic Cleaning Plants, *Trans. Am. Inst. Mining Met. Engrs.*, **94,** 351–354 (1931).
18. Ralston, Oliver C.: The Control of Ore Slimes, *Eng. Mining J.*, **101,** 763–769, 890–894 (1916).
19. Stewart, R. F., and E. J. Roberts: The Sedimentation of Fine Particles in Liquids, *Trans. Inst. Chem. Engrs. (London)*, **11,** 124–137 (1933).
20. Young, George J.: Slime Filtration, *Trans. Am. Inst. Mining Engrs.* **42,** 752–784 (1911).
21. Comings, E. W.: Thickening Calcium Carbonate Slurries, *Ind. Eng. Chem.*, **32,** 663–667 (1940).

CHAPTER XXI

AUXILIARY OPERATIONS

Auxiliary operations include storage, conveying, disposal of products, sampling, and weighing. As explained in Chapter I, these auxiliary operations perform no dressing operations proper; but dressing operations cannot be conducted without them. In detailed flow sheets, in total space occupied, and in capital costs, the devices for auxiliary operations represent a large proportion of the whole.

Storage is made necessary by one of the following considerations or by a combination of them: (1) intermittent reception of the ore, (2) intermittent disposal of the products, (3) adoption of intermittent mill operations, (4) existence of surges in crushed-ore flow or pulp flow, and (5) desirability of providing against possible breakdowns or against a possible interruption in the reception of ore or in the disposal of the products.

Conveying is necessary to accomplish one of the following objectives or a combination of them: (1) transport ore, water, pulp, etc. from one device to another, (2) feed broken ore or pulp, and (3) distribute broken ore or pulp from a single stream into several streams in parallel. Of these, the first aim is the most important. It can be subdivided into transport by gravity and mechanized transport.

Usually, concentrates are sold under contracts. Examples of contracts for base-metal concentrates have been given in Chapter I. Tailings may be sold or given away for road building, railroad ballast, or as sand; they may also be used as a cheap and convenient fill for empty mine stopes; but usually they are allowed to accumulate in tailing dumps.

Sampling and weighing are control operations designed to check up on mill operations and to permit the formulation of definite and accurate accounting.

STORAGE[21a]

Depending upon the nature of the material treated, storage is accomplished in stock piles, bins, tanks, or ponds.

Stock Piles. Stock piles are used to store outdoors coarse material of low value, especially if the duration of storage is extensive. An example is the storage of iron ore at Lake Superior mines. In designing stock piles, it is merely necessary to know the angle of repose of the broken ore, the volume occupied by the broken ore (cubic feet per ton), and the tonnage. The ore is dumped from a trestle so as to form a cone. Retrieving is done by shovels or draglines.

Bins. Bins are used more extensively than stock piles. They are applicable to dry, drained, or filtered material where the

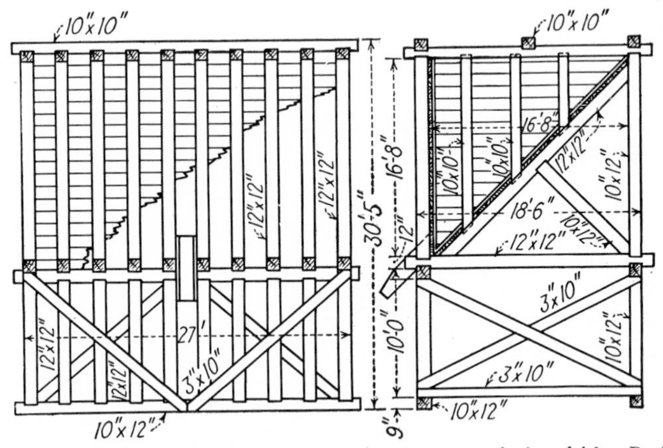

Fig. 251.—Slanting-bottom timber ore bin (150-ton capacity). (*After Barbour.*)

storage period is short. Bins are built of wood, concrete, or steel[2] in one of many shapes and are so designed that their contents can be discharged by gravity through a suitable gate, or by a suitable feeder.

Flat-bottomed bins are rectangular boxes, usually of wood. Because of this shape, the bins cannot be emptied completely and must retain a substantial tonnage of dead rock. This disadvantage is compensated by three advantages: (1) The dead rock can be barred down in times of emergency, (2) the first cost of flat-bottomed bins is lower than that of other types of bin, (3) there is no wear on the bottom of the bin.

Sloping-bottomed bins can be built to discharge on one side, two sides, or from below. They are usually built of wood (Fig. 251).

Steel bunkers may be built in the form of a circular or of an elliptical cylinder, or with a catenary profile; the latter is regarded as particularly suitable from an engineering standpoint.[10]

To design bins, it is required first to ascertain the tonnage for which provision is to be made. The second step requires application of the mechanics of semifluids as worked out by Coulomb, Rankine, Cain, Janssen, Airy,[1] and Ketchum to determine the loads and pressures at different points. The third step is the choice of suitable members to bear the loads and pressures. For the design of bins, the reader is referred to Taggart [21] or to Ketchum.[15]

The capacity for which a bin is designed depends, of course, on the service which is expected of it. A surge bin placed between a jaw crusher and a stamp mill need not have a capacity greater than is required to supply the stamps for a few minutes provided the jaw crusher is operating during the same period as the stamps. On the other hand, a bin expected to take up the slack between a crushing plant operating one shift per day and a grinding plant operating continuously should have a capacity not smaller than a 24-hr. ore supply. A bin expected to harmonize a mine working six days per week and hoisting during one shift only with a continuous mill should have a capacity equal to three days' ore supply: two days' capacity is required to meet the periodic nonproductive period of the mine, and one day's capacity is required to meet possible interruptions in the mill or at the mine. Generally speaking, the total bin capacity of a mill should not be less than two days' supply, and the bins should be kept about half full. Bin capacity depends, of course, on local conditions, and may have to be considerably in excess of the minimum.

Tanks. Tanks are used for storing suspensions of fine particles whenever it is desired to give some chemical reaction an opportunity to proceed. Provision must be made to keep the suspension from settling, to evacuate the "conditioned" pulp, and to vary the duration of the conditioning. Suitable agitator tanks are similar to thickeners, but of smaller cross section and greater depth. If oxygen is to participate in the reaction, violent aeration is used, the air being introduced by mechanical means or by an air lift. The equipment is that devised for hydrometallurgical operations.

Tanks are also used to minimize surges in the composition of pulped products (as of flotation middlings). To that end, a tank providing storage for a quantity of pulp equivalent to that flowing in 15 to 45 min. is usually sufficient.

Storage of Water. Ponds are used for storing water, a precious commodity for the mill man, without which he may well be compelled to cease operating. The water supply in many mining districts is inadequate, so the water in tailings and concentrates must be retrieved. In some districts, the water supply is alternately abundant and scarce. In other cases, the water is of inadequate quality and must be purified by settling and by the addition of chemicals. Storage and purification of water may reach considerable proportions and become a large item of expense in the operation of a mill.

It is customary to store water by disposing the tailing in the form of a dam. This operation, known as impounding, allows a pond to form behind the dam. The disposal of the tailing thus results in storage of water. Water is pumped from the pond to a small reservoir or tank built at a higher level than the mill, from which it flows by gravity and with constant head to wherever it is needed. It is a good practice* to design this reservoir for a capacity sufficient to care for mill requirements for 8 hr. or more. This is sufficient time for repairing tank supply lines or pumps.

Capacity of the pond should be ample to allow for the most adverse seasonal conditions, including evaporation losses, water retention in the tailing and concentrate, and temporary failure of the water supply.

Water requirements are extremely variable. Dry plants, of course, use no appreciable quantity of water, then come flotation plants with 3 to 5 tons of water per ton of ore, and gravity plants with 10 to 30 tons of water per ton of ore. If full reclamation is practiced, the net water makeup may be as low as 20 gal. per ton, as in some collieries, and may remain as high as 1.0 to 1.5 ton per ton of ore, as in some flotation plants.

CONVEYING

Railroad, aerial tramway, and truck transportation are not employed as an integral part of a mill, but they constitute the principal means of influx to and efflux from the mill.

* Private communication, George G. Griswold.

Within the mill, conveying is a continuous operation. Gravity transport is the flow of crushed solids or that of a suspension of solids in water in which the actuating force is gravity. Gravity is resisted by frictional forces within the stream of crushed solids or of pulp, and at the surface of the conveyance on which the stream is moving. Gravity transport is carried out in chutes and launders.

Mechanized transport requires the supply of energy. For crushed solids, it takes the form of bucket elevators, belt conveyors, pan conveyors, vibrating conveyors, screw conveyors; for suspensions in water, it takes the form of centrifugal or diaphragm pumps, bucket elevators, tailing wheels, and air lifts.

Chutes. Chutes are steeply inclined troughs of rectangular section for gravity transport of dry solids. The slope should be sufficient for free flow; this depends on the coefficient of friction between the rock and the material of which the chute is made. On bright steel,[14] a slope of 15 to 25° is sufficient to start sliding and 14 to 22° to keep particles sliding (dynamic friction). Steeper slopes are required for wood.[20] In practice, slopes for dry material are from 40 to 45° for ore on wood and 25 to 35° for coal on steel. Since most ores are more or less damp, steeper slopes are necessary. A slope of 45° is minimum for steel bottoms, and 55° is much safer.

Launders. Launders are gently sloping troughs of rectangular, triangular, or semicircular section for gravity transport of suspensions of ore or mineral in water. The solid is carried in suspension, by sliding, or by rolling. The slope must increase with particle size, with the solid content of the suspension, and with specific gravity of the solid. The effect of depth of water is complex: if the particles are carried in suspension, a deep flow is advantageous as it speeds up the pulp; if the particles are carried by rolling, a deep flow may be disadvantageous.[3,12,18,19]

Launders are most commonly built of wood with a rectangular cross section, but occasionally they are built of metal or concrete. The latter is adopted in large, fireproofed plants.

For conveying fine material (less than 1 mm. in diameter), launders are generally unlined. For coarse material, a lining is used. Wood, canvas belting, rubber belting, sheet iron, cast slag,[16] glass, and concrete have been used. Wooden linings are short-lived but cheap and light. Old rubber belting is also suitable.

For details as to slope and construction, reference should be made to a handbook.

Bucket Elevators. Bucket elevators[8,9,13,22] consist of an endless chain or belt to which a number of buckets are fastened (Fig. 252). The chain travels vertically, or at a steep angle, and is driven from the head pulley. Material is fed to the elevator near the base, or "boot," directly into the buckets, or it is fed into the boot where the buckets pick it up.

Discharge is at the top where the buckets turn from an upright to an upside-down position and is facilitated by the centrifugal throw of their contents on passage around the head pulley. In some bucket elevators, the effect of centrifugal force is more important than in others.

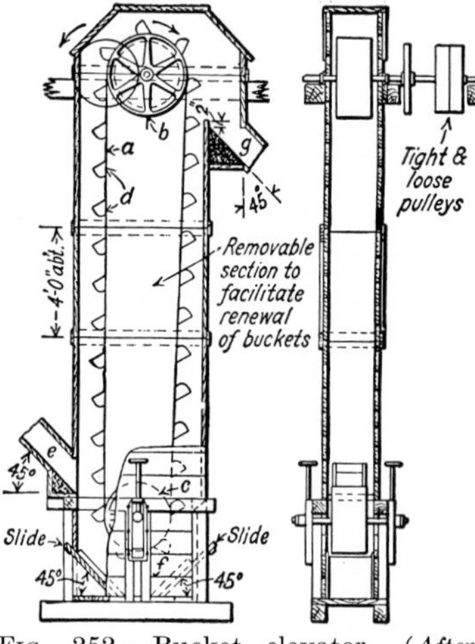

FIG. 252.—Bucket elevator. (*After Taggart.*)

Fast-moving elevators or those in which dust or splash occurs require a housing. This is usually made of wood, *e.g.*, of 1-in. planks. Slow elevators, or those lifting coarse dry material, may be open. Boots are usually made of cast iron. Buckets are usually made of iron or steel; they may have replaceable wearing edges. Replacement and upkeep of belt and buckets are the largest elements of cost.

Bucket elevators have the greatest range of usefulness of any elevating contrivance: they can handle coarse dry material, fine dry material, coarse pulps, or fine pulps. For the elevation of coarse dry material, belt conveyors are generally preferred; for the elevation of fine pulps, pumps are preferred. But for the elevation of coarse pulps, or of sticky, thickened concentrates, bucket elevators are best.

Belt Conveyors.[17] A belt conveyor consists of a belt passing around a head and a tail pulley and resting on troughing idlers for the carrying run and on return idlers for the return. Drive is through the head pulley if that provides enough wrap for the drive or through special driving pulleys arranged to give an angle of belt wrap much in excess of 180°. The idlers are of two kinds: cast iron with babbitted hubs and grease-cup lubrication, and cast or pressed steel with ball-bearing support. The latter are more expensive but save much power. Idlers on the carrying run are troughed as this increases capacity; they are spaced 3 to 5 ft. apart. Return idlers are not troughed, and the spacing is about 10 ft. Belts are of rubber-covered cotton duck, or of

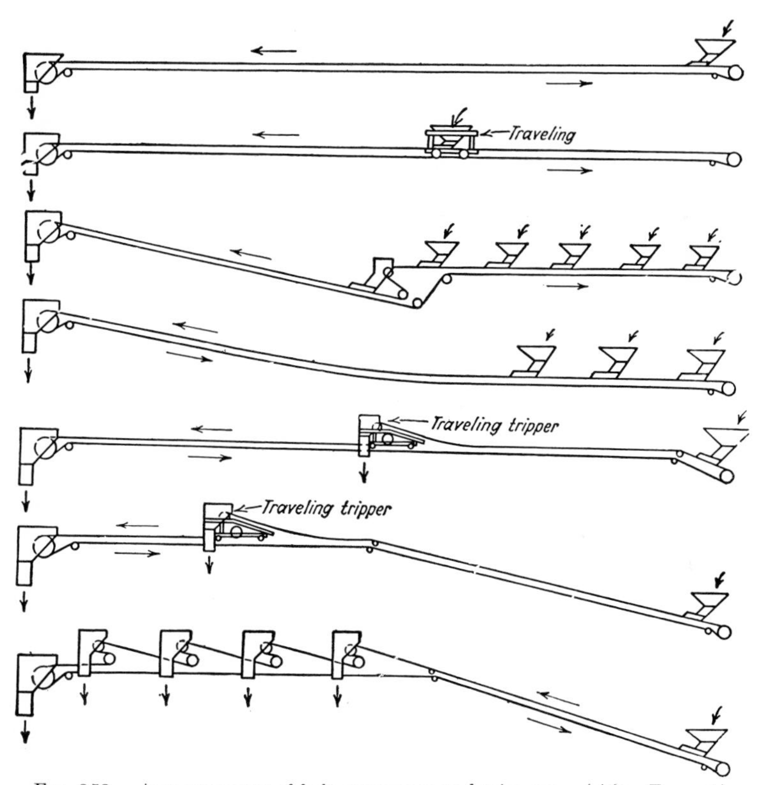

Fig. 253.—Arrangements of belt conveyors and trippers. (*After Taggart.*)

waterproofed but otherwise uncovered canvas (balata belt). The belt may have 3 to 10 or more plies of the cotton duck or canvas, and the rubber covering may be from $\frac{1}{16}$ to $\frac{1}{4}$ in. thick. Since much of the wear is in the center, belts can be built with extra-thick rubber or with more plies at the center. Most belts in the United States are of the rubber-covered type. Endless steel belts have been used in Germany.

Conveyor belts can be used for conveying horizontally, for conveying up or down a gentle incline; they can receive feed from a variable point, as from under a long multiple-discharge bin; they can distribute feed through traveling trippers along a long bin, etc. These arrangements are summarized in Fig. 253.

The maximum permissible slope depends upon the coefficient of friction between the belt and the material that is being transported. Usual maximum is 18 to 20°, but this may be increased to 24° to 25° in favorable cases, as with evenly crushed, dry, granular material at fairly high speeds. Again the maximum slope may be under 15° if somewhat wet, sloppy tailing is being transported.

The largest item of cost in the operation of belt conveyors is the upkeep, and of that perhaps one-third is the cost of the belt proper. Care of the belt is then good economy. One of the simple means to lengthen belt life is to prevent coarse, sharp abrasive material from coming in direct contact with the belt. This is easily done where the belt carries the discharge of a crusher; a by-passing screen or grizzly can always be installed so as to spout the undersize on the belt ahead of the oversize discharge. On standing, belts deteriorate faster than any other equipment in a mill; this is because of the aging of the rubber.

Approximate power requirement is given by Eq. [XXI.1] for conveyors with ordinary idlers and by Eq. [XXI.2] for conveyors with ball-bearing idlers.

$$HP = \left(\frac{0.02L}{100} + \frac{0.01H}{10}\right)T \qquad \text{[XXI.1]}$$

$$HP = \left(\frac{0.0087L}{100} + \frac{0.01H}{10}\right)T, \qquad \text{[XXI.2]}$$

in which L is the horizontal length in feet, H is the rise in feet, and T the tonnage per hour.

Pan and Apron Conveyors. Pan and apron conveyors consist of articulated steel pans carried on chains. They are driven at low speeds and are suitable for coarse, abrasive, or hot material that cannot be carried on belt conveyors. They are usually horizontal, but slopes up to 15° to 30°, depending on the shape of the pans, may be used. Their principal application in mineral dressing is in feeders.

Vibrating Conveyors. Vibrating conveyors operate on the same principle as some vibrating screens, *e.g.*, the Jeffrey-Traylor vibrating screen. They consist of a vibrator and a conveying surface. They are well suited for the transport, either horizontally, or at a gentle upward slope, of dry, granular solids, but may give trouble with damp feed. Wear is very low on this type of conveyor.

Screw Conveyors. Screw conveyors consist of a spiral blade attached to a revolving shaft; they operate on a principle similar to that of the Akins classifier. Screw conveyors are suitable for the conveyance of sandy materials, moist or dry, for short distances, and where space is lacking for the installation of other devices. The wear is high.

Centrifugal Pumps. Centrifugal pumps are used for two duties: (1) elevating water or clear solutions, and (2) elevating sandy or slimy suspensions of solids having a substantial content of abrasive solids.

Centrifugal pumps consist of a chamber inside of which an impellor revolves so as to centrifuge the fluid. The chamber is round but with gradually enlarging radius up to an effluent pipe. The velocity of the effluent is equivalent to a head; by it the fluid is pushed up to the desired level. The velocity head should equal the lift plus frictional losses in the pump and pipe. The net rise rarely exceeds 50 ft., but installations with a lift of 75 to 100 ft. have been made. Multilift pumps are used for larger lifts (largely on the main water supply).

The efficiency of centrifugal pumps depends upon the rate of fluid flow, the head, and the speed. Well-designed pumps show an efficiency of 50 to 75 per cent, according to their size, exclusive of pipe-line friction and power-transmission losses, when operated at the optimum capacity. Practical average efficiency, including pipe-line loss, power-transmission loss, and consideration for the fact that operation is not always possible at optimum capacity, may be as low as 30 per cent.

The principal items of expense, besides that for the electrical energy, are for the replacement of worn impellors (runners) and liners, especially where a sand is elevated. In elevating water, power is, of course, the principal item of expense.

Centrifugal pumps are of two types, *viz.*, the stuffing-box type and the glandless type (Fig. 254) typified by the Wilfley sand pump. The latter type eliminates leakage of grit into bearings and has made pumping a convenient alternative for the elevating or conveying of sandy materials.

Diaphragm Pumps. Diaphragm pumps are suction pumps limited in lift by atmospheric pressure. The maximum lift for water equals the atmospheric pressure (expressed in feet of water) minus the vapor pressure of water, minus the frictional losses. For pulp of a specific gravity higher than water, it is proportionately smaller.

FIG. 254.—Sectional view of Wilfley sand pump. (*A. R. Wilfley and Sons.*)

A diaphragm pump consists of a chamber with a valve at the bottom and a movable valved diaphragm at the top. The diaphragm reciprocates fifteen to one hundred times per minute. In each cycle, pulp is sucked through the bottom valve on the upstroke of the diaphragm; at the end of the upstroke, the lower valve closes and the upper valve opens; during the downstroke, pulp is passed from below to above the diaphragm; at the end of the downstroke, the diaphragm or upper valve closes and the lower valve opens, to start a new cycle.

Tailing (Sand) Wheels. Tailing wheels are large rotating wheels up to 60 ft. in diameter fitted with buckets. The buckets scoop up from a pit and discharge near the top. New lift is 70 to 80 per cent of the diameter of the wheel and power efficiency 40 to 50 per cent. First cost is high. Generally speaking, centrifugal sand pumps and bucket elevators are preferable because of lower first cost and reduced requirement for space.

Air Lifts. Air lifts are devices for elevating water, solutions, slime pulps, or even sand pulps. The expansion of compressed air is utilized as a source of energy; the compressed air is introduced into a pipe at a considerable submergence below the free level of the liquid. As the specific gravity of the column of pulp and air is less than that of pulp, the level in the pipe will rise higher than the free level of the liquid. If the discharge of compressed air is sufficiently rapid, and if furthermore it is in the form of small bubbles, there is not much tendency for the bubbles to rise without pushing the pulp ahead of them (slippage), and an efficient operation is obtained.

The advantage of air lifts over other methods of pulp elevation is the absence of wear. On the other hand, compressed air is required instead of the more widely available electricity, and deep wells for submergence are needed. Submerged length of pipe is from twice to two-thirds of the emerged length of lift, being proportionately more for low lifts.

FEEDING

Feeding is essentially a conveying operation in which the distance traveled is short and in which close regulation of the rate of passage is required. The rate of passage of the material treated in crushers, grinders, screens, classifiers, concentrating machines, etc., has considerable influence on the quality of the work done; and best results are obtained if a uniform rate of feed is maintained. Where the operations that follow each other are at the same rate, it is unnecessary to interpose feeders; but where principal operations are interrupted by a storage step, it is necessary to provide a feeder.

A typical feeder consists of a small bin (possibly attached to, or an integral part of, a large bin) with a gate and a suitable conveyor. Feeders of many types have been designed, notably chain feeders, apron feeders, pan feeders, belt feeders, roller feeders, rotary feeders, reciprocating-plate feeders, plunger feeders, revolving-disk feeders, and vibrating feeders.

Figure 255 represents an *apron feeder*. The rate of discharge is controlled by raising or lowering a gate (not shown in cut) or by modifying the speed of the feeder.

FIG. 255.—Apron conveyor and feeder, heavy type, with steel plate pans. (*Allis-Chalmers Manufacturing Co.*)

Chain feeders are used to regulate the feed to jaw crushers. They consist of a small bin with an opening blocked by heavy chains. Discharge is obtained by slow revolution of the chains, and the rate of discharge is controlled by varying the speed of the chains.

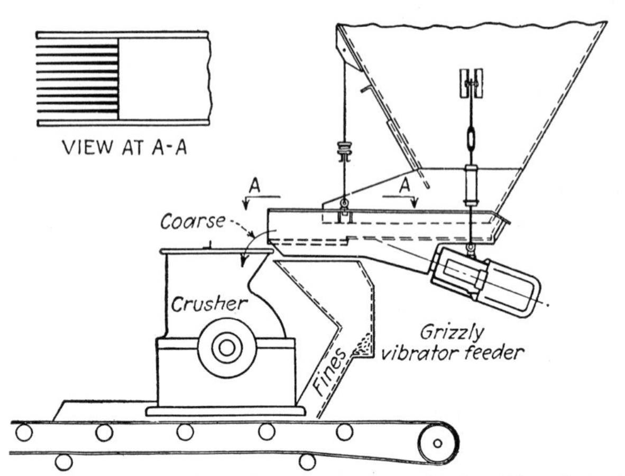

FIG. 256.—Jeffrey vibrating feeder and grizzly. (*Jeffrey Manufacturing Co.*)

Figure 256 represents a combined *vibrating feeder* and grizzly. It is operated by mechanism similar to that of electromagnetically actuated screens.

Figure 257 represents a *revolving-disk* type of *feeder* (Challenge feeder). This is widely used in conjunction with gravity stamps. Motion of the central stamp causes a jerklike circular progression of the revolving disk that forms the bottom of the hopper. The feeder is adjusted by changing the downward return of the lever *B* attached to the friction pawl *D* by means of the tension of the spring *N*.

Reagent feeders for use in flotation have already been mentioned.

Flow Distribution. Distribution of a stream of solids can be accomplished by placing one or more crude splitters across the flow. If more accurate splitting is desired, a device based on the principle of mechanical samplers can be used.

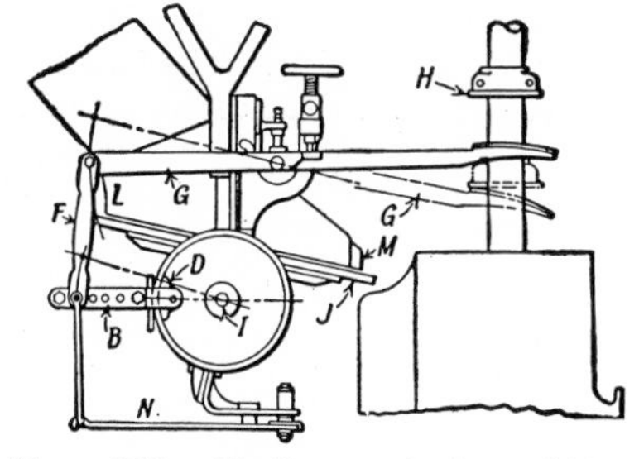

FIG. 257.—Challenge feeder. (*After Taggart.*)

Distribution of a stream of pulp is usually made by means of a revolving distributor. This consists of a spout revolving about a vertical axis in which the whole stream is received. Discharge from the spout falls into a compartmented annulus from which pipes lead to the machines in which the next operation is performed.

SAMPLING

Sampling is the art of securing in a small weight or *sample* a representative fraction of a relatively large lot. To be representative, a sample must contain an adequate number of particles, and the sample must have been selected in a fair way.[5,7,11]

Size of Sample. The minimum number of particles in a sample depends upon (1) the approximate content of the substance for which an assay or chemical analysis is to be made, (2) the accuracy of sampling that is sought, and (3) whether the material consists of free or of locked particles, and if locked, upon the nature and character of the locking.

By reasoning back from the theory of errors, it can be shown that the number of particles n (assumed to be all of the same grade, and free) required is

$$n = 0.45\frac{x}{y^2}. \qquad \text{[XXI.3]}$$

In this equation, x is the approximate volumetric content of the substance for which an assay is to be made, and y is the probable volumetric error of sampling.*

Equation [XXI.3] can be established as follows: if a sample of n particles is actually made up of nx free particles of mineral A, and of $n(1 - x)$ particles not containing mineral A, each individual particle deviates from the mean by $(1 - x)$ or by x, depending upon whether it is a particle of A, or some other particle. The frequencies of these deviations are, of course, nx and $n(1 - x)$. Hence the probable error of the sample is

$$y = \frac{0.6745}{\sqrt{n-1}\sqrt{n}}\sqrt{[(1-x)n]x^2 + [xn](1-x)^2}.$$

This result is obtained by substituting frequencies and deviations in the standard equation for deviation of the mean in the Theory of Errors.[21b] This equation can be simplified to

$$y = \frac{0.6745}{\sqrt{n-1}}\sqrt{x(1-x)}.$$

If x is small $(1 - x) \rightarrow 1$, and if n is large $\sqrt{n-1} \rightarrow \sqrt{n}$. Hence, as an approximation

$$y = 0.67\sqrt{\frac{x}{n}} \qquad \text{or} \qquad n = 0.45\frac{x}{y^2}.$$

In applying Eq. [XXI.3] to a crushed product finer than a limiting screen size, it is necessary to make some assumption as to the average size of the particles in the product. It is more than safe to assume the product to have the size of the openings in the limiting screen. However, it seems reasonable to take as average a volume mean,[21c] or $D = \left(\frac{\Sigma nd^4}{\Sigma nd^3}\right)$ which gives as answer a size approximately 0.7 times the limiting sieve size. Since, furthermore, crushed particles are of irregular shape, averaging in volume perhaps one-half the volume of cubes of equivalent size, the effective number of particles n' per gram is,

* The *permissible* error, that is, the largest error allowable, should be smaller than the probable error, and is usually one-third as large (Priv. Comm., O. Cutler Shepard).

as an acceptable approximation,

$$n' = \frac{6}{\Delta a^3}, \qquad \text{[XXI.4]}$$

in which Δ is the specific gravity of the particles and a the opening of the limiting screen, expressed in centimeters.

The following examples will illustrate the use of formulas [XXI.3] and [XXI.4].

Example 1. The assay of a gold ore reduced to pass a 150-mesh screen is about 0.1 oz. per ton. What weight of sample is required to assure correctness of sampling to 0.005 oz. per ton, if the gold is all free, and if the specific gravity of the rock is 3.0?

The weight proportion of gold to gangue is roughly 1 in 300,000; the volume proportion is then roughly 1 in 2,000,000. Hence x in Eq. [XXI.3] is 0.000,000,5. Likewise y is 0.000,000,025. From these data,

$$n = 0.45 \frac{0.000{,}000{,}5}{(0.000{,}000{,}025)^2} = 0.36 \times 10^9 \text{ particles.}$$

From Eq. [XXI.4], the number of particles per gram is

$$n' = \frac{6}{3 \times (0.0104)^3} = 1.8 \times 10^6.$$

The weight required is then $W = (0.36 \times 10^9)/(1.8 \times 10^6) = 200$ g. Clearly, the sample should weigh several "assay tons."

Example 2. A flotation tailing reduced to pass a 100-mesh screen contains about 0.3 per cent copper (as chalcopyrite). What weight of sample is required to assure correctness of sampling to 0.01 per cent Cu, assuming the chalcopyrite to be all free, and the specific gravity of the tailing to be 3?

The volume content of chalcopyrite is about $(0.3/0.345) \times (3/4.2)$, or 0.6 per cent.

Hence $x = 0.006$. Likewise, $y = 0.0002$ and

$$n = 0.45 \left[\frac{0.006}{(0.0002)^2} \right] = 0.67 \times 10^5 \text{ particles.}$$

From Eq. [XXI.4], the number of particles per gram is

$$n' = \frac{6}{3 \times (0.0147)^3} = 0.64 \times 10^6.$$

The weight required is $(0.67 \times 10^5)/(0.64 \times 10^6) = 0.1$ g.

Example 3. Gold in an ore of the same assay as that of No. 1 occurs as particles averaging 1 mm. in size. What is the minimum size of sample that will be accurate to 0.005 oz. per ton?

As in Example 1, the number of particles required is 0.36×10^9. Since the number of particles per gram is now only $2 \times \frac{1}{3} \times \frac{1}{(0.1)^3}$, or 666, the weight required is $W = 0.36 \times 10^9/666 = 540{,}000$ g., or about 0.5 ton.

Example 4. Gold in an ore of the same assay as that of No. 1 occurs as grains averaging 10 microns in size. What is the minimum size of sample (reduced to 150-mesh) that will be accurate to 0.005 oz. per ton?

As in Example 1, the number of particles required is $n = 0.36 \times 10^9$. Since the grain size is much smaller than the particle size (see Chapter IV), the effective size to use is the grain size. Here

$$n' = 2 \times \frac{1}{3} \times \frac{1}{(0.001)^3} = 0.67 \times 10^9$$

and

$$W = \frac{0.36 \times 10^9}{0.67 \times 10^9} = 0.5 \text{ g.}$$

These examples make it clear that the individual makeup of an ore exerts a tremendous influence on the minimum size of sample required to give results of any one particular accuracy. It has long been assumed that particle size is of great importance in this connection. That is true provided the particles are free. But if the particles are locked, the true yardstick is not particle size but grain size.

Special difficulty is encountered in precious-metal ores since y is very small. That is, the size of sample for a given accuracy becomes fairly large. In other cases, the samples that are taken are usually so generous that the problem can be dismissed as nonexistent. It is likely, indeed, that in dealing with ores other than gold ores the samples currently taken in practice are often hundreds or thousands of times larger than they need to be in relation to the analytical accuracy that is sought.

In dealing with gold ores in which part of the gold is coarse, a solution is afforded by the malleability of the metal: the gold can be screened from a large sample as flattened particles (*metallics*), and the remainder assayed as usual with the appropriate correction for "metallics" inserted in the final answer. But it is futile to make a correction for "metallics" unless a sufficiently large sample is crushed and screened. As shown in Example 3, this may become considerable under adverse circumstances.

Sampling Procedure.[4,6] The foregoing has indicated sufficiently what considerations must guide one in choosing the size of sample. In taking the sample either manually or by machine, the great danger lies in systematic errors which may be introduced either by the sampling procedure or by segregation in the ore stream or ore pile.

Grab samples are easy to take but inaccurate; generally the results are high.

Sampling by coning and quartering has been used widely; but it can be made systematically inaccurate in the hands of a skilled and unscrupulous operator.

Shovel sampling is another manual sampling method. Although not more accurate than coning or quartering, it lends itself less to tampering.

Several *mechanical samplers* have been designed; their use in mills is spreading as the demand for accurate knowledge of operating conditions is increasing.

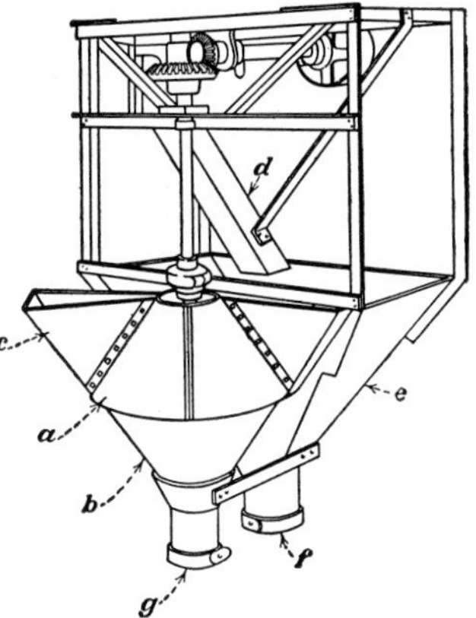

Fig. 258.—Vezin sampler. (*After Taggart.*)

An accurate mechanical sampler must be designed to cut each part of a stream of broken ore or pulp for the same proportion of the total time; otherwise systematic segregation errors will be introduced.

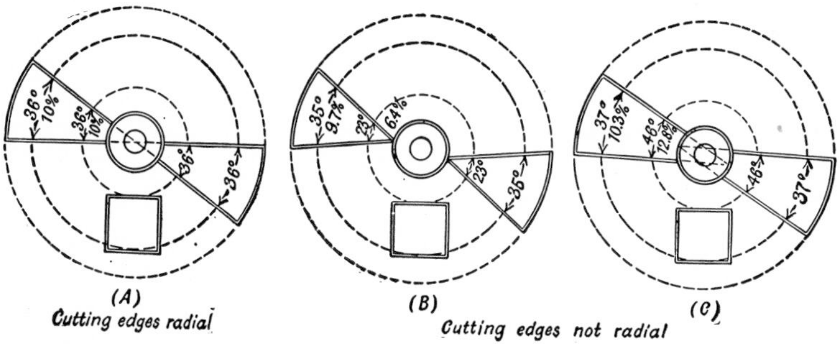

Fig. 259.—Proper (*A*) and improper (*B*, *C*) arrangement of cutting edges in Vezin sampler. (*After Taggart.*)

The Vezin sampler (Fig 258) is one of the best. It consists of a revolving cutter in the shape of a circular sector of such dimensions as to cut the whole stream of ore. When it is properly built (Fig. 259, *A*), *i.e.*, when the cutting sector or sectors are shaped with their center at the center of revolution of the sampler, an accurate sample is secured. In addition, the cutting sector must be at least three to four times as wide as the coarsest particles are large, in order to avoid clogging.

Usual cuts made by Vezin-type samplers are $\frac{1}{5}$ to $\frac{1}{20}$. By making a cut of the cut, a $\frac{1}{25}$ to $\frac{1}{400}$ sample is obtained.

For sampling wet pulps, *cutters* are used. The edges of the cutter should travel at right angles to the stream. Cutters can be actuated by any one of a variety of means, tilting boxes, eccentric gears, pistons, electrical devices, etc. Timing devices are also used to actuate the cutters at definite time intervals. One of the better wet samplers is the Geary-Jennings Sampler of the Galigher Company. This sampler cuts an accurate sample at time intervals that are adjustable. Timing and actuation of the sampler are electrical.

WEIGHING

Determination of the quantity of dry ore can be made in several ways:

1. By weighing ore cars or trucks and weighing of tare. This requires an operator.

2. By weighing the concentrate and multiplication of this number by the ratio of concentration. The ratio of concentration, in turn, is calculated from assays of feed, concentrate, and tailing.

3. By weighing the ore stream at some convenient point and by making an allowance for the moisture content.

The last method is the most accurate; it can be checked by the others. Continuous weighing devices suitable for that purpose are typified by the Merrick weightometer.

In the Merrick weightometer, a portion of a belt conveyor is suspended by means of rods from weighing levers. The weight is counterbalanced on the balance beam by means of an iron float in a mercury bath. A change in the weight on the belt causes a change in the position of the beam, and actuates a mechanical integrator. The other element that actuates the integrator is the speed of the belt. Allowance is made for the weight of the empty belt (by blank setting). Occasional calibration of the meter is required.

Literature Cited

1. AIRY, WILFRED: *Proc. Inst. Civ. Engrs. (London)*, **131** (1897), cited by KETCHUM, ref. 15.
2. BARBOUR, PERCY E.: Steel *vs.* Timber Ore Bins, *Eng. Mining J.*, **99,** 195–196 (1915).
3. BLUE, F. K.: Flow of Water Carrying Sand in Suspension, *Eng. Mining J.*, **84,** 536–539 (1907).
4. BRIDGMAN, H. L.: A New System of Ore Sampling, *Trans. Am. Inst. Mining Engrs.*, **20,** 416–442 (1891).
5. BRUNTON, D. W.: A New System of Ore Sampling, *Trans. Am. Inst. Mining Engrs.*, **13,** 639–645 (1884–1885).
6. BRUNTON, D. W.: The Theory and Practice of Ore Sampling, *Trans. Am. Inst. Mining Engrs.*, **25,** 826–844 (1895).
7. BRUNTON, DAVID W.: Modern Practice of Ore Sampling, *Trans. Am. Inst. Mining Engrs.*, **40,** 567–596 (1909).
8. GATES, ARTHUR O.: Belt and Bucket Elevators, *Eng. Mining J.*, **102,** 40–45 (1916).
9. GATES, ARTHUR O.: Wet Bucket Elevator Design, *Eng. Mining J.*, **96,** 725–727 (1913).
10. GEARY, J. F.: Design and Construction of Midvale Mill, *Eng. Mining J.*, **121,** 917–923 (1928).
11. GLENN, WILLIAM: Sampling Ores without Use of Machinery, *Trans. Am. Inst. Mining Engrs.*, **20,** 155–165 (1891).
12. GILBERT, G. K.: The Transportation of Debris by Running Water, *U. S. Geol. Surv., Prof. Paper* 86 (1914).
13. HETZEL, F. V.: "Belt Conveyors and Belt Elevators," John Wiley & Sons, Inc., New York (1922).
14. HOLBROOK, E. A., and THOMAS FRASER: Screen Sizing of Coal, Ores and Other Minerals, *U. S. Bur. Mines, Bull.* 234 (1925).
15. KETCHUM, MILO S.: "The Design of Walls, Bins, and Grain Elevators," McGraw-Hill Book Company, Inc., New York (1919).
16. LEDDELL, W. A.: Slag Lining for Launders, *Eng. Mining J.*, **102,** 644–646 (1916).
17. ROBINS, THOMAS, JR.: Notes on Conveying-belts and Their Use, *Trans. Am. Inst. Mining Engrs.*, **26,** 78–97 (1896).
18. SCHMITT, C. O.: "A Textbook of Rand Metallurgical Practice," Charles Griffin & Co., Ltd., London (1913).
19. SMART, EDGAR, HENRY FORBES JULIAN, and A. W. ALLEN: "Cyaniding Gold and Silver Ores," J. B. Lippincott Company, Philadelphia (1921).
20. STERLING, PAUL: The Preparation of Anthracite, *Trans. Am. Inst. Mining Engrs.*, **42,** 264–313 (1912).
21. TAGGART, ARTHUR F.: "Handbook of Ore Dressing," John Wiley & Sons, Inc., New York (1927); (*a*) Section 19, pp. 1033–1055; (*b*) p. 1373; (*c*) p. 1198, formula 10.
22. WIARD, E. S.: "The Theory and Practice of Ore Dressing," McGraw-Hill Book Company, Inc., New York (1915).

CHAPTER XXII

MINERAL DRESSING AS AN ART

Principles of physics and of physical chemistry lie at the root of every individual process used in mineral dressing. This has been implicitly assumed throughout. The synthesis of these various processes into a milling method and practice requires the use of other principles; those principles belong in the field of engineering economics. Just because they are considered at the end of this book does not mean that they have a lesser importance. Indeed, in many ways they are *the* most important factors that must be considered by the mineral technologist.

The most basic principle that must be kept in mind is that mineral dressing is adopted because it permits one to make a profit out of a venture of mineral exploitation where less profit, or no profit, would accrue without dressing. The same thought can be expressed in a socialistic instead of a capitalistic vein by saying that the efficiency of exploitation of a mineral deposit is usually enhanced by adoption of ore concentration.

The proper dressing method is that which will yield the greatest profit even if it is not that which yields the highest recovery or the cleanest concentrate. Where several methods may be considered in competition with each other, the correct method is determined by the yardstick of optimum profit, and no pet process or fashionable method should be allowed to warp the judgment of the engineer.

Within the scope of a single method, the best conditions as to fineness of grinding and specific procedure can be established. Basically, the problem of choosing the best conditions is the same in all instances. It consists in

1. Evaluating the net returns per ton of ore that can be obtained under the various conditions.

2. Evaluating the total cost that would be incurred in each instance.

3. Deducting (2) from (1).

Evaluating the net returns requires a knowledge of the grade of concentrate and tailing, composition of concentrate, smelter schedule, and schedule of transportation. The method is detailed elsewhere.[3a]

In most instances, the dressing works sell their product to another party, *e.g.*, to a smelter or to a domestic consumer. In the preparation of his product, the dressing engineer must keep in mind the wishes of the purchaser. Sometimes these wishes are expressed in the form of definite, financially perceptible bonuses and penalties, as in dealings with a smelter, but other times they are not expressed definitely and take the form of public favor or disfavor for a certain brand. In mineral dressing as in other commercial ventures, "the customer is always right."

Dressing Profit. The economics of a dressing operation can be subdivided into two sets of figures, one of which measures the additional gross return resulting from the dressing, and the other the expense incurred to accomplish this end. The difference is the profit made by the operation. Thus, if a silver-lead ore is worth $2.25 per ton (at the mill) in the unconcentrated condition, and $6.50 *per ton of ore* (at the mill) in the concentrated condition, the additional return resulting from dressing is $4.25 per ton of ore. If the cost of dressing is $1.50 per ton of ore, the dressing profit is $4.25 − 1.50, or $2.75 per ton of ore.

If the raw mineral is an ore of the common or precious metals, the increase in gross returns due to dressing may be measured readily by means of smelter schedules (either open schedules or private contracts), grades of products, recoveries, and freight schedules. Examples have already been given in Chapter I.

If the raw mineral is coal, the problem is complicated by the fact that purchase of coal by domestic and small industrial consumers is not on a quality basis: A schedule of prices, as a function of the ash, sulphur, volatile matter, or B.t.u. content has yet to be adopted.

Many coal companies have therefore accustomed themselves to regard a ton of coal as a ton of coal regardless of whether the coal is a raw product containing 12 per cent ash or a beneficiated product containing only 6 per cent ash. This view has justified an unusual method of accounting in which the beneficiation of the coal is penalized to the extent of the cost of mining the refuse. Thus if 10 tons of raw coal costing at the mill $1.50 per ton yield

9 tons of clean coal and 1 ton of refuse, the cost of dressing is usually made to include a charge of \$1.50/9, or \$0.166, per ton of coal to pay for the refuse: this is tantamount to viewing the waste as fuel! However unfair this accounting procedure does appear, it is justifiable in the eyes of the businessman by the relative lack of interest of many buyers in the quality of the coal that they buy.

The conditions are still less standardized in dealing with nonmetallic minerals. For example, the dressing works is often part of the scheme of operations of an interested user, so that proper evaluation of the increase in gross returns may be largely a matter of opinion. The trend is to better accounting and to a wider appreciation of the advantages of dressing.

Dressing Cost. The dressing cost is made up of two distinct elements: an operating cost and a capital cost. The operating cost can be determined very accurately, and there is little ground for disagreement between operators in ascertaining them. But the opposite is true of capital costs. The latter depend not only on the first cost of the plant—an item that can be ascertained very exactly—but also on proper charges for depreciation, obsolescence, and interest, items all of which are subject to opinion.

These two types of cost will be taken up in order.

Capital Cost. As has already been indicated, the capital cost consists of charges for depreciation, obsolescence, and interest.(1) Depreciation is the decrease in capital value that results from natural decay and from wear and tear. If adequate repairs are made constantly and charged to operation, the depreciation should be nil, as the plant at the end of x years should be as good as new. Unfortunately, this goal is difficult to reach. Sometimes overzealousness results in material improvement in plant: this is equivalent to hiding profits. At other times, a plant is allowed to run down: this is equivalent to claiming nonexistent profits. The difficulties of proper accounting are very great, and they are not diminished by the facts that both companies and tax collectors are interested parties.

Plant obsolescence is the decrease in value of the property that arises from the lack of need for that property either because there is no more material to treat, or because the material is better treated by some other process or with other machines than those of the plant. The second of these forms of obsolescence is general to all industry, but the first is more peculiarly acute in the mineral industry, the assets of which are being constantly depleted.

Both forms of obsolescence are proper capital charges against dressing. It is impossible to ascertain in advance how large they should be, and any estimate that comprises them implicitly includes a prophecy as to the life of the property and the likelihood of significant discoveries and inventions.

Interest on investment is sometimes not included as a cost, as it is reasoned by some people that the return on the capital invested is the reward of the enterpriser. However, if money is borrowed for the construction of a plant, interest has to be paid on the sum borrowed before a profit can be had by the enterpriser. Interest on investment equivalent to that which should be paid to a lender appears therefore a justifiable charge against plant operation.

The first cost of a dressing plant is usually expressed in dollars per ton of daily capacity. Thus a flotation plant may often be built for $600 to $800 per ton of daily capacity. A gravity plant is usually cheaper, especially if designed to treat coarse material.

Variations in unit cost are traceable primarily to complexity of treatment, secondly to rated tonnage, and thirdly to local conditions such as unit cost of labor, situation of the mill with reference to manufacturing centers, and topography of the mill site.

Consider a flotation mill that cost $800 per ton daily capacity, one designed to treat an ore the reserves of which are practically unlimited. The capital cost during the first year might well be as follows:

Interest on investment (5 per cent per annum).....	$40.00
Obsolescence (20-year life for equipment)..........	40.00
Capital cost per ton-year........................	$80.00
Capital cost per ton (300 working days per year)...	$ 0.267

After 20 years, the capital cost would be only half as large since the cost of the plant would have been written off and the item of interest on investment would have disappeared. It is likely, however, that in the intervening years new devices would have been installed, devices that are not a proper charge against operation, thus continuing to require amortization charges.

A contrasting case may be that of a flotation mill designed to re-treat a limited supply of old gravity tailings, a plant having cost but half as much, or $400 per ton; if it is estimated that the ore supply is to be depleted in 4 years, the capital cost the first year might well be as follows:

Interest on investment (5 per cent per annum)	$ 20.00
Obsolescence (4-year life for plant)	100.00
Capital cost per ton-year	$120.00
Capital cost per ton (250 working days per year)	$ 0.48

Comparison of the capital cost for these two plants shows the cheap plant to actually represent a higher capital cost. The cause for this effect is the difference in obsolescence. Clearly, capital costs are very much a function of the ore supply and of the number of working days per year.

The *salvage value* of a plant that has ceased to be useful is very small. The cost of the buildings and the labor expended in erection are wholly lost. If the remaining machinery is relatively new, a good secondhand price can be had, but usually its state of repair is so poor as to bring not over 15 to 20 per cent of the value of the machinery. All in all, the salvage value of a plant, figured on the first cost, rarely reaches 10 per cent.

Operating Cost. The operating cost consists of the following:

1. The wages of the operators.
2. The supplies of chemicals and spare machine parts.
3. The electric power or fuel.
4. General items such as the salaries of the staff, patent royalties, and a share in the expenses of the main office.

The operating cost can therefore be subdivided into four classes, as above. It can also be subdivided according to the type of operation involved, as, for example, into crushing, grinding, screening, classification, concentration, thickening, filtration, conveying, pumping, sampling and assaying, heating, lighting, and miscellaneous. Since this type of subdivision is different from the other, cross classification and cross subdivision of the operating cost is possible. This is a very useful and desirable method of keeping records. An example is afforded by Table 60.

TABLE 60.—SUMMARY OF OPERATING MILLING COSTS AT BRITANNIA PLANT, JULY 1931
(*After Munro and Pearse*[4])

	Operating				Repairs		Total
	Labor	Power*	Supplies	Reagents	Labor	Supplies	
Breaking to 6-in. size	$0.0023	$0.0002	$0.0020	....	$0.0005	$0.0002	$0.0052
Intermediate crushing	0.0178	0.0050	0.0158	....	0.0063	0.0037	0.0486
Fine grinding and classifying	0.0064	0.0148	0.0177	....	0.0040	0.0031	0.0460
Flotation	0.0030	0.0031	....	$0.0167	0.0028	0.0021	0.0277
Dewatering and handling concentrates	0.0027	0.0015	0.0012	....	0.0018	0.0022	0.0094
Sampling and assaying	0.0017	....	0.0003	....	....	....	0.0020
Superintendence, office and warehouse	0.0087	....	0.0005	....	....	....	0.0092
Miscellaneous	0.0074	0.0004	0.0072	......	0.0031	0.0008	0.0189
Flotation royalty	....	....	0.0100	....	....	....	0.0100
Totals	$0.0500	$0.0250	$0.0547	$0.0167	$0.0185	$0.0121	$0.1770

* Power cost was $0.0020 per kw.-hr. for July 1931. On account of July being a flush-water month, nearly all power consumed was generated by the company's own hydroelectric plant, hence the low unit cost for power. The average year-round cost for power is approximately double the preceding figure, which would then give an average milling cost of approximately $0.20 per ton of ore milled.

The operating cost depends upon the complexity of the plant, the size of the plant, the efficiency of the plant, the quantity of supplies needed, and the unit costs for labor and power. Great differences occur in operating cost as between different types of plants, from a few cents per ton in large coal washeries to as many dollars per ton for small, specialty plants treating finely ground minerals.

The great importance of plant capacity is generally not appreciated, but it should be clear that practically the same labor is required to operate any properly designed plant below a limiting size of perhaps 100 to 150 tons per day; the labor cost per ton for plants smaller than that critical size increases inversely as the size of the plant. Indeed, many metallurgists think that there is no place for a plant to dress metalliferous ores smaller than 30 to 50 tons per day, at least under the economic conditions that prevail in the United States. Some idea of the effect of mill capacity on operating cost may be acquired from Fig. 260.

Factors Involved in Deciding Whether a Mill Is Justified. To decide whether construction of a mill is justified, the following data, at least, are required:

1. Accurate sampling of the ore body.
2. Information as to the probable quantity of ore.
3. A fairly complete record of the results that might be obtained by dressing the ore.
4. Smelter schedules or other terms of sale, freight schedules and price ranges of metals or minerals.

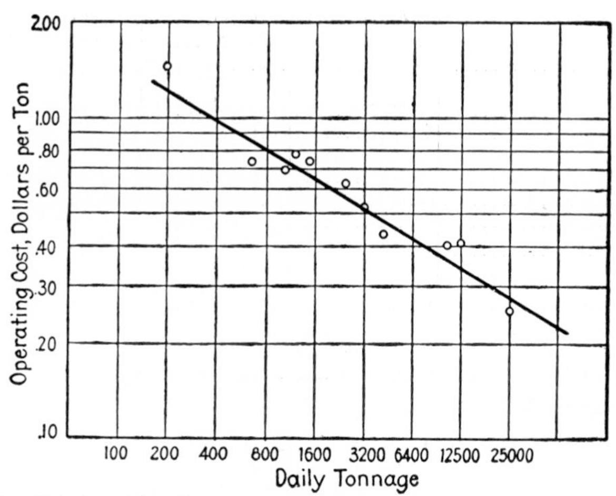

FIG. 260.—Relationship of operating cost to mill capacity in one-concentrate flotation plants.

It is not uncommon to find individuals or small groups of men who are anxious to build a mill on what is popularly known as a shoestring. This practice cannot be discouraged too strongly as it is wasteful of the capital of the prospector or small operator, as well as likely to give the stock-jobbing crook the golden opportunity to peddle his paper. No general limit may be set on ore tonnage, but it seems safe to say that few amalgamation mills can be made to pay under a total of 15,000 to 25,000 tons of ore, and that 50,000 tons or more are required for a flotation plant. As a general rule, 3 to 5 years' ore reserve should be the minimum on which a mill should be built with ore of average grade.

It is commonly found that in order to save a mining engineer's fee, tests are conducted on any kind of grab sample. Often the ore that the mill is actually called to treat is radically different from that on which the tests were based, with the result that a new mill is necessary. And it is also commonly found that a plant is built without any ore testing at all. No defense is possible for such foolishness.

The first two data involved in deciding whether a mill is justified are secured by the mining engineer, the other two by the mineral-dressing engineer. Of the mineral-dressing data, the test record is wholly technical, and the inquiries pertaining to the sale of prospective concentrates are often dealt in by a business executive.

Ore Testing. Ore testing is an art that comes naturally to one who is thoroughly familiar with mineral-dressing processes and practice. Specific directions can be given for some particular type of ore; but specific directions cannot be given that will fit all types of ores, since the aims and practices vary from one process or from one type of ore to another. Generally some, if not all, of the following steps are in order:

1. Cutting of a "head" or "feed" sample for analysis, and suitable analysis of that sample.

2. Determination of the minerals present, of the mineral associations, and of the fineness of the associations.

3. Ascertainment of the grinding characteristics, and of possible segregation of mineral constituents in various size ranges.

Completion of these determinations indicates which concentrating method or methods should be tested, *viz.*, whether, that is, the method should be a gravity-concentrating scheme, flotation, amalgamation, magnetic separation, or some of the less common methods. Table 61 summarizes usual findings in that respect.

The next objectives are, in order:

4. Determination of the behavior of the ore by the chosen method, and under standard conditions, after adopting a degree of crushing and grinding that is approximately suitable.

5. Behavior of the ore by the chosen method under conditions that deviate somewhat from the standard. This is designed to find out whether it is possible to improve on the results of the standard practice.

6. New departures in procedure.

7. Comparisons.

TABLE 61.—LIKELY ORE-TREATMENT PROCESS IN RELATION TO ORE CHARACTERISTICS

Principal valuable constituent	Auxiliary valuable constituents	Other constituents	Fineness of association	Likely process
Chromite	....	Silicates	Medium	Gravity concentration and flotation
Clay	....	Quartz	Fine	Dispersion and classification
Coal (anthracite)	....	Ash	Coarse	Heavy-fluid separation; hindered-settling classification
Coal (bituminous)	....	Ash, sulphur	Coarse	Heavy-fluid separation; jigging; pneumatic tabling; tabling
			Medium to fine	Tabling of sands and flotation of sludge
Copper (sulphides)	Nickel, gold, silver, zinc, lead	Pyrite, silicates	Medium or fine	Selective flotation
Copper (oxidized)	....	Iron oxides, silicates, carbonates	Medium or fine	Tabling of sands, and flotation of slimes; ammonia or acid leach
Diamond	....	Silicates	....	Disintegration by weathering; jigging; adhesion to grease
Gold (native gold)	....	Silicates	Coarse	Amalgamation, jigging, tabling
	....	Silicates, oxidized minerals	Fine	Cyanidation
	....	Pyrite, arsenopyrite, silicates	Fine	Collective flotation or cyanidation; collective flotation followed by cyanidation; cyanidation followed by collective flotation
	Copper (sulphides)	Pyrite, silicates	Fine	Selective flotation
Iron (hematite)	....	Silicates, gypsum	Medium	Heavy-fluid separation; hindered-settling classification; jigging and tabling
Iron (magnetite)	....	Silicates, apatite	Medium	Dry or wet magnetic separation
Iron (hematite)	....	Silicates, gypsum	Fine	Reducing roast followed by wet magnetic concentration
Kyanite	....	Other silicates	Medium	Gravity concentration; selective flotation

TABLE 61.—LIKELY ORE-TREATMENT PROCESS IN RELATION TO ORE CHARACTERISTICS.—(*Continued*)

Principal valuable constituent	Auxiliary valuable constituents	Other constituents	Fineness of association	Likely process
Lead (sulphide)	Silver, zinc, gold, copper	Pyrite, silicates	Medium or fine	Selective flotation
Lead (sulphide)	....	Carbonates and silicates	Coarse	Jigging of coarse particles and flotation of slimes
Lead (oxidized)	Gold, silver, copper	Carbonates and silicates	Fine	Tabling of sands and flotation of slimes
Lime-rock	....	Quartz, silicates	Fine	Classification and flotation
Magnesite	....	Silicates, calcite	Medium	Agglomeration; flotation
Manganese (dioxide)	....	Silicates	Medium	Dry magnetic separation; gravity concentration; flotation
Manganese (carbonate)	Lead and zinc sulphides	Silicates	Medium	Selective agglomeration; selective flotation
Phosphate	....	Quartz and clay	Medium	Washing and screening; agglomeration of screen undersize
Potash (sylvite or langbeinite)	....	Halite and magnesium salts	Medium	Flotation or agglomeration in natural brine
Silver (sulphide)	Gold	Pyrite, silicates	Fine	Selective or collective flotation
Silver (oxidized)	Gold, lead	Silicates, carbonates	Fine	Tabling of sands and flotation of slimes
Tin (oxide)	....	Silicates, carbonates	Medium	Jigging, tabling and vanning
	Tungsten	Silicates, carbonates, pyrite	Medium	Jigging, tabling and vanning. Concentrate retreated by flotation to eliminate pyrite, and magnetically to separate tin from tungsten
	....	Silicates	Fine	Vanning; flotation (experimental)
Tungsten (scheelite)	....	Silicates	Medium	Gravity concentration and flotation
Zinc (sulphide)	....	Silicates and carbonates	Coarse	Jigging and tabling of coarse particles and flotation of slimes
Zinc (sulphide)	Lead	Pyrite, marcasite, silicates	Medium or fine	Selective flotation
Zinc (franklinite)	....	Silicates	Medium	Magnetic separation (high intensity)
Zinc (carbonate or silicate)	....	Silicates	Medium	Gravity concentration

The microscope is an invaluable aid in ore testing, not only for the examination of polished pieces of uncrushed ore, but also for the inspection of products from the concentrating machine, either during its operation or afterwards.

Assays should be required of a precision commensurate with the needs, and the samples taken should be of proper magnitude.

Plant Design. The design of a plant involves a great many considerations; it requires a number of engineering decisions of the greatest importance as well as much detail work. Design of a plant includes the following points:

1. Is a plant justified from an economic or commercial standpoint?
2. Where should the plant be located?
3. What flow sheet should be adopted?
4. What machinery should be chosen?
5. Preparation of a detailed survey and topographic map of the mill site and of the site for tailing disposal.
6. Detailed arrangement of (*a*) essential, and (*b*) auxiliary equipment.
7. Design of structure, including layouts for plumbing, electricity, heating, lighting, ventilation, dust elimination.
8. Detailed estimate of cost, including bill of materials.

Relatively simple and inexpensive plants for the treatment of coarse minerals are usually designed to operate one shift per day. On the contrary, complex, expensive plants are almost always operated continuously, at least in the subdivision of the plant in which the mineral is fine.

Location of the Plant.(2) The general location of the plant must be selected in regard to the situation of the mine, the location of the market, facilities for tailing disposal, power supply, and water supply. The specific location to a large extent depends upon the local physiography.

Other things being equal, the most advantageous location of a dressing plant is at the mine. This eliminates haulage of waste and possible interruptions in ore supply. In the case of low-priced minerals such as coal, location of the dressing plant at the mine eliminates reshipment, an important item of cost.

If the user of the concentrate has unusual facilities, it may be advantageous, however, to locate the plant there. For example, many large flotation plants are located at smelters.

Location at the mine or quarry may be impracticable because of inadequate water supply, or because available space for tailing disposal does not exist. Either or both of these considerations may compel selection of a site other than one at the mine. Problems of water pollution may also enter into the selection of a suitable site.

In general, for small operations location at the mine is possible. For larger operations, the other factors increase in importance. For example, the Butte ores are treated at a mill 25 miles distant, located at the Anaconda Reduction Works. Among the reasons for this choice were the lack of water in Butte and the opportunity for more favorable labor conditions in Anaconda. The copper ores from Bingham, Utah, are treated at the twin Magna and Arthur mills, located some 20 miles away. The reasons for this choice were the lack of water and the lack of space for tailing disposal.

Flow Sheet. Selection of a flow sheet follows ore testing and is based on its findings and on general practice. Preparation of a qualitative flow sheet is a simple matter, but preparation of a quantitative flow is difficult. It requires (1) exact data on every phase of the proposed treatment scheme, (2) knowledge as to capacity of the machinery, not only according to the often optimistic predictions of manufacturers, but also according to the realities of past practice, and (3) broad judgment as to the most desirable solution wherever a choice between machines is required.

The flow sheet is necessarily drawn first so as to include principal equipment only, in a qualitative manner; then principal equipment in a quantitative manner; and finally so as to include both principal and auxiliary equipment in a quantitative manner. This last step is concerned largely with the transport of materials within the plant and as such requires knowledge of the spatial location of the equipment. It is irrevocably tied up with the detailed arrangement of the equipment.

Operation. Successful operation is predicated on continuity, accounting, and progressiveness. To secure continuity, it is necessary to ensure the supply of ore, the supply of spare parts, chemicals, water, and power, and to foster a cooperative attitude on the part of the workmen. Equipment must be kept lubricated and in good repair.

Successful operation is also predicated on accurate accounting as it is only by that means that economies can be effected. This accounting is twofold, technological and commercial. The technological accounting requires the ascertainment of the daily ore tonnage, and analyses or assays of feed and products.

The commercial accounting requires daily records of the man-shifts, of the requisitions against the commissary, or of the purchase of supplies, parts, and power.

The two sets of records permit the preparation of a daily metallurgical balance and a weekly or monthly cost sheet.

In addition, the control of operations by certain routine tests and laboratory examinations is well worth while as it permits the diagnosis and correction of unexpected difficulties. The proper check on operations requires not only a check on overall results, but also a check on each individual operation. Crushing, grinding, and classifying operations are checked by sizing tests in conjunction with tonnage and power records; gravity-concentrating operations are checked by float-and-sink tests; magnetic-separator operations are checked by magnetic-tube separations of separator products; and flotation operations are controlled by records of the pH, of the total electrolyte conductivity, of the reagent addition, and by routine laboratory tests on daily samples.

Microscopes are valuable for the control of some intermediate mill products, and for periodic checkups on the fineness of mineral association and degree of liberation achieved on the ore.

Finally, successful operation is also progressive operation: minor adjustments and improvements are constantly sought, first in parallel laboratory tests, then in the plant itself if the laboratory results are promising. These adjustments and improvements may have for object either improved metallurgical results or lowered costs.

Intermittent Operation. Where concentrating operations must be interrupted either as a routine or as an emergency event, a certain sequence must be followed: first the feed is cut off, then each machine in turn is stopped as the feed ceases to reach it, until the whole plant is stopped. In starting up, the last machine is started first, and the feed, at last, is turned on. Automatic switching can be designed for maintaining a proper sequence of stopping and starting.

Literature Cited

1. BARNES, FRANK E.: "Estimating Building Costs," McGraw-Hill Book Company, Inc., New York (1927).
2. CALLOW, JOHN M.: Design and Construction of Ore-treatment Plants, in TAGGART'S "Handbook of Ore Dressing," John Wiley & Sons, Inc., New York (1927), pp. 1263–1343.
3. GAUDIN, A. M.: "Flotation," McGraw-Hill Book Company, Inc., New York (1932), pp. 411–521; (*a*) pp. 466–486.
4. MUNRO, A. C., and H. A. PEARSE: Milling Methods and Costs at the Concentrator of the Britannia Mining and Smelting Co., Ltd., Britannia Beach, B. C., *U. S. Bur. Mines, Information Circ.* 6619 (1932).

INDEX